Anthony Wayne Vogdes

A classed and annotated Bibliography of the Paleozoic Crustacea

1698-1892

To which is added a Catalogue of the North American species

Anthony Wayne Vogdes

A classed and annotated Bibliography of the Paleozoic Crustacea 1698-1892
To which is added a Catalogue of the North American species

ISBN/EAN: 9783337176747

Printed in Europe, USA, Canada, Australia, Japan

Cover: Foto ©ninafisch / pixelio.de

More available books at **www.hansebooks.com**

A Classed and Annotated

BIBLIOGRAPHY OF THE PALÆOZOIC CRUSTACEA

1698 1892

TO WHICH IS ADDED A

Catalogue of North American Species.

BY

ANTHONY W VOGDES.

SAN ~~FRAN~~CISCO:
CALIFORNIA ACADEMY OF SCIENCES,
June, 1893.

Committee of Publication:

JOHN R. SCUPHAM.   GUSTAV EISEN.   FRANK H. VASLIT.

INTRODUCTION.

The object of this paper is to give the literature on the Palæozoic Crustacea and to aid students and palæontologists in their researches. It is the result of more or less constant work during the past decade. It gives a list of authors, and includes an index of the species described in each work.

I am well aware that the value of a work of this kind depends upon the care and labor given to its compilation. Every possible means of obtaining accurate information from a miscellaneous class of literature in many languages has been used, and it is hoped that there are but few errors and omissions.

I would thank those into whose hands this paper may fall for any information that would add to its value as a working hand-book.

A. W. Vogdes.

San Francisco, Cal., January 1, 1893.

A BIBLIOGRAPHY OF PALÆOZOIC CRUSTACEA
1698-1891.

A star (*) following a title indicates that the reference is recorded without opportunity for personal verification.

Agassiz (L.) In Murchison's Silurian System 1839, p. 605.

Pterygotus problematicus n. sp. *Onchus murchisoni* n. sp., p. 607, pl. 4, fig. 9-11. M'Coy refers this species to Ceratiocaris.

—— Monographie des poissons fossiles du vieux grès rouge ou système dévonien (old Red Sandstone) des iles Britanniques et de Russie.

Text, Soluere; atlas, Neuchâtel, 1844, pp. xix-xx, note; and atlas Tab. A. *Pterygotus anglicus* n. sp.

Alth (Alois von). Ueber die palæozoischen Gebilde Podoliens und deren Versteinerungen. Erste Abtheilung.

In Abhandl. k. k. geol. Reichsanst., vol. 7, pt. 1, Wien, 1874, 5 pl. *Eurypterus remipes* DeKay. *Stylomurus? Sphærexochus mirus* Beyr. *Ceraurus (Cheirurus) insignis* Beyr. *Illænus (Bumastus) barriensis* Murch. *Calymene blumembachi* Brong. *Dalmania caudata* Brünn. *Proetus podolicus* n. sp., *P. concinnus* Dalm., *P. dzieduszyckianus* n. sp. *Cyphaspis rugulosus* n. sp. *Beyrichia podolica* n. sp., *B. reussi* n. sp., *B. bilczensis* n. sp., *B. inornata* n. sp. *Primitia rectangularis* n. sp., *P. concinna* Jones, *P. oblonga* J. & H., *P. muta* Jones? *Leperditia baltica* His., *L. römeri* n. sp., *L. phaseolus* His. *Bairdia protracta* Eichw. *Leperditia tyraica* Schmidt. *L. römeri* is referred to this species on pl. 5, figs. 28-32, 34-36. *Pterygotus anglicus* Agas. *Encrinurus punctatus* Emm. *Illænus bouchardi* Barr.

Alton (Ed. de). See **Burmeister and Alton.**

Ami (Henry M.) Notes on *Triarthrus spinosus* Billings.

In Trans. Ottawa Field Nat. Club, No. 4, 1882-83, p. 88, 1 pl.

Angelin (N. P.) Palæontologia Scandinavica. Pars I. Crustacea formationis transitionis. Lipsiæ, 1852, 4to.

The first part of this work, entitled "Palæontologia Scandinavica," containing pp. 1-24 and pl. 1-24, was published by the Royal Swedish Academy of Sciences in 1852. The second portion in 1854, including a preface, pp. 1-9 and pp. 21-92, pl. 25-42.

Fasciculus 1, 1852, contains descriptions of the following species:
Olenus. Paradoxides tessini Broug. *P. forchhammeri* n. sp., *P. loveni*
n. sp. *Cryptonymus bellatulus* Dalm., *C. punctatus* Wahl., *C. obtusus* n.sp.,
C. lævis n. sp., *C. verrucosus* Dalm. *Eryx* n. g. (corrected by Angelin to
Elyx, the generic term *Eryx* having been used by Stephen in 1832 for a
genus), *E. laticeps* n. sp. *Acontheus* n. g. (corrected by the author to that
of *Aneucanthus*). *Aneucanthus acutangulus* n. sp. *Agnostus glandiformis*
n. sp., *A. bituberculatus* n. sp., *A. lævigatus* Dalm., *A. brevifrons* n. sp.,
A. glabratus n. sp., *A. lentiformis* n. sp., *A. pisiformis* Linné, *A. exsculptus* n. sp., *A. planicauda* n. sp., *A. reticulatus* n. sp., *A. punctuosus* n. sp.,
A. aculeatus n. sp. *Phacops bucculenta* Sjögr., *P. macroura* Sjögr., *P. conicophthalma* S. & B., *P. tumida* n. sp., *P. mucronata* Broug., *P. caudata* Brünn. (nec. Wahl.), *P. imbricatula* n. sp., *P. æquicostata* n. sp., *P. eucentra* n. sp., *P. sclerops* Dalm., *P. granulosa* n. sp., *P. breviceps* n. sp.,
P. 4-lineata n. sp. *Polytomurus euglyptus* n. sp. *Brachypleura 6-lineatus* n. sp., *B. 4-lineatus* n. sp. *Niobe* n. g., *N. lata* n. sp., *N. læviceps*
Dalm., *N. frontalis* Dalm., *N. emarginula* n. sp., *N. explanata* n. sp.
Megalaspis n. g., *M. gigas* n. sp., *M. heros* Dalm., *M. latilimbata* n. sp.,
M. multiradiata n. sp., *M. extenuata* Wahl., *M. rotundata* n. sp., *M. explanata* n sp., *M. stenorachis* n. sp., *M. planilimbata* n. sp., *M. limbata*
n. sp. *Nileus palpebrosus* Dalm., *N. armadillo* Dalm. *Ampyx nasutus*
Dalm., *A. tetragonus* n. sp., *A. carinatus* n. sp. *Proetus stokesi* Murch.,
P. concinnus Dalm., *P. elegantulus* Lovén, *P. lævis* n. sp., *P. ? limbatus*
n. sp., *P.? excavatus* n. sp., *P.? microphthalmus* n. sp., *P.? difformis* n.sp.
Calymene holometopa n. sp., *C. canaliculata* n. sp., *C. brachymetopa* n. sp.,
C. aculeata n. sp., *C. lejostraca* n. sp., *C. stenometopa* n. sp., *C. tuberculata*
Brünn. (*blumenbachii* Broug).

Fasciculus 2, 1854, contains descriptions of the following species:
Phaëthonides (Angelin uses this term for that of *Phæton* Barrande, 1846).
Phaëthonides stokesi Murch. *Forbesia concinna* Dalm., *F. conspersa* n. sp.
Goniopleura elegantula Lovén. *Celmus* n. g., *C. granulatus* n. sp. *Anomocare* n. g., *A. læve* n. sp., *A. limbatum* n. sp., *A. excavatum*, n. sp., *A. microphthalmum* n. sp., *A. difforme* n. sp., *A. aculeatum* n. sp., *A. acuminatum* n.sp. *Solenopleura* n. g., *S. holometopa* n.sp., *S. canaliculata* n.sp.,
S. brachymetopa n. sp. *Liostracus* n. g., *L. aculeatus* n. sp. *Solenopleura ? stenometopa* n. sp. *Calymene spectabilis* n. sp., *C. tuberculata* Brünn.
Homalonotus platynotus Dalm., *H. rhinotropis* n. sp. *Amphion. Pliomera fischeri* Eichw. *Chirurus ornatus* Dalm., *C. exsul* Beyr., *C. conformis* n. sp.
Cyrtometopus n. g., *C. clavifrons* Dalm. (ex. p.) *Acidaspis pectinata* n. sp.,
A. crenata Emm. *Trapelocera bicuspis* n. sp. *Pliomera mathesii* n. sp.,
P. actinura Dalm. *Cyrtometopus scrobiculatus* n. sp., *C. diacanthus* n. sp., *C. octacanthus* n. sp., *C. longispinus* n. sp. *Sphærexochus angustifrons* n. sp., *S. scabridus* n. sp., *S. latifrons* n. sp. *Acidaspis granulata*
Wahl. (ex. p.), *A. multicuspis* n. sp., *A. marklini* n. sp., *A. barrandei* n.sp.,
A. ? armata Boeck. *Trapelocera ? breviloba* n. sp. *Rhodope* n. g. (this

term was used by Siebold in 1848 for a genus of the Gasteropoda), *R. lineata* n. sp. *Dysplanus centaurus* Dalm., *D. centrotus* Dalm. *Bumastus lindströmi* n. sp. *Illænus crassicauda* Wahl. *Rhodope ? oblongata* n. sp. *Æglina ? oblongula* n. sp. *Olenus truncatus* Brünn., *O. attenatus* Boeck., *O. sphænopygus* n. sp., *O. aculeatus* n. sp., *O. gibbosus* Wahl., *O. aciculatus* n. sp., *O. ? acanthurus* n. sp. *Peltura scarabæoides* Wahl. *Parabolina spinulosa* Wahl. *Acerocare* n. g., *A. ecorne* n. sp. *Leptoplastus* n. g., *L. stenotus* n. sp., *L. raphidophorus* n. sp., *L. ovatus* n. sp. *Eurycare* n. g. (this genus is referred by J. W. Salter, Mem. Geol. Survey United Kingdom, Brit. Organic Remains, decade xi, pl. 8, p. 3, to the genus *Sphærophthalmus* Ang.), *E. brevicauda* n. sp., *E. angustatum* n. sp., *E. latum* Boeck., *E. camuricorne* n. sp. *Sphærophthalmus* n. g., *S. flagellifer* n. sp., *S. alatus* Boeck., *S. teretifrons* n. sp. *Anopocare* n. g. (this genus is referred to J. G. O. Linnarsson, Geol. Förening. Stockholm Förhdl., vol. 5, 1880-81, p. 136, to the genera *Peltura* and *Sphærophthalmus;* the specimen, according to Dr. Linnarsson, in the Royal Museum of Sweden, "marked by Angelin as the *abdomen till Stenom. pusillus*, contains several small heads of *Sphærophthalmus alatus*, and a small pygidium of *Peltura scarabæoides;*" Dr. Linnarsson refers to this specimen as Angelin's type for the genus *Anopocare* and also remarks "that Angelin intended to change the family name of *Sphærophthalmus alatus* and their nearest forms to that of *Stenometopus*, inasmuch as another specimen of *Sph. alatus* has, in Angelin's own handwriting, the generic name of *Stenometopus*.") *Anopocare pusillum* n. sp. *Megalaspis acuticauda* n. sp., *M. rudis* n. sp. *Ptychopyge* n. g., *P. applanata*. *Asaphus rimulosus* n. sp., *A. expansus* Linné (ex. p.) *A. raniceps* Dalm., *A. fallax* Dalm., *A. lævigatus* n. sp., *A. acuminatus* Boeck. *Ptychopyge glabrata* n. sp. *Megalaspis excavatozonata* n. sp., *M. zonata* n. sp. *Asaphus platyurus* n. sp. *Ptychopyge rimulosa* n. sp., *P. elliptica* n. sp., *P. multicostata* n. sp., *P. lata* n. sp., *P. angustifrons* Dalm., *P. limbata* n. sp., *P. media* n. sp., *P. aciculata* n. sp. *Bronteus laticauda* Wahl., *B. platyactin* n. sp., *B. polyactin* n. sp. *Holometopus* n. g., *H. aciculatus* n. sp., *H. ornatus* n. sp., *H. limbatus* n. sp. *Isocolus* n. g., *I. sjögreni* n. sp. *Corynexochus* n. g., *C. spinulosus* n. sp., *C.? umbonatus* n. sp. *Nileus ? lineatus* n. sp. *Symphysurus breviceps* n. sp. *Euloma* n. g., *E. læve* n. sp., *Pharostoma? ælandicum* n. sp. *Conocoryphe dalmani* n. sp. *Bumastus? glomerinus* Dalm. *Forbesia? brevifrons* n. sp. *Aneucanthus acutangulus* n. sp. *Trinucleus wahlenbergi* Rouault, *T. ceriodes* n. sp., *T. carinatus* n. sp., *T. coscinorinus* n. sp., *T. affinis*. *Sphærocoryphe* n. g., *S. dentata* n. sp. *Deiphon globifrons* n. sp. *Staurocephalus clavifrons* n. sp. *Lichas celorrhin* n. sp., *L. laciniatus* Wahl. (ex. p.), *L. affinis* n. sp., *L. polytomus* n. sp., *L. depressus* n. sp., *L. convexus* n. sp., *L. concinnus* n. sp., *L. rotundifrons* n. sp., *L. laticeps* n. sp., *L. latifrons* n. sp., *L. ælandicus* n. sp., *L. gibbus* n. sp., *L. pusillus* n. sp., *L. deflexus* Sjögr., *L. ornatus* n. sp. *Conocoryphe? glabrata* n. sp. *Dolichometopus* n. g., *D. svecicus* n. sp. *Lichas pachyrrhinus*

Dalman, *L. norvegicus* n. sp. *Platymetopus planifrons* n. sp. *Lichas conformis* n. sp., *L. cicatricosus* Lov., *L. G-spinus* n. sp., *L. dalecarlicus* n. sp., *L. gothlandicus* n. sp., *L. aculeatus* n. sp. *Platymetopus lineatus* n. sp. *Sphærexochus? clarifrons* Sars. (nec. Dalm. nec. His.), *S. ? deflexus* n. sp., *S. wegelini* n. sp., *S. conformis* n. sp., *S. granulatus* n. sp. *Sphærocoryphe granulata* n. sp. *Deiphon lævis* n. sp., *D. punctatus* n. sp, *Cyrtometopus speciosus* Dalm. (non Sternb. Sars. His.), *C. foveolatus* n. sp., *C. affinis* n. sp., *C. sarsi* n. sp., *C. tumidus* n. sp., *C. gibbus* n. sp. *Chirurus speciosus* His. (non Dalm. Sars. Sternb.), *C. glaber* n. sp., *C. punctatus* n. sp. *Ampyx costatus* Boeck., *A. foveolatus* n. sp., *A. mammillatus* Sars., *A. nasutus* Dalm., *A. ? aculeatus* n. sp. *Raphiophorus setirostris* n. sp., *R. tumidus* n. sp., *R. culminatus* n. sp., *R. depressus* n. sp., *R. scanicus* n. sp. *Lonchodomas rostratus* Sars., *L. crassirostris* n. sp., *L. affinis* n. sp., *L. jugatus* n. sp., *L. domatus* n. sp. *Trinucleus discors* n. sp., *T. seticornis* His. *Lichas 4-spinus* n. sp. *Trinucleus affinis* (cfr. t. 39, f. 5), *T. bucculentus* n. sp., *T. foveolatus* n. sp. *Harpes wegelini* n. sp., *H. costatus* n. sp., *H. scanicus* n. sp. *Arraphus* n. g., *A. corniculatus* n. sp. *Harpides rugosus* S. & B. H. *breviceps* n. sp. *Centropleura* n. g. (J. G. O. Linnarsson, Svenska Vetensk. Akad. Handl., vol. 8, 1869, No. 2, p. 71, cites Angelin's *Centropleura angusticauda* and *C. dicræura* under Dr. Owen's genus *Dikelocephalus*). *Centropleura (Paradoxides) loveni* type, *C. ? dicræura* n. sp., *C. serrata* n. sp., *C. angusticauda* n. sp. *Cryptonymus caudatus* n. sp. *Cybele dentata* Esm. *Cryptonymus striatus* n. sp. *Cybele brevicauda* n. sp. *Pliomera primigena* n. sp. *Liostracus costatus* n. sp. *Bronteus ? nudus*. *Telephus granulatus* n. sp., *T. bicuspis* n. sp., *T. wegelini* n. sp. *Ogygiocaris* n. g., *O. dilatata* Brünn. *Euloma ornatum* n. sp. *Liostracus muticus* n. sp. *Cyrtometopus ? decacanthus* n. sp. *Holometopus ? clatifrons* n. sp. *Bronteus marklini* n. sp.

——— Palæontologia Scandinavica. Part I. Crustacea formationis transitionis. Lipsiæ, 1854, 4to.

Second edition. This work contains 92 pp. and 41 plates, with a description of the genera and species described in parts one and two; omitting pp. 21-24 of Fasciculus No. 1, 1852, which includes a description of the following genera and species: *Proetus stokesi* Murch., *P. concinnus* Dalm., *P. elegantulus* n. sp., *P. lævis* n. sp., *P.? limbatus* n. sp., *P. ? excavatus* n. sp., *P. ? microphthalmus* n. sp. *P.? difformis* n. sp. *Calymene holometopa* n. sp., *C. canaliculata* n. sp., *C. brachymetopa* n. sp., *C. aculeata* n. sp., *C. lejostraca* n. sp., *C. stenometopa* n. sp., also *Calymene tuberculata* Brünn.

——— Palæontologia Scandinavica. Pars I. Crustacea formationis transitionis fasciculi 1 et 2 cum tabulis 48. Holmiæ, 1878 (3d ed.)

This work contains plates 1, 1a, 2, 3, 20, 22, with an appendix, pp. 93-96. Amended and revised by Prof. Gustaf Lindström.
The appendix contains descriptions and figures of the following species: *Paradoxides tessini*, *P. tessini* var. *wahlenbergii*, *P. tessini* var. *œlandicus*, *P. affinis*, *P. tuberculatus*, *P. forchhammeri*, *P. tumidus*. *Centropleura loveni*, *C. steenstrupi*. *Ogygiocaris dilatata*, *O. dilatata* var. *sarsi*, *O. dilatata* var. *strömi*, in addition to the genera and species described in the second edition.
This edition contains the omitted text, pp. 21-24 of the 2d edition, also the revision of the omitted text, pp. 21-29.

——— Palæontologia Scandinavica. Plates A and B.
Plate A was first issued with the second edition of Palæont. Scand., 1854. The work was afterwards revised and republished in 1860, accompanied by Plates A. and B, without text or descriptions.
Fig. 36 a-b, Plate A has received the name of *Beyrichia angelini* Barrande. Regio A, near Andrarum.
Fig. 9 a-c, *Leperditia primordialis* Linnarsson.
Fig. 1, *Leperditia baltica* His. (Barr.)
Plate B exhibits a fragment of *Ceratiocaais* from Regio E, Gothland, which shows 7 to 8 free segments.

Anthony (J. G.) New Trilobites.
In Am. Jour. Sci., 1st series, vol. 34, 1838, p. 379.
Ceratocephala ceralepta, figs. 1 and 2.

——— Description of a new fossil (*Calymene bucklandii*).
In Am. Jour. Sci., 1st series, vol. 36, 1839, p. 106.
In this article a specimen of *Ceraurus pleurexanthemus* Green was figured and described under the name of *Calymene bucklandii* as a new species.

Audouin (J. V.) Recherches sur les rapports naturels qui existent entre les Trilobites et les animaux articulés.
In Annales Gén. Sc. Phys. Nat., Bruxelles, vol. 8, 1821, pl. 126, p. 33; also Isis (oder Encyc. Zeitung), Oken, vol. 1, 1822, p. 87.
Calymene blumenbachii.
An important result of the investigations of this author was gained by the enunciation and establishment of the following principles:
1st. That trilobites differ only from the other articulata in points of minor importance, and that, beyond a doubt, they belong to this group.
2d. That Trilobites exhibit the greatest analogies with the Isopodes. Later investigations, with the discovery of ambulatory appendages, have not changed this classification. Dr. Henry Woodward remarks in the Ency. Britannica, article Crustacea, p. 659, that "there seems, however,

no good reason to urge against the conclusion that the Trilobita were an earlier and more generalized type of crustacea, from which the later and more specialized Isopoda have arisen."

3d. That the want of feet seems to be a necessary characteristic of their skeleton formation, although this point still remains problematical.

Bailey (W. H.) On two new species of Crustacea (*Belinurus* König) from the Coal Measures in Queen's County, Ireland; and some remarks on forms allied to them.

In Rept. 28th Meeting Brit. Assoc. Adv. Sci., 1858, p. 76.

The author rearranges the genus *Belinurus* as a subgenus to *Limulus*, describing two new species under the names of *Belinurus regina* and *B. arcuatus*.

—————— On the annuloid Crustacea of the Coal Measures.

In Jour. Geol. Soc. Dublin, vol. 8, 1858, p. 89.

In this paper a description of a fossil Crustacean from the Bilboa Colliery, Queen's County, was given, with a few remarks on the allied species from Coal Brook Dale, which had been included with it in the genus *Limulus*, and it was proposed to removed all these Coal-Measure Crustacea from that genus and group them into a new one under the name of *Steropis*.

—————— Remarks on some Coal-Measure Crustacea belonging to the genus *Belinurus* König, with descriptions of two new species from Queen's County, Ireland. Pl. 5.

In Annals. Nat. Hist., 3d series, vol. 2, 1863, p. 105. Translated by Milne-Edwards, Annales Gén. Sc. Pys. Nat., Bruxelles, 1863.

The author herein describes *Belinurus reginæ*, *B. arcuatus*, *B. bellulus* König, *B. rotundus*?

Charles König in his work (Icones Fossilium Sectiles, London, 1825) simply mentions the name of *Belinurus bellulus* in his explanation of pl. 18, fig. 230.

—————— Figures of characteristic British fossils, with descriptive remarks.

Pt. 1, pls. 1-15, Cambrian and Lower Silurian, London, 1867; pt. 2, pls. 11-20, 1869; pt. 3, pls. 21-30, 1871; pt. 4, pls. 31-42, 1875.

Part No. 1 contains description of *Calymene tuberculosa* and *Phacops caudatus*.

The following fossil Crustacea are cited from the Cambrian: *Palæopyge ramsayi* Salter. *Hymenocaris vermicauda* Salt. *Agnostus principeps*, Salt. *Paradoxides davidis* Salt. *Olenus micrurus* Salt., *O. cataractes*

Salt., *O. humilis* Phil. *Dikelocephalus ?celticus* Salt. *Conocoryphe invita* Salt.
Fossils of the Lower Llandeilo or Tremadoc Slates: *Conocoryphe depressaa* Salter. *Angelina sedgwickii* Salter. *Asaphus affinis* McCoy. *Niobe homfrayi* Salt. *Psilocephalus innotatus* Salt. *Cheirurus frederici* Salt.
Fossils of the Llandeilo: *Agnostus maccoyii* Salt. *Asaphus tyrannus* Murch. *Ogygia buchii* Brong. *Æglina binodosa* Salt. *Trinucleus fimbriatus* Murch., *T. lloydii* Murch. *Ampyx nudus*, Murch. *Calymene duplicata* Murch.
Part 2, 1869, contains descriptions of Fossils of the Caradoc or Bala: *Cythere? umbonata* Salt. *Primitia maccoyi* Salt. *Beyrichia complicata* Salt. *Agnostus trinodus* Salt. *Illænus bowmani* Salt. *Trinucleus concentricus* Eaton. *Lichas laxatus* McCoy. *Calymene brevicapitata* Port. *Homalonotus bisulcatus* Salt. *Sphærexochus mirus* Beyr. *Phacops brongniartii* Port.
Part 3, 1871—Fossils of the Wenlock: *Calymene blumenbachii* Brong. *Encrinurus punctatus* Brünn., *E. variolaris* Brong. *Illænus barriensis* Murch. *Phacops caudatus* Brünn., *P. downingiæ* Murch. *Homalonotus delphinocephalus* Green. *Proetus latifrons* McCoy. *Beyrichia kloedeni* McCoy.
Fossils of the Ludlow: *Homalonotus knightii* König. *Ceratiocaris papilo* Salter. *Pterygotus problematicus* Ag. *Eurypterus pygmæus* Salt. *Pterygotus bilobus* Salt.
Part 4, 1875—Devonian: *Phacops latifrons* Bronn. *Harpes macrocephalus* Gold. *Homalonotus armatus* Burm. *Bronteus flabellifer* Gold. *Estheria membranacea. Entomis serrato-striata* Sandb. *Pterygotus anglicus* Ag. *Stylonurus symondsii* Salter.
Fossils of the Carboniferous: *Brachymetopus ouralicus* Vern. *Phillipsia pustulata* ,Schl. *Griffithides globiceps* Phillips. *Leperditia subrecta* Port. *Entomoconchus scouleri* McCoy. *Belinurus reginæ* Baily, *B. trilobitoides* Buckland.

——— On Trilobites and other fossils from Lower or Cambro-Silurian strata in the County of Clare.
In Jour. Royal Geol. Soc. Ireland, vol. 17, pt. 1, 1886, p. 21.

Banks (R. W.) On the Tilestones or Downton sandstones, in the neighborhood of Kington, and their contents.
In Quart. Jour. Geol. Soc. London, vol 12, 1856, p. 93, plate.
Pterygotus sp.? *Eurypterus pygmæus* Salter. *Himantopterus banksii* Salter.

Barrande (Joachim). Notice préliminaire sur le système Silurien et les Trilobites de Bohême, Leipsic, 1846, 97 pp.

Étage A.
Étage B.
Étage C: *Paradoxides tessini, P. linnæi, P. rotundatus, P. pusillus.*
Conocephalus sulzeri, C. striatus, C. emmrichii, C. coronatus. Ellipsocephalus hoffi, E. nanus, E. tumidus. Arion n. g., *A. ceticephalus. Sao* n. g., *S. hirsuta. Trilobites decipiens. Battus integer, B. bibullatus, B. nudis, B. granulatus, B. orion, B. affinis, B. rex, B. cuneifer. Hydrocephalus* n. g., *H. carens, H. saturnoides. Monadina* n. g., *M. distincta, M. omicron. Trilobites desideratus.*

Étage D: *Phacops proævus, P. socialis, P. hawlei, P. sclerops, P. elongatus, P. dubius, P. phillipsii. Calymene pulchra, C. parvula, C. incerta. Odontopleura buchii, B. primordialis. Asaphus ingens, A. nobilis. Cheirurus claviger. Trinucleus ornatus, T. goldfussii, T. bucklandi. Caphyra* n. g., *C. radians. Dione* n. g., *D. formosa. Illænus perovalis, I. crassicauda. Trilobites lindaueri. Egle* n. g., *E. rediviva. Battus tardus.*

Étage E: *Phacops fæcundus, P. bulliceps, P. trapeziceps, P. glockeri, P. intercostatus. Tril. heteroclytus. Arethusa* n. g. *A. koninckii. Spærexochus mirus. Asaphus (Nileus) bouchardi. Cheirurus insignis. C. beyrichi, C. quenstedti. Calymene diadamata, C. baylei, C. beaumonti. Staurocephalus* n. g., *S. murchisoni. Lichas scabra, L. propinqua, L. palmata, L. simplex. Harpes tenui punctatus. Odontopleura prevosti. O. dufrenoyi, O. mira, O. verneuili, O. leonhardi, O. minuta, O. tricornis. Cyphaspis burmeisteri, C. depressa. Bronteus partschii, B. haidingeri, B. amhignus. Phæton* n. g., *P. archiaci, P. membranaceus, P. striatus. Proetus ryckholtii, P. intermedius, P. decorus, P. venustus.*

Étage F: *Phacops intemedius, P. breviceps, P. hausmanni, P. lævigatus. Bronteus campanifer, B. palifer, B. angusticeps, B. zippei, B. umbellifer, B. formosus. Lichas haueri, L. parvus. Proetus concinnus, P. lepidus, P. tuberculatus, P. myops, P. inæquicostatus. Harpes ungula. Cheirurus insignis, C. cordæ, C. gibbus. Odontopleura verneuili, O. hörnesii, O. lacerata. Cyphaspis clavifrons, C. cerberus. Trilobites orbitatus.*

Étage G: *Phacops hausmanni, P. spinifer, P. reussii, P. protuberans, P. bronnii. Bronteus brongniarti, B. porosus, B. pustulatus, B. formosus. Cheirurus sternbergi, C. gibbus. Cyphaspis clavifrons, Odontopleura derelicta. Proetus gracilis, P. sculptus, P. lovenii.*

—— Nouveaux Tribolites. Supplément à la notice préliminaire sur le système Silurien et les Trilobites de Bohême. Prague, 1846, 40 pp.

Étage C: *Sao nana.*

Étage D: *Cheirurus claviger, C. scuticauda, C. globosus, C. radiatus. Phacops parabola, P. deshayesii, P. solitarius. Tril. musca. Ampyx portlockii. Odontopleura keyserlingii. Illænus wahlenbergii. I. hisingeri.*

Étage E: *Arethusa nitida. Harpes crassifrons.*

Étage F: *Proetus unguloides, P. complanatus, P. fallax, P. discretus.*

Phæton planicauda, P. latens. Bronteus elongatus, B. brongniarti. Odontocephalus subterarmata.
Étage G: *Harpes d'orbignyanus. ˋCheirurus minutus. Phacops hoeninghausii.*

——— Ueber das Hypostoma und Epistoma, zwei analoge aber verschiedene Organe der Trilobiten.
In Neues Jahrbuch für Mineral., 1847, p. 385; pl. 8.
1849. The author illustrates the hypostomæ of *Phacops socialis, P. breviceps,* also that of *Cheirurus insignis.*
The generic term *Dione* is herein modified to that of *Dionide,* p. 391.

——— *Sao hirsuta* Barrande, ein Bruchstück aus dem "Système silurien du centre de la Bohême."
In Neues Jahrbuch für Mineral., 1849, p. 385, pl. 7.

——— Note on Trilobites.
In Ber. Mittheil. Freunde Naturw., Wien, vol. 4, 1849, p. 353.
Noticed by J. W. Salter.
In Quart. Jour. Geol. Soc. London, vol. 5, 1849, pt. 2, p. 34.

——— *Deiphon forbesii,* neuer Trilobit aus Böhmen.
In Ber. Mittheil. Freunde Naturw., Wien, vol. 5, 1850, p. 6.

——— Versuch einer Klassification der Trilobiten.
In Neues Jahrbuch für Mineral, 1850, p. 769.

——— Système silurien du centre de la Bohême. 1ʳᵉ partie. Recherches paléontologiques. Prague et Paris, 1852.
Vol. 1. Text, Crustacés, Trilobites, and descriptions of the following new genera:
Arionellus n. g. *Arethusina* (generic term changed from that of *Arethus* to *Arethusina*). *Dione* (generic term changed from *Dione* to *Dionide*). *Æglina* n. g. *Cromus* n. g. *Telephus* n. g. *Harpes ungula* Sternb., *H. vittatus* Barr., *H. venulosus* Cord., *H. montagnei* Cord., *H. reticulatus* Cord., *H. crassifrons* Barr., *H. d'orbignyanus* Barr. *Remopleurides radians* Barr. *Paradoxides bohemicus* Boeck.. *P. sacheri* Barr., *P. spinosus* Boeck, *P. rotundatus* Barr., *P. inflatus* Cord., *P. imperialis* Barr., *P. orphanus* Barr., *P. pussillus* Barr., *P. rugulosus* Cord., *P. desideratus* Barr. *Hydrocephalus carens* Barr., *H. saturnoides* Barr. *Sao hirsuta* Barr. *Arionellus ceticephalus* Barr. *Ellipsocephalus hoffi* Schlott, *E. germari* Barr. *Conocephalites sulzeri* Schlott, *C. coronatus* Barr., *C. striatus* Emm., *C. emmrichi,* Barr.
Proetus, Section 1 with 9 segments: *Pretus sculptus* Barr.
Section 2 with 10 segments: *Proetus ryckholti* Barr., *P. frontalis* Cord.,

P. superstes Barr., *P. myops* Barr., *P. unguloides* Barr., *P. orbitatus* Barr., *P. retroflexus* Barr., *P. micropygus* Cord., *P. ascanius* Cord., *P. serus* Barr., *P. lusor* Barr., *P. gracilis* Barr., *P. inæquicostatus* Barr., *P. fallax* Barr., *P. latens* Barr., *P. heteroclytus* Barr., *P. bohemicus* Cord., *P. neglectus* Barr., *P. tuberculatus* Barr., *P. loreni* Barr., *P. memnon* Çord., *P. natator* Barr., *P. insons* Barr., *P. mæstus* Barr., *P. eremita* Barr., *P. curtus* Barr., *P. complanatus* Barr., *P. intermedius* Barr., *P. lepidus* Barr., *P. venustus* Barr., *P. decorus* Barr., *P. astianax*, Cord., *P. archiaci* Barr., *P. planicauda* Barr. (*Proetus* (*Phaëtonellus*) *planicauda* Novák), *P. striatus* Barr. *Phillipsia parabola* Barr. *Cyphaspis halli* Barr., *C. burmeisteri* Barr., *C. barrandei* Cord., *C. cerberus* Barr., *C. davidsoni* Barr., *C. convexa* Cord., *C. novella* Barr., *C. humillima* Barr., *C. depressa* Barr. *Arethusina konincki* Barr., *A. nitida* Barr. *Phacops cephalotes* Cord., *P. sternbergi* Cord., *P. intermedius* Barr., *P. bœcki* Cord., *P. fecundus* Barr., *P. breviceps* Barr., *P. bronni* Barr., *P. miser* Barr., *P. signatus* Cord., *P. volborthi* Barr., *P. glockeri* Barr., *P. trapeziceps* Barr., *P. bulliceps* Barr. *Dalmania hausmanni* Broug., *D. auriculata* Dalm., *D. spinifera* Barr., *D. reussi* Barr., *D. rugosa* Cord., *D. cristata* Cord., *D. fletcheri* Barr., *D. maccoyi* Barr., *D. hawlei* Barr., *D. deshayesi* Barr., *D. dubia* Barr., *D. socialis* Barr., *D. solitaria* Barr., *D. phillipsi* Barr., *D. morrisiana* Barr., *D. orba* Barr. *Calymene blumenbachi* Broug., *C. diademata* Barr., *C. incerta* Cord., *C. declinata* Cord., *C. parvula* Barr., *C. baylei* Barr., *C. tenera* Barr., *C. pulchra* Barr. *Homalonotus bohemicus* Barr., *H. rarus* Cord. *Lichas scabra* Beyr., *L. palmata* Barr., *L. haweri* Barr., *L. ambigua* Barr., *L. heteroclyta* Barr., *L. simplex* Barr. *Trinucleus bucklandi* Barr., *T. ornatus* Sternb., *T. goldfussi* Barr., *T. ultimus* Barr. *Ampyx portlocki*, Barr., *A. rouaulti* Barr., *Dionide formosa* Barr. *Asaphus nobilus* Barr., *A. ingens* Barr. *Æglina rediviva* Barr., *A. speciosa* Cord., *A. pachycephalus* Cord. *Illænus hisingeri* Barr., *I. panderi* Barr., *I. wahlenbergianus* Barr., *I. salteri* Barr., *I. transfuga* Barr., *I. bouchardi* Barr., *I. tardus*. *Acidaspis primordialis* Barr., *A. keyserlingi* Barr., *A. tremenda* Barr., *A. verneuili* Barr., *A. vesiculosa* Beyr., *A. buchi* Barr., *A. leonhardi* Barr. *A. hörnesi* Barr., *A. geinitziana* Cord., *A. rœmeri* Barr., *A. dormitzeri* Cord., *A. minuta* Barr., *A. pectinifera* Barr., *A. derelicta* Barr., *A. ruderalus* Cord., *A. propinqua* Barr., *A. mira*, *A. prévosti* Barr., *A. dufrénoyi* Barr., *A. portlocki* Cord., *A. tricornis* Barr., *A. solitaris* Barr., *A. lacerata* Barr., *A. truncata* Cord., *A. subterarmata* Barr., *A. radiata* Goldf., *A. monstrosa* Barr., *A. laportei* Cord., *A. grayi* Barr. *Cheirurus clavinger* Beyr., *C. globosus* Barr., *C. insocialis* Barr., *C. tumescens* Barr., *C. scuticauda* Barr., *C. insignis* Beyr., *C. obtusatus* Cord., *C. hawlei* Barr., *C. beyrichi* Barr., *C. quenstedti* Barr., *C. gibbbus* Beyr., *C. sternbergi* Boeck., *C. cordai* Barr., *C. pauper* Barr., *C. bifurcatus* Barr., *C. minutus* Barr. *Placoparia zippei* Cord. *Sphaerexochus mirus* Beyr. *Staurocephalus murchisoni* Barr. *Deiphon forbesi* Barr. *Dindymene friderici-augusti* Cord., *D. haidingeri* Barr. *Amphion lindaueri* Barr. *Cromus intercostatus*

Barr., *C. beaumonti* Barr., *C. bohemicus* Barr., *C. transiens* Barr. *Bronteus elongatus* Barr., *B. sieberi* Cord., *B. thysanopeltis* Barr., *B. campanifer* Beyr., *B. dormitzeri* Barr., *B. zippei* Barr., *B. cælebs* Barr., *B. fosmosus* Barr., *B. oblongus* Cord., *B. kutorgai* Barr., *B. transversus* Cord., *B. viator* Barr., *B. furcifer* Cord., *B. palifer* Cord., *B. simulans* Barr., *B. planus* Cord., *B. brongniarti* Barr., *B. tenellus* Barr., *B. partschi* Barr., *B. angusticeps* Barr., *B. haidingeri* Barr., *B. nuntius* Barr., *B. spinifer* Barr., *B. umbellifer* Beyr., *B. edwardsi* Barr., *B. porosus* Barr., *B. brevifrons* Barr., *B. infaustus* Barr., *B. richteri* Barr., *B. pustulatus* Barr., *B. hawlei* Barr. *Telephus fractus* Barr. *Agnostus integer* Beyr., *A. nudus* Beyr., *A. bibullatus* Barr., *A. rex* Barr., *A. garanulatus* Barr., *A. tardus* Barr. *Trilobites inchoatus* Barr., *T. musca* Barr., *T. infaustus* Barr., *T. mutilus* Barr., *T. ferus* Barr., *T. orphanus* Barr. *Harpes naumanni* Barr. *Paradoxides lyelli* Barr., *P. expectans* Barr. *Dalmanites angelini* Barr. *Acidaspis desiderata* Barr. *Cheirurus neglectus* Barr. *Acidaspis rebellis* Barr. *A. hawlei* Barr. *Deiphon forbesi* Barr. *Harpides grimmi* Barr. *Phacops hœninghausi* Barr., *P. emarginatus* Barr. *Calymene interjecta* Cord. *Illænus distinctus* Barr.

—————— On *Ceratiocaris bohemicus*.

In Neues Jahrbuch für Mineral., 1853, pt. 3, pp. 335-347.
On p. 335 of this article there is a review of Dr. D. D. Owen's Rept. Geol. Survey Wisconsin, Iowa, and Minnesota.
The author gives a rough wood cut of *Ceratiocaris bohemicus*.

—————— Note sur quelques nouveaux fossiles découverts aux environs de la ville de Rokitzan, dans le bassin silurien du centre de la Bohème.

In Bull. Soc. Géol. France, 2d series, vol. 13, 1856, p. 532.
Mere mention of locality and period.

—————— Wiedererscheinung der Gattung *Arethusina* Barrande.

In Neues Jahrbuch für Mineral., 1858, p. 257, 1 pl.
Arethusina sandbergeri Barr., *A. konincki* Barr.

—————— Documents anciens et nouveaux sur la faune primordiale et le Système Taconique en Amérique.

In Bull. Soc. Géol. France, 2d series, vol. 18, 1861, pp. 203-321, pls. 4, 5.
Paradoxides asaphoides Emmons. *P. thompsoni* Hall. *P. macrocephalus* Emmons. *P. vermontana* Hall. *Peltura (Olenus) holopyga* Hall. *Atops trilineatus* Emmons. *A. punctatus* Emmons. *Triarthrus becki* Green. *Microdiscus quadricostatus* Emmons.

―――― I. Réapparition du genre *Arethusina* Barrande. II. Faune silurienne des environs de Hof, en Bavière. Prague et Paris, 1868.

Arethusina sandbergeri, *A. konincki*. *Conocephalites bavaricus* Barr., *Bavarilla* (n. sub. gen. *Conocephalites*), *B. hofensis* Barr. *Conocephalites? münsteri* Barr., *C. geinitzi* Barr., *C. wirthi* Barr., *C.? problematicus* Barr., *C. quæsitus* Barr., *C. innotatus* Barr., *C. extremus* Barr., *C. deficiens* Barr., *C. discrepans* Barr. *Olenus guembeli* Barr., *O. frequens* Barr., *O. expectans* Barr. *Agnostus bavaricus* Barr. *Asaphus wirthi* Barr. *Lichas primulus* Barr. *Calymene tristani* var. *bavarica* Barr. *Cheirurus gracilis* Barr., *C. discretus* Barr. *Trilobites prævalens* Barr., *T. corpulentus* Barr. *Trilobites V. X. Y. and Z.*

―――― Trilobites. I. Résumé général de nos études sur l'évolution des Trilobites. II. Distribution verticale des Trilobites dans le bassin silurien de la Bohême. III. Parallèle entre les Tribolites et Céphalopodes siluriens de la Bohême. IV. Epreuve des théories paléontologiques par la réalité. Prague et Paris, 1871.

―――― Système silurien du centre de la Bohême. 1re partie. Recherche paléontologiques. Supplément au vol. 1. Trilobites. Crustacés divers et Poissons. Prague et Paris, 1872, 32 pls.

Trilobites. *Carmon* n. g. *Telephus* n. g. *Triopus*.

Phyllopoda. *Ceratiocaris*. *Aptychopsis* n. g. *Cryptocaris* n. g. *Pterocaris* n. g.

Ostracodes. *Aristozoe* n. g. *Bolbozoe* n. g. *Callizoe* n. g. *Caryon* n. g. *Crescentilla* n. g. *Cytheropsis* n. g. *Elpe* n. g. *Hippa* n. g. *Nothozoe* n. g. *Orozoe* n. g. *Zonozoe*, n. g.

Cirrhipediæ. *Anatifopsis* n. g. *Plumulites* n. g. *Bactropus* n. g. *Dryalus* n. g.

Harpes benignensis Barr., *H. crassifrons* Barr., *H. naumanni* Barr., *H. primus* Barr., *H. transiens* Barr. *Remopleurides radians* Barr. *Paradoxides expectans* Barr., *P. pusillus*, *P. rugulosus* Cord. *Ellipsocephanus germari* Barr., *E. hoffi* Schlot. *Proetus comatus* Barr., *P. frontalis* Cord., *P. lusor* Barr., *P. micropygus* Cord., *P.? perditus* Barr., *P.? primulus* Barr., *P. superstes* Barr., *P. vicinus* Barr. *Phillipsia parabola* Barr. *Carmon mutilus* Barr., *C. primus* Barr. *Cyphaspis coronatus* Barr., *C. sola* Barr. *Harpides grimmi* Barr. *Phacops fecundis* Barr., var. *communis* Barr., also var. *degener* Barr., *P. fugitivus* Barr., *P. modestus* Barr. *Dalmanites angelini* Barr., *D. atavus* Barr., *D. hawlei* Barr., *D. mac-*

coyi Barr., *D. oriens* Barr., *D. perplexus* Barr., *D. spinifera* Barr. *Calymene arago* Rou., *C. bifida* Barr., *C. blumenbachi* Brong., *C. pulchra* Barr. *Homalonotus bohemicus* Barr., *H. inexpectatus* Barr., *H. medius* Barr., *H. minor* Barr., *H. rarus* Cord. *Lichas avus* Barr., *L. branikensis* Barr., *L. incola* Barr., *L. rudis* Barr. *Trinucleus bucklandi* Barr., *T. reussi* Barr., *T. ultimus* Barr. *Ampyx gratus* Barr., *A. portlocki* Barr., *A. tenellus* Barr. *Dionide formosa* Barr. *Asaphus alienus* Barr., *A. nobilis* Barr., *A. quidam* Barr. *Ogygia desiderata* Barr., *O. discreta* Barr., *O. sola* Barr. *Barrandia crassa* Barr. *Aeglina armata* Barr., *A. princeps* Barr., *A. prisca* Barr., *A. rediviva* Barr., *A. speciosa* Cord., *A. sulcata* Barr. *Illænus advena* Barr., *I. aratus*, *I. bohemicus* Barr., *I. bouchardi* Barr., *I. calvus* Barr., *I. hospes* Barr., *I. katzeri* Barr., *I. oblitus* Barr., *I. puer* Barr., *I. zeidleri* Barr. *Acidaspis buchi* Barr., *A. derelicta* Barr., *A. monstrosa* Barr., *A. orphana* Barr., *A. peregrina* Barr., *A. pigra* Barr., *A. rara* Barr., *A. sparsa* Barr., *A. spoliata* Barr., *A. tricornis ?* Barr., *A. ursula* Barr., *A. victima* Barr. *Cheirurus comes* Barr., *C. completus* Barr., *C. cordai* Barr., *C. fortis* Barr., *C. gryphus* Barr., *C. insocialis* Barr., *C. neglectus* Barr., *C. neuter* Barr., *C. pater* Barr., *C. pectinifer* Barr., *C. sternbergi* Boeck, *C. vinculum* Barr. *Areia bohemica* Barr., *A. fritschi* Barr. *Placoparia zippei*, *P. gandis* Cord. *Sphærexochus mirus* Beyr., *S. bohemicus* Barr, *S. latens* Barr., *S. ? ultimus* Barr. *Deiphon forbesi* Barr. *Dindymene bohemica* Barr., *D. frid. augusti* Cord. *Amphion senilis* Barr. *Cromus beaumonti* Barr., *C. bohemicus* Barr. *Bronteus acupunctatus* Barr., *B. asperulus* Barr., *B. billingsi* Barr., *B. binotatus* Barr., *B. clementinus* Barr., *B. expectans* Barr., *B. ivanensis* Barr., *B. magus* Barr. *B. palifer* Beyr., *B. perlongus* Barr., *B. pustulatus* Barr., *B. rhinoceros* Barr., *B. scharyi* Barr., *B. sosia* Barr., *B. tardissimus* Barr., *B. thysanopeltis* Barr., *B. umbellifer* Beyr. *Bohemilla stupenda* Barr. *Triopus draboviensis* Barr. *Agnostus caducus* Barr., *A. nudus* Beyr., *A. perrugatus* Barr., *A. similaris* Barr., *A. tardus* Barr. *Trilobites contumax* Barr., *T. expectatus* Barr., *T. incongruens* Barr. *Æglina gigantea* Barr. *Acidaspis prévosti* Barr. *Bronteus furcifer* Cord., *B. gervilleicans* Barr., *B. indocilis* Barr., *B. infaustus* Barr.

Crustaces divers non Trilobitiques. Phyllopodes: *Ceratiocaris bohemicus* Barr., *C. debilis* Barr., *C. decipiens* Barr., *C. docens* Barr., *C. inæqualis* Barr., also var. *decurtata* Barr., *C. primulus* Barr., *C. scharyi* Barr., *C. tardus* Barr. *Aptychopsis primus* Barr. *Atychopsis ?* *Cryptocaris bohemica* Barr., *C. contracta* Barr., *C. obsoleta* Barr., *C. pulcha* Barr., *C. solida* Barr., *C. suavis*, *C. tardissima* Barr., *C. ? rhomboidea* Barr. *Pterocaris bohemica* Barr.

Ostracodes: *Aristozoe amica*, *A. bisulcata* Barr., *A. inclyta* Barr., *A. jonesi* Barr., *A. lepida* Barr., *A. memoranda* Barr., *A. orphana* Barr., *A. perlonga* Barr., *A. regina* Barr. *Beyrichia bohemica* Barr., *B. hastata* Barr., *B. ? barbara* Barr. *Bolbozoe anomala* Barr., *B. bohemica* Barr., *B. jonesi* Barr. *Callizoe bohemica* Barr. *Caryon bohemicum* Barr. Cres-

centilla pugnax Barr. *Cythere ? bohemica* Barr. *C.? paradoxa* Barr. *Cytheropsis derelicta* Barr., *C.? melonica* Barr., *C. testis* Barr. *Elpe inchoata* Barr., *E. pinguis* Barr. *Entomis dimidiata* Barr., *E. migrans* Barr., *E. pelagica* Barr., *E. rara* Barr. *Hippa latens* Barr., *H. rediviva* Barr. *Leperditia desiderata* Barr., *L. fragilis* Barr., *L. rarissima* Barr., *L. solitaria* Barr. *Isochilina ? formosa* Barr. *Nothozoe pollens* Barr. *Orozoe mira* Barr. *Primitia consobrina* Barr. *P. debilis* Barr., *P. fugax* Barr., *P. fusus* Barr., *P. modesta* Barr., *P. monas* Barr., *P. perforata* Barr., *P. prunella* Barr., *P. socialis* Barr., *P. tarda* Barr., *P. timida* Barr., *P. transiens* Barr. *Zonozoe drabaviensis* Barr., *Z. complexa* Barr.

Eurypterides: *Pterygotus bohemicus* Barr., *P. comes* Barr., *P. cyrtochela* Barr., *P. expectatus* Barr., *P. kopaninensis* Barr., *P. mediocris* Barr., *P. nobilis*, Barr. *Eurypterus pugio* Barr.

Cirrhipedes: *Plumulites bohemicus* Barr., *P. compar* Barr., *P. contrarius* Barr., *P. delicatus* Barr., *P. discretus* Barr., *P. folliculum* Barr., *P. fraternus* Barr., *P. minimus* Barr., *P. regius* Barr., *P. squamatula* Barr. *Anatifopsis bohemica* Barr., *A. acuta* Barr., *A. longa* Barr., *A. prima* Barr.

Fossiles incertae sedis: *Bactropus longipes* Barr., *B. tenuis* Barr. *Dryalus obscurus* Barr.

—— and **Verneuil** (E. de). See **Verneuil** (E. de) and **Barrande** (Joachim).

Barrett (S. T.) The coralline or Niagara limestone of the Appalachian system as represented at Nearpass's Cliff, Montague, N. J.

In Am. Jour. Sci., 3d series, vol. 15, 1878, p. 370.
Proetus pachydermatus n. sp.

—— Description of a new Trilobite, *Dalmanites dentata*.

In Am. Jour. Sci., 3d series, vol. 11, 1876, p. 200, plate.

—— On *Dalmanites dentata*.

In Am. Jour. Sci., 3d series, vol. 12, p. 70.

Barris (W. H.) New fossils from the Corniferous formation at Davenport.

In Davenport Acad. Nat. Sci., vol. 2, 1879, p. 282, pls. 10 and 11.
Proetus davenportensis n. sp., pl. 11, fig. 8.

Barrois (Charles). Note sur les fossiles de Cathierville.

In Bull. Soc. Géol. France, 3d series, vol. 8, 1880, p. 266, plate 7.
Phacops fecundus, Dalmanites hausmani.

―――― Recherches sur les terrains anciens des Asturies et de la Galice.
In Mém. Soc. Géol. Nord, Lille, vol. 2. 1882, 630, pp. 20 pls.
Paradoxides pradoanus Barr., *P. barrandei* Barris. *Conocephalites sulzeri* Zenk., *C. ribeiro* Barr., *C. castroi* n. sp. *Arionellus ceticephalus* Barr. *Homalonotus pradoanus* Vern. *Phacops latifrons* Bronn. *Phillipsia brongniarti* Fischer, *P. castroi* n. sp., *P. derbyensis* Mart. *Entomis grand-uryi* n. sp. *Calymene tristani*. *Illænus hispanicus* Vern.

―――― Sur les faunes siluriennes de la Haute-Garonne. Pls. 6, 7.
In Annales Soc. Géol. Nord, Lille, vol. 10, 1883, p. 151.
Dalmanites gourdoni n. sp. *Cyphaspis belloci* n. sp. *Phacops fecundus* Barr. *Lichas* sp. *Ceratiocaris bohemicus* Barr.

―――― Les appendices des Trilobites.
In Annales Soc. Géol. Nord, Lille, vol. 11, 1883-84, p. 228.

―――― Mémoire sur le calcaire à polypiers de Cabrières (Hérault).
In Annales Soc. Géol. Nord, Lille, vol. 13, 1885-86, p. 94, plate.
Phacops latifrons Bronn., var. *occitanicus* Trom. Grass. *Bronteus meridionalis* Trom. Grass.

―――― Sur la faune de Hont-de-Ver (Haute-Garonne).
In Annales Soc. Géol. Nord, Lille, vol. 13, 1885-6, pls. 2, 3.
Lichas gourdonni n. sp. *Harpes pyrenaicus* n. sp. *Bronteus raphaeli* n. sp. *B. trutati* n. sp. *Dalmanites courdoni* Barrois. *Cyphaspis belloci* Barrois. *Phacops fecundis* Barr. *P. breviceps ?* Barr.

―――― Sur le calcaire dévonien de Chaudefonds (Maine-et-Loire).
In Annales Soc. Géol. Nord, Lille, vol. 13, 1885-86, p. 170.
Cheirurus gibbus Beyr. *Acidaspis vesiculosa* Beyr. *Bronteus canaliculatus* Gold. *Harpes macrocephalus* Goldf.

―――― Aperçu de la constitution géologique de la rade de Brest.
In Bull. Soc. Géol. France, 3d series, vol. 14, 1886, pls. 31-33.
Homalonotus le hiri Barrois.

―――― Faune du Calcaire d'Erbray (Loire Inférieure). Contribution à l'étude du terrain Devonien à l'ouest de la France. Lille, 1889, 346 pp., 17 plates.
Harpes venulosus Corda. *Bronteus gervillei* Barr. *Cheirurus sternbergi ?*

Boeck. *Cryphæus pectinatus* A. Roemer. *Phacops fecundus* Barr. *Proetus bohemicus* Corda. *P. fallax* Barr. *P. ligeriensis* n. sp. *P. gosseleti* n. sp. *P. cornutus* Goldf. *P. vicinus* Barr.

—————— Mémoire sur la Faune du Grès Armoricain.
Soc. Géol. du Nord Annales xix, 1891, p. 134, 5 plates.
Myocaris lutraria Salter. *Ceratiocaris* sp.? *Mesothyra oceani* Hall

Baumer (J. W.) Naturgeschichte des Mineralreichs mit besonderer Anwendung auf Thüringen. 2 vols. Gotha, 1763–64, 1767.
The author gives a brief description of "trigonella striata conchites trilobus striatus." vol. 1, p. 328, pl. xi, fig. 32 a-b. (Encrinurus sp.)

Bayle (E.) Explication de la carte géologique de la France. Vol. 4. Atlas, Paris, 1878.
Première partie. Fossiles principaux des terrains.
Plate 1: *Megalaspis desmaresti* Broug., figs. 1–3.
Plate 2: *Homalonotus gervillei* DeVer., figs. 1, 3 and 6. *H. brongniarti* Deslong., fig. 2. *H. vicaryi* Salter, fig. 4. *H. deslongchampsi* Tromelin, fig. 5.
Plate 3: *Calymene tristani* Broug., figs. 1–9. *C. aragoi* Marie Rouault, fig. 10.
Plate 4: *Cheirurus* sp., fig. 1. *Illænus* sp., fig. 2. *Phacops latifrons* Bronn., figs. 4 and 5. *P. potieri*, figs. 7–10. *Cryphæus michilini* Marie Rouault, figs. 18–21. *Goldfius gerville* Barb. *Proetus æhlerti*. *Phillipsia gemmulifera* Phil., fig. 22.
Plate 5: *Acidaspis buchi* Barr., fig. 1.

Bean (W.) Entomostracans from the Coal Measures of Newcastle-upon-Tyne.
In Mag. Nat. Hist. (London), vol. 9, 1836, p. 376.
Cypris arcuata now *Beyrichia arcuata*.

Beckmann (J.) Commentatio de reductione rerum fossilium ad genera naturalia protyporum.
In Novi Comm. Soc. Scient. Götting., vol. 3, 1873, p. 95.

Beecher (C. E.) *Ceratiocaridæ* from the Upper Devonian Measures in Warren County.
In Second Geol. Survey Pennsylvania, vol. PPP, 1884, 2 pls.
Elymocaris n. g. *Tropidocaris* n. g.
Author's edition issued separately, entitled: "*Ceratiocaridæ* from the Chemung and Waverly Groups at Warren, Pa.," with 2 pls., Harrisburg, 1884, 22 pp.
Elymocaris punctata Hall. *E. socialis* n. sp. *Elymocaris siliqua* n. sp. *Tropidocaris bicarinata* n. sp. *T. interrupta* n. sp. *T. alternata* n. sp.

Belt (Thomas). On some new Trilobites from the Upper Cambrian rocks of North Wales.
In Geol. Mag., vol. 4, London, 1867, p. 294, pl. 12.
Agnostus pisiformis Linné var. *obesus*, *A. nodosus* n. sp. *Olenus gibbosus*.

—— On the Lingula flags or Festiniog group of the Dolgelly district. Part iii.
In Geol. Mag., vol. 5, London, 1868, p. 5, pl. 2.
Conocoryphe ? williamsonii n. sp. *C. ? longispina* n. sp. *C. ? bucephala* n. sp. *Sphærophthalmus bisulcatus* Phil. *Agnostus obtusus* n, sp. *A. trisectus* Salt. *A. barlowii* n. sp.

Bergeron (M. T.) Etude paléontologique et stratigraphique des terrains anciens de la Montagne Noire.
In Bull. Soc. Géol. France, 3rd series, vol. 15, 1887, No. 5.
Harpes. Phacops. Cheirurus.

Bergeron (Jules). Sur une formes nouvelle de Trilobite de la famille des Calymenidæ (genre *Calymenella*).
In Bull. Soc. Céol. France, 3d series, vol. 18, 1890, p. 365, plate 5.
Calymene declinata Barr. *Homalonotus delphinocephalus* Green. *Calymenella boisseli* n. sp. *C. bayani* Trom. & Lebesc.
Diagnosis of *Calymenella*: Glabella slighly convex, rounded in front, having three furrows, of which the last two are well marked. The posterior furrow is inclined towards the rear, lobes slightly elevated, fixed cheeks wide, limb well developed in front of the glabella sometimes ending in a point. Pygidium the same as Calymene.
This genus includes within its limits the American species *Calymene rostrata* Vogdes, and *C. nasuta* Ulrich.

Beushausen (Louis). Beiträge sur Kenntniss des oberharzer Spiriferensandsteins und seiner Fauna.
Inaugural-Dissertation. Göttingen, 1884, p. 31.
Liste der aus dem Oberharzer Spiriferensandstein beschriebenen Arten ihrer Verbreitung im rheinischen Unterdevon.

Beyrich (Ernst). Ueber einige böhmische Trilobiten. Berlin, 1845, 1 pl.
Cheirurus n. g. *Sphærexochus* n. g. *Trochurus* n. g.
Cheirurus insignis n. sp. *C. claviger* n. sp. *C. sternbergii* Boek. *C. gibbus* n. sp. *Sphærexochus mirus* n. sp. *Lichias laciniata* (Lovén.) *L. scabra* n. sp. *L. boltoni* Green. *L. dissidens* n. sp. *Trochurus speciosus* n. sp. *Bronteus umbellifer* n. sp. *B. palifer* n. sp. *B. campanifer* n. sp. *B. laticauda* Wahl. *Battus integer* n. sp. *B. nudus* n. sp.

―――― Untersuchunger über Trilobiten. Zweites Stück als Fortsetzung zu der Abhandlung "Ueber einige böhmische Trilobiten." Berlin, 1846, 3 pls.

Harpides n. g. *Cheirurus gibbus* Beyr., *C. exsul* n. sp., *C. ornatus* Dalm. *Sphærexochus* n. sp.? *S. mirus* Beyr., (*S. beyrichi* Lindstom), pl. 1, fig. 9. *Lichas laciniata* n. sp., *L. angusta* n. sp., *L. tricuspidata* n. sp., *L. scabra* n. sp. *Aryes armatus* Gold., *A. speciosus* Beyr., *A. anglicus* n. sp. *Bronteus campanifer* Beyr., *B. palifer* Beyr., *B. pendulus*, *B. insignitus* Römer. *Odontopleura ovata* Emm., *O. mutica* Emm., *O. inermis* n. sp., *O. brightii* Murch., *O. cornuta* n. sp., *O. vesiculosa* n. sp. *Calymene diademata* Barr., *C. parvula* Barr., *C. pulchra* Barr. *Proetus concinnus*, *P. cornutus*. *Trinucleus ornatus* Sternb. *Harpes ungula* Sternb. *Harpides hospes* n. sp. *Staurocephalus murchisoni*.

―――― Ueber *Leaia leidyi*.

In Zeitsch. Deutsch. geol. Gessel. vol. 16, 1864, p. 363.

Bigsby (J. J.) A sketch of the Geology of the Island of Montreal.

In Annals Lyc. Nat. His. N. Y. vol. 1, 1825, p. 198.

The author figures a trilobite (*Trinucleus concentricus*) pl. 15, fig. 1.

Bigsby (J. J.) Description of a new species of Trilobite.

In Jour. Acad. Nat. Sc. Phila., vol. 4, pt. 2, 1825, p. 365, pl. 23.
Paradoxides boltoni (*Lichas boltoni*).

―――― Thesaurus Siluricus. The flora and fauna of the Silurian period. London, 1868.

Billings (E.) Fossil foot prints.

In Canadian Naturalist, vol. 1, 1856, p. 35, wood cuts.
Protichnites septem-notatus.

―――― On some of the characteristic fossils of the Lower Silurian rocks of Canada.

In Canadian Naturalist, vol. 1, 1856, p. 39, 11 wood cuts.

The author figures *Isotelus gigas*, *Calymene senaria*, and the hypostoma of *Isotelus gigas*.

―――― Fossils of the Upper Silurian rocks, Niagara and Clinton goups.

In Canadian Naturalist, vol. 1, 1856, p. 57, 1 pl.
Phacops limulurus Green.

―――― Description of fossils occurring in the Silurian rocks of Canada.
In Canadian Naturalist, vol. 1, 1856, p. 312, 10 wood cuts.
Homalonotus delphinocephalus Green.

―――― Report for the year 1856 of E. Billings, Esq., palæontologist, addressed to Sir Wm. E. Logan, etc.
In Geol. Survey Canada Rept. Progress, 1853-54-55-56, p. 247.
Triarthrus spinosus n. sp. *Acidaspis horani* n. sp. *Bronteus lunatus* n. sp.

―――― Fossils of the Calciferous sandrock, including those of a deposit of white limestone at Mingan, supposed to belong to the formation.
In Canadian Naturalist, vol. 4, 1859, p. 345, 12 wood cuts.
Bathyurus n. g., *B. extans* Hall (type), *B. amplimarginatus* n. sp., *B. conicus* n. sp., *B. cybele* n. sp. *Leperditia anna* Jones.

―――― Descriptions of some new species of Trilobites from the Lower and Middle Silurian rocks of Canada.
In Canadian Naturalist, vol. 4, 1859, p. 367, 18 wood cuts.
Illænus globosus n. sp., *I. bayfieldi* n. sp., *I. americanus* n. sp., *I. conradi* n. sp., *I. milleri* n. sp., *I. trentonensis* Emmons, *I. ovata* Con., *I. angusticollis* n. sp., *I. conifrons* n. sp., *I. clavifrons* n. sp., *I. arcturus* Hall., *I. orbicauda* n. sp., *I. grandis* n. sp., *Amphion canadensis* n. sp., *Triarthrus glaber* n. sp., *T. beckii* Green., *T. glaber* Billings.

―――― Fossils of the Chazy limestone, with descriptions of new species.
In Canadian Naturalist, vol. 4, 1859, p. 426, 28 wood cuts.
Bathyurus angelini n. sp. *Harpes antiquatus* n. sp.

―――― Description of some new species of fossils from the Lower and Middle Silurian rocks of Canada.
In Canadian Naturalist, vol. 5, 1860, p. 49, 16 wood cuts.
Dalmanites bebryx n. sp., *D. achates* n. sp. *Phacops orestes* n. sp. *Cheirurus icarus* n. sp. *Proetus alaricus* n. sp.

―――― On some new species of fossils from the limestone near Point Levi, opposite Quebec.
In Canadian Naturalist, vol. 5, 1860, p. 301, 30 figs.
Agnostus americanus n. sp., *A. orion* n. sp., *A. canadensis* n. sp. *Conocephalites zenkeri* n. sp. *Dickelocephalus magnificus* n. sp., *D. planifrons*

n. sp., *D. oweni* n. sp., *D. belli* n. sp., *D. negalops* n. sp., *D. cristatus* n. sp. *Arionellus cylindicus* n. sp., *A. sublavatus* n. sp. *Menocephalus sedgwicki* n. sp., *M. globosus* n. sp. *Bathyurus capax* n. sp., *B. dubius* n. sp., *B. bituberculatus* n. sp., *B. armatus* n. sp., *B. saffordi* n. sp., *B. oblongus* n. sp., *B. cordai* n. sp., *B. quadratus* n. sp. *Cheirurus apollo* n. sp., *C. eryx* n. sp. *Asaphus illænoides* n. sp., *A. goniurus* n. sp.

—— On some of the rocks and fossils occurring near Phillipsburg, Canada East.
In Canadian Naturalist, vol. 6, 1861, p. 310, 6 figs.
Bathyurus saffordi Bill., *B. cordai* Bill. *Menocephalus globosus* Bill.?
Amphion salteri n. sp.

—— On some new or little-known species of Lower Silurian fossils from the Potsdam group (Primordial zone). Montreal, 1861, 24 pp.
This pamphlet, with the addition of a description of *Archæocyathus profundus*, the suppression of that of *A. minganensis*, and the notes on pages 10, 11 and 12, was reprinted in the "Palæozoic Fossils of Canada," vol. 1, 1865, pp. 1-24, also in part in the "Geol. Vermont," vol. 2, 1861, pp. 942-960.
Olenellus thompsoni Hall., *O. vermontana* Hall. *Conocephalites miser* n. sp., *C. adamsi* n. sp., *C. teucer*, *C. vulcanus* n. sp., *C. arenosus* n. sp. *Bathyurus senectus* n. sp., *B. parvulus* n. sp. *Ampyx halli* n. sp.

—— Description of a new species of *Harpes* from the Trenton limestone, Ottawa.
In Canadian Naturalist, vol. 8, 1863, p. 36, wood cut.
Harpes dentoni n. sp.

—— Geological Survey of Canada. Report of progress from its commencement to 1863. Illustrated by 498 wood cuts in the text. Montreal, 1863. Atlas of maps and sections: contains illustrations of species under the following genera:
Harpes. Paradoxides. Conocephalites. Triarthrus. Bathyurus. Proetus. Phacops. Dalmanites. Calymene. Lichas. Trinucleus. Asaphus. Illænus. Acidaspis. Cheirurus. Amphion. Encrinurus. Bronteus. Beyrichia. Leperditia. Isochilina. Eurypterus. Cytheropsis.

—— Description of new species of *Phillipsia* from the Lower Carboniferous rocks of Nova Scotia.
In Canadian Naturalist, vol. 8, 1863, p. 209, 1 wood cut.
Phillipsia howi.

———— Description of some new species of fossils, with remarks on others already known, from the Silurian and Devonian rocks of Maine.
In Proc. Portland Soc. Nat. Hist., vol. 1, 1869, p. 104, plate.
Dalmanites epicrates n. sp.   *Cheirurus tarquinius* n. sp.   *Proetus junius* n. sp., *P. macrobius* n. sp.   *Bronteus pompilius* n. sp.   *Phacops trajanus* n. sp.   *Lichas.?*

———— Palæozoic fossils. Vol. 1. Containing descriptions and figures of new or little known species of organic remains from the Silurian rocks. 1861–65. 426 pp. 401 figures. Geol. Survey Canada.

The first portion of this work, consisting of 24 pages, was issued in November, 1861, and re-appears in this work, with the exception of the notes on pages 10, 11, and 12 of the original edition. The second part, consisting of pp. 25–56, was issued in January, 1862, and the third, pp. 57–168, in June, 1862; the fourth, pp. 169–344, in February, 1865.

Agnostus americanus, A. canadensis, A. fabius, A. galba, A. orion. Amphion barrande, A. canadensis, A. convexus, A. julius, A. insularis, A. salteri, A. westoni. Ampyx halli, A. læviusculus, A. normalis, A. ratilius, A. semicostatus. Arionellus cylindricus, A. subclavatus. Asaphus canalis, A. curiosus, A. goniurus, A. huttoni, A. illænoides, A. morrisii. A. pelops, A. quadraticaudatus. Bathyurus amplimarginatus, B. arcuatus, B. armatus, B. bituberculatus, B. breviceps, B. caudatus, B. capax, B. conicus, B. cordai, B. cybele, B. dubius, B. gregarius, B. minganensis, B. nero, B. oblongus, B. parvulus, B. perplexus, B. perspicator, B. quadratus, B. saffordi, B. senectus, B. smithii, B. solitarius, B. strenuus, B. timon, B.vetulus. Bathyurellus n.g., *B. abruptus, B. expansus, B. formosus, B. fraternus, B. litoreus, B. marginatus, B. nitidus, B. rarus, B. validus.   Beyrichia atlantica.   Cheirurus apolllo, C. glaucus, C. eryx, C. mercurius, C. perforator, C. polydorus, C. pompilius, C. prolificus, C. satyrus, C. sol, C. solitarius, C. vulcanus.   Conocephalites adamsii, C. arenosus, C. miser, C. teucer, C. vulcanus, C. zenkeri. Dikelocephalus affinis, D. belli, D. corax, D. cristatus, D. devinei, D. hisingeri, D. magnificus, D. megalops, D. missisquoi, D. oweni, D. pauper, D. planifrons, D. selectus, D. sesortris.   Dolichometopus convexus, D. gibberulus, D. rarus.   Endymion* n. g., see *Endymionia meeki.   Encrinurus mirus.   Harpes dentoni, H. granti, H. ottawœnsis.   Harpides atlanticus, H. desertus, H. concentricus.   Holometopus angelini.   Illænus americanus, I. arcuatus, I. consimilis, I. consobrinus, I. fraternus, I. incertus, I. simulator, I. tumidifrons, I. vindex.   Leperditia concinnula, L. turgida, L. ventralis.   Lichas jukesii, L. minganensis.   Loganellus logani.   Menocephalus globosus, M. salteri, M. sedgwicki.   Nileus affinis, N. macrops, N. scrutator.   Olenellus thompsoni, O. vermontana.   Olenus logani.   Remopleurides affinis, R. canadensis, R. panderi, R. schlotheimi.   Shumardia*

n. g., *S. glacialis*, *S. granulosa*. *Sphærexochus parvus*. *Telephus americanus*. *Triarthrus fischeri*.

——— Catalogues of the Silurian fossils of the Island of Anticosti, with descriptions of some new genera and species. 1866. 93 pp., 28 figs. Geol. Survey Canada.

Asaphus platycephalus Stokes, *A. notans* n. sp., *A. alacer* n. sp., *A. megistos* Locke. *Illænus orbicaudatus* Billings, *I. grandis* Billings. *Dalmanites calicephalus* Hall. *Cheirurus pleurexanthemus* Green. *C. icarus* Billings, *C. numitor* n. sp. *Proetus alaricus* Billings. *Harpes ottawænsis* Billings. *Calymene blumembachii* Brong. *Leperditia canadensia* Jones. *Cheirurus insignis* Beyr., *C. nuperus* n.sp. *Phacops orestes* Billings. *Dalmanites macrora ?* Ang., *D. anticostiensis* n. sp. *Encrinurus multisegmentatus* Portl., *E. punctatus* Brünn., *E. elegantulus* n. sp. *Sphærocoryphe salteri* n. sp. *Harpes consuetus* n. sp. *Sphærexochus canadensis* n. sp. *Lichas canadensis* n. sp. *Bronteus insularia* n. sp. *Dionide ? perplexa* n. sp. *Beyrichia decora* n. sp. *B. venusta* n. sp. *Leperditia anticostians* Jones.

——— Note on some specimens of Lower Silurian Trilobites.
In Quart. Jour. Geol. Soc. London, vol. 26, 1870, p. 479, pls. 31, 32.
Asaphus platycephalus Stokes, with some of the legs preserved.

——— Palæozoic fossils. Vol. 2, pt. 1. 1874.
In Geol. Survey Canada, 144 pp., 35 figs. and 9 pls.
Proetus phocion n. sp. *Agraulos socialis* n. sp., *A. strenuus* n. sp. *A. affinis* n. sp. *Solenopleura communis* n. sp. *Anapolenus venustus* n. sp. *Paradoxides tenellus* n. sp., *P. decorus* s. sp.

——— On some new or little known fossils from the Silurian and Devonian rocks of Canada.
In Canadian Naturalist, new series, vol. 7, 1884, p. 230.
Lichas superbus n. sp.

——— Appendix Geological Survey of Newfoundland Report of Progress for the year 1881. St. Johns, 1882, pp. 1–16, 2 plates.
Agraulos socialis n. sp., *A. strenuus* n. sp. *Anapolenus venustus* n. sp. *Paradoxides tenellus* n. sp., *P. decoris* n. sp.

Blumenbach (J. F.) Abbildungen naturhistorischer Gegenstände. Göttingen, 1810.
The author remarks on a species of *Calymene blumenbachii*, which he figures on pl. 50, that "according to all analogy it must have been without doubt the petrification of some species of insect without wings."

The specimen was from Dudley limestone, and afterwards named by Alex. Brongniart *Calymene blumenbachii.*

Boeck (C.)   Notitzer til Læren om Trilobiten.
In Nyt Mag. Naturvid., vol. 1, Christiania, 1827, p. 11.
1. *Trilobitern tessini (Bohemicus).*   2. *T. spinosus major.*   3. *T. spinosus minor.*   4. *T. gracilis.*   5. *T. bucephalus.*   6. *T. hoffi.*   7. *T. sulzeri.*   8. *T. sternbergi.*   9. *T. zippei.*   10. *T. hausmanni.*
Noticed in Bull. Sci. (Férussac), vol. 14, 1828, p. 45.

———— Uebersicht der bisher in Norwegen gefundenen Formen der Trilobiten-Familie.
In Gæa Norwegica (Keilhau), Christiania, 1838, pt. 1, p. 138.
The species described in this work under the general term of Trilobiten are referred to the following genera:
1. *Phacops elliptifrons.*   2. *Trilobiten elegans* to *Phacops latifrons* Bronn.   3. *T. scabes* to *Lichas norwegicus* Ang.   4. *Phacops conicophathalmus.*   5. *Phacops extensa.*   6. Not defined.   7. *Cybele dentata.*   8. *Dikelocephalus serratus.*   9. *T. Armatus.*   10. *Dalmanites hausmanni, Trilobiten semilunare* to *Dalmanites caudata.*   11. *D. mucronatus.*   12. *Trilobiten plicatus* to *Cybele dentata.*   13. *Encrinurius punctatus.*   14. *Trilobiten sphaericus* to *Cyphaspis ceratophthalmus.*   15. *T. clavifrons* to *Cyrtometopus sarsi* Ang., also to *Cheirurus speciosus.*   16. *Calymene blumembachi.*   17. *Amphion fischeri.*   18. *Ceratopyge forficula.*   19. *Cheirurus acicularis.*   20. *T. lyra* to *Ceratopyge forficula.*   21. *Megalaspis grandis.*   22. *Ptychopyge angustifrons.*   23. *Ogygia dilatata.*   24. *Asaphus expansus.*   25. *Asaphus acuminatus.*   26. *Melagaspis limbatus.*   27. *Asaphus striatus.*   28. *Trilobiten frontalis* to *Niobe insignis* Linrs. in part.   29. *Symphysurus palpebrosus.*   30. *Trilobiten depressa* to *Nileus armadello* var. *depressa.*   31. *T. laevis* and *T. intermedius.*   32. *T. oblongatus.*   33. *Symphysurus angustatus.*   34. *Illænus crassicauda.*   35. *Dysplanus centrotus.*   36. *Oyggia asellus.*   37. *T. gibbosus* to *Olenus truncatus. O. attenuatus.*   38. *Sphærophthalamus alatus.*   39. *Eurycare latum.*   40. *Harpides rugosa.*   41. *Trinucleus granulatus.*   42. *Trinucleus ornatus.*   43. *T. trinucleum.*   44. *T. bronni.*   45. *Ampyx nasutus, A. rostrata.*   46. *A. mammillatus.*   47. *A. nasutus.*   48. *Peltura scarabæoides. Trilobiten pusillus* Sars. to *Shumardia pusillus.*

———— Om Trilobiterne.
In Skand. Naturf. Förhand., vol. 2, 1841, p. 289; Neues Jahrbuch für Mineral., 1841, p. 724.

Boll (Ernst).   Beitrag zur Kenntniss der Trilobiten.
In Dunker and von Meyer's Palæontographica, vol. 1, Cassell, 1846-1851, p. 126, plate 17.
Ampyx bruckneri n. sp.   *Battus pisiformis.*

——— Ueber die Arten der Gattung *Beyrichia* in norddeutschen Silurian-Geröllen.

In Zeitschr. deutsch. geol. Gesell., Berlin, vol. 8, 1856, p. 321, 4 figs.
A communication from Herr Boll to Herr Beyrich.
Beyrichia jonesi, B. spinulosa, B. hians.

——— Die Beyrichien der norddeutscher silurischen Gerölle.

In Archiv. Vereins Freunde Naturg. Mecklenburg, 1862, p. 114, pl. 1.
Beyrichia tuberculata Boll., *B. kockii* n. sp., *B. protuberans* n. sp., *B. cincta* n. sp., *B. dalmaniana* Jones, *B. buchiana* Jones, *B. klödenii* M'Coy, *B. complicata* Salter, *B. nodulosa* n. sp., *B. spingera* n. sp., *B. jonesi* Boll., *B. maccoyana* Jones, *B. elegans* n. sp., *B. salteriana* Jones, *B. siliqua* Jones, *B. hians* Boll., *B. wilkensiana* Jones, *B. nundula*, *B. torosa* n. sp.

Bonnycastle (Capt. R. H.) On the Transition rocks of the Cataraqui.

In Am. Jour. Sci., 1st series, vol. 20, 1831, p. 74, plate. Continued from vol. 18, 1830, p. 104.

The author illustrates from the transition limestone of Kingston, Canada, under the old name of *Entomolithus paradoxus* Linné, a specimen of *Bathyurus extans* Hall.

Bolsche (W.) Ueber Prestwichia rontunda.*

Born (*Chev.* de). Lithophylacium Bornianum. Index fossilium quæ collegit, et in classes ac ordines disposuit. 2 vols. Pragæ, 1772.

In Crustacea, vol. 2, p. 5.

The author illustrates one Bohemian Paradoxides and enumerates six other species of tribolites, with citations to the illustrations of other authors.

Entomolithus paradoxus and Entom. incognitus, pl. 3, fig. 6, have been referred by Sternberg to *Tril. tessini*. Entom. expansus caput truncatus, to *Tril. hausmanni* Entom. paradoxi caput læve, to *Tril. sulzeri*.

The last reference is doubtful. Born cites Wilken's figure Tab. 7, fig. 36, which is a direct copy of Lhywd's illustration of *Trinucleus*. The Entomolithus from Andrarum is probably *Olenus truncatus*. Bromell's woodcut on p. 77 is given as a reference for this species. The Entom. expansus capite lævi tuberculis nullis (Acta Soc. Reg. Holm, 1759, Tab. 2, fig. 1), from Bohemia is a doubtful species.

Bornemann (J. G.) Palæontologisches aus dem cambrischen Gebiete von Camalgrande in Sardinien.

In Zeitschr. deutsch. geol. Gesell., Berlin, 1883, p. 270.

Olenellus zoppei. *Conocoryphe* sp.? *(Olenus armatus* Meng.) *Illænus meneghini.*

Bradley (Frank). Description of a new Trilobite from the Potsdam sandstone, with a note by E. Billings, of Montreal, Canada.

In Proc. Am. Assoc. Adv. Sci., 14th meeting, 1861, 2 woodcuts.
Conocephalites miuutus.
See, also, Am. Jour. Sci., 2d series, vol. 30, 1860, p. 241; Canadian Naturalist, vol. 5, 1860, p. 420.

Brady (M. P.) See **Jones (T. Rubert), Kirkby (J. W.) and Brady (M. P.).**

Brocchie (M. P.) Note sur un crustacé fossile receuilli dans les schistes d'Autun.

In Bull. Soe. Géol. France, 3d series, vol. 8. No. 1, 1880, p. 1, pl. 1.
Describes *Palæocaris typus* Meek & Worthen. *Gampsomyx fimbriatus* Jordan. *Nectotelson* n. g. *N. rochei* n. sp.

Brodie (P. B.) History of fossil insects in the Secondary rocks of England. London, 1845, p. 105, pl. 1, fig. 11.

This author describes and figures a fossil from Coalbrookdale resembling the caterpillar of the Emperor moth, which was afterwards described by J. W. Salter under the name of *Eurypterus (Arthropleura) ferox.*

This species is referred by Woodward to the genus *Euphoberia,* and has no relation with *Eurypterus.*

Brogger (W. C.) Fossiler fra Oxna og Kletten.

In Geol. Föreningens Stockholm Förhandl., vol. 2, 1875, 1 pl.
Paradoxides kjerulfi Linrs., *P. tessini* Brong. *Agnostus parvifrons* Linrs., *A. punctuosus* Ang., *A. fallax* Linrs., *A. laevigatus. Liostracus aculeatus* Ang., *L. microphtalmus* Ang.

—— Andrarums-kalk ved Breidengen i valders.

In Geol. Föreningens Stockholm Förbandl., vol. 3. 1876, 1 pl.
Agnostus lævigatus Dalm., *A. bituberculatus* Ang. *Paradoxides forchhammeri* Ang. *Liostracus microphtalmus* Ang. *Solenopleura bruchymetopa* Ang. *Agnostus aculeatus* Ang.

—— Om Trondhjemsfeldtets midlere Afdeling mellem Guldalen og Meldalen Christiania.

In Vidensk. Selskab. Förhandl., 1877, No. 2, p. 1, 2 pls.
Trilobites n· sp., p. 17, pl. 1, f. 1-7 (Dalmanites).

―――― Om Paradoxides skifrene ved Krekling.

In Nyt Mag. Naturvid., vol. 24, pt. 1, 1878, 6 pls.

Étage 1b: *Olenellus (Paradoxides) kjerulfi* Linrs. *Arionellus primaerus* n. sp.

Étage 1c: *Paradoxides tessini* Broug., *P. rugulosus* Cord. *Liostracus aculeatus* Aug., *L. linnarssoni* n. sp., *L. microphtalmus* Aug. *Conocoryphe sulzeri* Schloth. *Elyx latilimbatus* n. sp. *Conocephalites ornatus* n. sp. *Arionellus difformis* var. *aculeata. Agnostus gibbus* Linrs. also var. *hybrida, A. fallax* Linrs. also var. *tricuspis, A. punctuosus* Aug. also var. *affinis* and var. *bipunctata, A. incertus* n. sp., *A. nathorsti* n. sp., *A. parvifrons* Linrs., also var. *mammillata* and var. *nepos, A. truncatus* n. sp., *A. nudus* Beyr. var. *marginata, A. glandiformis* Aug. (?) A. *A. lævigatus* Dalm.

Étage 1d: *Paradoxides forchammeri. Dolichometopus suecicus. Liostracus microphthalmus. Solenopleura brachymetopa, S. holometopa. Anomocare excavatum, A. ? magnum* n. sp. *Ellipsocephalus circulus* n. sp. *Arionellus difformis* and var. *aculeata* var. *acuminata. Agnostus fallax, A. kjerulfi* n. sp., *A. nathorsti* n. sp., *A. aculeatus, A. brevifrons, A. glandiformis, A. lævigatus* also var. *armata, forfex* and *similis, A. bituberculatus.*

―――― Die Silurischen Étagen 2 und 3, im Kristianiagebiet und auf Eker. Kristiania, 1882, 376 pp., 12 plates.

Ostracoda: *Beyrichia nana* n. sp. *Isochilina ? socialis* n. sp.

Trilobitæ: *Agnostus pisiformis* Linné, also var. *socialis* Tullb., *A. sidenbladhi* Linrs., *A. reticulatus* Aug. *Ampyx domatus* Aug., *A. nasutus* Dalm. *Trinucleus. Nileus* subgenus *Symphysurus incipiens* n. sp., *S. angustatus* Sars. & Boeck., *S. palpebrosus* Dalm. Subgenus *Nileus* Dalm. (sens. strict), *N. limbatus* n. sp., *N. armadillo* Dalm., also var. *depressa* Sars. & Boeck. var. *oblongata* Boeck. *Niobe obsoleta* Linrs., *N. læviceps* Dalm., *N. insignis* Linrs., *N. emarginula* Aug. *Ptychopyge angustifrons* Dalm., *P. limbata* Aug. *Megalaspis polyphemus* n. sp., *M. stenorachis* Aug., *M. planilimbata* Aug.? *M. explanata* Aug., *M. limbata* Boeck var. *minor, M. limbata* Boeck var. *typica, M. extenuata* Wahl., *M. grandis* Sars., *M. gigas* Aug., *M. heroides* n. sp., *M. acuticauda* Aug., *M. heros* Dalm. *Asaphus expansus* Linné var. *incerta, A. raniceps* Dalm., *A. acuminatus* Boeck, *A. striatus* Sars., *A. striatus* Sars. & Boeck *form typica. Illænus* subgenus *Dysplanus centrotus* Dalm., *I. dalmani* Volborth. *Euloma ornatum* Aug. *Olenus* (sens. strict.), *O. attenuatus* Boeck, *O. aculeatus* Aug. Subgenus *Parabolina spinulosa* Wahl., *P. heres* n. sp. Subgenus *Parabolinella* n. subgenus *P. limitis* n. sp., *P. rugosa* n. sp. *Peltura* subgenus *Protopeltura* n. subgenus *P. acanthura* Aug. *Peltura* (sens. strict.), *P. bidentata* n. sp., *P. planicauda* n. sp., *P. scarabæoides* Wahl. also var. *acutidens, P. scarabæoides* form typica. Subgenus *Cyclog-*

nathus transiens n. sp., *C. costatus* n. sp., *C. micropygus* Linrs. Genus (subgenus?) *Triarthrus angelini* Linrs.
Leptoplastus Section 1 subgenus *Leptoplastus* subg. *Enrycare.*
Section 2 subgenus *Sphæropthalmus* subg. *Ctenopyge.* *Leptoplastus ovatus* Ang.? *L. stenotus* Ang.? *Eurycare latum* Boeck, *E. augustatum* Ang. *Sphærophtalmus alatus* Boeck., *S. majusculus* Linrs. *Ctenopyge spectabilis* n. sp., *C.? lobata* n. sp. *Boeckia* n. g., *B. hirsuta* n. sp. *Ceratopyge forficula* Sars. Genus? *Conophrys pusilla* Sars. *Dicelocephalus dicræurus* Ang., *D. serratus* Boeck, subsp. *D. angusticauda* Ang. *Harpides rugosus* Sars. & Boeck. *Remopleurides dubius* Linrs. *Holometopus? elatifrons* Ang. *Lichas celorrhin* Ang. *Cheirurus foveolatus* Ang., *C. clavifrons* Dalm., *C. verrucosus* n. sp. *Amphion primigenus* Ang., *A. fischeri* Eichw. *Cybele bellatula* Dalm. *Phacops sclerops* Dalm.

—— *Pardoxides ölandicus* niväet ved Ringsaker, Norge.
In Geol. Föreningens Stockholm. Förhandl, vol. 6, 1884, 1 pl.

—— Om alderen af Olenelluszone i Nordamerika.
In Geol. Foreningens Stockholm Forhandl, vol. 8, 1886, p. 182.
Contains important criticisms on American Primodial fossils.

—— Ueber die Ausbildung des Hypostomes bei einigen skandinavischen Asaphiden.
In Ofversigt Kongl. Svenska Vet.-Acad. Handl., vol. 11, 1886, 3 pls.
Asaphus. Isotelus. Ptychopyge. Megalaspis. Megalaspides n. g. *Niobe. Asaphellus. Ogygia. Nileus. Symphysurus.*

Bromell (M.) Lithographia Suecana.
In Acta Literaria Sueciæ, vol. 2, Upsaliæ, 1725.
This work was continued in several parts through this volume, but it is only at p. 493, in the section called "De lapidibus insectiferis Scanices et Gothicis," that the Trilobites are treated. We have there the *Oleni* of the alum slate and *Agnostus pisiformis.*
A separate edition of the work was printed at Stockholm and Leipzig, in 1740, bearing the same title as the original work.
This work contains rough woodcuts of *Olenus truncatus* Brünn, *Agnostus pisiformis* Linné and *Peltura scarabæoides* Wahl.

Brongniart (Alex.) Histoire naturelle des Crustacés fossiles, sous les rapports zoologique et géologique, savoir les Trilobites. Les Crustacés proprement dits par A.-G. Desmarest. 11 pls. Paris, 1822.
This author was the first systematic writer upon the Trilobites. He arranges twenty-one species under five new generic names, as follows:

First genus, *Calymene*, species 1. *C. de blumembach*, 2. *C. de tristan*, 3. *C.? variolaire*, 4. *C.? macrophtalme*. Second genus, *Asaphus*, species 1. *A.? cornigère*, 2. *A. de debuch*, 3. *A. de hausmann*, 4. *A. caudigère*, 5. *A. large queue*. Third genus, *Ogygia*, species 1. *O. de guettard*, 2. *O. de desmarest*. Fourth genus, *Paradoxides*, species 1. *P. de tessin*, 2. *P. spinulex*, 3. *P. scaraboïde*, 4. *P.? gibbeux*, 5. *P. lacinié*. Espèces de genre incertain: *Trilobite granulé*, *T. ponctué*, *T. bucéphale*, *T. tentaculé*. Fifth genus, *Agnostus*, species 1. *A. pisiforme*.

—— and **Desmasrest** (A.-G.) Histoire naturelle des Trilobites et des Crustacés.

In Jour. Mines, vol. 9, 1822, p. 273.

Brongniart (Charles). Note sur un nouveau genre d'Entomostracé fossile provenant du terrains Carbonifere des environs de St. Etienne.

In Annales Sci. Géol., vol. 6, 1876.

Bronn (Heinrich). Ueber zwei neue Trilobiten-Arten.

In Zeitschr. für Mineral., vol. 1, pt. 1, 1825, p. 317, pl. 2.
Calymene latifrons n. sp., *C. schlotheimi* n. sp.

—— Geognostische zoologische Untersuchungen in den russisch-baltischen Provinzen von Herrn Dr. Eduard Eichwald.

In Zeitschr. für Mineral., vol. for 1828, pt. 1, p. 104.

Bronn (H. G.) Lethæa geognostica, oder Abbildungen und Beschreibungen der für die Gebirgs-Formationen bezeichnendsten Versteinerungen. Stuttgart, 1835–38.

2 vols., and atlas with 47 pls.

Crustaceen vol. 1, 1835, *Eurypterus remipes* Dekay. *Eidotea* Scouler. *Calymene blumenbachii*, *C. macrophthalma*, *C. latifrons*, *C. schlotheimi* Bronn. *Trimerus delphinocephalus* Green. *Dipleura dekayi* Green. *Asaphus expansus* Linné. *Hemicrypturus* Green. *Nileus*, *Asaphus (Nileus) gigas*. *Asaphus platycephalus* Stokes. *Isotelus gigas* DeKay. *Asaphus gigas* Dalm. *Brongniarta isotela* Eaton. *Illœnus*, *Asaphus (Illœnus) crassicauda* Wahl. *Lichas*. *Ampyx*. *Asaphus (Ampyx) nasutus* Dahlm. *Triarthrus beckii* Green. *Cerurus pleurexanthemus* Green. *Cryptolithus tessclatus* Green. *Bronyniartia platycephalius* Eaton. *Homalonotus knightii* König. *Ogygia guettardi* Brong. *Paradoxides tessini*. *Conocephalus sulzeri* Schloth, *C. costatus* Zenk. *Ellipsocephalus hoffii* Schloth. *Otarion diffractum* Zenk. *Agnostus pisiformis* Linné.

Bronn and **Roemer** (F.) H. G. Bronn's Lethæa geognostica, oder Abbildungen und Beschreibungen der für die Gebirgs-Formationen bezeichnendsten Versteinerungen. 3d ed. Stuttgart, 1851–56.

3 vols. and atlas, 124 pls.
Vol. 1, part 2, Palæo-Lethæa by F. Roemer.
Crustacea vol. 1, p. 525.

Entomostraca: *Cythere baltica* His. *Bairdia curta* M'Coy. *Cypridina serrato-striata* Sandb. *Cyprella chrysalidea* DeKon. *Entomoconchus scouleri* McCoy. *Beyrichia tuberculata* Klöd. *Dithyrocaris testudineus* Scouler. *Ceratiocaris inornatus* M'Coy.

Trilobitæ: *Harpes ungula*, *H. venulosus*. *Harpides hospes*, *H. rugosus*. *Remopleurides radians*. *Paradoxides tessini*. *Hydrocephalus carens*. *Sao hirsuta*. *Arionellus ceticephalus*. *Ellipsocephalus hoffi*. *Olenus truncatus*. *Conocephalites sulzeri*. *Proetus cuvieri*, *P. bohemicus*. *Cyphaspis burmeisteri*. *Phillipsia derbyensis*, *P. globiceps*, *P. gemmulifera*. *Arethusina konincki*. *Phacops latifrons*. *Dalmania hausmanni*, *D. caudata*. *Cryphdeus calliteles*, *C. laciniatus*, *C. punctatus*, *C. stellifer*. *Calgmene blumenbachii*. *Homalonotus delphinoephalus*, *H. knightii*, *H. bisulcatus*, *H. rudis*, *H. platynotus*, *H. rhinotropis*, *H. armatus*, *H. crassicauda*, *H. (Dipleura) dekayi*. *Lichas scabra*. *Trinucleus goldfussi*. *Ampyx tetragonus*, *A. nasutus*. *Dionide formosa*. *Asaphus expansus*, *A. (Isotelus) platycephalus*. *Ogygia buchii*, *O. guettardi*. *Aeglina rediviva*. *Illænus crassicauda*. *Acidaspis prevosti*, *A. mira*. *Ceraurus insignis*. *Placoparia zippei*. *Sphærexochus mirus*. *Staurocephalus murchisoni*. *Deiphon forbesi*. *Dindymene haidingeri*. *Zethus verrucosus*. *Amphion fischeri*. *Encrinurus punctatus*. *Cromus intercostatus*. *Bronteus haidingeri*, *B. flabellifer*. *Telephus*. *Agnostus pisiformis*.

Poecilopoda: *Eurypterus remipes*, *E. scouleri*, *E. tetragonophthalmus*. *Adelophthalmus (Eurypterus) granosus* H. v. Meyer. *Eurypterus cephalaspis* McCoy. *Himantopterus* Salter, 1856. *Pterygotus anglicus*, *P. leptodactylus*. *Limulus rotundatus*. *Gampsonychus* Burmeister, 1855, *G. fimbriatus* Jordon. *Cyclus radialis*. *Bostrichopus* Goldfuss, 1839. *B. antiquus*. *Chonionotus lithanthracis* and *Arthropleura* H. v. Meyer. *Cyclophthalmus* Corda, 1835.

———— Note über die mit Homalonotus verwandten Trilobitengenera.

In Neues Jahrbuch für Mineral., 1840, p. 435, pl. 8, figs. 1 a, b.
Homalonotus delphinocephalus, and on the subgenera *Trimerus* and *Dipleura* Green.

———— Index palæontologicus, oder Uebersicht der bis jezt bekannten fossilen Organismen, bearbeitet unter

Mitwirkung der Herren H. R. Göppert und Herm. von Meyer von Dr. H. G. Bronn, Nomenclator palæontologicus in alphabetischer Ordnung. Stuttgart, 1848, 3 vols.

────── Ueber *Gampsonyx fimbricatus* Jordon aus der Steinkohlen-Formation von Saarbrücken und vom Murg-Thal.
<small>In Neues Jahrbuch für Mineral., 1850, p. 575.</small>

Brueckmann (F. E.) Epistolarum iteneriarum centuria. Wolfenbüttelæ, 3 vols.
<small>In Epist. 23, pl. 2, figs. 1-7, 1732, and Epist. 64, pl. 3, fig. 5, 1737.</small>
<small>This author compares the fossil Crustacea with the Mollusca and calls them *Armata veneris*.</small>

Brunnich (F. E.) Beskrivelse over Trilobiten, en Dyreslægt og dens Arter, med en nye Arts Aftegning.
<small>In Nye Samlig of det Kongelige Danske Videnskabers Selskabs Skriften, Kiöbenhavn. 1781, p. 384, 1 pl.</small>
<small>This author describes *Trilobus (Calymene) tuberculatus, T. (Dalmanites) caudatus* (figures of head and pygidium), *T. dilatatus, T. (Encrinurus) punctatus*.</small>

Buch (Leopold von). Beiträge zur Bestimmung der Gebirgs-Formationen in Russland.
<small>In Archiv. mineral. Geogn., etc., vol. 15, Berlin, 1840, p. 1, pl. 2.</small>
<small>The author's edition was published in Berlin, 1840, 128 pp., 3 plates and one map.</small>
<small>*Asaphus expansus. Illænus crassicauda. Calymene polytoma, C. blumenbachi*, including *Zethus varrucosus* Pander. *Phacops sclerops* (including *Calymene sclerops* Dalm., *C. macrophthalma* Pander, pl. 5, f. 5, a-c, also Brongniart's fig. pl. 1, f. 4, and *Calymene downingiæ* Murch, pl. 14, f. 3). *Phacops macrophthalma (Zethus uniplicatus* Pander, pl. 5, f. 7). *Illænus* or *Nileus armadillo*.</small>

Buckland (William). Geology and mineralogy considered with reference to natural theology. Bridgewater treatise. 2 vols. London, 1837, 87 pls. [4th ed., London, 1869, 2 vols., 90 pls.]
<small>Translated into German by L. Agassiz.</small>
<small>This work contains an important article on the structure of the eyes of</small>

Trilobites, also figures of *Limulus americanus*, *Phacops caudatus*, magnified view of a portion of the eye of *Calymene macrophthalmus*. Hypostoma of *Asaphus platycephalus* Stokes (plate 63, 4th edition). *Calymene blumembachii*, *C. macrophthalmus*. *Asaphus tuberculatus (Encrinus punctatus*, *A. (Ogygia) debuchii*. *Paradoxides tessini*. *Ogygia guettardii*. *Asaphus (Phillipsia) gemmuliferus*, *A. (Phacops) caudatus* (pl. 64, 4th edition). *Limulus trilobitoides* (pl. 66, 4th edition). This species is now referred to *Belinurus trilobitoides* Buckland.

Bureau (Louis). Note sur deux nouveaux *Dalmanites* des schistes ardoisiers d'Angers.

In Bull. d'Etudes Sci. D'Angers, n. serie, xviii, 1888, p. 183, plate.

Dalmanites edwardsi, *D. lapeyrei*.

Burmeister (H.) Die Organisation der Trilobiten, etc. Berlin, 1843, 6 pls.

The English translation is entitled: "The organization of Trilobites deduced from their living affinities, with a systematic review of the species hitherto described by Hermann Burmeister, edited from the German by Prof. Thomas Bell, F. R. S., and Prof. E. F. Forbes, F. R. S." London, Ray Society, 1846. There is a supplementary appendix by the editors.

Burmeister classifies the then known species under 24 genera arranged in the following order:

1. *Eurypterus remipes* DeKay, *E. lacustris* Harlan, *E. tetragonophthalmus* Fisch. 2. *Cytherina baltica* His., *C. phaseolus* His. 1. Trilobitæ: *Trinucleus caractaci* Murch., *T. granulatus* Wahl., *T. fimbriatus*, *T. ornatus* Sternb., *T. tessalatus* Green. 2. *Ogygia buchii* Brong., *O. guettardi* Brong. 3. *Odontopleura ovata* Emm., *O. elliptica*. 4. *Arges amatus* Goldf. 5. *Bronteus flabellifer* Goldf., *B. laticauda* Wahl. 6. *Paradoxides bohemicus* Boeck, *P. tessini*, *Entom. bucephalus* Wahl. is referred to the hypostoma of *P. bohemicus*, *P. spinulosus*. 7. *Olenus truncatus* Brünn, *O. forficulu* Sars., *O. scarabæoides* Wahl. 8. *Conocephalus sulzeri*, *C. striatus*. 9. *Ellipsocephalus hoffii*, Schloth. 10. *Harpes ungula* Sternb. Section 2 Trilobites having the power of rolling themselves into a ball. 11. *Calymene tristani* Brong., *C. polytoma* Dalm., *C. blumenbachii* Brong., *C. callicephala* Green. 12. *Homalonotus:* (A) *Dipleura dekayi* Green. (B) *Trimerus delphinocephalus* Green. *H. knightii* König., *H. herschelii* Murch., *H. armatus*. 13. *Phacops latifrons* Brong., *P. protuberans* Emm., *P. anchiops* Green, *P. sclerops* Dalm., *P. conophthalmus* Emm., *P. odontocephalus* Green, *P. macrophthalmus* Brong., *P. rotundifrons* Emm., *P. proævus* Emm., *P. hausmanni* Brong., *P. caudatus* Brünn., *P. mucronatus* Brong., *P. arachnoides* Goldf., *P. stellifer*. 14. *Cyphaspis* n. g., *C. ceratophthalma* Goldf. 15. *Proetus cuvieri* Steininger. 16. *Aeonia* n. g. The author refers the following species to this genus: *Calymene concinna*

Dalm. *Asaphus stokesii* Murch.; also *Aeonia verticalis* n. sp. *Calymene diops* Green. 17. *Archegonus* n. g. *(Phillipsia* and *Griffithides* Portl.) *A. æqualis* v. Meyer. 18. *Illænus crassicauda* Wahl., *I. giganteus* Guettard. This species was named *Ogygia desmaresti* by Brongniart in 1822. *Illænus (Bumastes) barriensis* Murch. 19. *Dysplanus* n. g., *D. centrotus* Dalm. 20. *Asaphus (Nileus) armadillo* Dalm., *A. palpebrosus* Dalm., *A. læviceps* Dalm., *A. expansus* Linné, *A. tyrannus* Murch., *A. raniceps* Dalm., *A. extenatus* Wahl., *A. (Isoteles) platycephalus* Stokes. 21. *Ampyx nasutus* Dalm., *A. mammillatus* Sars., *A. rostratus* Sars. The appendix contains notes on *Calymene rariolaris* Brong. *Tril. punctatus* Brünn. *Triarthrus beckii* Green, etc. 22. *Agnostus pisiformis*, *A. lævigatus* Dalm., *A. integer* Beyr., *A. nudus* Beyr., with other enumerated genera and species wholly unavailable for systematic arrangement.

The supplementary appendix by Thomas Bell and E. F. Forbes contains an alphabetical list of British Trilobites, also a copy of McCoy's generic description of the following genera: *Tiresia*, *Forbesia*, *Portlockia*, *Trinodus* (Synopsis Silurian Fossils Ireland, 1846); the generic descriptions of *Cheirurus*, *Sphærexochus*, *Lichas*, *Trochurus* (Beyrich, Ueber einige böhmische Trilobiten, 1845), and a list of the genera and species described by Barrande) (Notice préliminaire sur le systéme silurien et les Trilobites de Bohême, 1846).

Burmeister and **Alton** (Ed. de). Neue Beobachtungen über die Organisation der Trilobiten. (*)

In Zeitung für Zoologie, Zootomie und Palæozoologie, Alton und Burmeister, Leipsig, vol. 1, 1846, p. 67, pl. 1, figs. 13-19.

Callaway (Charles). On a new area of Upper Cambrian rocks in South Shropshire, with a description of of a new fauna.

In Quart. Journ. Geol. Soc. London, vol. 33, 1877, p. 652, plate.

Asaphus (Asaphellus n. subgen.*) homfrayi* Salt., *A. (Platypeltis* n. subgen.*) croftii* Callaway. *Agnostus dux* n. sp. *Conocoryphe monile* Salter. *Olenus salteri* Callaway, *O. triarthrus* n. sp. *Conophrys* n. g., *C. salopiensis* Callaway. *Lichapyge* n. g., *L. cuspidata* Callaway. *Primitia* sp.?

Castelnau (Laporte de). Communiqué à l'Académie des Sciences de Paris la découverte des pieds des Trilobites qu'il aurait vus dans lintérier d'un éxemplaire enroulé de l'Amérique du Nord.

In Inst. France, 1842, p. 74; Neues Jahrbuch für Mineral., 1843, p. 504; Comptes Rendus, vol. 14, 1842, p. 344.

(Phacops rana Green.)

———— Essai sur le système silurien de l'Amérique Septentrionale. Paris, 1843, 27 pls.
Asaphus micrurus, A. limulurus, A. cordieri, A. caudatus, A. edwardii, A. murchisoni. Homalonotus giganteus, H. herculaneus, H. atlas. Arctinurus n. g., *A. boltoni. Calymene bufo. Odontocephalus. Acantholoma. Aspidolites. Discranurus.*

The author illustrates plate 2, figs. 1-4, several sections of *Phacops bufo* Green, from Capon, Virginia.

The first figure shows a transverse section made in the rear of the eyes. Immediately below the axis there is a large cylindrical spot and another on the right side occupying the position of a posterior pair of cephalic legs. Fig. 4 gives a longitudinal section, with two spots corresponding to two thorax segments.

On p. ix we find the following remarks: "Upon a transverse section of the thorax of a specimen of Calymene from North America. We perceive a ferruginous spot which occupies the place where one ought to find the cylindrical trunk of the animal, and a little lower down on the right side we distinguish another spot of the same kind, but of different form, which resembles approximately the marks which might have been made by the print of a foliaceous foot similar to that of a Branchiopodæ. In another specimen we find upon a longitudinal fracture two spots of the same color, but straight and elongated, which seem to correspond to two distinct rings of the thorax, and which could very well be of the same nature as those seen upon the face of the preceding specimen."

Champernowne (A.) Note on a find of *Homalonotus* in Red Beds at Torquay.
In Geol. Mag., new series, decade 2, vol. 8, 1881, p. 487.

Chapman (E. J.) *Asaphus canadensis.*
In Canadian Jour., new series, vol. 1, 1856, p. 482.

———— A review of the Trilobites, their characters and classification.
In Canadian Jour., new series, vol. 1, 1856, p. 271, woodcuts.

———— *Trinucleus concentricus.*
In Canadian Jour., new series, vol. 3, 1858, p. 414, woodcut.

———— On some new Trilobites from Canadian rocks.
In Annals Mag. Nat. Hist., 3d series, vol. 2, 1858, p. 9, 2 figs.
Asaphus canadensis n. sp., *A. halli* n. sp.

———— On some new Trilobites from Canadian rocks.
In Canadian Jour., new series, vol. 3, 1858, p. 230, 2 woodcuts.
Asaphus canadensis, A. halli.

—— On the hypostoma of *Asaphus canadensis*, and on a third new species of Asaphus from the Canadian rocks.
In Canadian Jour., new series, vol. 4, 1859, p. 1, 2 figs.
Asaphus canadensis, A. hincksii n. sp.

On the probable nature of the supposed fossil tracks known as *Protichnites*, etc.
In Canadian Jour., new series, vol. 16, 1877, p. 7.

—— Some remarks on the classification of the Trilobites as influenced by stratigraphic relations; with outlines of a new grouping of these forms.
In Trans. Royal Soc. Canada, vol. ., 1890.
Retrospective view of the suborders Sections and Families.
Order Trilobita. Suborder I: Pussilliformes Fam. 1, Agnostidæ.
Suborder II: Latiformes. 1, Levati Fam. 2, Illænidæ Fam. 3, Asaphidæ. 2, Sulcati Fam. 4, Basilicidæ Fam. 5, Dikelocephalidæ. 3, Palmati Fam. 6, Lichasidæ Fam. 7, Bronteidæ. 4, Columnati Fam. 8, Phaetonidæ Fam. 9, Proetidæ.
Suborder III: Conifrontes. 1, Longiconi Fam. 10, Homalonotidæ Fam. 11, Calymenidæ Fam. 12, Triarthridæ. 2, Curticoni Fam. 13, Conocephalidæ Fam. 14, Conocoryphidæ. 3, Vittati Fam. 15, Olenidæ Fam. 16, Arethusinidæ Fam. 17, Harpesidæ.
Suborder IV: Frontones. 1, Annulati Fam. 18, Phacopsidæ Fam. 19, Encrinuridæ Fam. 20, Cheiruridæ. 2, Armati Fam. 21, Acidaspidæ Fam. 22, Paradoxidæ. 3, Globosi Fam. 23, Trinucleidæ Fam. 24, Ampyxidæ Fam. 25, Æglinidæ. 4, Oculosi Fam. 26, Bohemillidæ Fam. 27, Remopleuridæ.

Chtchegloff. Sur les Trilobites en général et en particulier sur ceux de Zarskoë-Selo.
In Jour. für neue Entdeckungen in der Phys. Chem. Natur und Technologie St. Petersburg, Nos. 1, 2, 1827.
Deucalion n. g., *D. brongniarti* n. sp., pl. 8, fig. 9b.

Clarke (J. M.) New Phyllopod Crustaceans from the Devonian of western New York.
In Am. Jour. Sci., 3d series, vol. 23, 1882, p. 476, plate.
Spathiocaris n. g. *Lisgocaris* n. g. *Estheria pulex* n. sp. *Spathiocaris emersonii* n. sp. *Lisgocaris lutheri* n. sp.

—— Cirriped Crustacean from the Devonian.
In Am. Jour. Sci., 3d series, vol. 24, 1882, p. 55.
Plumilites devonicus.

────── New discoveries in Devonian Crustacea.
In Am. Jour. Sci., 3d series, vol. 25, 1883, p. 120.
Dipterocaris n. g., *D. pennœ-dædali* n. sp., *D. procne* n. sp., *D. præ-cervæ* n. sp.

────── Ueber deutsche oberdevonische Crustaceen.
In Neues Jahrbuch für Mineral., 1884, pt. 1, p. 178, 1 plate.
Spathiocaris lata Woodward, *S. koeneni* Clarke, *S. ungulina* Clarke, *S. (Cardiocaris?) congener* Clarke. *Entomis variostriata* Clarke. *Dithyocaris kayseri* Clarke.

────── On the higher Devonian fauna of Ontario County, New York.
In Bull. U. S. Geol. Survey, No. 16, 1885, 3 pls.
Ceratiocaris beecheri n. sp., *C. longicaudus*, *C. simplex* n. sp. *Beyrichia dagon* n. sp. *Echinocaris whitfieldi* n. sp. *Spathiocaris emersoni* Clarke. Crustacean tracks?

────── The structure and development of the visual area in the Trilobite *Phacops rana* Green.
In Jour. Morphology, vol. 2, No. 2, Nov., 1888, Boston, p. 253, pl. 21.

────── The Hercynian question. A brief review of the development and present status, with a few remarks upon its relation to the current classification of American Palæozoic faunas.
In 42d Rep. N. Y. State Mus., 1889, p. 408. Author's edition, 32 pp.

────── The genus *Bronteus* in the Chemung rocks of New York.
In 42d Rep. N. Y. State Mus., 1889, p. 403, fig. Author's edition, 6 p.
Bronteus senescens n. sp.

────── As Trilobitas do Grez de Ereré e Maecurú Estado do Pará, Brazil.
In Archivos do Museu Nacional, vol. 9, 1890, 2 plates.
Homalonotus oiara n. sp., *H. derbyi* n. sp., *H. (Calymene) acanthurus* n. sp. *Phacops braziliensis* n. sp., *P. menurus* n. sp., *P. scirpeus* n. sp., *P. ? pullinus* n. sp., *P. (Dalmanites) macropyge* n. sp. *Dalmanites maecurua* n. sp.; also var. *D. australis* n. sp., *D. galea* n. sp., *D. infractus* n. sp., *D. tumilobus* n. sp., *D. gemellus* n. sp. *Dalmanites (Cryphaeus) paituna* H. & R., *D. gonaganus* n. sp. *Homalonotus lougicaudatus*. *Phacops anceps* n. sp.

———— The Hercyn-frage and the Helderberge Limestone in North America.
In American Geologist, vol. 7, 1891, p. 109.

———— Notes on the genus Acidaspis.
Author's ed., 1892. Communicated to the N. Y. State Geologist, Dec., 1890. In 44th Rep. N. Y. State Mus., 1892.
The author proposes the following scheme of classification:
Genus *Ceratocephala*, Warder, 1838.
1. Species having the occipital ring: *(a)* Smooth, or with a central tubercle. *Odontopleura* Emmrich, type *O. orata* Emm. *(b)* With a single large straight median spine. *Acidaspis* Murchison, type *A. brighti* Murch. syn. *Acantholoma* Conrad. *(c)* With two straight divergent spines. *Ceratocephala* Warder, sensu stricto. Type *C. goniata* Warder. *(d)* With two spirally recurved spines of great size. *Dicranurus* Conrad. Type *D. hamatus* (Conrad) Hall.
2. Species with confluent glabellar lobes, oblique thoracic pleuræ and spineless pygidium. *(e) Selenopeltis* Corda. Type *S. buchi* Barr. syn. *Polyeres* Rouault. *(f) Ancyropyge* n. subgn. Type *Acidaspis romingeri* Hall. *Ceratocephala goniata* Ward. *Acidaspis verneuilli* Barr. *Trapelocera hoernesi* Cord. *Dicranurus hamatus* Conrad. *Acidaspis monstrosa* Barr., *A. brighti* Murch., *A. monstrosa* Barr., *A. romingeri* Hall. *Odontopleura orata* Emm. *Selenopeltis stephani* Cord.

———— Note on *Coronura aspectans* Conrad. The *Asaphus diurus* Green.
Author's ed., 1892. Communicated to the N. Y. State Geologist, Dec., 1890. In 44th Rep. N. Y. State Mus., 1892, p. 105.
Coronura diurus Green, pl. 4.

———— Observations on the *Terataspis grandis* Hall. The largest known trilobite.
Author's ed., 1892. Communicated to the N. Y. State Geologist, Dec., 1890. In 44th Rep. N. Y. State Mus., 1892, p. 111.
Terataspis grandis Hall, plate.

Clarke (W. B.) On the occurrence of Trilobites in New South Wales.
In Quart. Jour. Geol. Soc. London, vol. 4, 1848, p. 63 (not descriptive).

Claypole (E. W.) Note on a large Crustacean from the Catskill group of Pennsylvania.
In Proc. Am. Philos. Soc., Phila., vol. 21, 1883, p. 236, plate.
Dolichocephala n. g., *D. lacoana* n. sp.

────── On the occurrence of the genus *Dalmanites* in the Lower Carboniferous rocks of Ohio.
In Geol. Mag., decade 3, vol. 1, London, 1884, p. 303.
Dalmanites ? cuyahogæ n. sp.

Coignou (Miss). On a new species of *Cyphaspis* from the Carboniferous rocks of Yorkshire.
In Quart. Jour. Geol. Soc., London, vol. 46, 1890, p. 421, 5 woodcuts.
Cyphaspis davidsoni n. sp., *C. megalops* McCoy, *C. cerberus* Barr., *C. coronata* Barr., *C. acanthina* n. sp.

Conrad (T. A.) Report on the Palæontological Department.
In Rept. Geol. Survey, N. Y., 1838, p. 107.
Platynotus n. g. *Paradoxides boltoni* Bigsby is taken for the type. This work also contains a catalogue of the organic remains of New York arranged into six groups.

────── Third Annual Report of the Palæontological Department of the Survey.
In Rept. Geol. Survey, N. Y., 1840, p. 199.
Odontocephalus n. g., *O. selenurus*. *Asaphus halli* n. sp. *Acidaspis tuberculatus*. *Acanthaloma* n. g.

────── On the Silurian system; with a table of the strata and characteristic fossils.
In Am. Jour. Sci., 1st series, vol. 38, 1840, p. 86.

────── Description of new genera and species of organic remains. Crustacea.
In Rept. Geol. Survey, N. Y., 1841, p. 48 1 pl.
Aspidolites n. g. *Dicranurus* n. g. *Asaphus ? acantholeurus*, *A. ? denticulatus*, *A. nasutus*, *A. aspectans*. *Calymene senaria*.
The plate accompanying this report was distributed with a few copies. Republished in the 15th Rept. N. Y. State Cab. Nat. Hist., pl. 11.

────── *Cytherina alta*.
In Geol. Survey, N. Y., pt. 3, comprising the survey of the 3d Geol. Dist. (Vanuxem), 1842, p. 112, fig. 26, No. 6.

────── Observations on the Silurian and Devonian systems of the United States, with descriptions of new organic remains.
In Jour. Acad. Nat. Sci., Phila., vol. 8, 1842, p. 228, 6 pls.
Asaphus corycæus, *A. ? trentonensis*. *Calymene spinifera*, *C. camerata*. *Dipleura*.

The author remarks in regard to *Calymene spinifera*: "Among the specimens of Lower Silurian fossils of Missouri in the collection of my friend J. N. Nicollet, Esq., I noticed the tail of this curious trilobite, which has on the back of one of the articulations of the middle lobe an erect curved spine, curved towards the caudal extremity, and more than one-third of an inch in length."

Conrad's species should replace *Cyphaspis girardeauensis* Shumard, 1855.

―――― Observations on the lead-bearing limestone of Wisconsin, and descriptions of a new genus of Trilobites and fifteen new Silurian fossils.

In Proc. Acad. Nat. Sci., Phila., vol. 1, 1843, p. 329.

Thaleops n. g., *T. ovatus* n. sp.

Corda (A. J. C.) and **Hawle** (Ignaz). Prodrom einer Monographie der böhmischen Trilobiten.

In Abhandl. böhm. Gesell. Wiss., Prag, vol. 5, 1847, pp. 129-292, 7 pls.

This work is generally attributed to A. J. C. Corda. The 58 new genera described in this work are arranged by Joachim Barrande, Système silurien du centre de la Bohême, vol. 1, p. 39, as follows:

I. DIVISION. TELEJURIDES.

1. Phlysacium } = Hydrocephalus Barr.
2. Phanoptes   }
3. Crithias \
4. Tetracnemis |
5. Goniacanthus |
6. Enneacnemis |
7. Acanthocnemis \ = Sao Barr.
8. Acanthogramma /
9. Endogramma |
10. Micropyge |
11. Selenosema /
12. Staurogmus
13. Herse   } = Arionellus Barr.
14. Agraulos }
15. Conocoryphe \
16. Ptychoparia   } = Conocephalus Zenk.
17. Ctenocephalus /
18. Selenopeltis = Acidaspis Murch.
19. Polytomurus = Dionide Barr.
20. Tetrapsellium = Trinucleus Lhwyd.
21. Phalacroma = Agnostus (in part) Brong.
22. Selenoptychus = Æglina Barr.
23. Mesospheniscus = Agnostus integer Beyr.

24. Diplorrhina = Agnostus integer.
25. Condylopyge } =Agnostus rex Barr.
26. Lejopyge
27. Microparia=Æglina Barr.
28. Plæsiacomia=Illænus Dalm.
29. Cyclopyge=Æglina (in part) Barr.
30. Alceste=Illænus (in part) Dalm.
31. Xiphogonium=Proetus Stein.
32. Conoparia=Cyphaspis Burm.
33. Goniopleura=Proetus elegantulus Ang.
34. Aulacopleura—Arethusina Barr.
35. Pharostoma=Calymene Brong.
36. Atractopyge } =Dalmania (in part) Emmrich.
37. Odontochile

(This generic term was used in 1834 for a genus of the Coleoptera.)

II. DIVISION. ODONTURIDES.

38. Amphytrion=Remopleurides Portlock.
39. Arthrorhachis=Agnostus tardus Barr.
40. Peronospsis=Agnostus tardus Barr.
41. Battus=Agnostus integer Barr.
42. Pleuroctenium=Agnostus granulatus Barr.
43. Thysanopeltis=Bronteus (in part) Gold.
44. Dindymene=Dindymene Corda.
45. Prionopeltis= { Proetus (in part) Stein.
                   Phætonides Barr.
46. Asteropyge } =Dalmania (in part) Dalm.
47. Metacanthus
48. Odontopyge=Olenus spinulosus.
49. Placoparia=Placoparia Corda.
50. Eccoptochile=Cheirurus Beyr.
51. Actinopeltis=Cheirurus Beyr.
52. Trochurus=Staurocephalus Barr.
53. Corydocephalus
54. Dicranopeltis
55. Acanthopyge   } =Lichas Dalm.
56. Dicranogmus
57. Trapelocera=Acidaspis Murch.
58. Ceratopyge=Olenus forficula Sars.

Telejurides: 1, *Phlysacium paradoxum* n. sp. 2, *Phanoptes pulcher* n. sp. 3, *Crithias minima* n. sp. 4, *Tetracnemis elegantula* n. sp., *T. spuria*, *T. dubia* n. sp., *T. selenophora* n. sp. 5, *Goniacanthus abbreviatus* n. sp., *G. partschii*. 6, *Enneacnemis* n. g., *E. lyellii* n. sp., *E. herschelii* n. sp. 7, *Herse neubergii*. 8, *Acanth. verrucosa*, *A. glabra*. 9, *Acanthogramma speciosa* n. sp., *A. verruculosa* n. sp. 10, *Endogramma salmii* n. sp. 11, *Micropyge bachofenii* n. sp. 12, *Ellipsocephalus gracilis* n. sp. 13, *Seleno-*

sema thunii n. sp. 14, *Conocoryphe sulzeri* Schloth., *C. latifrons* n. sp., *C. mutica* n. sp., *C. granulata* n. sp., *C. punctata* n. sp. 15, *Ptychoparia striata* Sternb., *P. pulchra* n. sp. 16, *Ctenocephalus* n. g., *C. barrandei* n. sp. 17, *Agraulos delphinocephalus* n. sp., *A. ceticephalus* n. sp., *A. lobulosus* n. sp., *A. carinatus* n. sp., *A. porosus* n. sp. 18, *Staurogmus muricatus* n. sp., *S. acuminatus* n. sp., *S. latus* n. sp. 19, *Paradoxides bohemicus* Boeck, *P. rotundatus* Barr., *P. dormitzeri* n. sp., *P. latus* Zenker, *P. pussilus* Barr., *P. rugulosus* n. sp., *P. inflatus* n. sp.

Selenopeltides: *Selenopeltis stephani* n. sp., *S. humbolti* n. sp., *S. buchii* n. sp., *S. beyrichii* n. sp.

Trinucleides: 1, *Polytomurus formosus* Barr., *P. speciosus* n. sp. 2, *Ampyx bohemicus* n. sp. 3, *Trinucleus ornatus* Sternb., *T. bucklandii* Barr., *T. senftenbergii* n. sp. (*T. ornatus* Beyr. part excl. icon.) *T. cribrosus* n. sp., *T. elegantulus* n. sp., *T. barrandei*, *T. minor* n. sp., *T. pragensis* n. sp. 4, *Tetrapsellium pulchrum* n. sp.

Phalacromides: 1, *Phalacroma priscum* n. sp., *P. quadrinotatum* n. sp., *P. gibbosum*, *P. bibullatum* Barr., *P. ellipticum* n. sp., *P. nudum* n. sp., *P. emarginatum* n. sp., *P. ovatum* n. sp., *P. carinatum* n. sp., *P. applanatum* n. sp., *P. scutiforme* n. sp., *P. laevigatum* n. sp. 2, *Selenoptychus rotundatus* n. sp. 3, *Mesospheniscus cuneifer* n. sp. 4, *Diplorrhina rotundata* n. sp., *D. triplicata* n. sp., *D. orion* n. sp., *D. umbonata* n. sp., *D. sirius* n. sp., *D. elliptica* n. sp., *D. asperula* n. sp., *D. selenophora* n. sp., *D. monas* n. sp., *D. affinis* n. sp., *D. cristata* n. sp. 5, *Condylopyge rex* n. sp. 6, *Lejopyge laevigata* His.

Illænides: 1, *Microparia speciosa* n. sp. 2, *Nileus? acuminatus* n. sp., *N. cyclurus* n. sp., *N. bouchardii* Barr. 3, *Symphysurus palpebrosus* Dalm. 4, *Dysplanus centrotus* Dalm. 5, *Illænus limbatus* n. sp., *I. attenuatus* n. sp., *I. laticeps* n. sp., *I. glaberrimus* n. sp., *I. dubius*, *I. asaphoides* n. sp., *I. subtriangularis* n. sp., *I. minutus* n. sp. 6, *Plæsiacomia rara* n. sp.

Bronteides: *Bronteus campanifer* Beyr., *B. zippei* Barr., *B. aulicus* n. sp., *B. transversus* n. sp., *B. asperulus* n. sp., *B. intermixtus* n. sp., *B. partschii* Barr., *B. sparsus* n. sp., *B. subtriangularis* n. sp., *B. planus* n. sp., *B. carinatus* n. sp., *B. gigas* n. sp., *B. palifer* Beyr., *B. sculptus* n. sp., *B. angusticeps* Barr., *B. sieberi* n. sp., *B. oblongus*, *B. furcifer* n. sp., *B. umbellifer* Beyr., *B. berkeleyanus* n. sp., *B. pendulus* Beyr., *B. aciculatus* n. sp., *B. porosus* Barr., *B. haidingeri* Barr., *B. pulcher* n. sp., *B. brongniartii* Barr. 2, *Cyclopyge megacephala* n. sp., *C. pachycephala* n. sp., *C. marginata* n. sp.

Phacopides: 1, *Alceste latissima* n. sp. 2, *Ogygia*. 3, *Asaphus ingens* Barr., *A. nobilis* Barr. 4, *Hemicrypturus expansus*. 5, *Archegonus æqualis*. 6, *Xiphogonium sieberianum* n. sp., *X. declive* n. sp., *X. planicauda* n. sp., *X. sculptum* n. sp. 6, *Proetus tuberculatus* Barr., *P. bohemicus* n. sp., *P. lepidus* Barr., *P. confusus* n. sp., *P. myops* Barr., *P. buchii* n. sp., *P. dufresnoyi* n. sp., *P. mucronatus* n. sp., *P. fischeri* n. sp., *P. boops* n. sp., *P. leiurus* n. sp., *P. frontalis* n. sp., *P. elegantulus* n. sp., *P. ryckholtii* Barr.,

P. ovalifrons n. sp., *P. reussi* n. sp., *P. platycephalus* n. sp., *P. lovenii* Barr., *P. convexus* n. sp., *P. angelini* n. sp., *P. fallax* Barr., *P. forchhammeri* n. sp., *P. micropygus* n. sp., *P. longulus* n. sp., *P. mancus*, *P. asaphoides* n. sp., *P. dubius*. 7, *Phillipsia derbyensis*. 8, *Griffithides globiceps*. 9, *Goniopleura elegantulus* Lovén. 10, *Cyphaspis barrandei* n. sp. 11, *Conoparia cornigera* n. sp., *C. cerberus* n. sp., *C. macrocephala* n. sp., *C. misera* n. sp., *C. rugosa* n. sp., *C. asper* n. sp., *C. convexa* n. sp., *C. verrucosa* n. sp., *C. glabra* n. sp., *C. burmeisteri*. 12, *Olenus gibbosus*. 13, *Aulacopleura koningkii* n. sp., *A. augusticeps* n. g. 14, *Calymene parvula* Barr., *C. diademata* Barr., *C. affinis* n. sp., *C. interjecta*, *C. incerta* Barr., *C. baylei* Barr., *C. declinata* n. sp. 15, *Pharostoma pulchrum* Beyr. 16, *Homalonotus delphinocephalus*. 17, *Atractopyge verrucosa* Dalm. 18, *Cybele bellatula*. 19, *Encrinurus punctatus*. 20, *Amphion beaumonti*. 21, *Odontochile hausmanui* Brong., *O. applanata* n. sp., *O. rugosa* n. sp., *O. cristata* n. sp., *O. auriculata*, *O. aspera* n. sp., *O. subdepressa*, *O. laticauda*, *O. tuberculata* n. sp., *O. spinifera* Barr. 22, *Phacops socialis* Barr., *P. goldfussii* n. sp., *P. procevus* Emm., *P. sieberi* n. sp., *P. ovoideus* n. sp., *P. quadratus* n. sp., *P. ruderalis* n. sp., *P. pentagonus* n. sp., *P. eichwaldii* n. sp., *P. phillipsii* Barr., *P. deshayesii* Barr., *P. hawlei* Barr., *P. cultrifrons* u. sp., *P. glockeri* Barr., *P. horridus* n. sp., *P; intermedius* Barr., *P. breviceps* Barr., *P. hoseri* n. sp., *P. laticauda* n. sp., *P. maximus* n. sp., *P. verrucifer* n. sp., *P. decorus* n. sp., *P. reclinatus* n. sp., *P. signatus* n. sp., *P. hoeninghausii* Barr., *P. bronui* Barr., *P. protractus* n. sp., *P. sternbergii* n. sp., *P. boeckii* n. sp., *P. oculatus* n. sp. *P. cephalotes* n. sp., *P. foecundus* Barr., *P. exasperatus* n. sp., *P. asper* n. sp.

Part II. Odonturides. Remopleurides: 1, *Amphitryon murchisonii* n.sp. *Remopleurides laterispina*.

Battoides: 1, *Arthrorhachis tarda* Barr. 2, *Peronopsis integra* Beyr. 3, *Battus pisiformis* Dalm., *B. beyrichii* n. sp. 4, *Pleuroctenium granulatum* n. sp., *P. minutum* n. sp.

Thysanopeltides: *Thysanopeltis speciosa* n. sp. Pelturides. *Dindymene friderici angusti* n. sp., *D. speciosa* n. sp. 2, *Prionopeltis priamus* n. sp., *P. hector* n. sp., *P. paris* n. sp., *P. archiaci* Barr., *P. polydorus* n. sp., *P. polymnestor* n. sp., *P. troilus* n.sp., *P. æneas* n. sp., *P. striatus* n. sp., *P. ascanius* n. sp., *P. astyanax* n. sp., *P. memon* n. sp. 3, *Asteropyge arachnoides*. 4, *Metacanthus stellifer*. 5, *Peltura scarabæoides*.

Cheirurides: 6, *Placoparia zippei* n. sp., *P. macroptera* n. sp., *P. grandis* n. sp. 7, *Eccoptochile clavigera* n. sp., *E. aspera* n. sp., *E. perlata* n. sp., *E. curta* n. sp. 8, *Artinopeltis caroli-alexandri* n. sp. 9, *Chirurus insignis* Beyr., *C. quenstedti* Barr., *C. obtusatus* n. sp., *C. beyrichi* Barr., *C. gibbus* Beyr., *C. sternbergii* Beyr., *C. cruciatus* n. sp., *C. verrucosus* n. sp., *C. cordai* Barr., *C. affinis* n.sp., *C. globifrons* n. sp. 10, *Trochurus speciosus* Beyr. 11, *Sphærexochus mirus* Beyr.

Lichades: 1, *Corydocephalus flabellatus*, *C. verrucosus* n. sp., *C. interjectus* n. sp., *C. propinquus* n. sp. 2, *Lichas laciniata*. 3, *Dicranopeltis scabra, D. granulosa* n. sp., *D. aspera* n. sp., *D. parva* n. sp., *D. simplex* n. sp. 4, *Acanthopyge leuchtenbergii* n. sp., *A. pulchra* n. sp., *A. speciosa.* 5, *Dicranogmus pustulatus* n. sp.

Odontopleurides: 1, *Odontopleura siemangii* n. sp., *O. prevostii* Barr., *O. tenuispina* n. sp., *O. ruderalis* n. sp., *O. derelicta* Barr., *O. mira* Barr., *O. bronni* n. sp., *O. hoseri* n. sp., *O. dormitzeri* n. sp., *O. reichenbachii* n. sp., *O. neumanni* n. sp., *O. haushoferii* n. sp., *O. zenkeri* n. sp., *O. steenstrupi* n. sp., *O. beyrichii* n. sp., *O. dufresnoyi* Barr., *O. zenonis* n. sp., *O. lindackeri* n. sp., *O. dumortieri* n. sp., *O. minuta* Barr., *O. geinitziana* n. sp., *O. tricornis* Barr., *O. crassicornis*, *O. tenuicornis* n. sp., *O. portlockii* n. sp., *O. germari* n. sp., *O. imperfecta* n. sp., *O. laportii* n. sp., *O. quinqueloba* n. sp., *O. rotundata* n. sp., *O. truncata* n. sp., *O. burmeisteri* n. sp., *O. impar*, n. sp., *O. primordialis* Barr. 2, *Trapelocera rhabdophora* n. sp., *T. verneuillii, T. vesiculosa, T. hoernesii, T. leptodonta.* 3, *Arges armatus.* 4, *Ceratopyge forficula* Sars. 5, *Ceraurus pleurexanthemus* Green.

Harpides: *Harpes ungula* Sternb., *H. concavus* n. sp., *H. crassifrons* Barr., *H. sculptus* n. sp., *H. reuulosus* n. sp., *H. reticulatus* n. sp., *H. carinatus* n. sp., *H. convexus* n. sp., *H. ruderalis* n. sp., *H. montagnei* n. sp., *H. d'orbignyannus* Barr. *Harpides hospes* Beyr. *Conocoryphe mutica* n. sp. *Bronteus elongatus* Barr.

Cozzens (Issachar). Description of three new fossils from the Falls of the Ohio.
In Annals Lyceum Nat. Hist., N.Y., vol. 4, 1846, p. 157, 1 pl.
Piliolites n. g.
The author therein describes a species of *Proetus* under the name of *Piliolites ohioensis.*

Da Costa (E. M.) A letter concerning the fossil from Dudley.
In Philos. Trans. Royal Soc., vol. 48, pt. 1, No. 42, 1753, p. 286, pl. 1, figs. 6 8.
Dr. Da Costa herein gives the name of *Pediculus marimus major trilobus* to then called Dudley fossil *(Calymene blumenbachii)*, and declares it to be a crustaceous animal nearly related to the living *Isopodes.*

——— Description of a curious fossil animal.
In Gentleman's Mag., vol. 25, London, 1754, p. 24, pl. —, fig. 3.
(Calymene blumenbachii.)

Dalman (J. W.) Nagra petrifacter fundne i Ostergötlands ofvergangskalk afterknade och beskrifne.
In Svenska Vetensk. Akad. Handl., 1824, p. 368, pl. 4.
Entomostr. actinurus n. sp.

—— Om Palæaderna eller de sa kallade Trilobiterna. Stockholm, 1826, 6 pls.

Svenska Vetensk. Akad. Handl., 1826, p. 226.

The German translation bears the following title: Ueber die Palæaden, oder die sogenannten Trilobiten von J. D. Dalman, von Friedrich Engelhart, mit 6 Kupfertafeln, Nürnberg, 1828.

Dalman arranges 41 species of Trilobites under 5 genera and 4 subgenera, as follows:

Division I. Oculated: 1st genus, *Calymene;* 2d genus, *Asaphus.*

Division II. Typhlini: 3d genus, *Ogygia;* 4th genus, *Olenus (Paradoxides* Br.) Section 2. Battoides: 5th genus, *Battus (Agnostus* Br.)

Subgenera: 1, *Asaphus (Nileus);* 2, *Asaphus (Illænus);* 3, *Asaphus (Ampyx);* 4, *Lichas* n. g.

1, *Calymene blumembachii* Broug. var. (a) *tuberculata,* (b) *pulchella.* 2, *Calymene bellatula* n. sp. 3, *C. polytoma* n. sp. 4, *C. actinura* Dalm. 5, *C. sclerops* n. sp. 6, *C. punctata* Brünn. 7, *C. concinna* n. sp.

II. *Asaphus.* Sectio 1, genuini: 1, *Asaphus mucronatus* Brong. 2, *A. caudatus* Brünn. Subdivisio 2. Cornigeri ecaudati: 3, *A. extenuatus* Wahl. 4, *A. granulatus* Wahl. Divisio II. Mutici: *Asaphus angustifrons* n. sp. 6, *A. expansus.* 7, *A. frontalis* n. sp. 8, *A. læviceps* n. sp. 9, *A. palpebrosus* n. sp. Sectio II. Nileus: 10, *Asaphus (Nileus) armadillo* n. sp. Sectio III. Illænus. Divisio 1, Cornigeri: 11, *Asaphus (Illænus) centrotus* n. sp. Divisio II. Mutici: 12, *Asaphus (Illænus) crassicauda* Wahl. 13, *A. (Illæuus) laticauda* Wahl. Sectio IV. Lichas: 14, *Asaphus laci viatus* Wahl. Sectio V. Ampyx: 15, *Asaphus (Ampyx) nasutus* n. sp.

III. 1, *Olenus tessini* Brong. 2, *O. bucephalus* Wahl. 3, *O. spinulosus* Wahl. Divisio II. Mutici: 4, *Olenus gibbosus* Wahl., *O. scarabæoides* Wahl. IV. *Battus (Agnostus)* Brong. *B. pisiformis* Linn.

In the second part of this work the genera and species are arranged as follows:

I. Calymene. Sectio 1. Capitis angulis posticis elongatis attenuatis. 1, *Calymene variolaris* Brong. Sectio 2. Capitis angulis posticis rotundatis. 2, *Calymene blumembachii* Brong. 3, *C. tristani* Brong. 4, *C. bellatula* Dalm. 5, *C. polytoma* Dalm. 6, *C. actinura* Dalm. 7, *C. sclerops, C. macrophthalma* (Brong. tab. 1, figs. 5, a, b, c). 8, var. *B.* (Brong. tab. 1, figs. 4, a, b). 9, *C. protuberans.* 10, *C. schlotheimii* Bronn. 11, *C. latifrons* Bronn. 12, *C. punctata* Brünn. 13, *C. concinna* Dalm.

II. Asaphus. Sectio 1. Genuini Divisio 1. Cornigeri subdivisio 1. Caudati. 1, *Asaphus macronatus* Brong. 2, *A. caudatus* Brünn. 3, *A. auriculatus* Dalm. 4, *A. hausmanni* Brong. Subdivisio 2. Cornigeri ecandati. 5, *Asaphus granulatus* Wahl. 6, *A. extenuatus* Wahl. Divisio 2. Mutici. 7, *Asaphus angustifrons* Dalm. 8, *A. dilatus* Brünn. 9, *! A. buchii* Brong. 10, *A. expansus* Linn. 11, *A. frontalis* Dalm. 12, *A. læviceps* Dalm. 13, *A. gigas* DeKay. 13b, *A. gigas, planus* DeKay. 14, *A. palpebrosus* Dalm. Sectio II (Nileus). 15, *Asaphus (Nileus) armadillo*

Dalm. Sectio III (Illænus). 16, *Asaphus (Illænus) centrotus* Dalm.
Divisio 2. Mutici. 17, *Asaphus (Illænus) crassicauda* Wahl. 18, *A. (Illænus) laticauda* Wahl. Sectio IV? (Lichas). 19, *Asaphus laciniatus* Wahl. Sectio V (Ampyx). 20, *Asaphus ? (Ampyx) nasutus* Dalm.
III. Ogygia. 1, *O. guettardi* Brong. 2, *O. desmarestii* Brong.
IV. Olenus. Divisio 1. Cornigeri. 1, *Olenus tessini* Brong. 1 var. *B. ? (Paradoxides bohemicus)*. 2, *O. spinulosus* Wahl. 3, *O. bucephalus* Wahl. Divisio II. Mutici. 4, *Olenus gibbosus* Wahl. Subdivisio 2. 5, *Olenus scarabæoides* Wahl.
V. Battus (Agnostus Brong.) 1, *B. pisiformis* Linn.
Addenda. 1, *sulzeri* Schloth. 2, *hoffi* Schloth. 3, *Calymene ? speciosa* Dalm. 4, *Calymene ? verrucosa* Dalm. 5, *Asaphus ? schroeteri* Schloth. 6 ?, *relatus* Schloth. 7, *pustulatus*. 8 ?, *granum* Schloth.
1, *Trilobites spærocephalus* Schloth. 2, *T. problematicus* Schloth. 3, *T. bituminosus* Schloth. 4, *T. tentaculatus* Schloth.

———— Arsberättelse om nyare zoologiska arbeten och uwptäckter.

This work forms a part of the annual report of the keepers of the Swedish State Museum for 1828. It contains (p. 134) under the title of "Nya Svenska Palæades" a description of the following Trilobites: *Calymene ornata, C. verrucosa, C. clavifrons, C. ? centrina. Asaphus heros, A. platynotus. Nileus glomerinus. Ampyx pachyrrhinus. Battus lævigatus, B. pisiformis* var. *spiniger*.

———— Nouvelles espèces de Trilobites ou Palæades de Suède.

In Bull. Férussac, vol. 17, 1829.
A notice of the preceding reference.

Dames (W.) Ueber *Hoplolichas* und *Conolichas*, zwei Untergattungen von *Lichas*.

In Zeitschr. Deutsch. geol. Gesell., Berlin, 1877, vol. 29, pt. 1, p. 793, 3 pls.

The type used for the new genus *Hoplolichas* is *Lichas tricuspidatus* Beyr., and for *Conolichas*, *Lichas æquiloba* Steinh. *Hoplolichas tricuspidata* Beyr., *H. proboscidea* n. sp., *H. conico-tuberculata* Nieszk. *Conolichas æquiloba* Steinh., *C. triconica* n. sp., *C. schmidtii* n. sp.

———— Geologische Reisenotizen aus Schweden.

In Zeitschr. Deutsch. geol. Gesell., Berlin, 1881, vol. 33, p. 405.
Paradoxides œlandicus, P. tessini, P. forchhammeri, P. kjerulfi. Ellipsocephalus. Lichas. Megalaspis. Asaphus? Agnostus, etc.
Not descriptive.

—— Erste Abhandlung. Cambrische Trilobiten von Lian Tung, China (Richthofen). Vol. 4. Berlin, 1883, pls. 1, 2.

The author suggests the new generic term *Dorypyge* for certain Trilobites which have the pygidium armed with strong spines, the penultimate pair being the longest, including in the genus *Dikelocephalus quadriceps* Hall and Whitfield and *D. gothicus* Hall and Whitfield.

Under *Anamocare* Angelin he includes the following American species: *Conocephalites hamulus* Owen, *C. wisconsinensis* Owen, and *C. pattersoni* Hall.

Conocephalites frequens, *C. quadriceps*, *C. typus*, *C. sabquadratus*. *Anomocare latelimbatum*, *A. minus*, *A. planum*, *A. nanum*, *A. majus*, *A. subcostatum*. *Liostracus talingensis*, *L. megalurus*. *Dorypyge* n. g., *D. richthofeni*. *Agnostus chinensis*.

—— E. W. Claypole: On the occurrence of the genus Dalmanites in the lower Carboniferous rocks of Ohio. Geol. Mag. London, 1884.

In Neues Jahrbuch für Mineral,, 1885, vol. 1, p. 102 (abstract).
Dalmanites. *Phillipsia*. *Proetus*.

—— C. D. Walcott: Appendages of the Trilobites. Science, vol. 3, 1884, p. 279.

In Neues Jahrbuch für Mineral., vol. 1. p. 102 (abstract).
Asaphus. *Calymene*. *Ceraurus*.

—— T. R. Jones and J. W. Kirkby: On some Carboniferous Entomostraca from Nova Scotia. Geol. Mag. London, 1884.

In Neues Jahrbuch für Mineral., vol. 1, 1885, p. 106 (abstract).
Leperditia. *Beyrichia*.

—— T. R. Jones and H. Woodward: Notes on Phyllopdiform Crustacea, referable to the genus *Echinocaris* from the Palæozoic rocks. Geol. Mag. London, 1884.

In Neues Jahrbuch für Mineral., vol. 1, 1884, p. 110 (abstract).

—— Chas. E. Beecher: Ceratiocaridæ from the Chemung and Waverly group of Warren County. Second Geol. Survey, Penn. Rep. PPP.

In Neues Jarbuch für Mineral., vol. 1, 1884, p. 110 (abstract).

———— J. Mickleborough: Locomotory appendages of Trilobites. Geol. Mag. London, 1884.

In Neues Jahrbuch für Mineral, 1885, p. 477.

———— F. Schmidt and T. R. Jones: On some Silurian *Leperditia*. Annals Nat. Hist., series 5, vol. 9, 1882, p. 168.

In Neues Jahrbuch für Mineral., 1885, vol. 1, p. 105 (abstract).

Dana (James D.) Trilobites in the Potsdam sandstone.

In Edinburgh New Philos. Jour., vol. 6, 1857, p. 350.

———— On the supposed legs of Trilobites *(Asaphus platycephalus).*

In Annals Nat. Hist., 4th series, vol. 7, London, 1871, p. 366; Am. Jour. Sc., 3d series, vol. 1, 1871, p. 320.

———— Manual of geology, treating of the principles of the science with special reference to American geological history. 2d. ed. New York, 1874.

This work contains illustrations of the following Crustacea: *Dalmanites hausmanni. Paradoxides harlani. Conocoryphe mathewi. Agnostus acadia. Conocoryphe minuta. Dicellocephalus minnesotensis. Conocoryphe iowensis. Agnostus rex. Olenus micrurus. Sao hirsuta. Hymenocaris vermicauda. Bathyurus saffordi. Bathyurellus nitidus. Amphion barrandei. Asaphus gigas. Calymene blumenbachii. Lichas trentonensis. Trinucleus concentricus. Agnostus lobatus. Leperditia fabulites. Triarthrus beckii. Illænus davisii. Ampyx mudus,* etc., etc.

D'Archiac (A.) and **Verneuil** (E. de). On the fossils of the older deposits in the Rhenish Provinces, preceded by a general survey of the fauna of the Palæozoic rocks, and followed by a tabulated list of the fossils of the Devonian system of Europe.

In Bull. Soc. Géol. France, vol. 13, 1841–42; Trans. Geol. Soc. Lond., vol. 6, pt. 2, 1842, p. 303.

————, **Fischer** (de Waldheim G.) and **Verneuil** (E. de). Paléontologie de l'Asie Mineure. Paris, 1866, atlas, 21 pls.

Homalonotus gervillei DeVern, *H. longicaudatus* Murch. *Cryphæus asiaticus* DeVern. *Phacops latifrons* Bronn. *Homalonotus salteri* DeVern. *Cryphæus callitelus* Green, *C. stellifer* Burm., *C. pectinatus* Römer, *C. abdullahi* DeVern.

Dawson (J. W.) On the Lower Coal Measures as developed in British America.
In Quart. Jour. Geol. Soc. London, vol. 15, 1859, p. 62.
Tracks of a Crustacean.

───── Note on the fossils of Nova Scotia.
In Canadian Naturalist, vol. 5, 1860, p. 297 (wrongly paged 197 in the volume).
Homalotonous dawsoni Hall.

───── On the footprints of *Limulus* as compared with the *Protiehnites* of the Potsdam sandstone.
In Canadian Naturalist, vol. 7, 1862, p. 271; Am. Jour. Sci., 2d series, vol. 34, 1862, p. 416.

───── Acadian geology. The geological structure, organic remains and mineral resources of Nova Scotia, New Brunswick and Prince Edward Island, etc. London.
Three editions, 1855-78. (3d ed. contains the supplement to the 2d ed.)
Mr. C. F. Hartt's descriptions of the trilobites first appeared in the third edition of this work. The original MSS. described *Microdiscus dawsoni* under the new generic name of *Dawsonia*, but on Prof. E. Billing's authority the author has referred it to *Microdiscus dawsoni*.
Agnostus acadicus Hartt, *A. similis* Hartt. *Beyrichia* (c). *B. jonesii* n. sp., *B. pustulosa* Hall, *B. equilatera* Hall. *Calymene blumembachii* var. *Cythere, Bairdia, Cytherella inflata. Dipostylus dawsoni* n. sp. et n. g. *Conocephalites baileyi* Hartt, *C. matthewi* Hartt, *C. robbii* Hartt, *C. orestes* Hartt, *C. elegans* Hartt, *C. ouangondianus* Hartt, *C. tener* Hartt, *C. aurora* Hartt, *C. thersites* Hartt, *C. gemini-spinosus* Hartt, *C. halli* Hartt, *C. neglectus* Hartt, *C. quandratus* Hartt, *C. formosus* Hartt. *Dalmania logani* Hall. *Estheria. Eurypterus pulicaris* Salter. *Amphipeltis paradoxus. Homalonotus dawsoni* Hall. *Leperditia sinuata* Hall, *L. okeni. Leaia leidyi. Microdiscus dawsoni* Hartt. *Paradoxides lamellatus* Hartt, *P. micmac* Hartt. *Phillipsia howi* Bill., *P. vindobonensis* Hartt.

───── Supplement to the 2d edition of Acadian geology. Containing additional facts as to the geological structure, fossil remains and minaral resources of Nova Scotia, New Brunswick and Prince Edward Island. London, 1878.
Homalonotus dawsoni. Protichnites carbonarius. Anthracopalæmon (Palæocarbus) hillianum.

4

—— Note on two Palæozoic Crustaceans from Nova Scotia.
In Geol. Mag., decade 2, vol. 4, London, 1877, p. 56. (See also Supplement Acadian Geol., 1878, p. 53, fig. 10.) *Anthracopalæmon (Palæocarbus) hillianum. Homalonotus dawsoni.*

—— Description of *Pterygotus canadensis* n. sp.
In Canadian Naturalist, new series, vol. 9, 1881, p. 103, fig.

D'Orbigny (A.) Geologie du voyage dans l'Amerique meridonale Palæontologie, Paris, 1842, 22 plates.
Calymene verneuili n. sp., pl. 1, figs. 4-5.

Day (F. H.) On the fauna of the Niagara and Upper Silurian rocks, as exhibited in Milwaukee county, Wisconsin, and in counties contiguous thereto.
In Trans. Wisconsin Acad. Sci., vol. 4, 1876-77, p. 113.

Dekay (J. E.) Observations on the structure of Trilobites, and description of an apparently new genus, with notes on the geology of Trenton Falls, by James Renwick. Read Nov. 22, 1824.
In Annals Lyceum Nat. Hist., N. Y., vol. 1, 1824, p. 174, 2 pls.; Isis (oder Encycl. Zeitung), Oken, 1825 and 1832.
Isotelus n. g., *I. gigas* n. sp.

—— Observations on a fossil crustaceous animal of the order Branchiopoda. Read Dec. 12, 1825.
In Annals Lyceum Nat. Hist., N. Y., vol. 1, pt. 2, p. 375, pl. 29. Republished in Harlan's Med. Phys. Researches, 1835, p. 297.
Eurypterus n. g., *E. remipes* n. sp.

—— Report on several fossil multilocular shells from the State of Delaware, with observations on a second specimen of the new fossil genus Eurypterus. Read Oct. 2, 1827.
In Annals Lyceum Nat. Hist., N. Y., vol. 2, 1828, p. 273.
Eurypterus remipes Dekay.

Deslongchamps (E.) Mémoire sur les corps organisés fossiles du grès intermédiaire du Calvados.
In Soc. Linn. Calvados, vol. 2, 1825, p. 291, 2 pls.
The author herein describes two species of the genus *Homalonotus* under the names *Asaphus bronguiarti* and *A. brevicaudatus*. *Asaphus incertus*.

Dethleff (von) and **Boll** (Ernst). Die Trilobiten Mecklenburgs.
In Archiv. Vereins Freunde Naturg. Mecklenburg, 1858, p. 155.
This article contains a catalogue of species under the following genera: *Remopleurides, Paradoxides, Ellipsocephalus, Olenus, O. (Sphærophthalmus), Proetus, Cyphaspis, Harpides, Phacops, Calymene, Homalonotus, Lichas, Trinucleus, Ampyx, Asaphus, Ogygia, Illænus, Nileus, Acidaspis, Cheirurus, Sphærexochus, Sphærocoryphe, Amphion, Encrinurus, Bronteus, Telephus, Holometopus, Dolichometopus, Agnostus. Lichas arenswaldii (Metopias verrucosus* Quenst.) n. sp. *Cheirurus korhii* n. sp. *Agnostus nasutus* n. sp., *A. neobrandenburgensis* Boll. =*(Baltus pisiformis* Boll., in Dunk. et Meyer Palæont., 1, xvii, 7).

Devine (T.) Description of a new Trilobite from the Quebec group.
In Canadian Naturalist, vol. 8, 1863, p. 95, 2 figs.
Loganellus n. g.
Similar forms have been referred by Corda to the genus *Ptychoparia*.
Olenus ? logani n. sp.
The author also uses for the same species the new name of *Loganellus quebecensis*.

—— Description of a new Trilobite from the Quebec group.
In Canadian Naturalist, vol. 8, 1863, p. 210, fig.
Menocephalus salteri n. sp.

Dewalque (G.) Prodrome d'une description géologique de la Belgique. Bruxelles, 1868, 2d ed., 1880.
Contains important list of fossils, and observations.

Dumont (A. H.) Mémoire sur la constitution géologique de la province de Liège.
In Mem. Cour. Acad. Roy. Sci., Bruxelles, vol. 8, 1832, p. 353. Table of fossils.

Durocher (J.) Sur le test des Trilobites et des animaux fossils de la Bretagne en général.
In Bull. Soc. Géol. France, 2d ser., vol. 7, p. 307; also vol. 8, 1850-51, p. 160.

Dwight (W. B.) Recent explorations in the Wappinger Valley limestone of Dutchess county, New York.
In Am. Jour. Sci., 3d ser., vol. 27, 1884, p. 349, pl. 7.
Bathyurus ? crotalifrons n. sp.

—— Recent explorations in the Wappinger Valley limestone and other formations of Dutchess county, New York.
In Am. Jour. Sci., 3d ser., vol. 28, 1889, p. 139, plate.
Fossiliferous strata of the paradoxides zone at Stissing *Leperditia ebenina* n. sp. *Olenoides stissingensis* n. sp.

Eaton (Amos). Geological Equivalents.
Description of *Ogygies latissimus* and *Cancer trilobioides*.
In Am. Jour. Sci., 1st ser., vol. 21, 1832, pp. 135, 136, note.

—— Trilobites.
Am. Jour. Sci., 1st ser., vol. 22, 1832, p. 165.
Brongniartia n. g.
We learn from reference to Dr. Amos Eaton's description and figures given in his "Geological Text-book," p. 33, pl. 1, fig. 3 (2d ed., 1832), that the original of *Brongniartia carcinodea* Eaton, was the same as that afterwards described by Dr. Jacob Green (Monthly Am. Jour. Geology, vol. 1, June, 1832; also Mon. Tril. N. A., p. 87, pl. 1, fig. 6, cast 34), as *Triarthrus beckii*.

—— Geological text-book for aiding the study of North American geology; being a systematic arrangement of facts collected by the author for his pupils, etc. 2d ed., 5 pls. June 15, 1832.
The following species are described in this work, pp. 31-34: Genus "*'alamena*," *C. blumenbachii* (*Calymene senaria* Conrad), pl. 2, fig. 19; genus *Asaphus*, *A. hausmannii*, *A. caudatus* (*Dalmanites limulurus* Green), pl. 2, fig. 18; *A. selenourus* (*Dalmanites selenurus* Eaton), pl. 1, fig. 1; genus *Brongniartia*, *B. platycephala* (*Homalonotus dekayi* Green), pl. 2, fig. 20; *B. isotelea* (*Asaphus platycephalus* Stokes), pl. 2, fig. 22; *B. carcinoida* (*Triarthrus becki* Green), pl. 1, fig. 3; genus *Nuttainia*, *N. concentrica* (*Trinucleus concentrica* Eaton), pl. 1, fig. 2; *N. sparsa* (*Homalonotus dekayi* Green).

Eichwald (Eduard). Observationes geognostico-zoologicæ per Ingriam marisque Baltici Provincias nec non de Trilobites. Casani, 1825, 4 pls.
The author herein describes and figures eight species of Lower Silurian Trilobites under the new generic term *Cryptonymus*. In his later work ("Lethæa Rossica," p. 1449) the first three are referred to the genus *Asaphus*, viz.: *Cryptonymus schlotheimii* to *Asaphus schlotheimii*, *Cryp. weissii* to *Asaphus weissii*, *Cryp. panderi* to *Asaphus expansus* Wahl. The last five are classed as follows: *Cryp. lichtensteinii* under *Niobe lichtensteinii*, *Cryp.*

rosenbergii to *Illœnus rosenbergii*, *Cryp. wahlenbergii* to *Illœnus wahlenbergii*, *Cryp. rudolphii* to *Illœnus rudolphii*, *Cryp. parkinsonii* to *Illœnus parkinsonii*.

This work also contains a description and a figure of *Asaphus fischeri* (*Amphion fischeri* Eich.)

The description of two new Carbonic Trilobites are added by Dr. G. Fischer, pp. 53–55, pl. 4, figs. 4 and 5: *Asaphus brongniarti*, pl. 4, fig. 5, and *A. eichwaldi*, pl. 4, fig. 4. Valerian von Möller ("Ueber die Trilobiten der Steinkohlen-formation des Ural," 1867, p. 7) refers these species to *Phillipsia eichwaldi* Fisch., applying the term to the non-mucronate pygidium. The Russian mucronate species, figured by Eichwald ("Lethœa Rossica," pl. 54, fig. 10) as *Griffithides eichwaldi*, was assigned by the same author to *Phillipsia mucronata* McCoy.

——— Zoologia specialis, quam expositis animalibus tum vivis tum fossibilibus potissimum Rossiæ in Universum et Poloniæ. Vilnæ, vols. 1–3, 1829–31.

Vol. 2, 1830, contains descriptions of the following Trilobites: *Calymene blumenbachii* Brong., *C. fischeri* Erchw. *Asaphus mucronatus* Dalm.

Cryptonymus: 1, *C. expansus* Wahl. (*cornigerus* Brong. *lichtensteinii* Echw.) 2, *C. weissii* Echw., *C. schlotheimii* Echw. 4, *C. crassicauda* Wahl. 5, *C. rosenbergii*. *Nileus armadillo* Dalm. *Ampyx nasutus* Dalm.

——— Die Thier und Pflanzenreste des alten rothen Sandsteins und Bergkalks in Novogrodschen Gouvernement.

In Bull. Acad. Petropol. St. Petersburg, vol. 7, 1840.
Otarion eichwaldi.
This species is referred to *Phillipsia mucronata* by Möller.

——— Ueber das silurische Schichten-System von Esthland. St. Petersburg, 1840.

Krustazeen, p. 64.

Calymene odini, *C. blumenbachii* Brong., *C. downingia* Murch., *C. macrophthalma* Murch., *C. bellatula* Dalm., *C. zembnitzkii*. *Amphion (Asaphus) fischeri*. *Zethus verrucosus* Pander. *Cryptonymus punctatus* (Wahl.) Eichwald, *C. variolaris*, *C. wörthii*, *Asaphus expansus* Wahl., *A. weissii*, *A. schlotheimii*, *A. latus* Pander, *A. laceniatus* Dalm., *A. dilatatus* Dalm., *A. devexus*, *A. tyrannus* Murch., *A. vulcani* Murch. *Illœnus crassicauda* Dalm., *I. rosenbergii*, *I. cornutus* Pander, *I. perovatis* Murch., *I. centrotus* Dalm. *Trinucleus sparkskii*. *Homalonotus herschelii* Murch. *Nileus armadillo* Dalm. *Agnostus nasatus* Dalm.

—— Die Urwelt Russlands durch Abbildungen erläutert.

4 pts., St. Petersburg, 1840-48, 14 pls.

In the first part of this work, in the list of fossils given on p. 22, Eduard von Eichwald refers three species to *Cryptonymus;* a generic name which he used in 1825 for an entirely different series of fossil crustacea. In the present work he substitutes the name for such species as *Calymene variolaris, C. punctatus, C. wörthii,* etc. The name is used without a generic description, and only indicated by the above named species.

Metopias n. g., p. 60, pl. 3, fig. 4. (Part 2, 1842.)
Metopias hübneri n. sp., *M. verrucosus* n. sp., *M. aries* n. sp. *Asaphus obsoletus* Phil. *Agnostus boeckii* n. sp. (part 3, p. 68).

—— Beiträge zur Geologie und Palæontologie Russlands. Moscou, 1854.

Bunodes lunula n. sp.

—— Die Grauwacken Schichten von Liev- und Esthland.

In Bull. Soc. Imp. des Naturalistes Moscou, vol. 27, 1854, No. 1, p. 3.
Cypridina minuta n. sp., *C. balthica* His. *Eurypterus remipes* Dekay. *Pterygotus anglicus* Agass. *Bunodes lunula* Eichw.

—— Ueber die Gattungen *Cryptonymus* und *Zethus*.

In Bull. Soc. Imp. des Naturalistes Moscou, vol. 28, 1855.
Separate ed., Moscou, 1855, 23 pp.
The author maintains that the genus *Cryptonymus* should be retained for several Russian and Swedish Trilobites as defined by N. P. Angelin ("Palæont. Scand.," 1852, p. 2), with *Calymene punctatus* for its type.

—— Beiträge zur geographischen Verbreitung der fossilen Thiere Russland's.

In Bull. Soc. Nat., Moscou, vols. 28-30, 1855-57 (vol. 30, 1857).
Bull. No. 4, 1857, p. 305, contains descriptions of the following Crustacea:
Ostracoda: *Cytherina pyrrha* n. sp., *C. phaseolus* His., *C. minuta* Eichw., *C. (Cyclas) fos* n. sp., *C. ovata* n. sp. *Cypridina baltica* His., *C. grandis* Schrenk, *C. laevigata* n. sp., *C. microphthalma* n. sp. *Bairdia qualeni* n. sp., *B. curta* M'Coy, *B. æqualis* n. sp., *B. excisa* n. sp., *B. distracta* n. sp. *Beyrichia gibbirosa* n. sp., *B. umbonata* n. sp., *B. striolata* n. sp., *B. colliculus* n. sp. Poecilopoda. *Agnostus pisiformis* Broug., *A. paradoxus* n. sp., *A. brevifrons* Ang.
Trilobiten: *Harpes (Trinucleus) spaskii* Eichw. *Trinucleus issedon* n. sp. *Ampyx nasutus*. *Lonchodomas affinis* Ang. *Raphiophorus conulus* n. sp. *Lichas hübneri* Eichw., *L. luciniatus* Wahl., *L. lævis* Eichw., *L. coniceps* Leuchtb. *Cerurus (Cheirurus) speciosus* His., *C. gladiator* n. sp., *C. scut-*

iyer, C. aculeatus, C. sembnitzkii. Sphaerexochus clavifrons Dalm. *Sphaerocoryphe uries* Eichw. *Staurocephalus (Sphaerexochus) cranium* Kut. *Amphion fischeri. Cryptonymus* Eichw. *(Encrinurus* Emm.*), C. parallelus* Eichw., *C. wörthii, C. punctatus, C. verrucosus* Pand. *Calymene blumenbachi* Brong., *C. verticalis* Burm. *Odontochile sclerops* Dalm. *Phacops latifrons* Burm., *P. macrophthalma* Brong. *Chasmops odini, C. powisii* Sow. *Proetus concinnus* Dalm. *Zethus* Pand. *(Cyphaspis* Burm.?*), Z. uniplicatus* Pand., *Z. biplicatus, Z. triplicatus. Cyphaspis (Platymetopus) planifrons* Ang. *Griffithides* Port. *(Phillipsia* Port., *Otarion* Zenk. ?*) G. derbyensis, G. pustulatus* Schloth, *G. truncatula, G. eichwaldi* Fischer, *G. brongniarti* Fischer, *G. uralicus* Vern., *G. jonesii* DeKon. *Ogygia buchii, O. dilatata* Brün: *Homalonotus elongatus. Megalaspis heros* Dalm., *M. longicauda* Leucht., *M. extenuata* Wahl., *M. remigium* n. sp., *M. centron* Leucht. *Asaphus expansus, A. raniceps* Dalm., *A. weissii* Eichw., *A. expansus* var. *cornuta, A. lichtensteinii* n. sp., *A. latus* Pand., *A. devexus, A. hyorrhinus* Leucht. *Ptychopyge angustifrons* Dalm., *P. applanta* Ang., *P. globifrons. Dysplanus rosenbergii* n. sp., *D. centrotus* Dalm. *Illænus crassicauda, I. parkinsonii* n. sp., *I. wahlenbergii, I. rudolphii, I. atavus, I. tauricornis* Kut., *I. perovalis* Murch., *I. bouchardi* Barr. aff. *I. barriensis* Murch. *Nileus armadillo* Dalm. *Bronteus flabellifer* Gold., *B. insularis. Odontochile (Dalmania) exilis* n. sp.

Copepodarien: *Eurypterus tetragonophthalmus* Fischer. *Pterygotus anglicus* Agass.

Xiphosuren: *Eidothea (Limulus) oculata* Kut. *Bunodes lunula* Eichw.

——— Lethæa Rossica, ou paléontologie de la Russie. Stuttgart, 1852–69. 3 vols., with atlas, 113 pls.

The Crustacea are arranged in the following order: First order, Ostracopoda, first family, Cypridina, genus 1, *Leperditia*. 2, *Bairdia*. 3, *Beyrichia*. Second order, Pœcilopoda, second family, Agnostidæ, genus 4, *Agnostus*. Third order, Copépoda, third family, Eurypteridæ, genus 5. *Eurypterus*. 6, *Pterygotus*. Fourth order. Xiphosuridæ, fourth family, Limulidæ, genus 7, *Campylocephalus* n. g, Fifth order, Isopodes, Trilobites, family Harpidæ, genus 9, *Harpes*. 10, *Trinucleus*. 11, *Ampyx*. 12, *Lonchodomas*. 13, *Raphiophorus*. Sixth family, Lichidæ, genus 14, *Lichas*. Seventh family, Cheiruridæ, genus 15, *Acidaspis*. 16, *Ceraurus*. 17, *Sphærexochus*. 18, *Zethus*. 19, *Sphærexocoryphe*. Eighth family, Amphionidæ, genus 20, *Amphion*. 21, *Homalonotus*. 22, *Cryptonymus*. Ninth family, Calyminidæ, genus 23, *Calymene*. 24, *Acaste*. 25, *Phacops*. 26, *Chasmops*. 27, *Proetus*. 28, *Griffithides*. 29, (not given). 30, *Cyphasis*. Tenth family, Remopleuridæ, genus 31, *Bunodes*. 32, *Pseudonisrus*. 33, *Remopleurides*. Eleventh family, Asaphidæ, genus 34, *Asaphus*. 35, *Megalaspis*. 36, *Niobe*. 37, *Ptychopyge*. 38, *Ogygia*. 39, *Illænus*. 40, *Rhodope*. 41, *Dysplanus*. 42, *Actinobolus* n. g. 43, *Nileus*. Twelfth family, Brontidæ, genus 44, *Bronteus*.

Leperditia baltica His., *L. grandis* Schrenk, *L. ornata* n. sp., *L. phaseolus* His., *L. ovulum* Eichw., *L. minuta* Eichw., *L. foveolata* n. sp., *L. microphthalma* Eichw., *L. recta* DeKeys. *Bairdia protracta* n. sp., *B. curta* M'Coy, *B. qualeni* Eichw., *B. æqualis* Eichw., *B. distracta* Eichw., *B. excisa* Eichw., *B. lævigata* Eichw., *B. scapha* n. sp., *B. pyrrhæ* Eichw., *B. ovata* n. sp., *B. cyclas* DeKeys. *Beyrichia tuberculata* Kloed., *K. kloedeni* M'Coy, *B. umbonata* Eichw., *B. striolata* Eichw., *B. colliculus* Eichw., *B. gibberosa* Eichw., *B. stricta* DeKeys, *B. schrenkii* DeKeys. *Agnostus pisiformis, A. paradoxus* Eichw., *A. nodiger* n. sp. *Euryglerus fischeri* n. sp. *Pterygotus anglicus* Ag. *Campylocephalus oculatus* Kut.

Trilobites: *Harpes spaskii* Eichw. *Trinucleus issedon* Eichw. *Ampyx nasutus* Dalm. *Lonchodomas longirostris* n. sp. *Raphiophorus convolus* Eichw. *Lichas macrocephala* n. sp., *L. eichwaldi* Nieszk., *L. angusta* Beyr., *L. hübneri* Eich., *L. ornata* Ang., *L. coniceps* Leuchtb., *L. verrucosa* Eichw., *L. lævis* Eichw., *L. laciniata* Wahl., *L. concinna* Aug., *L. dalecarlica* Ang., *L. oelandica* Ang. *Acidaspis crenata* Emm. *Ceraurus gladiator* Eichw., *C. scutiger* Eichw., *C. aculeatus* Eichw., *C. exsul* Beyr., *C. macrophthalmus* Kut., *C. glaber* Aug., *C. speciosus* Dalm., *C. zembnitzkii* Eichw., *C. affinis* Aug., *C. tumidus* Aug., *C. approximatus* n. sp. *Sphærexochus clarifrons* His. *Zethus uniplicatus* Paud., *Z. biplicatus* Eichw., *Z. triplicatus* Eichw. *Sphærocoryphe dentata* Ang., *S. aries* Eichw. *Amphion fischeri* Eichw. *Homalonotus elongatus* n. sp. *Cryptonymus punctatus* Brünn., *C. bellatulus* Dalm., *C. parallelus* Eichw., *C. würthii* Eichw., *C. rex* Niesk. *Calymene blumembachii* Broug., *C. denticulata* n. sp. *Acaste exilis* Eichw., *A. caudata* Brünn., *A. truncato-caudata* Port., *A. sclerops* Dalm., *A. macrouhthalma* Broug. *Phacops latifrons* Bronn. *Chamops odini* Eichw., *C. macrourus* Aug., *C. conicophthalmus* Boeck. *Proctus concinnus* Dalm. *Griffithides eichwaldi* Fischer, *G. brongniardi* Fischer, *G. jonesii* Port., *G. uralicus* DeVern., *G. biserialis* n. sp., *G. derbyensis* Mart., *G. truncatulus* Phil. *Cyphaspis elegantulus* Ang., *C. planifrons* n. sp. *Remopleurides nanus* Leuchtb. *Bunodes lunula* Eichw. *Pseudoniscus aculeatus* Nieszk. *Asaphus expansus, A. schlotheimii* Eichw., *A. weissii* Eichw., *A. raniceps* Bock. *(A. expansus* var. *raniceps* Dalm.) *A. acuminatus* Boeck., *A. angustifrons* Dalm., *A. hyorrhinus* Leuchtb., *A. centron* Leuchtb., *A. latus* Paud., *A. devexus* Eichw., *A. platyurus* Ang., *A. platycephaus* Stock. *Megalaspis longicauda* Leuchtb., *M. remigium* Eichw., *M. rudis* Aug., *M. heros* Dalm., *M. extenuata* Wahl. *Niobe lichtensteinii* Eichw., *N. læviceps* Dalm., *N. lata* Aug. *Ptychopyge lata* Aug., *P. applanata* Aug., *P. rimulosa* Aug., *P. globifrons* Eichw., *P. limbata* Ang. *Ogygia buchii* Broug. *Illanus crassicauda* Wahl., *I. whalenbergii* Eichw., *I. laticlavius* n. sp., *I. parkinsonii* Eichw., *I. davisii* Salter, *I. cornutus* Eichw., *I. oblongatus* Ang., *C. rudolphii* Eichw., *I. rosenbergii* Eichw., *I. bouchardi* var. *minuta* Cord., *I. barriensis* Murch. *Rhodope lata* Ang. *Dyplauus centrotus* Dalm. *Actinobolus atavus* Eichw. *Nileus armadillo* Dalm. *Bronteus granulatus* Goldf., *B. insularis* Eichw.

―――― Beiträge zur nähern Kenntniss der in meiner Lethæa Rossica beschriebenen Illænus, etc.
In Bull. Soc. Imp. des Naturalistes Moscou, 1863, No. 4, p. 372.
Illaenus parkinsonii Eichw., *I. oblongatus* Eichw., *I. laticlavius* Eichw., *I. wahlenbergii* Eichw., *I. rudolphii* Eichw., *I. rossenbergii* Eichw., *I. cornutus* Eichw. *Actinobolus atavus* Eichw.

Emerson (B. K.) On the geology of Frobisher Bay and Field Bay.
Appendix 3. Narrative of the second Arctic expedition made by Charles F. Hall, 45th Cong., 3d sess., Senate Doc. 27, Washington, 1879.
Leperditia alta Conrad, *L. canadensia* Jones. *Primita muta* Jones, *P. frobisheri* n. sp. *Beyrichia symmetrica* n. sp. *Triarthrus beckii* Green. *Calymene senaria* Conrad. *Trilobites* sp. ? *Cyphaspis* ? *frobisheri* n. sp. *Asaphus* sp. ? *Phacops* sp. ?

Emmons (Ebenezer). Geology of New York. Pt. 2. Comprising the survey of the Second Geological District. Albany, 1842.
Isotelus gigas, fig. 99 (1); *Bumastus trentonensis*, fig. 100 (1); *Calymene senaria*, fig. 100 (2); *Illænus trentonensis*, fig. 100 (3); *Ceraurus pleurexanthus*, fig. 100 (6); *Trinucleus tessellatus*, fig. 100 (7); *Triarthrus beckii*, fig. 110 (1); *Trinucleus caractaci*, fig. 112 (1).

―――― The Taconic system, based on observations in New York, Massachusetts, Maine, Vermont and Rhode Island. Albany, 1844; 6 pls. and map.
Atops n. g., *A. trilineatus* n. sp., fig. 8; *Elliptocephala* n. g., *E. asaphoides* n. sp. (not *Ellipsocephalus* Zenker, 1833), fig. 9.

―――― Agriculture of New York, comprising an account of the classification, composition and distribution of the soils and rocks, etc. Vol. I. Albany, 1846; 21 pls. and maps.
Atops trilineatus, fig. 8; *Elliptocephala asaphoides*, fig. 9.

―――― On the identity of the *Atops trilineatus* and the *Triarthrus beckii* Green, with remarks upon the *Elliptocephalus asaphoides*.
In Proc. Am. Assoc. Adv. Sci. First meeting, Philadelphia, 1848, p. 16.
Atops trilineatus, *Triarthrus beckii*, *Elliptocephalus asaphoides*.

—— American Geology, containing a statement of the principles of the science, with full illustrations of the characteristic American fossils. Vol. 1, Parts I, II and VI. Albany, 1855–57, atlas, 17 pls.

This incomplete work only appeared in three parts, as enumerated above.

Trilobites. Taconic System: *Elliptocephalus asaphoides* Emm. *Olenus gibbosus* Dalm. *Paradoxides bohemicus. Atops trilineatus* Emm. *Microdiscus* n. g., *M. quadricostatus* n. sp.

Crustacea: *Conocephalus striatus. Harpes. Acidaspis. Trinucleus concentricus. Calymene senaria* Conrad. *Trianthrus beckii* Green. *Phacops callicephalus. Illaenus crassicauda, I. trentonensis. Isotelus gigas. Ogygia vetusta. Acidaspis trentonensis. Ceraurus pleurexanthemus, C. vigilans. Beyrichia simplex* Jones, *B. regularis* n. sp., *B. ciliata* n. sp. *Agnostus lobatus. Cytherina subelliptica* n. sp., *C. crenulata* n. sp., *C. subcylindrica* n. sp. *Dikelocephalus minnesotensis* Owens. *Illaenus arcturus* Hall. *Asaphus marginalis. Isotelus canalis. Asaphus obtusus. Calymene conradi* n. sp. *Lichas trentonensis*, etc.

—— Manual of geology, designed for the use of colleges and academies. New York, 1860.

Paradoxides? quadrispinosus. fig. 57; *P. macrocephalus* fig. 70; *Atops punctatus*, fig. 71; *Microdiscus quadricostatus*, fig. 73; *Calymene senaria*, fig. 87; *Illaenus trentonensis*, fig. 89 (a); *I. crassicauda*, fig. 89 (b); *Triarthrus beckii*, figs. 89 (c) and 91; *Isotelus gigas*, fig. 89 (d); *Beyrichia, Cytherina, Cyproides*, fig. 90; *Trinucleus caractaci*, fig. 93 (1); *Hemicrypturus*, fig. 97 (2); *Illaenus barriensis*, fig. 98 (3); *Dalmania selenurus*, fig. 118 (1); *Phacops bufo*, fig. 124 (6); *Dalmania calliteles*, fig. 124 (7); *Phacops nupera*, fig. 138 (2); *Cyphaspis girardeauensis*, fig. 145 (2); *Homalonotus dekayi*, figs. 134, 135.

Note A, p. 280, the author remarks: " We now know the following trilobites, all of which belong to a slate beneath the Calciferous, viz,: *Atops punctatus, Elliptocephalus (Paradoxides) asaphoides, Paradoxides thompsoni. P. vermonti, P. macrocephalus, P. (Pagura) quadrispinosus,* and *Microdiscus quadricostatus.*"

Dr. Emmons misprints *Peltura* and uses it in a subgeneric sense to *Paradoxides (Pagura) quadricostatus.*

Emmrich (H. F.) De Trilobitis Dissertatio petrefactologica quam consensu et auctoritate amplissimi philosophorum ordinis, etc. Berolini, 1839, 1 pl., 56 pp.

Classifies 65 species and 9 genera and 6 subgenera, as follows:
1. Phacops.
2. Asaphus { Sec. 1, Ogygia.
           2, Asaphus.
           3, Illœnus.
           4.
3. Calymene { Sec. 1, Calymene.
             2, Trimerus.
4. Dipleura.
5. Conocephalus.
6. Ellipsocephalus.
7. Ampyx.
8. Paradoxides.
9. Odontopleura.
   Cyphæus.

Emmrich includes *Cyphœus* Green in his classification of the genera, but without enumeration.

1, *Phacops macrophthalmus* Brong. 2, *P. protuberans* Dalm. 3, *P. sphæricus* Esm. 4, *P. variolaris* Park. 5, *P.? odontocephalus* Green. 6, *P. clavifrons* Dalm. 7, *P. conophthalmus* Boeck. 8, *P. sclerops* Dalm. 9, *P. anchiops* Green. 10. *P. roundifrons* n. sp. 11, *P. mucronatus* Brong. 12, *P. caudatus* Brünn. 13, *P. hausmanni* Brong. 14 *P. proævus* n. sp.

II. Asaphus Sec. 1, *Ogygia*. 1, *Asaphus guettardi* Brong. 2, *A. desmarestii* Brong. 3, *A. corndensis* Murch. 4, *A. buchii* Brong. 5, *A. dilatatus* Brünn. 6, *A. tyrannus* Murch. 7, *A. frontalis* Dalm.

Subg. II. Asaphus genuini: 8, *A. extenuatus* Wahl. 7, *A. angustifrons* Dalm. 10, *B. expansus* Wahl. 11, *A. weissii* Eich. 12, *A. gigas* DeKay. 13, *A. læviceps* Dalm. 14, *A. palpebrosus* Dalm. 15, *A. armadillo* Dalm.

Sec. III. Illænus: 16, *Asaphus barriensis* Murch. 17, *A. crassicauda* Wahl. 18, *A. centrotus* Dalm.

Sec. IV. *Asaphus concinnus* Dalm. 20. *A. globiceps* Phil. 21, *A. dalmani* Gold.

Appendix: 22, ? *Asaphus laticauda* Wahl. 23, ? *A. flabelliformis* Gold.

III. *Calymene.* 1, *C. bellatula* Dalm. 2, *C. polytoma* Dalm. 3, *C. blumenbachi*, 4, *C. tristani* Brong. 5, *C. callicephala* Green. 6, *C. selenocephala* Green. 7, *C. delphinocephala* Green. 8, *C. ludensis* Murch.

IV. *Dipleura dekayi* Green.

V. *Conocephalus sulzeri* Schloth. 2, *C. striatus* n. sp.

VI. *Ellipsocephalus hoffi* Schloth.

VII. *Paradoxides*. 1, *P. gibbosus* Wahl. 2, *P. acuminatus* n. sp. 3, *P. latus* Boeck. 4, *P. tetragonocephalus* Green. 5, *P. spinulosus* Wahl. 6, *P. scarabœoides* Wahl. Sec. 2, *P. longicaudatus* Zenk., *P. tessini*, *P. harlani* Green.

VIII. *Ampyx nasutus* Dalm., *A. mammillatus* Sars., *A. rostratus* Sars. Sec. 3, *Cryptolithus*. *A. granulatus* Wahl. *(Cryptolithus) tessellatus* Green,

A. (Cr.) caractaci Murch., *A. fimbriatus* Murch., *A. radiatus* Murch., *A. lloydii* Murch.
IX. *Odontocephalus* n. g. *O. ovata* n. sp. *Calymene arachnoidea* Goldf. *Cryphæus* Green.
The author figures *Phacops rotundifrons, Conocephalus striatus, Odontopleura ovata, Cryptolithus caractaci* and *Asaphus globiceps.*

—— Zur Naturgeschichte der Trilobiten. Programm zur öffentlichen Prüfung, welche mit den Zöglingen der Realschule in Meiningen Donnerstag dem 28, und Freitag den 29. März 1844 im grossen Hörsaale der Anstalt abgehalten werden soll, etc. Meiningen, 1844, 1 pl.

Pt. 1. Zur Morphologie der Trilobiten, pp. 4-13.
Pt. 2. Ueber die Trilobitengattungen, pp. 13-18.
Pt. 3. Ueber die Verbreitung der Trilobiten in den Gebirgsschichten, pp. 18-28.

The genera are arranged into 25 families in the following order:
A—Phacops in the restricted sense (Calymene Broug. in part). Contractile. Pleuræ rounded at the extremity, with an acutely-sharpened anterior margin. Axis of pygidium furnished with 11 or fewer joints.
P. macrophthalmus Broug., *protuberans* Dalm., *cryptophthalmus* Emm. *(Calymene lævis* Münst.?), *sclerops* Dalm. *(powisii* Murch.), *conophthalmus* Boeck., *downingiæ* Murch., *proævius* Emm., *rotundifrons* Emm.

B—Dalmania (Asaphus Brong. in part, Asaphus Goldf.) Not contractile. Pleura at the anterior margin scarcely sharpened, frequently acute. Glabella always clavate, distinctly lobed. Pygidium with more than 11, sometimes as many as 22, joints to the axis, frequently elongated at the extremity of the tail. *Phacops caudatus* Brünn., *mucronatus* Broug., *hausmanni* Broug., *odontocephalus* Green., *truncato-caudatus* Port.

II. Phillipsia Port. Facial suture extending to the posterior margin. Glabella almost cylindrical. Thorax 9 to 10 jointed.
(a) Nine-jointed species: *Phacops kelli, jonesii, ornata.*
(b) Ten-jointed species: *Asaphus dalmani* Goldf. *Calymene æqualis* v. Mey.

Family II. Asaphi in the restricted sense. Contractile species. Pleuræ always sharpened at the anterior margin, generally obtuse at the extremity.
III. Griffithides Port. Glabella gibbous, clavate. Thorax 9-jointed. Pygidium furnished with 10 joints and upwards. *G. globiceps, longispinus* Port., *obsoletus* Phil.
IV. Gerastos Goldf. (Proetus Stein.) Glabella truncate, cylindrical, indistinctly lobed. Thorax 10-jointed. Tail less and indistinctly jointed. *G. concinnus* Dalm.

V. Asaphus Brong. Glabella well defined, clavate until it disappears; the articulation of the tail is likewise distinct or indistinct until it disappears completely. Thorax 8-jointed.

VI. Illaenus Dalm. emend. Without glabella and without articulation of the tail. Thorax (9) 10-jointed. Joints smooth. Eyes distant from the axis of the head, located nearer to the margin. *I. casaicauda, centrotus* and *Bumastus barriensis*.

VII. B. Ogygia Brong. emend. Body 7? 8-jointed. *O. guettardi, desmarestii, buchii, dilatata* Brun. *Asaphus tyrannus* does not belong here, but to Asaphus.

VIII. Brontes Goldf. Caudal shield clypeoid, with radiating ribs. Thorax 10-jointed. *B. flabellifer* Goldf.

IX. Nuttainia Eaton. Ribs of the tail distinct, connected only by their margins, terminating in triangular points. Thorax 10-jointed. *N. hibernica* Port.

Family III. Calymene.

X. Encrinurus n. g. Possessing the power of contraction. Eyes smooth. Glabella clavate. Pygidium with a many-jointed axis and few ribs. *Entomostracites punctatus* Wahl. *Amphion multisegmentatus* and *Ogygia rugosa* Port.

XI. Amphion Pander. Eyes smooth. Glabella of equal width, lobed. Thorax many-jointed (20-jointed). Pygidium small and distinctly articulated. *A. frontilobus* Pand.

XII. Calymene Brong. emend. With *oculi hiantes*. Glabella narrowing anteriorly, lobed. Thorax 13-jointed. Pygidium articulated. Ribs mostly furcated. *C. blumembachii* Brong.

XIII. Homalonotus König (Trimerus Green). The same as XII, but the glabella is not lobed. *H. knightii* König. *Herschelii* Murch., *delphinocephalus* Green.

B. Paradoxides. Not contractile.

XIV. Conocephalus Zenk. Thorax 14-jointed. Eyes smooth.? In other respects like XI. *C. sulzeri* Schloth, *striatus* Emm.

XV. Ellipsocephalus. *E. hoffi* Schloth.

XVI. Anthes Goldf. *A. scrabaeoides* Wahl.

XVII. Paradoxides Bron. emend. (Olenus Goldf.) *P. gibbosus*.

Family IV. Odontpleura. Facial suture terminating at the posterior margin more or less distant from the angles of the head. Eyes smooth. Glabella narrow anteriorly, divided into lobes at the sides. Margin of the head dentated. Thorax 8-9-jointed, and upwards. Pygidium considerably smaller than the head, with longer or shorter spiny appendages. The pleuræ of the thorax are likewise frequently elongated into spines. Odontopleura n. g. Glabella semi-circular. Facial suture, see above. Eyes smooth. Thorax 9-jointed. Pygidium furnished with a 2-jointed axis, and dentated or spiny margin. *O. bispinosa* Emm., *O. ovata* Emm., *O. mutica* Emm., *O. cremata* Emm., *O. centrina* Dalm., *O. dentata* Goldf.

XIX. Arges. Eyes ? Facial suture straight posteriorly. Cheeks ? Thorax 8-jointed. Pygidium with an indistinct 4-jointed axis. All the pleuræ are serrated.

Family V. Trinuclei. Head with a broad even reticulate foveolate margin, which is lengthened into horns posteriorly. Facial suture terminating at the posterior margin. Eyes generally unknown.

XX. Cryptolithus Green (Trinucleus Murch.) Glabella clavate, divided into lobes at the sides. Thorax with 6 segments. Pygidium articulated. Eyes? *C. granulatus* Green, *C. caractaci* Murch.

XXI. *Ampyx nasutus, rostratus* Sars., *mammillatus* Sars. Harpes Goldf. differs from Trinucleus by the greater number of rings of the thorax.

Section II. All the segments of the body at the posterior part of the cephalic shield are homologous and movable, only the last joint is without fins.

Family VI. Oleni.
XXI. *Olenus tessini*.
XXIII. *Remopleurides colbii, longicaudatus, lateri-spinifer*.
XXIV. Ceraurus Green.
XXV. Agnostus Brong. (Battus Dalm.) *A. pisiformis* Brong.

———— Ueber die Trilobiten.

In Neues Jarhbuch für Mineral., 1845, p. 18, 1 pl.
For list of genera, see preceding reference.

———— On the morphology, classification, and distribution of the Trilobites.

Scientific Memoirs, edited by Richard Taylor, vol. 4, 1846, p. 253, 1 pl.
Translation of the preceding reference.

Esmark (H. M. T.) Om nogle nye Arter af Trilobiter.

In Mag. for Naturvidensk., Anden Rækkes. Christiana, vol. 1, 1833, p. 268, pl. 8.

Trilobiter axillus n. sp., *T. elliptifrons* n. sp., *T. sphærius* n. sp., *T. semilunaris* n. sp., *T. dentatus* n. sp.

Etheridge (Robert). Description of the Palæozoic and Mesozoic fossils of Queensland.

In Quart. Jour. Geol. Soc. London, vol. 28, 1872, p. 317.
Griffithides dubius n. sp., pl. 18, fig. 7.

———— Observations on a few Graptolites from the Lower Silurian rocks of Victoria, Australia, with a further note on the structure of Ceratiocaris.

In Annals Mag. Nat. Hist., London, 4th series. vol. 14, 1874, p. 1.

———— Palæontology of the coasts of the Arctic lands visited by the late British expedition under Capt. Sir George Nares, etc.
In Quart. Jour. Geol. Soc. London, vol. 34, 1878, p. 568.
Bronteus sp.? allied to *B. hibernicus* Port., *B. flabellifer* Goldf., *Asaphus* like *A. tyrannus* Murch. *Calymene senaria* Conrad? n. sp. *Encrinurus lævis* Ang. *Proetus* sp.

———— Address of the President.
In Quart. Jour. Geol. Soc. London, vol. 37, 1881, Proceedings, p. 37.

———— Reports of the committee, consisting of Mr. R. Etheridge, Dr. Henry Woodward, and Prof. T. Rupert Jones, secretary, on the fossil Phyllopoda of the Palæozoic rocks.
See reports of the secretary, recorded under Jones (T. Rupert). See, also. Woodward (Henry) and Etheridge (Robert); Huxley (T. H.) and Etheridge (Robert, Jr.)

Etheridge (Robert, Jr.) Memoir Geological Survey of Scotland. Edinburgh. Explanation Map, 23, p. 93.
Ceratocaris.

———— On the remains of *Pterygotus* and other Crustaceans from the Upper Silurian series of the Pendland Hills.
In Trans. Geol. Soc. Edinburgh, vol. 2, 1874, part 3, pp. 314-316.
Fragmentary remains of *Pterygotus* and a supposed Crustacean, probably a Phyllopod. Species of *Dictyocaris* are described, but without specific names.

———— Notes on further localities for *Acanthospongia smithii* Young, and *Estheria dawsoni* Jones.
In Geol. Mag. decade 2, vol. 3, 1876, p. 576.
Estheria dawsoni Jones recorded from the Red Sandstone group, the lowest member of the Calciferous series at the base of the British Carboniferous formation.

———— On the remains of a large Crustacean probably indicative of a new species of Eurypterus, or allied genus *(Eurypterus? stevensoni)*, from the Lower Carboniferous (Cement stone group) of Berwickshire.
In Quart. Jour. Geol. Soc. London, vol. 33, 1876, pp. 223-228.

—— On the occurrence of a Macrurous Decapod (*Antrapalæmon? woodwardi*) in the Red Sandstones, or lowest group of the Carboniferous Formation, in the Southeast of Scotland.

In Quart. Jour. Geol. Soc. London, vol. 33, 1876, pp. 863-878, plate 27.

—— On Pinnocaris Lapworthi.

In Proc. Roy. Phys. Soc., 1878, vol. 4, p. 164.

Turrilepas scotia n. g. *Pinnocaris* n. g. *Solenocaris solenoides* Young. *Pinnocaris lapworthi* n. sp. *Lichas* sp. *Salteria primæva* Wy. Thomson. *Acidaspis grayæ* n. sp. *Cheirurus trispinosus* Young. *Agnostus trinodus* Salter.

—— A Catalogue of Australian Fossils (including Tasmania and the Island of Timor), Stratigraphically and Zoologically arranged. Cambridge, 1878, pp. viii and 226.

This work notes species of *Bronteus*, *Calymene*, *Cheirurus*, *Cromus*, *Encrinurus*, *Entomis*, *Forbesia*, *Harpes*, *Homalonotus*, *Hymenocaris*, *Illænus*, *Lichas*, *Phacops*, *Proetus*, *Staurocephalus*, *Trinucleus*, *Phillipsia*, *Griffithides* and *Brachymetopus*.

—— On the occurrence of the genus *Dithyrocaris* in the Lower Carboniferous or Calciferous Sandstone series of Scotland, and on that of a second species of *Anthrapalæmon* in those beds.

In Quart. Jour. Geol. Soc. London, vol. 35, 1879, pp. 464-474, plate 23.

Dithyrocaris testudineus Scouler, *D. tricornis* Scouler, and two indicated species *Anthrapalæmon woodwardi* Etheridge, *A. macconochii* n. sp.

—— On the occurrence of a small new Phyllopodous Crustacean, referable to the genus *Leaia*, in the Lower Carboniferous rocks of the Edingburgh neighborhood.

In Ann. Mag. Nat. His., 5th series, vol. 3, 1879, pp. 257-263.

Notes on a collection of fossils from the Palæozoic rocks of New South Wales, Part 1.

In Jour. Rev. Soc. N. S. W., vol. 14, 1880, pp. 247-258, plate.

Encrinurus punctatus Brünn.

—— Description of the remains of Trilobites from the Lower Silurian rocks of the Mersey River District, Tasmania.

In Papers and Proc. Royal Soc. Tasmania for 1882 (1883), pp. 150-163, plate.
Conocephalus stephensi Etheridge, Jr., *Dikelocephalus tasmanicus* Etheridge, Jr., are described from the Cambrian beds of Caroline Creek, also an indicated species of the genus Asaphus.

—— On the occurrence of the genus *Turrilepas* and the jaws of annelids in the Upper Silurian rocks of New South Wales.
In Geol. Mag. 1890, vol. 7, pp. 337-340.
Turrilepas mitchelli Etheridge, Jr.

—— and Woodward (Henry). On some specimens of *Dithyrocaris* from the Carboniferous Limestone series, East Kilbride, and from the Old Red Sandstone of Lanarkshire.
Geol. Mag. 1773, vol. 10, p. 482-486, plate 16. Continued in vol. 1, 1874, pp. 107-111, plate 5.

—— On the identity of *Bronteus partschii* DeKoninck (non Barrande) from the Upper Silurian rocks of New South Wales.
In Proc. Linn. Soc. N. S. Wales, 1890, part 3, pp. 501-504, plate 18.
Bronteus jenkensi n. sp.

—— and Mitchell (John). The Silurian Trilobites of New South Wales with references to those of other parts of Australia, Part 1.
In Proc. Linn. Soc. New South Wales, series 2, vol. 6, parts 1-3, 1891-92, p. 311, plate.
Family Proetidæ: 1, *Phæton* Barr. 2, *Forbesia* McCoy. 3, *Xiphogonium* Corda. 4, *Celmus* Ang.
Proetus browingensis Mitchell, *P. rattei* n. sp. *(P. ascanicus* Ratte), *P. australis* n. sp.

—— and Nicholson (H. A.) See Nicholson (H. A.) and Etheridge (Robert).

—— and Salter (J. W.) See Salter (J. W.) and Etheridge (Robert).

Faber (Charles L.) Remarks on some fossils of the Cincinnati group.
In Jour. Cincinnati Soc. Nat. Hist., vol. 9, 1886, p. 14.
Lepidocoleus n. g., pl. 1, fig. 1 a-f, *L. jamesi* (Hall & Whitfield) Faber.

Feistmantel (Ottokar). Ueber ein neues Vorkommen von nordischen silurischen Diluvialgeschieben bei Lampersdorf in der Grafschaft Glatz (Zeitschrift Lotos, Dec., 1874, p. 10, Prag).
Phacops downingiæ. Leperditia marginata. Beyrichia tuberculata, B. salteriana. Leperditia phaseolus.

Fischer (G. de Waldheim). Description of *Asaphus eichwaldi* and *A. brongniarti*. In Eichwald's Observationes Geognostico. Zoologigæ per ingriam Marinsque Baltici, etc., 1825, p. 54, pl. 4, figs. 4 and 5.

—— Notice sur l'Eurypterus de Podolie et le Chirotherium de Livonie. Moscou, 1829, 2 pls.
Eurypterus remipes Dekay, *E. lacustris?* Har., *E. tetragonophthalmus* n.sp.

—— Oryctographie du Gouvernment de Moscow. Moscow, 1830–37.
The author refigures the two pygidia described under the names of *Asaphus brongniarti* Fisch. and *A. eichwaldi*, Fisch. in 1825, and states that he considers them "as one and the same species for which he retains the name of Eichwald; the more so as another Trilobite already has the name of Brougniart."

Fitch (Asa). Fossils of Washington County, New York.
In Trans. New York Agric. Soc., vol. 9, 1850, p. 862.
Atops trilineatus Emm. *(Calymene becki* Gr.) *Olenus asaphoides. Trinucleus concentrica.*

Fletcher (W. F.) Observations on Dudley Trilobites.
In Quart. Jour. Geol. Soc. London, vol. 6, 1850, p. 235, pls. 27, 27 bis.
Lichas bucklandi Milne Edw., *L. hirsutus* n. sp., *L. grayii* n.sp., *L. salteri* n. sp., *L. barrandei* n. sp.

—— Observations on Dudley Trilobites. Part 2.
In Quart. Jour. Geol. Soc. London, vol. 6, 1850, p. 402, pl. 32.
Cybele punctata Wahl., *C. variolaris* Broug.

Foerste (A. F.) The Clinton group of Ohio.
In Bull. Denison Univ., vol. 1, 1885, p. 63, 2 pls.
Acidaspis. Bathyurus ?. Arionellus ?. Illænus daytonensis H. & W., *I. madisonianus* Whitf., *I. ambiguus* n. sp. *Calymene blumenbachii ?. Lichas breviceps* Hall. *Dalmanites werthneri* n. sp.

—— The Clinton group of Ohio. Part 2.
In Bull. Dennison Univ., vol. 2, 1887, p. 89, pl. 8.
Ceraurus. Acidaspis ortoni n. sp. *Proetus determinatus* n. sp. *Illænus draytonensis, I. madisonianus, I. ambiguus* Foerste. *Calymene vogdesi* n. sp. *Lichas breviceps* Hall. *Phacops pulchellus* n. sp. *Dalmanites werthneri* Foerste. *Encrinurus thresheri* n. sp.

—— Note on Illæni.
In Fifteenth Rept. Geol. Nat. Hist. Survey Minnesota, 1886, p. 478.
Illænus (Nileus) minnesotensis n. sp., *I. herricki* n. sp., *I. ambiguus* Foerste, *I. insignis* Hall.

—— Note on Palæozoic fossils. Part 2.
In Bull. Denison Univ., vol. 3, pt. 2, 1888, p. 117, pl. 13.
Microdiscus punctatus Salt. *Lichas halli* n. sp. *Sphærexochus mirus* Beyr. *Encrinurus bowningi* n. sp., *E. mitchelli* n. sp. *Phacops serratus* n. sp. *Lichas trentonensis* Conrad, *breviceps* Hall.

—— Notes on Clinton group fossils, with special reference to collections from Indiana, Tennessee and Georgia.
In Proc. Boston Soc. Nat. Hist., vol. 24, 1889, p. 261.
Calymene blumembachii var. *vogdesi* n. var., *C. rostrata* Vogd. *Illænus ambiguus* Foerste, *I. ioxus* |Hall. *Phacops pulchellus* Foerste. *Encrinurus punctatus* Wahl. *Lichas boltoni* var. *occidentalis ?* Hall. *Cyphaspis clintoni* n. sp.

Forbes (Edward). Description of *Ampyx nudus*.
In Mem. Geol. Survey United Kingdom, decade 2, London, 1849, pl. 10.
Ampyx nudus Murch., *A. tumidus* n. sp. *Brachampyx* n. g.
The author makes the following remark: "If, in the end, all the five ringed species should prove to have long heads, and those with 6 thoracic segments to have short and rounded ones, the latter section may be conveniently distinguished from the former as a subgenus, under the name of *Brachampyx*.
The author erroneously included the type of the genus, *Ampyx nasutus* Dalm., in his new subgenus; the term is misplaced and should be abandoned.

Ford (S. W.) Description of some new species of Primordial fossils.

In Am. Jour. Sci., 3d series, vol. 3, 1872, p. 419.
Agnostus nobilis n. sp.

———— Descriptions of new species of fossils from the Lower Potsdam group at Troy, New York.

In Am. Jour. Sci., 3d series, vol. 6, 1873, p. 137.
Microdiscus speciosus, fig. 2 a, b, n. sp. *Leperditia troyensis* n. sp.

———— Note on the discovery of a new locality of Primordial fossils in Rensselaer County, New York.

In Am. Jour. Sci., 3d series, vol. 9, 1875, p. 204.
Conocephalites (Atops) trilineatus Emmons.

———— On additional species of fossils from the Primordial of Troy and Lansingburg, Rensselaer County, New York.

In Am. Jour. Sci., 3d series, vol. 11, 1776, p. 369.
Microdiscus meeki n. sp.

———— Note on *Microdiscus speciosus*.

In Am. Jour. Sci., 3d series, vol. 13, 1877. p. 141.

———— On some embryonic forms of Trilobites.

In Am. Jour. Sci., 3d series, vol. 13, 1877, p. 265, pl. 4.
Metamorphoses of *Olenellus (Elliptocephalus) asaphoides* Emm.

———— Description of two new species of Primordial fossils.

In Am. Jour. Sci., 3d series, vol. 15, 1878, p. 124.
Solenopleura nana n. sp.

———— Note on the development of *Olenellus asaphoides*.

In Am. Jour. Sci., 3d series, vol. 15, 1878, p. 129.

———— Life of the Silurian age.

New York Tribune Extra, September 2, 1879.
In this paper Mr. S. W. Ford gives generic figures of Trilobites, including figures of *Microdiscus speciosus* Ford.

———— Note on the Trilobite *Atops trilineatus* Emmons.

In Am. Jour. Sci., 3d series, vol. 19, 1880, p. 152.

———— On additional embryonic forms of Trilobites from the Primordial rocks of Troy, New York, with observations on the genera *Olenellus, Paradoxides,* and *Hydrocephalus.*

In Am. Jour. Sci. 3d series, vol. 22, 1881, p. 250, 13 figs.

Olenellus asaphoides Emm. *Paradoxides spinosus* Boeck, *P. pusillus* Barr., *P. inflatus* Cord., *P. tessini* Brong., *P. kjerulfi* Linrs. *Hydrocephalus carens* Barr. *Olenellus thompsoni* Hall, *O. vermontanus* Hall.

———— and **Dwight** (W. B.) Preliminary report of S. W. Ford and W. B. Dwight upon fossils obtained in 1885 from metamorphic limestones of the Taconic series of Emmons, at Canaan, New York. Pl. 7.

In Am. Jour. Sci., 3d series, vol. 31, 1886, p. 248.

Genal spine of Trilobite, perhaps *Asaphus megistos*.

Frech (Fritz). Ueber das Devon der Ostalpen.

In Zeitschr. Deutsch. geol. Gesell., Berlin, vol. 39, 1888, p. 659.

Phacops grimburgi n. sp. *Cheirurus quenstedti* Barr., *C. quenstedti* mut. nov. *præcursor*. *Encrinurus novaki* n. sp. *Encrinurus* sp.? *Arethusina haueri* n. sp.

———— Die paläozoischen Bildungen von Cabrières, etc.

In Zeitschr. Deutsch. geol. Gesell., Berlin, vol. 39, 1887, p. 360, pl. 24.

Bronteus meridionalis Thom. Grass., *B. rouvillei* n. sp., *B. subcampanifer* n. sp., *B. dormitzeri ? Phacops fecundus* Barr., *P. fecundus* mut. nov. *suprade ronica, P. occitanicus* Trom. & Grass., *P. escoti* n. sp., *P. cf. lateseptatus. Cheirurus gibbus. Lichas meridionalis* n. sp. *Proetus cuvieri ?, P. complanatus* var. *P.* aff. *planicaudæ. Harpes rouvillei* n. sp.

Gehler (J. C.) De quibusdam rarioribus agri Lipsiensis petrifactis. Spec. I. Trilobites f. *Entomolithus paradoxus* Linné. Lipsiæ, 1793 (1 pl.)

The author gives a sketch of the bibliography of the Trilobites, and figures several specimens of *Calymene blumenbachii.*

Geinitz (H. B.) Grundriss der Versteinerungs-Kunde. Dresden, 1846, 28 pls.

The Trilobites are arranged into 19 families, as follows:

1. Calymene Brong. *(Amphion* and *Zethus* Pander). *C. blumembachii, C. tristani, C. polytoma, C. callicephala.*

2. Homalonotus König *(Trimerus* and *Dipleura* Green.) A—*Dipleura dekayi, B. Trimerus. H. knightii* König., *H. delphinocephilus, H. armatus, H. herschelii.*
3. Cyphaspis Burm. *C. clavifrons* Dalm.
4. Phacops Emm. (Pleuracanthus & Peltura, Milne Edwards). *P. latifrons, P. protuberans, P. anchiops, P. rotundifrons, P. macrophthalmus, P. odontocephalus, P. sclerops, P. hausmanni, P. caudatus, P. mucronatus, P. archnoides, P. stellifer.*
5. Aeonia Burm. (Calymene Dalm., Asaphus Emm.) *A. diops, A. concinna* Dalm., *A. cornuta* Goldf.
6. Illaenus Burm. (Bumastus Aut.) A—*Illaenus crassicauda, I. giganteus* Burm. B—Bumastus. *I. barriensis.*
7. Archegonus Burm. *(Illænus* Dalm. *Asaphus* Emm. *Calymene* H. v. Mey. *Phillipsia* Port.) A—*Dysplanus centrotus.* B—*Archegonus æqualis, A. globiceps.*
8. Asaphus Brong. (*Asaphus & Nileus* Dalm. *Isotelus* Dekay. *Hemicrypturus* Green). A—*Nileus* Dalm. *A. (N.) armadillo* Dalm. B—*A. palpebrosus, A. expansus, A. tyrannus, A. raniceps, A. extenatus, A. platycephalus.*
9. Ampyx Dalm., *A. nasutus* Dalm., *A. mammillatus* Sars., *A. rostratus* Sars.
10. Trinucleus Murch. *(Cryptolithus* Green). *T. cataracti, T. granulatus, T. fimbriatus, T. ornatus, T. tessellatus.*
11. Ogygia Brong. *O. buchii, O. guettardi.*
12. Odontopleura Emm. (Acidaspis Murch.) *O. ovata* Emm., *O. elliptica* Burm.
13. Arges Goldf. *A. armatus* Goldf.
14. Bronteus Goldf. (Goldius DeKon.) *B. flabellifer* Goldf., *B. signatus* Phil., *B. luticauda* Wal.
15. Paradoxides Brong. (Olenus Dalm.) *P. bohemicus, P. spinulosus.*
16. Olenus Burm. (Paradoxides and Olenus Aut.) *O. gibbosus, O. scrabaeoides, O. forficula* Sars.
17. Conocephalus Zenk. *C. sulzeri* Schloth, *C. striatus* Emm.
18. Ellipsocephalus Zenk. *E. Hoffi* Schloth.
19. Harpes Goldf. *H. ungula.*
Pœcilopoda 3. *Belinurus* König, *B. trilobitoides* Buckl.
Family 1. Eurypteridœ: *Eurypterus* Dekay, *E. remipes, E. lacustris* Harlan, *E. tetragonopthalmus.*
Family 2. Cytherinidœ: *Cytherina* Lam., *C. balthica* Hisinger, *C. phaseolus* His.

——— Carbonformation und Dyas in Nebraska. Dresden, 1866, 5 pls.

Phillipsia sp.? *Cythere nebrascensis* n. sp., *C. cyclas?* Keys var.

—— Ueber Thierfährten und Crustaceen-Reste in das unteren Dyas, oder dem unteren Rothliegenden der Gegend von Hohenelbe.
In Isis, 1862, 2 plates.
Dalmanites ? kablikae n. sp. *Kablikia* n. g., *K. dyadica* n. sp. *Branchiopus stagnalis* n. sp.

Genzmar. Beschreibung einer versteinerten Muschel mit dreifachen Rücken. (*)
In Gesell. der Ober-Lausitz Lobau, vol. 2, 1758; also, vol. 3, p. 185, figs. 17-25; Neues, Mag., Hamburg, 1772.
The author gives a description of the Trilobites of Mecklenburg, especially those of his own collection.

Gerstaecker (A.) In Bronn's Die Klassen und Ordnungen des Thier-Reichs wissenschaftlich dargestellt in Wort und Bild. Leipsig und Heidelberg, 1866–79, p. 1142, pls. 43–49.
The author illustrates *Eurypterus remipes* Dekay. *Pterygotus anglicus* Agas. *Belinurus reginæ* Baily. *Prestwichia rotundata* Woodw. *Hemiaspis limuloides* Woodw. *Trinucleus bucklandi* Barr., *T. goldfussi* Barr., *T. ornatus* Sternb. *Lichas palmata* Barr. *Cheirurus quenstedti* Barr., *C. gibbus* Beyr. *Ampyx rouaulti* Barr. *Cyphaspis burmeisteri* Barr., *C. barrandi* Cord., *C. cerberus* Barr. *Dionide formosa* Barr. *Proetus intermedius* Barr. *Arethusina konincki* Barr. *Harpes venulosus* Cord., *H. nugula* Sternb. *Phacops sternbergi* Cord. *Arionellus ceticephalus* Barr. *Acidaspis dormitzeri* Cord., *A. verneuilli* Barr. *Paradoxides bohemicus* Boeck. *Dalmanites socialis* Barr. *Phacops hoeninghausi* Barr., *P. fecundus* Barr. *Conocephalites sulzeri* Schloth. *Calymene pulchra* Barr., *C. parvula* Barr. *Remopleurides radians* Barr. *Dindymene haidingeri* Barr. *Placoparia zippei* Cord. *Cromus beaumonti* Barr. *Illænus wahlenbergianus* Barr. *Bronteus edwardsi* Barr., *B. brongniarti* Barr. *Asaphus nobilis* Barr. *Aeglina rediviva* Barr. *Hydrocephalus carens* Barr.
Sao hirsuta Barr. *Æglina prisca* Barr. *Sphærexochus mirus* Beyr. *Areia fritschi* Barr. *Staurocephalus murchisoni* Barr. *Deiphon forbesi* Barr. *Phillipsia parabola* Barr. *Homalonotus bohemicus* Barr. *Amphion senilis* Barr. *Ogygia desiderata* Barr. *Agnostus integer* Barr., *A. granulatus* Barr., *A. nudus* Beyr., *A. rex* Barr. *Dalmanites hausmanni* Brong. *Ampyx nasutus* Dalm. *Deiphon forbesi* Barr. *Staurocephalus murchisoni* Barr. *Hydrocephalus carens* Barr. *Cheirurus gibbus* Beyr. *Homalonotus dekayi* Green. *Phacops fecundus* Barr. *Phacops volborthi* Barr. *Asaphus platycephalus* Stokes. *Calymene blumenbachi* Brong. *Nileus armadillo* Dalm. *Acidaspis buchi* Barr. *Phacops sternbergi* Cord. *Zethus bellatulus* Dalm. *Phillipsia globiceps* Phill. *Spærexochus mirus* Beyrich.

Ellipsocephalus hoffi Schloth. *Asaphus extenuatus* Wahl. *Acidaspis verneuilli* Barr., *A. leonhardi* Barr. *Dalmanites socialis* Barr. Eye of *Bronteus brongniarti*. *Illænus tauricornis*, etc.

Giebel (C.) Die silurische Fauna des Unterharzes.
In Abhandl. Nat. Vereins Sachsen u. Thüringen, vol. 1, 1858, p. 1, 7 pls.
Harpes bischofi A. Roem. *Proetus pictus* n. sp. *Cyphaspis hydrocephala*. Römer. *Phacops angusticeps* n. sp., *P. ? sternbergi* n. sp. *Dalmanites tuberculatus* n. sp. *Lichas sexlobatus* Römer. *Acidaspis vulcana* Römer, *A. hercyniæ* n. sp. *Bronteus bishofi* n. sp.

Goldfuss (A.) Observation sur la place qu'occupent les Trilobites dans le règne animal.
In Annales Sci. Nat. Paris, vol. 15, 1828, p. 82, pl. 2.
Asaphus. *Calymene*.

—————— **Catalogue des Trilobites.**
In Handbuch der Geognosie von de la Beche, 1832.
This list was copied by Dr. Jacob Green in his "Monograph of the Trilobites of North America," p. 20.

—————— **Beiträge zur Familie der fossilen Crustaceen.**
In Nova Acta Physico Med., vol. 19, 1839, Breslau, p. 353, pl. 33.
Harpes n. g. *Arges* n. g. *Bostrichopus* n. g., *B. antiquus* n. sp. *Arges armatus* n. sp. *Harpes macrocephalus* n. sp. *Brontes flabellifer* n. sp. *Illænus ? triacanthus* n. sp.

—————— **Systematische Uebersicht des Trilobiten und Beschreibung einiger neuen Arten.**
In Neues Jahrbuch für Mineral., 1843, p. 537, pls. 4–6.
New genera *Anthes, Gerastos* and *Symphysurus*.
The author classifies 204 species under 31 genera principally on the characteristics of the eyes as follows:

I. Without eyes. a—without segments. 1, *Agnostus* Brong. b—with segments. (a) segments of the pygidium effaced. 2, *Ampyx*. 3, *Cryptolithus*. 4, *Arges*. 5, *Olenus*. (b) segments of the pygidium distinct. 6, *Zethus*. 7, *Otarion*.

II. Eyes smooth or finely reticulated. A—Thorax and pygidium indistinctly separated. (a) The last segment without appendages. 8, *Paradoxides*. 9, *Amphion*. 10, *Harpes*. (b) The last segment with appendages. 11, *Bronteus*. 12, *Ellipsocephalus*. 13, *Ceraurus*.
B—Pygidium clypiform, axis and sides not segmented. a—without dorsal grooves. The axis and sides indistinct. 14, *Nileus*. 15, *Bumastus*. b—Dorsal grooves faint. Grooves of the sides invisible on the pygidium. 16, *Dipleura*. c—Dorsal grooves prolonged even to the pygidium. 17, *Symphysurus*. 18, *Illænus*. 19, *Isotelus*. Pygidium clypiform, axis seg-

mented, sides smooth. 20, *Cryptonymus.* 21, *Ogygia.* Pygidium with distinct segments on the axis and sides. 22, *Odontopleura.* 23, *Conocephalus.* 24, *Gerastos.* 25, *Calymene.* 26, *Homalonotus.*
  III. Eyes reticulated. a—Front lobed. 27, *Asaphus.* 28, *Acaste.* b— Front not lobed. 29, *Phacops.*
  IV. Eyes missing. *Arges.*
  V. *Anthes.*

The author places *Olenus scarabæoides* Wahl. and *O. forficula* Sars. under this new genus.

Dr. Goldfuss gives brief descriptions of the following species:
  1. *Agnostus pisiformis* Linné, *A. spiniger* His., *A. lævigatus* His., *A. tuberculatus* Klöd., *A. gigas* Klöd., *A. granum* Schlot.
  2. *Ampyx nasutus* Dalm., *A. mammillatus* Sars., *A. rostratus* Sars., *A. pachyrrhinus* Dalm., *A. incertus* Delong.
  3. *Cryptolithus granulatus* Dalm., *C. tessellatus* Green., *C. caractaci* Murch., *C. fimbriatus* Murch., *C. radiatus* Murch., *C. lloydii* Murch., *C. rudus* Murch., *C. asaphoides* Murch., *C. bronnii* S. & B., *C. bigsbyi* Green., *C. concintricus* Eaton, *C. gracilis* Murch. *C. wilkensii* Münst., *C. ellipticus* Münst., *C. lævis* Münst., *C. nilssonii* Münst.
  4. *Arges armatus* Gold., *A. bimucronatus* Murch., *A. quadrimucronatus* Murch., *A. radiatus* Gold. tab. 4, fig. 1.
  5. *Anthes* Gold. (*Olenus* Dalm.) *A. scrabæoides* Wahl., *A. forficula* Sars.
  6. *Olenus gibbosus* Dalm., *O. acuminatus* Emm., *O. attenuatus* Boeck, *O. latus* Boeck, *O. rugosus* Boeck, *O. triarthrus* Harlan, *O. arcuatus* Harlan.
  7. *Zethus uniplicatus* Pand., *Z. verrucosus* Pand.
  8. *Otarion diffractum* Zenk., *O. squarrosum* Zenk.
  9. *Paradoxides tessini* Broug., *P. longicaudatus* Zenk., *P. pyramidalis* Zenk., *P. latus* Zenk., *P. spinulosus* Wahl., *P. boltoni* Green, *P. harlani* Green, *P. actinurus* Dalm., *P. spinosus* Boeck.
  10. *Amphion frontilobus* Pand., *A. odontocephala* Green.
  11. *Harpes macrocephalus* Gold., *H. speciosus* Münst.
  12. *Bronteus alutaceus* tab. 6, fig. 1, *B. granulatus* Gold. tab. 6, fig. 2, *B. flabellifer* Gold. tab. 6, fig. 3, *B. intermedius* Gold. tab. 6, fig. 4, *B. scaber* Gold., tab. 6, fig. 5, *B. canaliculatus* Gold. tab. 6, fig. 6, *P. signatus* Phill., *B. costatus* Münst.
  13. *Ellipsocephalus ambiguus* Zenk.
  14. *Ceraurus pleurexanthemus* Green, *C. acicularis* S. & B., *C. lyra* S. & B.
  15. *Nileus armadillo* Dalm., *N. chiton* Pand., *N. depressus* S. & B., *N. glomerinus* Dalm.
  16. *Bumastus barriensis* Murch., *B. franconicus* Münst., *B. planus* Münst.
  17. *Dipleura dekayi* Green.
  18. *Symphysurus læviceps* Dalm., *S. palpebrosus* Dalm., *S. lævis* Boeck, *S. intermedius* Boeck, *S. oblongatus* Boeck, *S. brevicaudus* Deslong.

19. *Illænus crassicauda* Dalm., *I. perocalis* Murch., *I. centrotus* Dalm., *I. asellus* Boeck, *I, triacanthus* Gold. is referred to *Phacops macrocephalus*.
20. *Isotelus gigas, I. megalops* Green, *I. stegops* Green, *I. cyclops* Green, *I. planus* Green, *I. platycephalus* Stokes.
21. *Cryptonymus expansus* Dalm., *C. extenuatus, C. limbatus* Boeck, *C. striatus* Boeck, *C. acuminatus* Boeck, *C. platynotus* Dalm., *C. schroeteri* Schlot.
22. *Ogygia guettardi* Brong., *O. desmarestii* Brong., *O. buchii* Brong.. *O. corndensis* Murch., *O. tyrannus* Murch., *O. subtyrannus* D'Arch. & Vern., *O. dilatata* Brünn., *O. angustifrons* Dalm., *O. frontalis* Dalm., *O. grandis* Sars., *O. pusilla* Münst., *O. grandæva* Gold., *O. sillimani* Broug. (De la Beche Man Geol.)
23. *Odontopleura ovata* Emm., *O. dentata* Gold. tab. 4, fig. 2, *O. brightii* Murch.
24. *Conocephalus sulzeri* Schloth, *C. striatus* Emm., *C. zippei* Sternb.
25. *Gerastos lævigatus* Gold. table 4, fig. 3, a-b. *Proetus curieri* Stein., *G. granulosus* Gold. tab. 4, figs. 4, a-b, *G. cornutus* Gold, tab. 4, figs. 1, *G. concinnus* Dalm., *G. globiceps* Phill., *G. schusteri* Roem., *G. sphæricus* Esm., *G. (Asaphus) brongniarti* Fischer.
26. *Calymene bellatula* Dalm.,*C. polytoma* Dalm.,*C. blumembachii* Brong. also var. *tuberculosa* Dalm. and var. *pulchella* Dalm, *C. tristani* Brong., *C. callicephala* Green, *C. selenocephala, C. platys* Green, *C. diops* Green. *C. punctata* Dalm., *C. propinqua* Münst., *C. articulata* Münst., *C. ornata* Dalm.
27. *Homalonotus delphinocephalus* Murch.,*H. ludensis* Murch.,*H. knightii* Murch., *H. herschelii* Murch., *H. greenii* Gold., *H. gigas* Roem., *H. ahrendii* Roem., *H. punctatus* Roem.
28. *Asaphus clavifrons* Dalm., *A. mucronatus* Brong., *A. semilunaris* Esm., *A. plicatus* B. & S., *A. longicaudatus* Murch., *A. caudatus* Brong., *A. tuberculato-caudatus* Murch., *A. hausmanni, A. speciosus* Dalm., *A. heros* Dalm., *A. stokesii* Murch., *A. dalmani* Gold., *A. arachnoides* tab. 5, fig. 3, *A. brongniartii* Delong., *A. proavus* Emm., *A. quadrilimbus* Phill., *A. obsoletus* Phill., *A. truncatulus* Phill., *A. micrurus* Green, *A. limulurus* Green, *A. wetherillii* Green, *A. pleuroptyx* Green, *A. laticostatus* Green, *A. selenurus* Green, *A. subcaudatus* Murch., *A. cawdori* Murch., *A. semiiferus* Phill., *A. gemmuliferus* Phill., *A. duplicatus* Murch., *A. raniceps* Phill., *A. astragalotes* Green, *A. crypturus* Green, *A. brevis* Münst., *A. eichwaldi* Fisch.
29. *Acaste downingiæ* Murch., *A. rotundifrons* Emm., *A. conophthalmus* Boeck., *A. extensus* Boeck, *A. sclerops* Dalm., *A. tetragonocephalus* Green. *A. jordani* Roem., *A. microps* Green.
30. *Phacops macrophthalmus* Brong., *P. protuberans* Dalm., *P. anchiops* Green, *P. variolaris* Brong., *P. ceratophthalmus* Gold. tab. 5, fig. 2, a-b, *P. intermedius* Münst., *P. subvariolaris* Münst., *P. granulatus* Münst., *P. lævis* Münst., *P. globiceps* Phill., *P. elegans* Boeck, *P. scaber* Boeck., *P. elliptifrons* Esm.

The author also gives a list of unclassified Trilobites.

Goodchild (J. G.) and Postlethwaite (J.) See Postlethwaite (J.) and Goodchild (J. G.)

Green (Jacob). Synopsis of the Trilobites of North America, in which some new genera and species are proposed.
Monthly Am. Jour. Geology. Phila., vol. 2, 1832, p. 558, pl. 14.
Republished in Dr. Jacob Green's "Monograph of the Trilobites of North America," Phila., 1832. List given in Dr. R. Harlan's "Med. Phys. Researches," Phila., 1835, p. 300.
Calymene blumenbachii Brong., *C. callicephala, C. selenocephala, C. platys, C. microps, C. anchiops, C. diops, C. bufo. Asaphus caudatus* Brünn, *A. selenurus, A. pleuroptyx, A. wetherilli, A. micrurus, A. limulurus, A. laticostatus. Trimerus* n. g., *T. delphinocephalus* (pl. 14, fig. 1). *Cryptolithus* n. g., *C. tessellatus* (pl. 14, fig. 4). *Dipleura* n. g., *D. dekayi* (pl. 14, figs. 8, 9). *Isotelus gigas, I. planus, I. cyclops* (pl. 14, fig. 7). *Triarthrus* n. g., *T. beckii* (pl. 14, fig. 6). *Paradoxides boltoni* (pl. 14, fig. 5). *Ceraurus* n. g., *C. pleurexanthemus* (pl. 14, fig. 10). *Calymene macrophthalma* Brong.

────── A monograph of the Trilobites of North America, with colored models of the species. Philadelphia, 1832, 94 pp., 1 pl. (Oct. 1, 1832).
Calymene blumenbachii Brong., Cast No. 1; *C. callicephala*, Cast No. 2. The cast of this species from Hampshire county, Va., exhibits a Calymene with a broad frontal margin. The other specimens referred to by Dr. G., from Ohio and Indiana, belong to *C. senaria* Conrad.
Calymene selenocephala, Cast No. 3; *C. platys*, Casts Nos. 4–5; *C. microps*, Cast No. 6; *C. anchiops*, Cast No. 7; *C. diops*, Cast No. 8; *C. macrophthalma* Brong., Cast No. 9; *C. bufo*, Cast No. 10, var. *rana*, Casts Nos. 11 and 12.
Asaphus laticostatus, Cast No. 13; *A. selenurus* Eaton, Casts Nos. 14 and 15; *A. limulurus*, Cast No. 16; *A. caudatus* Brünn., Cast No. 17; *A. hausmanni* Brong., *A. pleuroptyx*, Cast No. 18; *A. micrurus*, Cast No. 19; *A. wetherilli*, Cast No. 20. *Paradoxides boltoni* Bigsby. *Ogygia—Isotelus gigas* Dekay, Casts Nos. 21 and 22; *I. planus* Dekay, Cast No. 23; *I. cyclops*, Cast No. 24; *I. megalops*, Cast No. 25; *I. stegops*, Casts Nos. 26 and 27. *Cryptolithus tessellatus*, Cast No. 28. *Dipleura dekayi*, Casts Nos. 30 and 31. *Trimerus delphinocephalus*, Cast No. 32. *Ceraurus pleurexanthemus*, Cast No. 33. *Triarthrus beckii*, Cast No. 34. *Nuttainia sparsa* Eaton, Cast No. 35. *Brogniartia platycephala* Eaton. (This species is now referred to *Homalonotus delphinocephalus* Green.)
Dr. Green refers to the figure given by Bisgby, Ann. Lyc. Nat. His., N.Y., 1825, as *Cryptolithus bigsbyi* n. sp., should it prove to be a distinct species from *C. tessellatus* Green.

———— A new Trilobite. Letter to the editor, Dec. 26, 1833.
In Am. Jour. Sci., 1st series, vol. 23, 1833, p. 397.
Asaphus myrmecophorus n. sp.

———— Description of a new Trilobite from Nova Scotia.
In Trans. Geol. Soc., Penn., vol. 1, pt. 1, 1834, p. 37, pl. 4.
Asaphus ? crypturus n. sp.

———— Description of some new North American Trilobites.
In Am. Jour. Sci., 1st series, vol. 25, 1834, p. 334.
Calymene odontocephalus n. sp. *Asaphus astragalotes* n. sp., *A. tetragonocephalus* n. sp. *Paradoxides harlani* n. sp.

———— A supplement to the monograph of the Trilobites of North America, with colored models of the species. Philadelphia, 1835, 24 pp. (May 1, 1835).
Calymene odontocephalus, Cast No. 36. *Asaphus astragalotes*, Cast No. 37; *A. tetragonocephalus*, Cast No. 38. *Paradoxides harlani*, Cast No. 39. *Asaphus myrmecophorus*, Cast No. 40; *A. ? crypturus*, Cast No. 41; *A. micropleurus*, Cast No. 41 (42).

———— Description of a new Trilobite.
In Am. Jour. Sci., 1st series, vol. 32, 1837, p. 167.
Calymene phlyctænoides n. sp. *Asaphus platypleurus* n. sp.

———— Description of two new species of Trilobites. Read Jan. 24, 1837.
In Jour. Acad. Nat. Sci., Phila., vol. 7, 1837, p. 217, 2 woodcuts.
Gryphæus n. g., *C. boothii* n. sp. *Asaphus trimblii* n. sp.
J. W. Salter (Mon. Brit. Tril., p. 15) gives the following diagnosis of the genus, and uses it as a subgenus to *Phacops*: "Form of moderate size, depressed. Glabella depressed, not much expanded in front; all the lobes distinct, the front ones not greatly enlarged, genal angles long-spined. Pleura truncate, and the hinder ones often produced into spines. Tail large, of many segments; the margin spinose."

———— Description of several new Trilobites.
In Am. Jour. Sci., 1st series, vol. 32, 1837, p. 343, 2 woodcuts.
Gryphæus boothii Green, *C. callitelus* n. sp. *Trimerus jacksonii* n. sp. *Asaphus trimblii* Green.

—— Some remarks on the genus *Paradoxides* Brong., and the necessity of preserving the genus *Triarthrus*, proposed in the Monograph of the Trilobites of North America.
In Am. Jour. Sci., 1st series, vol. 33, 1838, p. 341.

—— Description of a new Trilobite.
In Am. Jour. Sci., 1st series, vol. 33, 1838, p. 406.
Calymene rowii n. sp.
This Trilobite was figured by Mr. Row in the Poughkeepsie Telegraph, Nov. 22, 1837.

—— Description of *Asaphus polypleurus*.
In Am. Jour. Sci., 1st series, vol. 34, 1838, p. 380.

—— Description of a new Trilobite.
In Annals Mag. Nat. His., 1st series, vol. 1, 1838, p. 79.
Calymene rowii Green.

—— Description of a new Trilobite.
In Am. Jour. Sci., 1st series, vol, 37, 1839, p. 40.
Asaphus diurus n. sp.
In a recent article on *Coronura aspectans* Conrad, by J. M. Clarke, the author refers this species to *Asaphus diurus* Green.
Professor Clarke remarks that Dr. Green's citation of locality "in Green county, Ohio, in the neighborhood of Xenia," is evidently erroneous. The species is cited by this author from the Corniferous limestone under *Coronura diurus* Green.

—— The inferior surface of the Trilobite discovered.
In The Friend, a weekly journal, Philadelphia, March 16, 1839.
Calymene bufo.

—— The inferior surface of the Trilobite discovered. Illustrated, with colored models. Philadelphia, 1839.
Calymene bufo.

—— Remarks on the Trilobites.
In Am. Jour. Sci., 1st series, vol. 37, 1839, p. 25.
This article appears to be a republication of the two preceding references, with the addition of the remarks on p. 38.

—— An additional fact, illustrating the inferior surface of *Calymene bufo*.
In Am. Jour. Sci., 1st series, vol. 38, 1840, p. 410.

Grewingk (C.) Geologie von Liv- und Kurland.
In Archiv Natur. Liv-, Ehst- u. Kurl., Dorpat, vol. 2, pt. 3, 1861, p. 479, 4 pls.
Contains list of fossils.

Griffith (Richard).
Two important works on the paleontology and geology of Ireland were prepared under the direction of Mr. Richard Griffith, by Frederick McCoy, who will be counted as the author, although Mr. Richard Griffith was the collector of the specimens and the financial supporter of the publications.
The contributions on the Paleozoic Crustacea are given under McCoy (Frederick) in this work.

Grote (A. R.) and **Pitt (W. H.)** New specimens from the Water Lime group at Buffalo, N. Y.
In Proc. Am. Assoc. Adv. Sci., 26th meeting, Nashville, 1877, p. 300.
Pterygotus cummingsi n. sp.

—— Description of a new Crustacean from the Water Lime group at Buffalo.
In Bull. Buffalo Soc. Nat. Sci., vol. 3, 1875, p. 1.
Eusarcus n. g., *E. scorpionis* n. sp.

—— On new species of *Eusarcus* and *Pterygotus* from the Water Lime group at Buffalo.
In Bull. Buffalo Soc. Nat. Sci., vol. 3, 1875, p. 17, plate.
Eusarcus scorpionis G. & P., *E. grandis* n. sp. *Pterygotus cummingsi* G. & P.

Gruenwaldt (M.) Beiträge zur Kenntniss der sedimentären Gebirgsformationen des Ural.
In Mém. Acad. Sci. St.-Pétersbourg, 7th series, vol. 2, No. 7, 1860, 6 pls.
Phillipsia derbyensis Mart.? *P. indeterminata* n. sp. *Leperditia biensis* n. sp.

Guettard (E.-T.) Mémoire sur les ardoisières d'Angers.
In Mém. Acad. Sci. Paris, 1757, p. 52, pls. 7-9. Reprint: Amsterdam, 1768.
The author in his memoir on the geology of Angers gives descriptions and figures of several Trilobites. These were afterwards described by Alex. Brougniart under the genus *Ogygia* and named by this author *Ogygia guettardii* and *O. desmaresti*.
The author describes three Trilobites which are referred to the modern genera of *Ogygia guettardi* Broug., *O. desmaresti* Broug. *(Illænus giganteus* Burm.), and *Calymene tristani* Broug.

Haldeman (S. S.) Report on the supposed identity of *Atops trilineatus* (Emmons) with *Triarthrus beckii*.
In American Jour. Agl. Sci., vol. 6, 1847, p. 194.
These comparisons were drawn from imperfect specimens of *Atops trilineatus*, the pygidium not being entire.

—— On the supposed identity of *Atops trilineatus* with *Triarthrus beckii*.
In Am. Jour. Sci., 2d series, vol. 5, 1848, p. 107.

Hall (Charles E.) Contribution to palæontology from the Museum of the Second Geological Survey of Pennsylvania.
In Proc. Am. Philos. Soc. Phila., vol. 16, 1877, p. 621.
Eurypterus pennsylvanicus n. sp., *E. (Dolichopterus) mansfieldi* n. sp.

Hall (James). Descriptions of two species of Trilobites, belonging to the genus *Paradoxides*.
In Am. Jour. Sci., 1st series, vol. 33, 1838, p. 139, 2 figs.
Read before the Yale Natural History Society, March 21, 1837.
Paradoxides becki Green, *P. eatoni* n. sp.

—— Geology of New York, Part IV, comprising the Survey of the Fourth Geological District. Albany, 1843.
Asaphus limulurus Green. *Bumastus barriensis. Calymene niagraensis* n. sp. *Homalonotus delphinocephalus* Green. *Calymene crassimarginata* n. sp. *Odontocephalus selenurus. Calymene bufo. Cryphæus callitelus. Dipleura dekayi. Calymene nupera* n. sp.

—— Trilobites of the inferior strata.
(Extracted and published in advance of the "Palæontology of New York," vol. 1, Albany, 1846, pp. 225-260, pls. 60-67.)
Ogygia? vetusta n. sp. *Asaphus? extans* n. sp. *Calymene multicosta* n. sp. Trilobites of the Trenton Limestone: *Illænus crassicauda* Wahl., *I. trentonensis* Emmons, *I. latidorsata* n. sp. *Isotelus gigas* DeKay. *Platynotus trentonensis* Conrad. *Calymene beckii* Green (including *Atops trilineatus* Emmons) *C. senaria* Conrad. *Acidaspis trentonensis* n. sp., *A. spiniger* n. sp. *Ceraurus pleurexanthemus* Green, *C. vigilans* n. sp., *C.? pustulosus* n. sp. *Phacops callicephalus* n. sp., *P.? laticaudus* n. sp. *Asaphus? nodostriatus* n. sp. *Trinucleus concentricus* Eaton.
Trilobites of the Utica Slate and Hudson River group: *Calymene beckii. Asaphus ? latimarginata* n. sp. *Isotelus gigas.*

Trilobites of the Hudson River group: *Trinucleus concentricus. Olenus asaphoides* Emmons, *O. undulostriatus* n. sp. *Thaleops (Illœnus) ovatus* Conrad. *Agnostus lobatus* n. sp.

——— Palæontology of New York. Vol. 1. Containing descriptions of the organic remains of the lower divisions of the New York system. Albany, 1847, 100 pls.

Trilobites from the Chazy Limestones: *Illœnus arcturus* n. sp., *I. crassicauda ? Asaphus ? obtusus* n. sp., *A. marginalis* n. sp. *Isotelus gigas ?, I. canalis* Conrad Ms. *Ceraurus* sp.? *Cytherina* sp.?
For other Crustacea see Trilobites of the inferior strata.

——— Remarks on the observations of S. S. Haldeman "On the supposed identity of *Atops trilineatus* with *Triarthrus beckii*."

In Am. Jour. Sci., 2d series, vol. 5, 1848, p. 322, 10 woodcuts.

——— Descriptions of new species of fossils, and observations upon some others previously not well known, from the Trenton limestones.

In Third Rept. N. Y. State Cab. Nat. Hist., 1850, p. 167, 5 pls. (2 eds.)
Asaphus extans Hall.
This species was used by Mr. E. Billings for the type of *Bathyurus*.

——— Description of new, or rare species of fossils, from the Palæozoic series.

In Rept. Geol. Lake Superior Land Dist. (Foster and Whitney), 1851, pt. 2, p. 203, pls. 22-35.
Harpes escanabiœ n. sp. *Dikelocephalus. Proetus. Phacops anchiops* Green, *P. callicephalus* Hall. *Asaphus barrandi* n. sp. Tracks of a Crustacean.?

——— Palæontology of New York. Vol. 2. Containing descriptions of the organic remains of the lower divisions of the New York system. Albany, 1852, 104 pls.

Trilobites of the Clinton Group.—*Homalonotus delphinocephalus. Cybele punctata* Brünn. *Calymene clintoni* Vanuxem, *C. blumenbachii* var.? *senaria* Conrad. *Acidaspis* sp.? *Phacops trisulcatus* Hall. *Ceraurus insignis* Beyr. *Beyrichia lata* Conrad.
Trilobites of the Niagara Group.—*Bumastus barriensis* Murch. *Phacops limururus* Green. *Ceraurus insignis* Beyr. *Calymene blumenbachii* var. *niagarensis* Hall. *Homalonotus delphinocephalus* Green. *Lichas boltoni* Bigsby. *Bronteus ? niagarensis* n. sp. *Arges phlyctainodes* Green. *Proetus*

corycoeus Conrad, *P. ? stokesii* Murch. *Beyrichia symmetrica* n. sp. *Cytherina spinosa* n. sp. *Onchus deweii* (the posterior spine of a Crustacea of the genus *Ceratiocaris ?*). *Calymene camerata* Conrad, from the Coralline limestone.

——— Notes on some fossils of the so-called Taconic system described by Doctor Emmons.
In Am. Jour. Sci., 2d series, vol. 19, 1855, p. 434.
Olenus (Elliptocephalus) asaphoides.

——— Description of new species of fossils from the Carboniferous limestone of Indiana and Illinois.
Read in 1856 and published separately by the author. See, also, Trans. Albany Inst., vol. 1, pt. 1, 1857, p. 1.
Cythere carbonaria n. sp.
Descriptions of these fossils, with additional notes by R. P. Whitfield, were published in the Bull. Am. Mus. Nat. Hist., New York, vol. 1, No. 3, 1882.
The original descriptions, with additional observation by James Hall, with the plates from R. P. Whitfield's memoir, were published in the 12th Annual Rept. Geol. Nat. Hist. Indiana, 1882, pp. 321-375, pls. 29-32.

——— The Trilobites of the shales of the Hudson River group.
In Twelfth Rept. New York State Cab. Nat. Hist., 1859, p. 59, 3 woodcuts. See, also, Palæont. New York, vol. 3, 1859; Geol. Vermont, vol. 1, 1861; 13th Rept. New York State Cab. Nat. Hist., 1860.
Olenus thompsoni n. sp., *O. vermontana* n. sp. *Peltura (Olenus) holopyga* n. g.

——— Geological Survey New York. Palæontology. Vol. 3. Containing descriptions and figures of the organic remains of the Lower Helderberg group and the Oriskany sandstone. 1855–59.
Pt. 1, text, 1859; pt. 2, plates, 1861, 120 pls.
Trilobites from the Lower Helderberg group: *Homalonotus vanuxemi* n. sp. *Bronteus barrandi* n. sp. *Proetus protuberans* n. sp. *Phacops hudsonicus* n. sp.; *P. logani* n. sp. *Dalmania pleuroptyx* Green, *D. micrurus* Green, *D. tridens* n. sp., *D. nasutus* Conrad. *Lichas bigsbyi* n. sp., *L. pustulosus* n. sp. *Acidaspis tuberculatus* Conrad, *A. hamata* Conrad. *Leperditia jonesi* n. sp., *L. alta* Conrad., *L. hudsonica* n. sp., *L. parasitica* n. sp., *L. parvuta* n. sp. *Beyrichia granulata* n. sp., *B. oculini* n. sp., *B. notata* n. sp. also var. *ventricosa*, *B. trisulcata* n. sp.

Waterlime group: *Eurypturus remipes* Dekay, *E. microphthalmus* n. sp., *E. lacustris* Harlan also var. *robustus*, *E. dekayi* n. sp., *E. pachycheirus* n. sp., *E. pustulosus* n. sp., *E. (Dolichopterus* subgen.), *D. macrocheirus* n. sp. *Pterygotus cobbi* n. sp., *P. macrophthalmus* n. sp., *P. osborni* n. sp. *Ceratiocaris maccoyanus* n. sp., *C. acuminatus* n. sp. *C. aculeatus* n. sp.

Trilobites from Georgia, Vermont: *Olenus thompsoni* Hall, *O. vermontana* Hall. *Peltura (Olenus) holopyga* Hall.

────── Description of new species of fossils from the Silurian rocks of Nova Scotia.
In Canadian Naturalist, vol. 5, 1860, p. 144.
Homalonotus dawsoni Hall. *Calymene blumenbachii*. *Dalmania logani* Hall. *Beyrichia pustuosa* Hall, *B. æquilatera* Hall. *Leperditia sinuata* Hall.

────── Contributions to palæontology, 1858 and 1859. Notes and observations upon the fossils of the Goniatite limestone in the Marcellus shale of the Hamilton group in the eastern and central parts of the State of New York, and those of the Goniatite beds of Rockford, Indiana; with some analogous forms from the Hamilton group proper.
In Thirteenth Rept. New York State Cab. Nat. Hist., 1860, p. 95.
Proetus doris n. sp.

────── Note upon the Trilobites of the shales of the Hudson River group in the town of Georgia, Vt.
In Thirteenth Rept. New York State Cab. Nat. Hist., 1860, p. 113.
Barrandia n. g.
This term was used by Frederick McCoy, in 1844, for a genus of Trilobites.
Bathynotus n. g., *B. holopyga* Hall. *Barrandia thompsoni* Hall, *B. vermontana* Hall.

────── New species of fossils from the Hudson River group of Ohio and the Western States.
In Thirteenth Rep. New York State Cab. Nat. Hist., 1860, p. 119.
Calymene christyi n. sp. *Proetus pariusculus* n. sp.

────── Geological Survey of Wisconsin. Description of new species of fossils from the investigations of the survey, etc. Report of the Superintendent of the Geo-

logical Survey, exhibiting the progress of the work, January 1, 1861, p. 49.
Illænus imperator n. sp., *I. taurus* n. sp. *Calymene mammilata* n. sp. *Dalmanites vigilans* n. sp.

——— Preliminary notice of the Trilobites and other Crustacea of the Upper Helderberg, Hamilton and Chemung groups.
Author's edition, published in 1861, 11 pls., 170 pp.; 15th Rept, New York State Cab. Nat. Hist., 1862, p. 82.
Olenellus n. g. *Calymene platys* Green. *Dalmania anchiops* Green also var. *armata*, *D. selenurus* Eaton, *D. ægeria* n. sp., *D. coronota* n. sp., *D. macrops* n. sp., *D. aspectans* Conrad, *D. myrmecophorus* Green, *D. helena* n. sp., *D. calyso* n. sp., *D. pleione* n. sp., *D. erina* n. sp., *D. bifida* n. sp., *D. boothi* Green. *Phacops bufo* Green, *P. rana* Green, *P. cristata* n. sp., *P. bombifrons* n. sp., *P. cacapona* n. sp. *Proetus conradi* n. sp., *P. angustifrons* n. sp., *P. hesione* n. sp., *P. clarus* n. sp., *P. crassimarginatus* Hall, *P. canaliculatus* n. sp., *P. verneuili* n. sp., *P. haldemani* n. sp., *P. rowi* Green, *P. marginalis* n. sp., *P. macrocephalus* n. sp., *P. macrocephalus* var. a, *P. auriculatus* n. sp., *P. occidens*, *P. longicaudus* n. sp. *Lichas armatus* n. sp., *L. grandis* n. sp. *Acidaspis* sp.? *Beyrichia punctulifera* n. sp. *Leperditia cayuga* n. sp., *L. spinulifera* n. sp., *L. seneca* n. sp. *Homalonotus dekayi* Green.

——— Note on the genus Cypricardites.
In Fifteenth Rept. New York State Cab. Nat. Hist., 1861, p. 192. Plate illustrating certain genera and species described by T. A. Conrad in the 5th Annual Rept. Pal. Dept. New York Geol. Survey, 1841, pl. 11 (reprint).
Dicranurus, pl. 11, fig. 1. *Aspidolites*, pl. 11, fig. 2. *Acidaspis tuberculatus*, pl. 11, fig. 3. *Asaphus ? acantholerus* Conrad, *A. ? denticulatus* Conrad.

——— Note upon the Trilobites of the Hudson River group, in the town of Georgia, Vermont.
In Geol. Vermont, vol. 1, 1861, p. 367, pl. 13.
Barrandia thompsoni Hall, *B. vermontana* Hall. *Bathynotus holopyga* Hall.

——— Palæontology of Wisconsin. Remarks upon the condition of the fossils in the rocks of the several formations. Catalogue of fossils known in the Palæozoic formations of Wisconsin, with observations upon some of the known species and descriptions of several new forms.

Geol. Survey Wisconsin, vol. 1, 1862, p. 425.
Dikelocephalus minnesotensis, D. pepinensis. Asaphus barrandi, A. (Isotelus) gigas. Dalmania vigilans, D. meta n. sp. *Calymene mammillata* n. sp. *Acidaspis danai* n. sp.

—— On a new Crustacean from the Potsdam sandtone. A letter addressed to Principal Dawson, dated Albany, Oct. 31, 1862.

In Canadian Naturalist, vol. 7, 1862, p. 443, figure. See The Geologist, vol. 6, London, 1863, p. 247, figure.

Aglaspis n. g. *(A. barrandi* Hall.)

—— On the occurrence of Crustacean remains of the genera *Ceratiocaris* and *Dithyrocaris*, with a notice of some new species from the Hamilton group and Genesee slate.

In Sixteenth Rept. New York State Cab. Nat. Hist., Albany, 1863, p. 71, plate.

Ceratiocaris armatus n. sp., *C. longicaudus* n. sp., *C. ? punctatus* n. sp. *Dithyrocaris neptuni* n. sp.

—— Preliminary notice of the fauna of the Potsdam sandstone, with remarks upon the previously known species of fossils, and description of some new ones from the sandstones of the Upper Mississippi Valley.

In Sixteenth Rept. New York State Cab. Nat. Hist., 1863, p. 119, 6 pls. Supplementary note, p. 110.

Triarthrella n. g. *Chariocephalus* n. g. *Ptychaspis* n. g. *Illænurus* n. g. *Pemphigaspis* n. g. *Aglaspis* n. g.

Descriptions of *Dikelocephalus minnesotensis* Owen also var. *limbatus, D. minnesotensis* var. *D. pepinensis* Owen, *D. spiniger* n. sp., *D. misa* n. sp., *D. osceola* n. sp. *Conocephalites minor* Shumard, *C. minutus* Bradley, *C. eos* n. sp., *C. perseus* n. sp., *C. shumardi* n. sp., *C. nasutus* n. sp., *C. oweni* n. sp., *C. eryon* n. sp., *C. anatinus* n. sp., *C. patersoni* n. sp., *C. ? binodosus* n. sp., *C. winona* n. sp., *C. iowensis* Owen., *C. wisconsensis* Owen, *C. hamulus* Owen, *C. diadematus* n. sp. *Arionellus bipunctatus* Shumard. *Ptychaspis miniscaensis* Owen, *P. granulosa* Owen. *P.? sp. Chariocephalus whitfieldi* n. sp. *Illænurus quadratus* n. sp. *Triarthrella auroralis* n. sp. *Agnostus josepha* n. sp., *A. parilis* n. sp., *A. disparilis* n. sp. *Aglaspis barrandi* n. sp. *Pemphigaspis bullata* n. sp. *Amphion ? matutina* n. sp. *Conocephalites ? (Arionellus ?) dorsalis* n. sp., *C. optatus* n. sp. *Lichas grandis* (subgenus ? *Terataspis)* L. *armatus, L. boltoni, L. nereus* n. sp.

―――― Notice of some new species of fossils from a locality of the Niagara group in Indiana, with a list of identified species from the same place.

In Trans. Albany Inst., vol. 4, 1864, p. 195. Author's edition published May 2, 1863, pp. 1-34.

Plates of the species 28th Rept. New York State Mus. Nat. Hist, 1879.

Dalmania verrucosa Hall. *Cyphaspis christyi* Hall. *Lichas breviceps* Hall, *L. boltoni* var. *occidentalis* Hall. *Calymene niagarensis* Hall. *Homalonotus delphinocephalus* Green. *Illænus barriensis* Murch.

―――― Account of some new or little known species of fossils from rocks of the age of the Niagara group.

In Twentieth Rept. New York State Cab. Nat. Hist., 1867, p. 305, 21 pls. Originally printed in advance for the 18th Rept. New York State Cab. Nat. Hist., 1864. Revised edition, Albany, 1870, with 22 pls.

The following Crustacea are described in the edition of 1867.

Illænus armatus n. sp., *I. insignis* n. sp.. *I. imperator* n. sp., *I. (bumastus) ioxus* Hall, *I. cumiculus* n. sp. *Bronteus acamas* n. sp. *Acidaspis danai* Hall. *Lichas breviceps ?* Hall, *L.* sp.? *L. pugnax* W. & M., *L. decipiens* W. & M. *Sphærexochus romingeri* Hall. *Calymene niagarensis* Hall. *Encrinurus nereus* n. sp. *Dalmania vigilans*. *Ceraurus niagarensis* n. sp. *Leperditia fonticola* n. sp.

The revised edition contains descriptions of the following species not in the edition of 1867, viz.:

Lichas pugnas W. & M., *L. obvius* n. sp.

―――― Geological Survey of the State of Wisconsin, 1859-63. Palæontology. Pt. 3. Organic remains of the Niagara group and the associated limestones. 94 pp., 18 pls. Albany, 1871.

The same was published in 1864 in advance for the 18th Rept. New York State Cab. Nat. Hist., 1867, under the title of "Account of some new or little known species of fossils from rocks of the age of the Niagara group." See 20th Rept. New York State Cab. Nat. Hist., rev. edition, 1870, p. 347.

―――― Descriptions of new species of Crinoids and other fossils from strata of the age of the Hudson River group and Trenton limestone.

In Twenty-fourth Rept. New York State Mus. Nat. Hist., 1872, p. 205, 3 pls.

Dalmania breviceps n. sp. *Proetus parviusculus* n. sp.

Advance sheets of this article were printed and distributed in November, 1866, except the first four species of Crinoidea mentioned in the foot note on p. 205, which were first published in 1871 in a pamphlet entitled "New species of fossils from the Hudson River group in the vicinity of Cincinnati, Ohio."

———— Descriptions of new species of fossils from the Hudson River group in the vicinity of Cincinnati, Ohio.

In Twenty-fourth Rept. New York State Mus. Nat. Hist., Albany, 1872, p. 225, 1 pl.

Published Oct., 1871, in advance of the State Museum Report.

Leperditia (Isochilina) cylindrica n. sp., *L. (Isochilina) minutissima* n. sp. *Beyrichia tumifrons* n. sp., *B. oculifer* n. sp.

———— The fauna of the Niagara group in Central Indiana.

Documents only of Twenty-eighth Rept. New York State Mus. Nat. Hist., 1876, 34 pls.

Description of Plate XXXII. *Leperditia faba* n. sp. *Beyrichia granulosa* n. sp. *Cyphaspis christyi* Hall. *Calymene niagarensis* Hall, *Sphærexochus romingeri* ? Hall. *Homalonotus delphinocephalus* ? Green. *Illaenus armatus* ? Hall.

Description of Plate XXXIII. *Dalmania vigilans* Hall, *D. verrucosa* Hall, *D. bicornis* n. sp.

Description of Plate XXXIV. *Lichas breviceps* Hall, *L. boltoni* var. *occidentalis* Hall. *Lichas* sp. ? *Dalmanites verrucosus* Hall.

———— Illustrations of Devonian fossils: Gasteropoda, Pteropoda, Cephalopoda, Crustacea, and Corals of the Upper Helderberg, Hamilton and Chemung groups, etc. Albany, 1876, 7 pp., 133 pls.

Crustacea—*Calymene platys* Green, *C. niagarensis* Hall. *Homalonotus dekayi* Gr. *Phacops cristata* Hall, *P. rana* ?, *P. bombifrons* Hall, *P. rana* Green, *P. cacapona* Hall, *P. bufo* Green, *P. nupera* Hall. *Dalmanites anchiops* Green, also var. *armatus*. *Dalmanites* sp.?, *D. denticulatus* Conrad, *D. emarginatus* n. sp., *D. concinnus* n. sp., *D. regalis* n. sp., *D. ægeria* Hall, *D. selenurus* Eaton, *D. coronatus* Hall, *D. calypso* Hall, *D. crina* Hall, *D. aspectans* Conrad, *D. helena* Hall, *D. myrmecophorus* Green, *D. boothii* Green, *D. pleione* Hall. *Acidaspis (Terataspis) grandis* Hall. *Acidaspis* sp.?, *A. (Terataspis) eriopis* Hall, *A. (Terataspis)* sp.? *Lichas pustulosus* Hall. *Dalmanites acantholeurus* Conrad. *Proetus angustifrons* Hall, *P. conradi* Hall, *P. canaliculatus* Hall, *P. clarus* Hall, *P. hesione* Hall. *Phillipsia minuscula* n. sp. *Proetus verneuili* Hall, *P. crassimarginatus* Hall, *P. longicaudus*. *Phillipsia ? (Brachymetopus ?) ornata* n. sp. *Proetus rowi* Green, *P. haldemani* Hall, *P. macrocephalus* Hall, *P. occidens*

Hall, P. *marginalis* Conrad. *Phillipsia levis* n. sp.? *Dithyrocaris neptuni* Hall. *Ceratiocaris armatus* Hall, *C. (Aristozoe) punctatus* Hall.

—— The fauna of the Niagara in Central Indiana.
In Twenty-eighth Rept. New York State Mus. Nat. Hist., 1879, p. 99, 34 pls.
Crustacea—*Leperditia faba. Beyrichia granulosa. Calymene niagarensis. Homalonotus delphinocephalus* Green. *Cyphaspis christyi. Illaenus armatus. Ceraurus (Cheirurus) niagarensis.*
Notes on the genus *Dalmanites* and *Odontocephalus*. *Dalmanites vigilans, D. verrucosus, D. bicornis. Lichas breviceps, L. boltoni* var. *occidentalis. Lichas* sp.?, *L. emarginatus* Hall.

—— Description of new species of fossils from the Niagara formation at Waldron, Indiana.
In Trans. Albany Inst., vol. 10, 1883, p. 57. Read March 18, 1879, and published separately in 1881.
Acidaspis fimbriata n. sp. *Illænus (Bumastus) ioxus* Hall.

—— Description of the species of fossils found in the Niagara group at Waldron, Indiana.
In Eleventh Ann. Rept. for 1881, Dept. Geol. Nat. Hist. Survey, Indiana, 1882, p. 217, 36 pls.
Leperditia faba. Beyrichia granulosa. Calymene niagarensis. Homalonotus delphinocephalus. Cyphaspis christyi. Acidaspis fimbriata. Illænus armatus, I. (Bumastus) ioxus. Ceaurus niagarensis. Note on the genus *Dalmanites* and *Odontocephalus. Dalmanites vigilans, D. verrucosus, D. bicornis. Lichas breviceps, L. boltoni* var. *occidentalis*.

—— Note on *Eurypteridæ* of the Devonian and Carboniferous formations of Pennsylvania, with a supplementary note on the *Stylonurus excelsior*.
In Proc. Am. Assoc. Adv. Sci., 33d meeting, Phila., 1884, p. 420.

—— *Eurypteridæ* from the Devonian and Carboniferous formations of Pennsylvania. With 6 pls.
Extracted Rept. Prog. PPP., 2d Geol. Survey Pennsylvania, 1884.
Eurypterus (Anthraconectes) mazonensis M. & W., *E. beecheri* n. sp., *E. pennsylvanicus* C. E. Hall, *E. mansfieldi* C. E. Hall, *E. stylus* n. sp., *E. potens* n. sp.

—— Note on the *Eurypteridæ* of the Devonian and Carboniferous formations of Pennsylvania.
In Second Geol. Survey Pennsylvania, Rept. Prog. PPP., 1884, 6 pls.
For list of species see the above reference.

────── Description of a new species of *Stylonurus* from the Catskill group.
In Thirty-sixth Rept. New York State Mus. Nat. Hist., 1884, p. 76, pl. 5.
Stylonurus excelsior Hall.
There was a short notice of this species published by D. S. Martin (Trans. New York Acad. Sci., vol. 2, 1882, p. 8). This notice was based upon a cast of the carapace in the New York State Museum, which had been labeled with name and locality by Prof. James Hall.

────── and **Whitfield** (R. P.) Descriptions of new species of fossils from the vicinity of Louisville, Ky., and the Falls of the Ohio.
In Twenty-fourth Rept. New York State Mus. Nat. Hist., 1872, p. 181, pl. 13, figs. 20 and 21.
Illænus cornigerus H. & W.

────── Geological Survey of Ohio. Vol. 2. Geology and palæontology. Pt. 2. Palæontology. 1875, p. 67, pls. 1–12.
Leperditia (Isochilina) cylindrica Hall, *L. (Isochilina) minutissima* Hall. *Beyrichia tumifrons* Hall, *B. oculifera* Hall, *B. chambersi* Miller, *B. quadrilirata* n. sp. *Plumulites jamesi* n. sp. *Calymene christyi* Hall. *Dalmania breviceps* Hall. *Proetus parvisculus* Hall. *Calymene niagarensis* Hall. *Encrinurus ornatus* n. sp. *Lichas breviceps* Hall.

────── Palæontology.
In U. S. Geol. Expl. 40th Parallel, vol. 4, pt. 2, 1877, 7 pls.
Crepicephalus (Loganellus) haguei n. sp., *C. (L.) nitidus* n. sp., *C. (L.) granulosus* n. sp., *C. (L.) unisulcatus* n. sp., *C. (L.) simulator* n. sp., *C. (L.) anytus* n. sp., *C. (Bathyurus ?) angulatus* n. sp. *Pterocephalus: Conocephalites (Pterocephalus) laticeps* n. sp. *Ptychaspis pusulosa* n. sp. *Chariocephalus tumifrons* n. sp. *Dikellocephalus bilobatus* n. sp., *D. flabellifer* n. sp. *Agnostus communis* n. sp., *A. neon* n. sp., *A. prolongus* n. sp., *A. tumidosus* n. sp. *Conocephalus subcoronatus* n. sp. *Crepicephalus ? (Loganellus) quadrans* n. sp. *Dikellocephalus quadriceps* n. sp., *D. wahsatchensis* n. sp., *D. ? gothicus* n. sp. *Bathyurus pogonipensis* n. sp. *Ogygia producta* n. sp., *O. parabola* n. sp. *Proetus peroccidens* n. sp., *B. loganensis* n. sp. *Crep. (L.) maculosus* n. sp. *Dikellocephalus multicinctus* n. sp.

────── assisted by **Clarke** (John M.) Geological Survey of the State of New York. Palæontology. Vol. VII. Text and plates, containing descriptions of the Trilobites and other Crustacea of the Oriskany, Upper

Helderberg, Hamilton, Portage, Chemung, and Catskill groups. Albany, 1888, 236 pp., 36 pls.

Trilobita: *Dalmanites* subgen. *Hausmanni* n. s. g. *Coronura* n. s. g. *Mesothyra* n. g. *Rhinocaris* n. g. *Schizodiscus* n. g. *Strobilepsis* n. g. *Calymene platys* Green. *Homalonotus major* Whitfield, *H. dekayi* Green. *Bronteus tullius* Hall. *Phacops cristata* Hall, also var. *pipa* Hall, *P. rana* Green, *P. bufo* Green, *P. cacapona* Hall, *P. nupera* Hall. *Dalmanites (Hausmania) pleuroptyx* Green, *D. (H.) concinnus* Hall. *Dalmanites (Hausmania) concinnus* var. *serrula* Hall, *D. (H.) phacoptyx* Hall, *D. (H.) meeki* Walcott, *D. (Coronura) aspectans* Conrad, *D. (C.) myrmecophorus* Green, *D. (C.) emarginatus* Hall, *D. (Cryphæus) comis* Hall, *D. (C.) pleione* Hall, *D, (C.) boothi* Green also var. *callitelus* Green, *D. (C.) barrisi* Hall, *D. (Odontocephalus) selenurus* Conrad, *D. (O.) bifidus* Hall, *D. (O.) ægeria* Hall, *D. (O.) coronatus* Hall, *D. (Corycephalus) regalia* Hall, *D. (C.) pygmaeus* Hall, *D. (Chamops) anchiops* Green also var. *armatus* Hall, *D. (C.) anchiops* var. *sobrinus* Hall, *D. (C.) calypso* Hall, *D. (C.) erina* Hall, *D. (C.) macrops* Hall. *Acidaspis callicera* Hall, *A.* sp., *A. romingeri* Hall. *Lichas (Terataspis) grandis* Hall, *L. (C.) hispidus* Hall, *L. (C.) eriopis* Hall, *L. (C.) hylaeus* Hall, *L. (Arges) contusus* Hall, *L. (Ceratolichas) gryps* Hall, *L. (C.) dracon* Hall. *Proetus conradi* Hall, *P. angustifrons* Hall, *P. hesione* Hall, *P. curvimarginatus* Hall, *P. latimarginatus* Hall, *P. cassimarginatus* Hall, *P. folliceps* Hall, *P. clarus* Hall, *P. canaliculatus* Hall, *P. verneuili* Hall, *P. microgemma* Hall, *P. stenopyge* Hall, *P. ovifrons* Hall, *P. delphinulus* Hall, *P.? planimarginatus* Meek, *P. tumidus* Hall, *P. haldemani* Hall, *P. macrocephalus* Hall, *P. rowi* Green, *P. jejunus* Hall, *P. phocion* Billings, *P. prouti* Shumard, *P. nevadæ* Hall, *P. occidens* Hall, *P.? longicaudus*. *Phaëthonides arenicolus* Hall, *P. varicellus* Hall, *P. gemmæus* Hall, *P.? denticulatus* Meek. *Cyphaspis minuscula* Hall, *C. stephanophora* Hall, *C. diadema*, *C. hybrida* Hall, *C. ornata* Hall also var. *baccata* Hall, *C. craspedota* Hall, *C. lævis* Hall.

Xiphosura: *Protolimulus eriensis* Williams. *Eurypterus beecheri* Hall. *Stylonurus excelsior* Hall, *S.? (Echinocaris?) wrightianus* Dawson.

Phyllocardia: *Ceratiocaris longicauda* Hall, *C. beecheri* Clarke, *C.? simplex* Clarke. *Echinocaris punctata* Hall, *E. whitfieldi* Clarke, *E. condylepis* Hall, *E. socialis* Beecher, *E. sublævis* Whitfield, *E. pustulosa* Whitfield, *E. multinodosa* Whitfield. *Elymocaris capsella* Hall, *E. siliqua* Beecher. *Tropidocaris hamiltonensis* Hall, *T. bicarinata* Beecher, *T. interrupta* Beecher.

Pinacaridæ: *Mesothyra oceani* Hall, *M. neptuni* Hall, *M. spumæa* Hall. *Dithyrocaris veneris*, *D. belli* Woodward.

Rhinocaridæ: *Rhinocaris columbina* Clarke, *R. scaphoptera* Clarke.

Discinocaridæ: *Spathiocaris emersoni* Clarke. *Dipterocaris pennæ-dædali* Clarke, *D. procne* Clarke, *D. pes-cervæ* Clarke.

Decapoda: *Palæopalæmon newberryi* Whitf.

Phyllopoda: *Estheria pulex* Clarke. *Schizodiscus capsa* Clarke.

Cirripedia: *Protobalanus hamiltonensis* Whitfield. *Palæocreusia devonicus* Clarke.

Lepadidæ: *Strobilepsis spinigera* Clarke. *Turrilepas flexuosus* Hall, *T. cancellatus* Hall, *T. devonicus* Clarke, *T. squama* Hall, *T. nitidulus* Hall, *T. foliatus* Hall, *T. tener* Hall, *T.? newberryi* Whitfield.

Genera and species not Devonian described in the volume: *Calymene niagarensis* Hall. *Homalonotus vanuxemi* Hall. *Phacops logani* Hall. *Dalmanites (Hausmania) pleuroptyx* Green, *D. (Corycephalus) dentatus* Barrett. *Lichas (Dicranogmus) ptyonurus* Hall, *L. (Conolichas) pustulosus* Hall, *L. (C.) bigsbyi* Hall. *Proetus missouriensis* Shumard. *Phaëthonides cyclurus* Hall, *P. macrobius* Billings. *Cyphaspis coelebs* Hall. *Phillipsia* sp.? *Eurypterus prominens* Hall, *E. approximatus* Hall. *Tropidocaris alternata* Beecher.

—— Report of the Director.
In 42d Rep. N. Y. State Mus. 1889, p. 1, plate.
Crustacean tracks.

Hancock (Abbany). Remarks on certain vermiform fossils found in the Mountain limestone districts of the north of England.
In Rept. 28th Meeting Brit. Assoc. Adv. Sci., 1858, Tran. of Sec., p. 80.
Tracks of Trilobites.

Hare (Sid. J.) Trilobites of the Upper Coal Measure Group of Kansas City, Missouri.
In the Kansas City Scientist, vol. 5, no. 3, 1891, plate.
The author describes *Phillipsia nodacostatus* n. sp., *P. major* Shumard, *P. cliftonensis* Shumard.
He refers *Proetus longicaudatus* Hall to *Phillipsia major*, remarking that the upper coal measures extend westward to near West Counday, Greenwood County, Kansas, whereas the trilobite in question was found some 25 miles east of this line at Madison.

Harlan (Richard). Critical notices of various organic remains hitherto discovered in North America.
In Trans. Geol. Soc. Pennsylvania, vol. 1, pt. 1, 1834, p. 46, 1 pl.
The author herein describes two species of *Eurypterus*, and gives a list of North American Trilobites.
Eurypterus remipes Dekay, *E. lacustris* Harlan.

—— Notice of nondescript Trilobites from the State of New York, with observations on the genus *Triarthrus*, etc:
In Trans. Geol. Soc. Pennsylvania, vol. 1, pt. 2, 1835, p. 263, pl. 15.
Paradoxides triarthrus Har., *P. acuatus* Har., *P. scarabæoides* Brong.

—— Medical and physical researches, or original memoirs in medicine, surgery, physiology, geology, zoology, etc. Phila., 1835, 36 pls.
Contains a republication of the above articles under Harlan, pp. 253-313 and 400-403.

Harley (J.) On the Ludlow bone bed and its crustacean remains.
In Quart. Jour. Geol. Soc. London, vol. 17, 1861, p. 542, pl. 17.

The author refers to the class Crustacea certain minute bodies of various forms from the Ludlow bone beds, including the *Conodonts* described by Pander. Twenty species are described and illustrated under the provisional genus *Astacoderma*.

Hartt (C. F.) Fossils of the Primordial or Acadian group at St. John.
In Acadian Geology, the geological structure, organic remains, and mineral resources of Nova Scotia, by J. W. Dawson.

—— Appendix B. List of New Brunswick fossils.
See Bailey's "Observation on the Geol. of Southern New Brunswick, etc.," 1865, p. 143.

—— On *Phillipsia vindobonensis* from the Carboniferous Limestone of Nova Scotia.
In Canadian Naturalist, new series, vol. 3, 1868, p. 214.

—— and **Rathbun** (Richard). Morgan Expedition, 1879-71.) On the Devonian Trilobites and Mollusks of Ereré, Province of Pará, Brazil.
In Annals Lyceum Nat. Hist., New York, vol. 11, 1875, p. 110.
Dalmania paituna n. sp. *Homalonotus oiara* n. sp.

Haswell (W.) Silurian Formation of the Pentland Hills, 1865.
Entomis impendens, n. sp. p. 38, pl. 3, fig. 11.

Hauer (F. von). Ueber Barrande's Versuch einer Classifikation der Trilobiten.
In Sitzungsber. d. Wien Akad. Math.-naturw. Classe, V, p. 304, 1850.

Haulp (K.) Die Fauna der Graptholitengest.
In Lausitzische Mag., vol. 44, Görlitz, 1878, p. 75, pl. 5.
Calymene blumembachii Brong. *Dalmania caudada* Brünn, *Phacops*

sp.? *P. variolaris.* *Odontopleura marklini* Ang. *Asaphus* sp.? *Raphisophorus culminatus* Ang. *Ampyx ?* sp. *(brevinasutus), A.* sp.? *Beyrichia kloedeni, B. tuberculata* Kloden, n. sp.? *Cytherina (Leperditia)* 6 species.

Hawle (Ignatz). See **Corda** (A. J. C.) and **Hawle** (Ignatz).

Hayden (F. V.) See **Meek** (F. B.) and **Hayden** (F.V.)

Heidenhain (F.) Ueber Graptholitenführende Diluvial-Geschiebe der norddeutschen Ebene.

In Zeitschr. Deutsch. geol. Gesell., vol. 21, 1869, p. 143, plate.
Calymene blumenbachii. Dalmanites caudata. Odontopleura ovata, O. nutica. Cyphaspis. Homalonotus. Beyrichia klæden, B. maccoyana, B. tuberculatus.

Hellman (A.) Die Petrefacten Thüringens. Cassel, 1866, 5 pts., 24 pls.
Dalmania tuberculata. Acidaspis buchi. Paradoxides spinosus.

Hermann (L. D.) Maslographia Brigæ, Massel, 1711, pl. 9, fig. 50; pl. 11, fig. 44; pl. 12, fig, 31.
The author herein compares a fragment of the pygidium of an *Encrinurus* to a scollop shell.

Herrick (C. L.) A sketch of the geological history of Licking county, accompanying an illustrated catalogue of Carboniferous fossils from Flint Ridge, Ohio.

In Bull. Denison Univ., vol. 2, pt. 1, 1887, p. 5, 7 pls.
Phillipsia sawallovi Shumard, *P. shumardi* Herrick, *P. meramecensis* Shumard, *P. missouriensis* Shumard, *P. major* Shumard, *P. cliftonensis* Shumard, *P. sangamonsis* Meek & Worthen, *P. scitula* M. & W., *P. doris* (Hall) Winchell, *P. rockfordensis* Winchell, *P. insignis* Winchell, *P. howi* Billings, *P. vindobonensis* Hartt, *P. trinucleata* n. sp., *P. leei* Wood.
P. gemmulifera Phill., *P. truncatula* Phill., *P. eichwaldi* var. *mucronata.* *P. colei* M'Coy. *P. derbyensis* Mart. *Griffithides globiceps* Phill., *G. longispinus* Portl., *G. longiceps* Portl., *G. seminiferus* Phill., *G. moriceps, G. acanthiceps* Wood, *G. roemeri* Möller, *G. gruenwaldti* Möller, *G. calcaratus* M'Coy. *Brachymetopus ouralicus* DeVer., *B. maccoyi* Port. *Proetus. Dalmanites? cuyahoga* Claypole.

—— Appendix 2. A Waverly Trilobite.
In Bull. Denison Univ., vol. 2, pt. 1, 1887, p. 69, pl. 7, fig. 14.
Phillipsia shumardi Herrick.

—— The Geology of Licking County, Ohio. Pt. IV. The Subcarboniferous and Waverly groups.
In Bull. Denison Univ., vol. 3, 1888, p. 13, pls. 2, 11 and 12.
Phillipsia meramecensis Sh., *P. shumardi* Her., *P. præcursor* Her.

—— Geology of Licking County. Pt. 4.
In Bull. Denison Univ., vol. 4, 1889, 1 pl.
Crustacea, p. 49.—*Phillipsia serraticaudata* n. sp., *P. ? consors* n. sp., *P. meramecensis* Shumard, *P. (Proetus) præcursor* Herrick. *Proetus ?* cf. *haldermani* Hall, *P. minutus* n. sp. *Phaëthonides occidentalis* n. sp., *P. spinosus* n. sp., *P. ? immaturus* n. sp., *P. ? lodiensis* Meek. *Cythere ohioensis* n. sp.

—— The Cuyahoga Shales and the problem of the Ohio Waverly.
In Bull. Geol. Soc. America, vol. 2, 1891, p. 31, plate.
Phaëthonides spinosus Herr. *Phillipsia? consors* Herr., *P. meramecensis* Shumard. *Cytherella unioniformis* n. sp.
Other species are mentioned, but not described in this article.

—— Notes upon the Waverly group in Ohio.
In Am. Geologist, vol. 3, 1889, p. 94, 4 pls. (not descriptive).

Hibbert (Dr.) On the fresh-water limestone of Burdiehouse, in the neighborhood of Edinburgh, belonging to the Carboniferous group of rocks; with supplementary notes on other fresh-water limestones.
In Trans. Royal Soc. Edinburgh, vol. 13, 1836, p. 169, pl. 9.
The author gives in his memoir several small woodcuts of his *Cypris scotoburdigalensis* and *Daphnoidia*. The latter received the name of "*Hibberti*" in Morris's Catalogue; the former is considered as a sufficiently distinct variety of *Lerperditia okeni*.
The author gives reduced figures of *Eurypterus remipes* Dekay and *E. lacustris* Harlan, taken from Dr. Richard Harlan's article on "Organic Remains of North America."

Hicks (Henry). Note on the genus *Anopolenus*.
In Quart. Jour. Geol. Soc. London, vol. 21, 1865, p. 477, figures.

—— Description of new species of fossils from the Longmynd rocks of St. David's.
In Quart. Jour. Geol. Soc. London, vol. 27, 1871, p. 399, pl. 15.
Plutonia sedgwicki n. sp. *Paradoxides harknessi* n. sp. *Conocoryphe lyellii* n. sp, *C. solvensis* n. sp. *Microdiscus sculptus* n. sp, *Agnostus antiqua* n. sp. *Leperditia ? cambrensis* n. sp.

—— On some undescribed fossils from the Menevian group, with a note on the Entomostraca by Prof. T. Rupert Jones.
In Quart. Jour. Geol. Soc. London, vol. 28, 1872, p. 173, pls. 5–7.
Cerausia n. g. *Agnostus davidis* Salter, *A. eskriggei* n. sp., *A. scutalis* Salter, *A. scarabæoides* Salter, *A. barrandei* Salter. *Arionellus longicephalus* n. sp. *Erinnys venulosa* Salter. *Carausia menevensis* n. sp. *Holocephalina inflata* n. sp. *Conocoryphe homfrayi* Salter, *C. coronatus* Barr. *Anopolenus impar* n. sp., *A. salteri* n. sp
Notes on the *Entomostraca* by Prof. T. Rupert Jones.
Leperditia hicksii Jones. *Entomis buprestis* Salter.

—— On the Tremadoc rocks in the neighborhood of St. David's, South Wales, and their fossil contents.
In Quart. Jour. Geol. Soc. London, vol. 29, 1873, p. 39, 3 pls.
Neseuretus n. g., *N. quadratus* n. sp., *N. recurvatus* n. sp., *N. ramsayensis* n. sp., *N. ? elongatus* n. sp. *Niobe menapiensis* n. sp., *N. solvensis* n. sp.

—— Descriptions of new species of fossils from the Arenig group of St. David's.
In Quart. Jour. Geol. Soc. London, vol. 31, 1875, p. 182, 3 pls.
Ampyx salteri Hicks. *Trinucleus etheridgei* n. sp., *T. ramsayi* n. sp. *Illaenus hughesii* n. sp. *Illaenopsis ? acuticaudata* n. sp. *Aeglina boia* Hicks, *A. ostusicaudata* n. sp. *Barrandea homfrayi* n. sp. *Phacoparia cambriensis* n. sp. *Phacops llanvirnensis* n. sp. *Calymene hopkinsoni* n. sp.

—— Appendix to Fossiliferous Cambrian shales, near Cœrnarvon, by J. E. Marr.
In Quart. Jour. Geol. Soc. London, vol. 32, 1876, p. 135.
Caryocaris marrii n. sp. *Aeglina hughesii* n. sp.

Hisinger (W.) Anteckningar i physik och geognösi under resor uti Sverige och Norrige. Stockholm, 8 pts., 1828–40.
The first three parts contain scarcely anything on Crustacea; the remaining parts contain only enumerations of fossils, without descriptions.
Cytherina phaseolus His., pt. 5, pl. 8, fig. 3.

—— Tableau des pétrifications de la Suède distribuées en ordre systématique. Stockholm, 1829.
This work enumerates 40 species of Silurian Crustacea previously described by J. W. Dalman.

The second edition of this work bears the title "Esquisse d'un tableau des pétrifications de la Suède," Stockholm, 1831.
This work contains 43 pages and a systematic table of the fossils, in which 46 species of fossil Crustacea are mentioned.

——— Lethæa Svecica, seu petrificata Sveciæ iconibus et characteribus illustrata duobus supplementis. Holmiæ, 1837–41, 4to, 42 pls.

The first part of this work was published in 1835 under the title "Icones petrifica orum Sveciæ. Fas. I. Animalia articulata et mollusca Cephalopoda." The plates, 1 to 10, were sent by the author only to his friends. In 1837 Lethæa Svecica was published; it contained 124 pages and 39 plates. The first three are marked A, B, C. The fossil Crustacea described in this work are those of J. W. Dalman's Palæaderna, with copies of his descriptions and figures. The second supplement, published at Stockholm in 1840, contains 11 pages and 3 plates. Figures and descriptions of *Calymene clavifrons, Asaphus (Trinucleus) secticornis, A. cyllarus (Trinucleus)* and *Calymene (Cheirurus) speciosa.* "Lethæa Svecica, seu petrificata Sveciæ, supplementa secundi continuatio," was published at Stockholm in 1841; it contains 6 pages and 3 plates (60-62). Figures and description of a *Calymene.*

Cytherina phaseolus His., *C. balthica* His. *Calymene blumenbachii* Brong. (a) *tuberculosa*, (b) *pulchella, C. bellatula* Dalm., *C. ornata* Dalm., *C. polytoma* Dalm., *C. actinura* Dalm., *C. sclerops* Dalm., *C. punctata, C. concinna* Dalm., *C. ? centrina* Dalm., *C. ? speciosa* (Dalm.) His., *C. ? pustulata.*
Asaphus mucronatus Brong., *A. caudatus, A. heros* Dalm., *A. extenuatus, A. granulatus, A. angustifrons* Dalm., *A. expansus, A. frontalis* Dalm., *A. platynotus* Dalm., *A. læviceps* Dalm., *A. palpebrosus* Dalm., *A. ? sulzeri. Nileus armadillo* Dalm., *N. glomerinus* Dalm. *Illaenus centrotus* Dalm., *I. centaurus* Dalm., *I. crassicauda, I. laticauda. Lichas laciniatus. Ampyx nasutus* Dalm., *A. pachyrrhinus* Dalm. *Olenus tessini, O. bucephalus, O., spinulosus, O. gibbosus, O. scarabæoides. Battus pisiformis,* also var. *spinger* Dalm., *B. lævigatus* Dalm.
Supplement II. *Calymene clavifrons* Dalm. *Asaphus seticornis* n. sp., *A. cyllarus* n. sp. *Calymene* sp. *Calymene sclerops ?* Dalm.

Hoeninghaus (F. W.) Beschreibung über Abbildungen von *Calymene arachnoides.* Crefeld, 1835, I pl.

——— Ueber die Versteinerungen des Uebergangs-Thonschiefers von Weisenbach im Dillenburgschen.

In Isis (oder Encycl. Zeitung), Oken, 1830, p. 96; Neues Jahrbuch für Mineral., 1831, p. 341.
Calymene macrophthalma Broug.

—— Trilobites der Geognostischen Sammlung von F. W. Hoeninghaus Geordnet der Systematischen uebersicht der Trilobiten von Herrn Professor Dr. Goldfuss. (No place or date.)

The author follows Goldfuss' classification placing the species under 30 genera.

—— *Harper reflexus.* Crefeld, 1847, 1 p., 1 pl.

Hoffman (E.) Sämmtliche bis jetzt bekannte Trilobiten Russlands.

In Verhandl. russ. k. mineral. Gesell. zu St. Petersburg, 1857-68, p. 21, 7 pls.

Lichas verrucosus Eichw., *'L. 'hybneri* Eichw., *L. eichwaldi* Kaiserling, *L. sexpunctatus* n. sp. *Cheirurus zembnitskii* Eichw., *C. macrophthalmus* Kut. *Sphærexochus cranium* Kut., *S. hemicranium* Kut., *S. platycranium* Kut., *S. enurus* Kut., *S. aries* Eichw., *S. coniceps. Zethus bellatulus* Dalm., *Z. verrucosus* Pand., *Z. uniplicatus* Pand. *Amphion fischeri* Eichw. *Encrinurus punctatus* Emm. *Bronteus granulatus* Goldf. *Proetus concinnus* Dalm. *Phillipsia eichwaldi* Fisch., *P. uralica* DeVern. *Calymene blumenbachii* Brong. *Phacops sclerops* Dalm., *P. macrophthalmus* Burm. *Chamops odini* Eichw. *Trinucleus. Ampyx nasutus* Dalm. *Asaphus expansus* Linné, *A. latus* Pand., *A. cornutus* Pand., *A. acuminatus* n. sp., *A. rotundifrons* n. sp., *A. hyarrhinus* Leucht, *A. centron* Leucht, *A. tyrannus* Murch., *A. buchii* Brong. *Illænus crassicauda* Wahl., *I. tauricornis* Kut. *Nileus armadillo* Dalm. *Agnostus pisiformis* Linné, *A. lentiformis* Aug. *Lichas lacinatus* Wahl.

Holl (F.) Handbuch der Petrefactenkunde, etc. Dresden, 1841, p. 155.

The author gives short descriptions of the Trilobites described by Alex. Brongniart, J. W. Dalman, E. Eichwald, von Schlotheim, and others.

Eurypterus remipes Dekay. *Calymene blumenbachii* Brong. varieties (a) *tuberculata*, (b) *tuberculosa*, (c) *pulchella*. *Calymene variolaris* Brong., *C. tristani* Brong., *C. macrophtalma* Brong., *C. bellatula* Dalm., *C. polytoma* Dalm., *C. sclerops* Dalm., *C. punctata*, *C. concinna* Dalm., *C. sulzeri* Schloth., *C. hoffii* Schloth. *Asaphus mucronatus* Brong., *A. caudatus*, *A. hausmanni* Brong., *A. extenuatus* Dalm., *A. granulatus* Wahl., *A. angustifrons* Dalm., *A. lichtensteinii* Eichw., *A. expansus*, *A. weissii* Eichw., *A. buchi* Brong., *A. dilatatus* Brünn., *A. frontalis* Dalm., *A. gigas* Dekay, *A. læviceps* Dalm., *A. palpebrosus* Dalm. *Nileus armadillo* Dalm. *Illænus centrotus* Dalm., *I. crassicauda*, *I. laticauda* Wahl., *I. wahlenbergii* Eichw., *I. rudolphii* Eichw. *Lichas laciniatus. Ampyx nasutus* Dalm. *Ogygia guettardi* Brong., *O. desmaresti* Brong. *Olenus tessini*, *O. bucephalus*

Wahl., *O. spinulosus, O. gibbosus, O. scarabæoides, O. fischeri (Asaphus)* Eichw. *Battus pisiformis* Dalm *Trilobite schroeteri, T. sphærocephalus, T. tentaculatus, T. problematicus, T. bitumnosus. Asaphus velatus, A. pustulatas. Tril. granum. Cryptonymus panderi; Asaphus brongniarti. Calymene clavifrons. Asaphus centaurus.*

Holl (H. B.) See **Jones (T. Rupert) and Holl (H. B.)**

Hollier (E.) On a specimen of *Homalonotus delphinocephalus* found at Dudley.

Trans. Manchester Geol. Soc., vol. 9, 1869, p. 28.

Holm (Gerhard.) Anteckningar om Wahlenberg's *Illænus crassicauda*.

In Ofversigt af K. Svenska Vetensk. Akad. Förhandl., vol. 37, 1880, No. 4, p. 3, Stockholm, pl. 5.

Illænus crassicaudu Wahl., *I. dalmani* Volb.

—— Bemerkungen über *Illænus crassicauda* Wahl.

In Zeitschr. deutsch. geol. Gesell., vol. 32, pt. 2–4, 1880, p. 559, pl. 23.

—— Ueber einige Trilobiten aus dem Phyllograptusschiefer Darlekarliens.

In Bihang till Kongl. Svenska Vet. Akad. Handl., vol. 6, No. 9, 1882, pl. 5. *Pliomera törnquisti. Megalaspis dalecarlicus. Ampyx pater. Agnostus törnquisti. Trilobites brevifrons. Niobe læviceps* Dalm.

—— De Svenska arterna af Trilobitslägtet *Illænus* Dalman.

In Bihang till Kongl. Svenska Vet. Akad. Handl., vol. 7, No. 3, 1882, 6 pls. Group I with 10 thorax segments: *Illænus esmarki* Schloth., *I. sphæricus* n. sp., *I. gigas* n. sp., *I. scrobiculatus* n. sp., *I. oblongatus* Ang., *I. fallax* n. sp., *I. chiron* n. sp., *I. crassicauda* Wahl., *I. tuberculatus* n. sp. Group II with 9 thorax segments: *I. centrotus* Dalm., *I. linarssonii* n. sp. Group III with 8 thorax segments: *I. megalophthalmus* Linrs., *I. parvulus* n. sp., *I. lineatus* Ang. Eyes wanting: *I. leptopleura* n. sp., *I. angelini* n. sp. Subgenus *Bumastus. Illænus barriensis* Murch., *I. insignis* Hall.

—— Om. thoraxledernas antal hos *Paradoxides tessini* Brongn.

In Geol. Föreningens, Stockholm, Förhandl., vol. 9, 1887, p. 408.

—— Om förekomsten af en Cruziana i öfversta Olenidskiffern vid Knifvinge i Vresta Kloster Socken i Ostergötland.

In Geol. Föreningens, Stockholm, Förhandl., vol. 9, 1887, p. 412.

—— Om *Olenellus kjerulfi* Linrs.
In Geöl. Föreningens, Stockholm, Förhandl., vol. 9, 1887, p. 493, pls. 14, 15.

—— Die Ostbaltischen Illaeniden.
In Mem. Acad. Imp. Sci., St. Petersburg, 7th series, vol. 33, No. 8, 1886, 12 plates.
Group with 10 thorax segments: *Illanus esmarkii* Schloth., *I. sphæricus* Holm., *I. jerensis* Holm., *I. laticlavius* Eichw., *I. chiron* Holm., *I. intermedius* Holm., *I. crassicauda* Wahl., *I. tauricornis* Kut., *I. ariensis* Holm., *I. sulcifrons* Holm., *I. plautini* Holm., *I. revaliensis* Holm., *I. dalmani* Volb., *I. oculosus* Holm., *I. chudleighensis* Holm., *I. sinuatus* Holm., *I. schmidti* Niesk., *I. ladogensis* Holm., *I. oblongatus* Aug., *I. roemeri* Volb., *I. angustifrons* Holm., *I. atavus* Eichw., *I. masckei* Holm.
II Group with 9 thorax segments: *Illænus centrotus* Dalm., *I. linnarssoni* Holm., *I. proles* Holm., *I. livonicus* Holm.
III Group with 8 thorax segments: *Illænus triquetrus* Volb., *I. caecus* Holm. (Eyes wanting.) Subgenus *Bumastus*. *Illænus barriensis* Murch.

Honeyman (D.) On new localities of fossiliferous Silurian rocks in eastern Nova Scotia.
In Canadian Naturalist, vol. 5, 1860, p. 293.
Dr. J. W. Dawson adds a note to this paper, describing *Homalonotus dawsoni* Hall, fig. 1, with notes on other fossils.

—— On the geology of Arisaig, Nova Scotia, with a note by Prof. T. Rupert Jones.
In Quart. Jour. Geol. Soc. London, vol. 26, 1870, p. 490.
The note by Prof. T. R. Jones, on page 492, on some Entomostraca from Arizaig, mentions the following genera:
Beyrichia tuberculata Klöden., *B. wilkensiana* Jones, *B. maccoyiana* Jones. *Primitia concinna* Jones.

—— Notes of examination by Prof. James Hall of the Silurian collection of the Provincial Museum, by the Rev. D. Honeyman.
In Proc. and Trans. Nova Scotian Inst. Nat. Sci., vol. 7, pt. 1, 1886, p. 14.
Asaphus ditmarsiæ Honeyman.

—— The Giant Trilobite of Moose River Iron Mine, Nova Scotia.
In Proc. and Trans. Nova Scotian Inst. Nat. Sci., vol. 7, pt. 1, 1888, p. 63.
In this article Dr. D. Honeyman refers *Asaphus ditmarsiæ* to Dr. Jacob Green's *Asaphus? crypturus*, giving a copy of the original description of that species from Trans. Geol. Soc. Pennsylvania, vol. 1, 1834, p. 37.

How (Henry). Notice of the occurrence of a Trilobite in the Lower Carboniferous limestone of Hants County.
In Proc. and Trans. Nova Scotian Inst. Nat. Sci., vol. 1, 1867, pt. 1, p. 87.
Phillipsia howi Billings.
This species is closely allied to *P. meramecensis* and *P. insignis*.

Hutton (F. W.) On a Trilobite from New Reefton, new to Australia.
In Proc. Linn. Soc. New South Wales, section 2, vol. 2, 1887-88, p. 257.
Homalonotus species not named.

Huxley (T. H.) Observations on the structure and affinities of *Himantopterus*.
In Quart. Jour. Geol. Soc, London, vol. 12, 1856, p. 34.

Huxley (T. H.) and **Salter** (J.W.) Monograph of the fossil Crustacea forming the genus *Pterygotus*, with its anatomy and affinities.
In Mem. Geol. Survey United Kingdom. Description of Brit. organic remains. Monograph No. 1. London, 1859, atlas of 16 pls.
Pterygotus bilobus Salt., *P. peronatus* Salt., also var. *plicatissimus*, *P. banksii* Salt., *P. stylops*, *P. acuminatus* Salt. *Eurypterus lanceolatus* Salt. *Pterygotus anglicus* Agas., *P. ludensis* n. sp., *P. gigas* n. sp., *P. problematicus* Agas., *P. arcuatus* n. sp., *P. punctatus* n. sp.

—— and **Etheridge** (Robert). A catalogue of the collection of fossils in the Museum of Practical Geology, with an explanatory introduction. London, 1865.

—— A catalogue of the Cambrian and Silurian fossils in the Museum of Practical Geology. London, 1878.
This catalogue was drawn up by Mr. E. T. Newton, under Prof. T. H. Huxley's superintendence.
The specimens were named by Mr. Robert Etheridge, palæontologist to the Geological Survey of Great Britain.

Jackson (Charles T.) Nouveau gisement de Trilobites découvert dans environs de Boston.
In Comptes Rendus Acad. Sci. Paris, vol. 43, 1856, p. 883.
Paradoxides tessini Brong., *P. harlani* Green, *P. spinulosus* Boeck.

—— Sur un moule du *Paradoxides harlani*.
In Comptes Rendus Acad. Sci. Paris, vol. 46, 1858, p. 254.

James (J. F.) Catalogue of fossils of Cincinnati group, published by order of the Committee on Palæontology. Cincinnati, 1871.

James (U. P.) Palæontology. Catalogue of Lower Silurian fossils of Cincinnati group, found at Cincinnati and vicinity, etc. Cincinnati, April, 1875.

—— Supplement to Catalogue of Lower Silurian fossils of the Cincinnati group, etc. Cincinnati, July 10, 1879.

—— Descriptions of new species of fossils from the Cincinnati group, Ohio and Kentucky.
In the Palæontologist, Cincinnati, No. 7, April 16, 1883.
Tracks of Crustacean (?), pl. 2, fig. 4.

Jenkins (Charles). On the geology of Yaas Plains. (Second paper.)
In Proc. Linnean Soc. N. S. Wales, 1878, plate.
Cheirurus insignis. Bronteus sp.? *Homalonotus* sp.? *Acidaspis brightii. Phacops* sp.?

Johnstrup (Fr.) Oversigt over de palæozoiske Dannelser paa Bornholm meddelt paa det 11te Skandinaviske Naturforskermede i Kjöbenhavn, 1873, p. 299.
This work contains important list of the fossils occurring at Bornholm, in the Cambrian and Lower Silurian formations.

Jones (T. Rupert). On Permian Entomostraca.
In Monograph of Permian Fossils of England (King). Palæont. Soc. London, vol. 3, 1849-54, p. 58.
The author describes and illustrates under the head of Crustacea, p. 54, plate xviii, the following species:
Cythere morrisiana Jones, *C. geinitziana* Jones, *C. elongata* Münst., *C. kuborgiana* Jones, *C. (Bairdia) curta* McCoy, *C. (Bairdia) gracilis* McCoy, *C. (Bairdia) acura* Jones, *C. (Cythereis ?) biplicata* Jones, *C. (Cytherella ?) inornata* McCoy, *C. (Cytherella ?) nuciformis* Jones. *Dithyrocaris permiana* Jones, *D. glypta* Jones.

—— In Journal of a Voyage in Baffin's Bay and Barrow Straits in the years 1850-51, etc., 2 vols. London, 1852. Appendix.
Leperditia balthica His. var. *arctica* n. sp., p. ccxxi, pl. 5, f. 13.

—— Notes on the Entomostraca.
In Quart. Jour. Geol. Soc. London, vol. 9, 1853, p. 160, pl. 7, figs. 5–7.

This article forms Appendix D to a paper on the Carboniferous and Silurian formations of the neighborhood of Bussaco, in Portugal, by Señor Carlos Ribeiro, with notes and a description of the animal remains by D. Sharpe, J. W. Salter and T. R. Jones, etc. See same journal, p. 135, pls. 7–9.

Beyrichia bussacensis n. sp., *B. simplex* n. sp.

—— Notes on the Palæozoic bivalved Entomostraca. No. 1. Some species from the Upper Silurian limestone of Scandinavia.
In Annals Mag. Nat. Hist., 2d Series, London, vol. 16, 1855, p. 81, pl. 5.

(Jugosæ.) 1, *Beyrichia buchinana;* 2, *B. tuberculata* Klöd., also var. *nuda antiquata;* 3, *B. dalmaniana;* 4, *B. maccoyiana;* 5, *B. salteriana.* (Corrugatæ.) 6, *Beyrichia wilckensiana* var. *plicata;* 7, *B. siliqua.* (Simplices.) 8, *Beyrichia mundula.*

—— Notes on Palæozoic bivalved Entomostraca. No. 2. Some British and foreign species of *Beyrichia*.
In Annals Mag. Nat. Hist., 2d series, London, vol. 16, 1855, p. 163, pl. 6.

Beyrichia complicata Salt., also var. *decorata, B. klödeni* McCoy, also var. *antiquata,* var. *torosa, B. lata* Vanuxem, *B. bussacensis* Jones, *B. ribeiriana* n. sp., *B. affinis* n. sp., *B. barrandiana* n. sp., *B. strangulata* Salt. var. (a), var. (b), var. (c), *B. bicornis* n. sp., *B. seminulum* n. sp., *B. simplex* Jones, *B. simplex ?* var.?, *B. mudula* Jones.

The author gives a table of the Beyrichiæ, arranged in three subdivisions as far as the species described will permit. The Lower Silurian Beyrichiæ are arranged under the groups *Simplices, Corrugatæ,* and *Jugosæ;* the Upper under *Corrugatæ,* and *Simplices.*

—— Notes on the Palæozoic bivalved Entomostraca. No. 3. Some species of *Leperditia.*
In Annals Mag. Nat. Hist., 2d series, London, vol. 17, 1856, p. 81, pls. 6–7.

Leperditia balthica His., *L. arctica* Jones, *L. alta* Conrad, *L. britianica* Rouault?, *L. gibbera* n. sp., *L. marginata* Keyserl. *L. solvensis* n. sp.

—— Notes on the Palæozoic bivalved Entomostraca. No. 4. North American species.
In Annals Mag. Nat. Hist., 3d series, London, vol. 1, 1858, p. 241, pls. 9, 10. See also Rogers's Final Rept. Geol. Survey Pennsylvania, 1858, pt. 2, p. 834, figs. 695–699.

(From Beechey Island.)—1, *Beyrichia rugulifera;* 2, *B. sigillata;* 3, *B. clathrata;* 4, *B. plagosa.*

(From Canada.)—1, *Beyrichia logani* (a) var. *reniformis*, (b) *leperditioides*. 2, *Leperditia canadensis*, (a) var. *labrosa*, (b) *L. canadensis* ?; 3, *L. anna*. *Isochilina*, subgenus of *Leperditia*. 4, *L. (Isoch.) ottawa*; 5, *L. (Isoch.) gracilis*. 6, *Cytheropsis concinna*; 7, *C. siliqua*; 8, *C. rugosa*.

(From the United States.)—1, *Leperditia alta* Conrad; 2, *L. gibbera* var. *scalaris*; 3, *L. pennsylvanica*; 4, *L. ovata*. 5, *Beyrichia maccoyiana* Jones; 6, *B. pennsylvanica*. 7, *Leperditia (Isochilina) cylindrica* Hall.

(Beechey Island, additional.)—*Cytheropsis concinna*.

—————— Additional notes on Palæozoic Entomostraca from Canada.

In Annals Mag. Nat. Hist., 3d series, London, vol. 1, 1858, p. 340.
Not descriptive.

—————— On the Palæozoic Entomostraca of Canada.

In Geol. Survey Canada, figure and description of Canadian organic remains, decade 3, Montreal, 1858, p. 91, pl. 2; also Annals Mag. Nat. Hist., 3d series, London, vol. 1, 1858, p. 244, pl. 9.

Isochilinia n. g. *Beyrichia logani* Jones. *Leperditia canadensis* Jones var. *labrosa*, var. *L. louckiana*, var. *pauquettiana*. var. *josephiana*, var. *anticostiana*, *L. anna* Jones, *L. amygdalina* Jones. *Isochilina* n. subg. *Leperditia (Isochilina) ottawa* Jones, *L. (I.) gracilis* Jones. *Cytheropsis concinna* Jones, *C. siliqua* Jones, *C. rugosa* Jones.

—————— Description of Entomostraca from the mountain limestone of Berwickshire and Northumberland, by Prof. T. Rupert Jones, F. G. S., etc., with notes on the strata in which they occur, by George Tate, F. G. S., etc.

In Proc. Berwickshire Nat. Club, 1864, p. 83.
Estheria striata Münst., also var. *tateana* Jones. *Candona (?) tateana* Jones. *Beyrichia tatei* n. sp.

—————— On fossil *Estheria* and their distribution.

In Quart. Jour. Soc. London, vol. 19, 1863, p. 140; also Nat. Hist. Rev., 1863, p. 262.
Estheria membranacea Pacht., *E. striata* Münst. also var. *tateana* and *leinertiana*, *E. tenella* Jord., *E. exigua* Eichw., *E. portlockii* Jones, *E. minuta* Alberti also var. *brodieana*, *E. mangaliensis* Jones, *E. ovata* Lea, *E. kotahensis* Jones, *E. concentrica* Bean, *E. murchisoniæ* Jones, *E. elliptica* Dunker also var. *subquadrata*, *E. forbesii* Jones, *E. middlendorfii* Jones. *Leaia leidyi* Lea also var. *salteriana* and *williamsoniana*.

—————— and **Kirkby** (J. W.) Notes on the Palæozoic bivalved Entomostraca. No. 5. Münster's species from the carboniferous limestone.

In Annals Mag. Nat. Hist., 3d series, London, vol. 15, 1865, pl. 20, p. 404.
Leperditia okeni Münster also var. *acuta, L. oblonga* n. sp., *L. parallela* n. sp., *L. suborbiculata* Münster. *Cytherella ? inflata* Münster. *Bairdia elongata* Münster, *B. hisingeri* Münster, *B. subcylindrica* Münster. *Cythere bilobata* Münster, *C. intermedia* Münster, *C. münsterina* n. sp.

——— and **Holl** (H. B.) Notes on the Palæozoic bivalved Entomostraca. No. 6. Some Silurian species (*Primitia*).
In Annals Mag. Nat. Hist., 3d series, London, vol. 15, 1865, p. 414, pl. 13.
Primitia n. g., *P. strangulata* Salter also var. (a), *P. salteriana* n. sp., *P. semicordata* n. sp., *P. simplex* Jones, *P. logani* Jones also vars. *reniformis* and *leperditioides, P. matutina* n. sp., *P. seminulum* Jones, *P. sigillata* Jones, *P. variolata* n. sp. also var. *paucipunctata, P. rugulifera* Jones, *P. renulina* n. sp., *P. mundula* Jones, *P. anna* n. sp., *P. bicornis* Jones, *P. umbilicata* n. sp., *P. cristata* n. sp., *P. tersa* n. sp., *P. trigonalis* n. sp., *P. beyrichiana* n. sp., *P. roemeriana* n. sp., *P. obsoleta* n. sp., *P. oblonga* n. sp., *P. ovata* n. sp., *P. semicircularis* n. sp., *P. pusilla* n. sp., *P. concinna* Jones, *P. muta* n. sp.

——— and **Kirkby** (J. W.) Notes on Palæozoic bivalved Entomostraca. No. 7. Some Carboniferous species.
In Annals Mag. Nat. Hist., 3d series, London, vol. 18, 1868, p. 32.
The author's give the following list of the species described in McCoy's Synopsis Carb., Foss., Ireland:
1, *Entomoconchus scouleri. Daphinia primæva (Cypridina primæva).* 3, *Bairdia curtus.* 4, *B. gracilis (B. subcylindrica* Münst.) 5, *Cythere amygdalina (Leperditia amygdalina.)* 6, *C. arcuata (Leperditia okeni* Münst. var. *subrecta* Portl.) 7, *C. bituberculata (Beyrichia bituberculata).* 8, *C. costata (Kirkbya costata).* 9, *C. cornuta (Leperditia okeni* Münst. var. *subrecta).* 10, *C. elongata.* 12, *C. excavata* are also referred to *Leperditia okeni* var. *subrecta.* 13, *C. hibbertii (Beyrichia ?)* 14, *C. inflata (Entomoconchus ?* vel. *Cypridella ?)* 15, *C. inornata (Leperditia okeni).* 16, *C. orbicularis (Cypridella ?)* 17, *C. pusilla (Entomoconchus ?* vel. *Cypridella ?)* 18, *C. scutulum (Leperditia okeni).* 19, *C. oblonga.* 20, *C. spinigera ?)* 21, *C. trituberculata.* 22, *C. gibberula.* 23, *C. subrecta* are also referred to *Leperditia okeni.* They comprise the varieties *subrecta* Portl. and *scotoburdigalensis* Hibbert, *Kirkbya annectens.*

——— and **Holl** (H. B.) Notes on Palæozoic bivalved Entomostraca. No. 8. Some Lower Silurian species from Chair of Kildare, Ireland.
In Annals Mag. Nat. Hist., 4th series, London, vol. 2, 1868, p. 54, pl. 7.

Primitia maccoyii Salter. *P. sancti-patricii* n. sp. *Cythere wrightiana* n. sp., *C. jukesiana* n. sp., *C. bailyana* n. sp., *C. harknessiana* n. sp. *Bairdia murchisoniana* n. sp., *B. griffithiana* n. sp., *B. salteriana* n. sp. *Cythere aldensis* M'Coy. *Cytheropsis* has been applied to this and other Palæozoic Entomostraca.

———— and **Holl** (H. B.) Notes on Palæozoic bivalved Entomostraca. No. 9. Some Silurian species.
In Annals Mag. Nat. Hist., 4th series, London, vol. 3, 1869, p. 211, pls. 14, 15.
Thlipsura n. g. *Cytherellina* n. g. *Kirkbya* n. g. *Moorea* n. g. *Æchmina* n. g.
Cythere corbuloides n. sp., *C. grindrodiana* n. sp. *Bairdia phillipsiana* n. sp. *Thlipsura corpulenta* n. sp., *T. tuberosa* n. sp.. *T. v-scripta* n. sp. *Cytherellina siliqua* also var. *grandis*, vars. *tersa* and *ovata*. *Æchmina cuspidata* n. sp., *Æ. clavulus* n. sp. *Beyrichia intermedia* n. sp. *Primitia lenticulari* n. sp., *P. umbilicata* J. & H., *P. bipunctata* Salt., *P. excavata* n. sp. *Kirkbya fibula* n. sp. *Moorea silurica* n. sp.,

———— On ancient Water-fleas of the Ostracodous and Phyllopodous tribes (bivalved Entomostraca).
In Month. Microsc. Jour., vol. 4, London, 1870, p. 184, pl. 61.
Bairdia curta M'Coy. *Thlipsura corpulenta* J. & H. *Cythere jukesiana* J. & H. *Cytherellina siliqua* Jones. *Æchmina cuspidata* J. & H. *Carbonia aynes* Jones. *Cypridina phillipsiana* n. sp. *Cypridella koninckiana* n. sp. *Cyprella subannulata* n. sp. *Entomoconchus scouleri* M'Coy. *Entomis divisa* Jones. *Primitia renulina* J. & H. *Kirkbya urei* Jones. *Moorea silurica* J. & H. *Leperditia balthica* His. *Isochilina gracilis* Jones. *Beyrichia wilckensiana* Jones, *B. complicata* Salt. *Leaia leidyi* Lea. *Estheria membranacea* Pacht, *E. tenella* Jones.

———— Explanation map 32. The geology of the neighborhood of Edinburgh, by H. H. Howell and A. Geikie; with appendix and list of fossils, by J. W. Salter.
In Mem. Geol. Survey Gr. Brit., London, 1861, p. 137, pl. 2, fig. 5.
Prof. T. Rupert Jones herein uses the name of *Entomis* for certain fossils. For a full generic description, see Annals Mag. Nat. Hist., 4th series, London, vol. 11, 1873, p. 413.

———— A monograph of the fossil *Estheria*.
In Palæont. Soc. London, vol. 14, 1862, 134 pp., 5 pls.
The following Palæozoic *Estheriæ* and *Leaiæ* are described in this work.
Estheria membranacea Pacht, *E. striata* Münster also vars. *tateana*, *binneyana*, *beinertiana*. *Leaia leidyi* Lea also vars. *salteriana*, *williamsoniana*. *Estheria tenella* Jordon, *E. exigua* Eichw., *E. portlockii* n. sp.

——— Note on *Estheria niddendorfii*.
In Quart. Jour. Geol. Soc. London, vol. 19, 1863, p. 73.

——— On some bivalved Entomostraca from the Coal Measures of South Wales.
In Geol. Mag., vol. 7, 1870, p. 214, pl. 9.
Cytheroid Entomostraca.
Carbonia evelinæ n. sp., *C. agnes* n. sp. var. *subrugulosa* (a), *ruglosa* (b). *Leaia leidyi*. *Daphnioid ?* Entomostraca. *Estheria adamsii* n. sp., *E. dawsoni* n. sp., *E. tenella* Jones, *E. pachtii* n. sp. *Anthracomya phillipsii* Williamson.

——— On the Palæozoic bivalved Entomostraca.
In Proc. Geologist Assoc. 1870.*

——— Note on the Entomostraca from the Cambrian rocks of St. David's.
In Quart. Jour. Geol. Soc. London, vol. 28, 1872, p. 183, pl. 5.
This paper forms a part of Henry Hick's article on some undescribed fossils from the Menevian Group. Same Jour., p. 173.
Leperditia hicksii Jones. *Entomis buprestis* Salter.

——— Notes on the Palæozoic bivalved Entomostraca. No. 10.
In Annals Mag. Nat. Hist., 4th series, London, vol. 11, 1873, p. 413.
Entomidella n. g. *Sulcuna* n. g. *Entomis concentrica* DeKon., *E. seraratostriata* Sandb., *E. nitida* F. A. Römer, *E. fragilis* F. A. Römer, *E. globulos* Richter, *E. gyrata* Richter, *E. teniata* Richter, *E. tuberosa* Jones, *E. impendens* Haswell, *E. biconcentrica* Jones, *E. aciculata* Jones, *E. dimidiata* Barr., *E. migrans* Barr., *E. pelagica* Barr., *E. rara* Barr. *Entomidella divisa* Jones, *E. buprestis* Salt.

——— On some bivalved Entomostraca chiefly Cyridenidæ of the Carboniferous formations.
In Quart. Jour. Geol. Soc. London, vol. 29, 1873, p. 409.
Sulcuna lepus, *S. cuniculus*. *Cyprella chrysalidea* DeKon., *C. annulata* DeKon. *Daphina primæva* M'Coy is herein referred to *Cypridina* proper.

——— Ancient Water-fleas, etc. Pt. 2.
In Monthly Microsc. Jour., vol. 10, London, 1873, p. 71.
Contains notes on the genera, but not descriptive of the species.

——— Notes on some forms of British Entomostraca from the Silurian rocks of Peeblesshire.
In Trans. Edinb. Geol. Soc., vol. 2, part 3, 1874, p. 321.
1, *Bairdia browniana* Jones. 2, *Beyrichia impendens* Jones. 3, *B. protenta* n. sp. 4, *Entomis aciculata* n. sp.

———, **Kirkby** (J. W.) and **Brady** (M. P.) Monograph of British fossil bivalve Entomostraca from the Carboniferous formations.
In Palæont. Soc. London, 1874-1884.
The Cypridinadæ and their allies, part 1, pp. 1-56, plates 1-5.
Cypridina primæva McCoy. 2, *C. radiata* n. sp. 3, *C. wrightiana* n. sp. 4, *C. bradyana* n. sp. 5, *C. brevimentum* n. sp. 6, *C. grossartiana* J. & K. 7, *C. youngiana* n. sp. 8, *C. phillipsiana* Jones. 9, *C. hinteriana* n. sp. 10, *C. thompsoniana* J. & K. 11, *C. pruniformis* n. sp. 12, *C. scoriacea* J. & K. 13, *C. oblonga* n. sp.
Cypridinella cummingii n. sp. 2, *C. superciliosa* n. sp. 3, *C. clausa* n. sp., 4, *C. bosqueti* n. sp. 5, *C. maccoyiana* n. sp. 6, *C. monitor* n. sp. 7, *C. romer* n. sp.
Cypridellina clausa n. sp. 2, *C. burrorii* n. sp. 2, *C. burrovii* var. *longnoriensis* n. sp. 3, *C. intermedia* n. sp. 4, *C. elongata* n. sp., *C. elongata* var. *hibernica*. 5, *O. galea* n. sp. 6, *C. romer* n. sp. 7, *C. alta* n. sp. 8, *C. bosqueti* n. sp.
Cypridella edwardsiana DeKon. 2, *C. koninckiana* Jones. 3, *C. obsoleta* n. sp. 4, *C. wrightii* n. sp. 5, *C. quadrata* n. sp. 6, *C. cyprelloides* n. sp.
Sulcuna lepus n. sp. 2, *S. cuniculus* n. sp. *Cyprella chrysalidea* DeKon. 2, *C. annulata* DeKon. *Bradycinetus rankinianus* J. & K. *Philomedes bairdiana* n. sp. *Rhombina hibernica* n. sp., *R. belgica* n. sp. *Entomoconchus scouleri* McCoy. 2, *E. orbicularis* n. sp. 3, *E. globosus* n. sp., *E.* sp. *Offa barrandiana* n. sp. *Polycope burrovii* n. sp. *P. simplex* J. & K., *P. youngiana* J. & K.

Part 1, no. 2, pp. i-iii, 57-92, plates 6-7.
1, *Cytherella valida* n. sp. 2. *C. benniei* n. sp. 3, *C. recta* n. sp. 4, *C. concinna* n. sp. 5, *C. hibernica* n. sp. 6, *C. brevis* Jones. 7, *C. obliquata* Brady MS. 8, *C. muciformis* Jones. 9, *C. æqualis* n. sp. 10, *C. tatei* Jones. 11, *C.? inflata* Münst. 12, *C. simplex* J. & K. 13, *C. obesa* n. sp. 14, *C.* sp.? 15, *C.* sp.? 16, *C.? rotundata* n. sp. 17, *C. scrobiculata* n. sp., *C. benniei* var. *intermedia* n. sp. 18, *C. regularis* n. sp., *C. regularis* var.? 19, *C. subreniformis* n. sp., *C. concinna* ? 20, *C. impressa* n. sp. 21, *C. richteriana* n. sp. *Philomedes elongata* n. sp.
Entomis concentrica DeKon. 2, *E. biconcentrica* Jones. 3, *E. burrovii* n. sp. 4, *E. koninckiana* n. sp. 5, *E. obscura* n. sp. *Beyrichia gigantea*. *Kirkbya costata* ?

——— and **Kirkby** (J. W.) Notes on the Palæozoic bivalved Entomostraca. No. 11. Some Carboniferous Ostracoda from Russia.
In Annals Mag. Nat. His., 4th series, London, vol. 15, 1875, p. 52, pl. 6.
Leperditia okeni Münst. and vars. *inornata* and *obliqua*. *Beyrichia in-*

termedia J. & H. *Primitia eichwaldi* n. sp. *Bairdia aequalis* Eichw., *B. ampla* Reuss, *B. plebeia* Reuss var. *rhombica* Jones also var. *munda* n. var. *Cythere (Potamocypris ?) bilobata* Münst. *Cytherella murchisoniana* n. sp. *Beyrichia gibberosa* Eichw., *B. colliculus* Eichw. *Kirkbya umbonata* Eichw., *K. striolata* Eichw. *Leperditia okeni* Münst. also var. *microphthalma* Eichw. *Bairdia excisa ?* Eichw., *B. distracta* Eichw., *B. qualeni* Eichw.

────── Notes on some fossil bivalved Entomostraca.
In Geol. Mag., decade 2, vol. 5, 1878, p. 100, pl. 3.
Estheria grevii n. sp., *E. dawsoni* Jones.

This paper contains also descriptions of other fossil Entomostraca from the Triassic, and descriptions of some fossil Ostracoda from the Ironstone of Shotover, near Oxford, involving remarks on all the known Wealden species; and lastly notes on some species found in the Purbeck strata of the Subwealden Boring, near Battle.

────── and **Kirkby** (J. W.) Description of the species of the Ostracodous genus *Bairdia* McCoy, from the Carboniferous strata of Great Britain.
In Quart. Jour. Geol. London, vol. 35, 1879, p. 565, 5 pls.
Bairdia curta M'Coy. 2, *B. plebeia* Reuss. 3, *B. hisingeri* Münst. 4, *B. ampla* Reuss. 5, *B. grandis* n. sp. 6, *B. mucronata* Reuss. 7, *B. submucronata* n. sp. 8, *B. sublongata* n. sp. 9, *B. subgracilis* Geinitz. 10, *B. brevis* J. & K. 11, *B. siliquoides* n. sp. 12, *B. amputata* Kirkby. 13, *B. praecisa* n. sp. 14, *B. nitida* n. Sp. 15, *B. circumcisa* n. sp. 16, *Bairdia* sp.?

The authors give a table of all the Palæozoic *Bairdia* known to them, omitting such as appear to be but varieties or synonyms.

────── Notes on the Palæozoic bivalved Entomostraca. No. 12. Some carboniferous species belonging to the genus *Carbonia* Jones.
In Annals Mag. Nat. Hist., 5th series, London, vol. 4, 1879, p. 28, pls. 2 and 3.
Carbonia n. g., *C. fabulina* J. & K. also var. *humilis*, var. *inflata*, var. *subangulata*. 2, *C. rankiniana* J. & K. 3, *C. subula* J. & K. 4, *C. scalpellus* n. sp. 5, *C. secans* J. & K. 6, *C. pungens* J. & K. 7, *Cythere ? (Carbonia ?) bairdoides* n. sp.

────── Notes on the Palæozoic bivalved Entomostraca. No. 13. *Entomis serrato-striata* and others of the so-called Cypridinen of the Devonian schist of Germany.
In Annals Mag. Nat. Hist., 5th series, London, vol. 4, 1879, p. 182, pl. 11.

The author remarks in regard to the genus *Richteria* (Neues Jahrb. für Mineral., 1874, p. 183): "I cannot find any reason to support the establishment of my proposed genus *Richteria* for the typical specimens show merely the conformation and characters of *Entomis*."
Entomis serratostriata Sandb., *E. tenella* Richter, *E. labyrinthica* Richter, *E. gyrata* Richter.

—————— Notes on some Palæozoic Entomostraca.
In Geol. Mag., decade 2, vol. 8, 1881, p. 337, 2 pls.

Cyprosis n. g. *Cyprosina* n. g. *Cypridina ? Cyprosis hawellii* n. sp. *Cyprosina whidbornei* n. sp. *Polycope devonica* n. sp. *Leperditia ? dorsalis* Richter. *Entomis calcarata* Richter. *Primitia armata* Richter? *P. ? cylindrica* Richter. *Beyrichia hollii* n. sp., *B. tuberculata* Klöden, *B. kloedeni* McCoy, var. *antiquata* Jones, *B. kloedeni* McCoy also *B. kloedeni* (typica) var. *tuberculata*, var. *antiquata*, var. *pauperata*, var. *intermedia*, var. *subtorosa*, var. *torosa*. *Beyrichia colwallensis* Holl, MS.

—————— Notes on some Palæozoic Entomostraca. No. 14. Some Cambrian and Silurian *Leperditia* and *Primitia*.
In Annals Mag. Nat. Hist., 5th series, London, vol. 8, 1881, p. 332, pls. 19, 20.
Characteristic features of the Leperditiadæ.
Leperditia balthica His. also var. *contracta*, *L. hisingeri* Schmidt, *L. phaseolus* His. var. *marginata*, *L. canadensis* Jones and *L. fabulites* Conrad with varieties, *L. amygdalina* Jones, *L. billingsii* n. sp., *L. alta* Conrad, *L. hicksii* Jones. *Isochilina grandis* Schrenk. *Primitia simplex* Jones, also varieties *sanctojohannesiana*, *lloydiana* and *milneana*.

—————— and **Schmidt** (F) On some Silurian *Leperditia*.
In Annals. Mag. Nat. Hist., 5th series, London, vol. 9, 1882, p. 168.
This article contains some criticisms on Prof. Jones' Notes on the Palæozoic Bivalved Entomostraca published in Ann. Mag. Nat. Hist., ser. 5, vol. 8, Nov., 1881.

—————— Notes on some Palæozoic bivalved Entomostraca. No. 15.
In Annals Mag. Nat. Hist., 5th series, London, vol. 10, 1882, p. 358, figs. 1 a, b.
Primitia allied to *Primitia barrandiana*.

—————— Notes on the Palæozoic bivalved Entomostraca. No. 16. 1. Some Palæozoic and other bivalved Entomostraca from Siberian Russia, pl. 6. 2. Some Palæozoic bivalved Entomostraca from Spitzbergen, pl. 9.

In Annals Mag. Nat. Hist., 5th series, London, vol. 12, 1883, p. 243, pls. 6 and 9.
Part 1. Some Palæozoic and other bivalved Entomostraca from Siberian Russia, plate vi.
Estheria minuta Alberti var. *karpinskiana*. 2, *Entomis serratostriata* Sand., *E. gyrata* Richter. 3, *Estheria striata* Münt.
Part 2. Some Palæozic Entomostraca from Spitzbergen, plate ix.
Leperditia isochilinoides n. sp. *Estheria nathorsti* n. sp.

―――― Notes on the late Mr. George Tate's specimens of the Lower Carboniferous Entomostraca from Berwickshire, Northumberland.
In Proc. Berw. Nat. Club, vol. 10, 1884, 1 pl.
Candona tateana Jones. *Bernix tatei* Jones. *Cythere (Macrocytris?) kirkyana* n. sp. *Carbonia fabulina* J. & K. *Leperditia scotoburdigalensis* Hibbert, *L. subrecta* Portl. *Beyrichia crinita* J. & K. MS. *Kirkbya spirilis* J. & K. MS. *Cytherella tatei* n. sp. *Estheria tenella* Jordan. *Darwinella? berniciana* n. sp.

―――― Notes on the Palæozoic bivalved Entomostraca. No. 17. Some North American *Leperditia* and allied forms.
In Annals Mag. Nat. Hist., 5th series, London, vol. 14, 1884, p. 339.
Leperditia canadensis Jones, *L. louckiana* Jones, *L. josephiana* Jones, *L. anticostiana* Jones, *L. fabulites* Conrad, *L. amygdalina* Jones, *L. alta* Conrad, *L. jonesi* Hall, *L. cylindrica* Hall. *Isochilina ottawa* Jones. *Primitia leperditoides* Jones.

―――― Report of the committee, consisting of Mr. R. Etheridge, Dr. H. Woodward and Prof. T. Rupert Jones (Secretary), on the fossil *Phyllopoda* of the Palæozoic rocks.
In Rept. 53d Meeting Brit. Asso. Adv. Sci., 1883, p. 215.
Hymenocaris vermicauda Salter, *H.? (Caryocaris) salteri* M'Coy. *Caryocaris marri* Hicks. *Lingulocaris lingulæcomes* Salter, *L. siliquiformis* n. sp. *Hymenocaris major* Salter. *Caryocaris wrightii* Salter.

―――― Second report of the committee consisting of Mr. R. Etheridge, Dr. H. Woodward, and Prof. T. Rupert Jones (Secretary), on the fossil *Phyllopoda* of the Palæozoic rocks.
In Rept. 54th Meeting Brit. Assoc. Adv. Sci., 1884, p. 75; Geol. Mag., n. s., decade 2, vol. 10, 1883, p. 461.

A. Shield not sutured along the back. Posterior margin entire.
Discinocaris bowniana H. W., *D. dubia* F. A. Roemer, *D. lata* H. W., *D. triastica* Reuss, *D.* n. sp., *D. congener* Clarke, *D. gigas* H. W. *Spathiocaris emersonii* Clarke, *S. ungulina* Clarke. *Pholadocaris leei* H. W., *P.* n. sp. *Lisgocaris lutheri* Carke. *Ellipsocaris dewalquei* H. W., *E.* n. sp.

B. Posterior margin truncate intended or slightly notched.
Cardiocaris roemeri H. W., *O. bipartita* H. W., *C. veneris* H. W., *C. koeneni* Clarke.

C. Posterior margin deeply notched.
Dipterocaris pes-cervæ Clarke, *D. vestusta* D'Arch. and DeVern., *D. procne* Clarke, *D. pennæ-dædali* Clarke, *E. etheridgei* Jones. *Pterocaris bohemica* Barr. *Crescentilla pugnax* Barr.

D. Sutured along the back. Nuchal suture angular.
Aptychopsis prima Barr., *A.* var. *secunda* nov., *A. wilsoni* H. W., *A. lapworthi* H. W., *A. glabra* H. W., *A.* n. sp., *A.* n. sp., *A. salteri* H. W., *A.* n. sp. *Peltocaris aptychoides* Salt., *P.* n. sp., *P. anatina* n. sp., *P. harnessi* Salt. *Pinnocaris lapworthi* R. E. jr.

—— and **Kirkby** (J. W.) On some Carboniferous Entomostraca from Nova Scotia.
In Geol. Mag., n. s., decade 3, vol. 1, 1884, p. 356, pl. 12.
Leperditia okeni Münst., also var. *scotoburdigalensis* Hibbert. *Beyrichia* or *Primitia ?* sp. *Carbonia fabulina* J. & K., *C.? bairdioides ?* J. & K. *Candona ? elongata* n. sp. *Cythere* sp. indet. *Estheria dawsoni* Jones. *Leaia leidyi* var. *salteriana* Jones.

—— and **Woodward** (Henry). On some Palæozoic Phyllopoda.
In Geol. Mag., n. s., decade 3, vol. 1, 1884, p. 348.
For lists of species see Second Report Committee on the Fossil Phyllopoda, 1884.

—— and **Woodward** (Henry). Notes on Phyllopodiform Crustaceans, referable to the genus *Echinocaris* from the Palæozoic rocks.
In Geol. Mag., n. s., decade 3, vol. 1, 1884, p. 393, plate.
Echinocaris wrightiana Dawson, *E. sublævis* Whitf., *E. pustulosa* Whitf.

—— Notes on the Palæozoic bivalved Entomostraca. No. 18. Some species of the Entomididæ.
In Annals Mag. Nat. Hist., 5th series, London, vol. 14, 1884, p. 391, pl. 15.
Entomis tuberosa Jones, *E. depressa* Salter, MS., *E. marstoniana* n. sp. *E. haswelliana* n. sp., *E. angelini* n. sp., *E. reniformis* Kolm., *E. globulosa* Jones, *E. impendens* Haswell. *Bolbozoe scotica* n. sp., *B. divisa* Jones. *Entomidella marrii* Hicks.

—— and **Woodward** (Henry). Notes on the British species of Ceratiocaris.

In Geol. Mag., n. s., decade 3, vol. 2, 1885, p. 385, pl. 10.

Ceratiocaris murchisoni Agas. and its variety *leptodactylus* M'Coy, *C. lundensis* H. W., *C. papilio* Salt., *C. stygia* Salt.

—— and **Woodward** (Henry). Notes on the British species of Ceratiocaris.

In Geol. Mag., n. s., decade 3, vol. 2, 1885, p. 460.

Ceratiocaris inorta McCoy, *C. oretonensis* H. W., *C. truncata* H. W., *C. solenoides* McCoy, *C. gobiiformis* n. sp., *C. salteriana* n. sp., *C. cassia* Salt., *C. n. sp., C. robusta* Salt., also var. *longa, C.* n. sp., *C. decora* Phillips, *C.? ensis* Salt., *C.? lata* Salt., *C.? insperta* Salt., *C.? perornata* Salt., *C.? elliptica* McCoy. *Physocaris vesica* Salt.

—— and **Kirkby** (J. W.) Notes on Palæozoic bivalved Entomostraca. No. 19. On some Carboniferous species of the Ostracodous genus *Kirkbya*. •

In Annals Mag. Nat. Hist., 5th series, London, vol. 15, 1885, p. 174, pl. 3.

Kirkbya permiana Jones, *K. umbonata* Eichw. and var. *radiata, K. oblonga* J. & K., *K. annectens* J. & K., *K. plicata* J. & K., *K. spiralis* J. & K., *K. spinosa* J. & K., *K. costata* M'Coy, also var. *moreana, K. scotica* J. & K., *K. rigida* J. & K., *K. urei* Jones.

—— Third Report of the committee consisting of Mr. E. Etheridge, Dr. H. Woodward and Prof. T. Rupert Jones (Secretary), on the Fossil Phyllopoda of the Palæozoic Rocks.

In Rep. 55th Meeting Brit. Assoc. Adv. Sci., 1885, p. 326.

1, *Ceratiocaris murchisoni* Agas. and its var. *leptodactylus* McCoy. 2, *C. ludensis* H. W. 3, *C. papilio* Salt. 4, *C. stygia* Salt. 5, *C. inornata* McCoy. 6, *C. oretonensis* H. W. 7, *C. truncata* H. W. 8, *C. solenoides* McCoy. 9, *C. gobiiformis* J. & W. 10, *C. salteriana* n. sp. 11, *C. cassia* Salt. 12, *C.* n. sp. 13, *C. robusta* Salt. also var. *longa* n. v. 14, *C.* n. sp. 15, *C. decora* (Doubtful genera. 16, *C. ensis.* 17, *C. lata* n. sp. 18, *C.? insperata.* 19, *C.* n. sp. 20, *C.? perornata.* 21, *C. elliptica.*) 22, *Physocaris vesica* Salt. 23, *Acanthocaris scorpioides, elongata* and *attenuata* n. sp. 24, *C.? longicauda* Sharp. 25, *C. deweii.* 26, *C. maccoyiana.* 27, *C. acuminata* n. sp. 28, *C. aculeata.* 29, *C. noetlingi* n. sp. 30, Barrande's species of Ceratiocaris: *Ceratiocaris docens, C. decipiens, C. scharyi, C. bohemica, C. inæqualis, C. debilis, C. tarda, C. primula.* 31, *Aristozoe amica, A. bisulcata, A. inclyata, A. lepida, A. memoranda, A. orphana, A. perlonga, A. regina, A.? jonesi. Orozoe mira. Callizoe bohemica. Nothozoe pollens.* 32,

Aristozoe regina and its abdominal appendages. 33, *Echinocaris* and its allies: *E. armata (punctata)* Hall, *E. sublævis* Whitf., *E. pustulosa* Whitf., *E. mutinodosa* Whitf., *E. socialis* Beecher. *Elymocaris siliqua* Beecher. *Tropidocaris bicarinata* Beecher, *T. interrupta* Beecher, *T. alternata* Beecher. *Echinocaris wrightiana* Dawson. *Colpocaris sinuata* M. & W.

—— and **Holl** (H. B.) Notes on the Palæozoic bivalved Entomostraca. No. 20. On the genus *Beyrichia* and some new species.

In Annals Mag. Nat. Hist., 5th series, London, vol. 17, 1886, p. 337, pl. 12.

Bollia n. g. *Kladenia* n. g. *Beyrichia tuberculata* Klöden var. *gibbosa* Reuter, *B. klödeni* M'Coy, also var. *granulata* Jones, var. *intemedia* Jones, and sub. var. *subspissa* nov., *B. klödeni* var. *subtorosa* and var. *tuberculata*, also var. *scotica* B. *concinna* n. sp., *B. maccoyiana* Jones, *B. jonesii* Boll., *B. admixta* n. sp., *B. lacumata* n. sp. *Bollia bicollina* n. sp., *B. uniflexa* n. sp. *Kladenia intermedia* J. & K. var. *marginata* n. sp.

—— Notes on the Palæozoic bivalved Entomostraca. No. 21. On some Silurian genera and species.

In Annals Mag. Nat. Hist., 5th series, London, vol. 17, 1886, p. 403, pls. 13, 14.

Strepula n. g. *Placentula* n. g. *Strepula concentrica* n. sp., *S. irregularis* n. sp., *S. beyrichioides* n. sp. *Bollia vinei* n. sp., also var. *mitis*. *Placentula excavata* J. & H. *Primitia lenticularis* J. & H., *P. roemeriana* J. & H., *P. fabulina* n. sp., *P. variolata* J. & H., *P. paucipunctata* J. & H., *P. humilis* n. sp., *P. valida* n. sp., also var. *breviata* and *augustata*, *P. tersa* J. & H., *P. umbilicata* J. & H., *P. cristata* J. & H., *P. ornata* J. & H., *P. cornuta* n. sp., *P. æqualis* n. sp., *P. diversa* n. sp. *P. seminulum* Jones, *P. furcata* n. sp.

—— and **Kirkby** (J. W.) Notes on the distribution of the Ostracoda of the Carboniferous formation of the British Isles.

In Abstracts Proc. Geol. Soc. London, session 1885-86, No. 486, p. 86. Not descriptive.

—— On Carboniferous Ostracoda from the Gayton borings, Northamptonshire.

In Geol. Mag., n. s., decade 3, vol. 3, London, 1886, p. 248, pl. 7.

Kirkbya variabilis n. sp., var. *a*, var. *b* and var. *c*, *K. plicata* J. & K. *Bythocypris sublunata* J. & K. *Macrocypris jonesiana* Kirkby. *Cytherella extuberata* J. & K., *C. attenuata* J. & K.

―――― Notes on the Palæozoic bivalved Entomostraca. No. 22. On some undescribed species of British Carboniferous Ostracoda.

In Annals Mag. Nat. Hist., 5th series, London, vol. 18, 1886, p. 249, pls. 6-9.

Bythocypris phillipsiana J. & H. var. *carbonica* n. var., *B. ? cuneola* J. & K., *B. cornigera* J. & K., *B. ? pyrula* n. sp., *B. ? moorei* n. sp., *B. thraso* J. & K., *B. lunata* n. sp. *Cythere ? gyripunctata* J. & K. *Leperditia armstrongiana* J. & K., *L. bosquetiana* n. sp., *L. youngiana* J. & K., *L. scotoburdigalensis* Hibbert, *L. parallela* J. & K., *L. obesa* J. & K., *L. compressa* J. & K., *L. lovicensis* n. sp., *L. acuta* J. & K. *Beyrichia radiata* J. & K., *B. longispina* n. sp., *B. fodicata* n. sp., *B. tuberculospinosa* n. sp., *B. multiloba* J. & K., *B. varicosa* n. sp., *B. ? bicæsa* n. sp. *Primitia ? holliana* n. sp. *Beyrichiella ? reticosa* n. sp., *B. ? ventricornis* J. & K. *Kirkbya tricollina* J. & K. *Moorea obesa* n. sp., *M. tenuis* n. sp. *Cytherella? reticulosa* J. & K., *C. valida* J. & K. var. *affiliata*, *C. ? elongata* n. sp. *Bythocythere antiqua* n. sp., *B. youngiana* n. sp. *Argillæcia æqualis* J. & K. *Aglaia? cypridiformis* J. & K. *Xestoleberis ? subcorbuloides* J. & K. *Macrocypris carbonica* Brady. *Carbonia wardiana* n. sp. *Cythere superba* J. & K., *C. ? obtusa* n. sp. *Bairdia legumen* J. & K., *B. subelongata* J. & K. var. *major*.

―――― On some fringed and other Ostracoda from the Carboniferous series.

In Geol. Mag., n. s., decade 3, vol. 3, London, 1886, p. 434, pls. 11, 12.

Beyrichiopsis n. g. *Beyrichiella* n. g. *Beyrichiopsis fimbriata* n. sp., *B. fortis* n. sp., *B. cornuta* n. sp., *B. crinita* J. & K., *B. subdentata* n. sp., *B. simplex* n. sp. *Beyrichiella cristata* J. & K. *Beyrichia arcuata* Bean., *B. fastigiata* J. & K., *B. bradgana* n. sp., *B. craterigera* Brady.

―――― On Palæozoic Phyllopoda.

In Geol. Mag., new series, decade 3, vol. 3, London, 1886, p. 456.

This paper contains the substance of the Fourth Report of a Committee on the Fossil Phyllopoda of the Palæozoic Rocks, 1886.

―――― and **Kirkby** (J. W.) Notes on the distribution of the Ostracoda of the Carboniferous formations of the British Isles.

In Quart. Jour. Geol. Soc. London, vol. 42, 1886, p. 496.

Beyrichiopsis. Phreatura n. g. *Youngia* n. g.

This generic name was used for a new genus of Trilobites by Lindström in 1885.

Note.—Since this paper was read many of the species have been described and figured in the Annals Mag. Nat. Hist. for October, 1886, and in the Geol. Mag. of the same date. Others in the Proc. Geologist Assoc., vol. 9, 1886.

Beyrichiopsis fimbricata J. & K. *Phreatura concinna* J. & K. *Youngia retidosalis* J. & K.

This valuable paper contains a classification of the Carboniferous strata of England and Scotland, with tabulated lists of the distribution of the Ostracoda.

——— Fourth report of the committee, consisting of Mr. R. Etheridge, Dr. H. Woodward, and Prof. T. Rupert Jones (Secretary), on the fossil Phyllopoda of the Palæozoic Rocks, 1886.

In Rep. 56th Meeting Brit. Assoc. Adv. Sci., 1886, p. 229.

1, *Ceratiocaris leptodactylus* McCoy. 2, *C. murchisoni* Agas. 3, *C. gigas* Salt. 4, *C. valida* n. sp. 5, *C. attenuata* n. sp. *(C.? tyrannus* Salt.) 6, *C. canaliculata* n. sp. 7, *C. halliana* n. sp. 8, *C. pardoëana* La Touche. 9, *C. ludensis* H. W. 10, *C. robusta* Salt., *C. lata, angusta* and *minuta* n. sp. 11, *C. papilio* and *C. stygia* Salt. 12, *C. laxa* n. sp. 13, *C. salteriana* J. & W. 14, *C. cassia* Salt., *C. cassioides* n. sp. 15, *C. compta* n. sp. 16, *C. inornata* McCoy. 17, *C. ruthveniana* n. sp. 18, *C. oretonensis* and *C. truncata* H. W. 19, *C. solenoides* McCoy and *C. gobiiformis* J. & W. 20, *Emmelezoe elliptica* McCoy, *E. tenuistriata* n. sp., *E. maccoyiana* n. sp. 21, *Xiphocaris ensis* Salt. 22, *Physocaris vesica* Salt. 23, *C.* sp. 24, *C. longicauda* Sharpe. 25, *Ptychocaris simplex* and *P. parvula* Ovák. 26, *Cryptozoe problematica* Packard. 27, Geological localities of Mr. S. M. Clarke's fossil Phyllopods. 28, List of British Palæozoic Phyllocarida described in the Third and Fourth Reports.

——— and **Kirkby** (James W.) A list of the genera and species of bivalved Entomostraca found in the Carboniferous formations of Great Britain and Ireland. With notes on the genera and their distribution.

In Proc. Geologist Assoc., vol. 9, 1887, p. 495.

I. Cypridinidæ: 1, *Cypridina*. 2, *Cypridinella*. 3, *Cypridellina*. 4, *Cypridella*. 5, *Sulcuna*. 6, *Cyprella*. 7, *Bradycinetus*. 8, *Philomedes*. 9, *Rhombina*.

II. Entomoconchidæ: 1, *Entomoconchus*. 2, *Offa*.

III. Polycopidæ: 1, *Polycope*.

IV. Entomididæ: 1, *Entomis*.

V. Cytherellidæ: 1, *Cytherella*.

VI. Leperditidæ: 1, *Leperditia*. 2, *Bernix*. 2, *Beyrichia*. 4, *Beyrichiella*. 5, *Beyrichiopsis*. 6, *Kirkbya*. 7, *Moorea*. 8, *Phreatura*.

VII. Cyprididæ: 1, *Aglaia*. 2, *Candona*. 3, *Argillœcia*. 4, *Macrocypris*. 5, *Bythocypris*. 6, *Bairdia*.

VIII. Dawinulidæ: 1, *Dawinula*.

IX. Cytheridæ: 1, *Cythere*. 2, *Xestoleberis*. 3, *Bythocythere*. 4, *Carbonia*. 5, *Youngia*.

The authors give generic descriptions with a catalogue of the genera and species of bivalved Entomostraca found in the Carboniferous formations of Great Britain and Ireland.
Figures of the following are given:
Beyrichiopsis fimbriata J. & K.   *Phreatura concinna* J. & K.   *Youngia rectidorsalis* J. & K.

———— Fifth report of the committee, consisting of Mr. R. Etheridge, D. H. Woodward, and Prof. T. Rupet Jones (Secretary), on the fossil Phyllopoda of the Palæozoic rocks, 1887.
In Rep. 57th Meeting Brit. Assoc. Adv. Sci., 1887, p. 60.
Ceratiocaris tyrannus Salt., and *C. patula* n. sp., *C. angelini* n. sp., *C. bohemica* Barr., also varieties *C.* n. sp., *C. concinna* n. sp., *C. scharyi* Barr., *C. pectinata* n. sp. *Phasganocaris pugio* Barr. var. *serrata*. *Dithyrocaris* Upper and Lower Carboniferous species, also Devonian and Silurian species. List of the known species of *Leaia*. Palæozoic species of *Estheria*.

———— Notes on the Palæozoic bivalved Entomostraca. No. 23. On some Silurian genera and species (continued).
In Annals Mag. Nat. Hist., 5th series, London, vol. 19, 1887, p. 117, pls. 4–7.
Prof. T. Rupert Jones remarks in this paper: "The genus *Cythere* comprises very few of the smooth subovate forms. and none that have toothless hinges. Hence we find that a real *Cythere* is scarcely known in Palæozoic strata."
Macrocypris vinei n. sp., *M. elegans* n. sp., *M. siliquoides* n. sp., *M. symmetrica* n. sp., *M.? alta* n. sp., *M.? crassula* n. sp. *Pontocypris mawii* n. sp., *P. gibbera* n. sp., *P. smithii* n. sp. *Bythocypris hollii* n. sp., *B.? reniformis* n. sp., *B.? botelloides* n. sp., *B. testacella* n. sp., *B. symmetrica* n. sp. and var., *B. concinna* n. sp. and var., *B. phillipsiana* and var. *major*, *B. pustulosa*, *B.? seminulum* n. sp., *B. acina* n. sp., *B. phaseolus* n. sp. *Cythere hollii* n. sp., *C. vinei* n. sp., *C.? subquadrata* n. sp. *Cytherella smithii* n. sp. and var. *Primitia punctata* n. sp., *P. valida* n. sp. *Incertæ sedis*.
Notes on Dr. Lindström's collection.

———— Notes on the Palæozoic bivalved Entomostraca. No. 24. On some Silurian genera and species (continued).
In Annals Mag. Nat. Hist., 5th series, London, vol. 19, 1887, p. 400, pls. 12, 13.
Thlipsura corpudenta J. & K., *T. tuberosa* J. & H., *T. angulata* n. sp., *T. plicata* n.sp., also var. *unipunctata, bipunctata, T. v-scripta* J. & H.

Octonaria octoformis n. sp., also vars. *intorta, simplex, informis, bipartita, persona* and *monticulata, O. undosa* n. sp., *O.? paradoxa* n. sp. *Bollia auricularis* n. sp., *B. interrupta* n. sp. *Primitia obliquipunctata* n. sp. *Moorea smithii* n. sp. *Xestoleberis corbuloides* J. & H. *Æchmina cuspidata* J. & H., *Æ. bovina* n. sp., *Æ. depressicornis* n. sp., *Æ. brevicornis* n. sp.

———— Notes on some Silurian Ostracoda from Gothland. Stockholm, 1887, 8 pp.

Primitiopsis n. g. *Bursulella* n. g. *Beyrichia clavata* Kolmodin, *B. klædeni* McCoy, also var. *antiquata* Jones and var. *tuberculata* Salt., *B. bolliana-umbonata* Reuter. *Leperditia grandis* Schrenk. *Entomis lindströmi* n. sp. *Primitia lævis* n. sp., *P. valida* J. & H., *P. grandis* n. sp., *P. reticristata* n. sp., *P. seminulum* Jones. *P. inæqualis* n. sp. *Primitopsis planifrons* n. sp., also var. *ventrosa* n. var. *Macrocypris vinei* Jones. *Bythocypris concinna* Jones., *B. hollii* Jones. *Thlipsura v-scripta* J. & H., *Æchmina bovina* Jones. *Bursulella triangularis* n. sp., *B. semiluna* n. sp., *B. unicornis* n. sp. Table of stratigraphical distribution by G. Lindström.

———— Notes on the Palæozoic bivalved Entomostraca. No. 25. On some Silurian Ostracoda from Gothland.

In Annals Mag. Nat. Hist., 6th series, London, vol. 1, 1888, p. 395.

Bursulella n. g. *Macrocypris vinei* Jones. *Pontocypris mawii* Jones. *Bythocypris symmetrica* Jones, *B. concinna* Jones, *B. hollii* Jones. *Klædenia apiculata* J. & H. *Beyrichia clavata* Kolmodin, *B. klædeni* McCoy, var. *granulata* Jones, *B. klædeni* McCoy var. *verruculosa*, also vars. *antiquata* and *tuberculata*, *B. bolliana-umbonata* Reuter, *B. tuberculata* and var. *(senex)*, var. *spicata*, var. *foliosa*. *Leperdita grandis* Schrenk. *Thlipsura v-scripta* J. & H. var. *discreta*. *Primitia lævis* Jones, *P. stricta* n. sp., *P. valida* J. & H., *P. grandis* Jones, *P. resticristata* Jones, *P. seminulum* Jones. *Primitiopsis planifrons* Jones, also var. *ventrosa* Jones. *Entomis lindstroemi* Jones, *E. inæqualis* Jones. *Æchmina bovina* Jones. *Bursulella triangularis* Jones, *B. semiluna* Jones, *B. unicornis* Jones.

———— Notes on the Palæozoic bivalved Entomostraca. No. 26. On some new Devonian Ostracoda, with a note on their geological position, by the Rev. G. F. Whitborne.

In Annals Mag. Nat. Hist., 6th series, London, vol. 2, 1888, p. 295, 1 pl.

Kyamodes n. g., *K. whidbornei* n. sp., also vars. *elliptica* and *obsolescens*.

———— and **Woodward** (Henry). A monograph of the British Palæozoic Phyllopoda.

Part 1, Ceraticaridæ, pp. 1-72, plates 1-12, Palæont. Soc., London, 1888.

1, *Ceratiocaris leptodactylus* McCoy. 2, *C. murchisoni* Agas. 3, *C. valida* J. & W. 4, *C. tyrannus* Salt. MS. 5, *C. gigas* Salt. MS. 6, *C.*

halliana J. & W.   7, *C. pardoëana* La Touche.   8, *C. canaliculata* J. & W.   9, *C. ludensis* H. W.   10, *C. papilio* Salt.   11, *C. stygia* Salt.   12, *C. longa* n. sp.   13, *C. robusta* Salt.   14, *C. patula* n. sp.   15, *C. angusta* J. & W.   16, *C. minuta* J. & W.   17, *C. inornata* McCoy.   18, *C. ruthveniana* J. & W.   19, *C. oretonensis* H. W.   20, *C. truncata* H. W.   21, *C. solenoides* McCoy.   22, *C. gobiiformis* J. & W.   23, *C. salteriana* J. & W.   24. *C. laxa* J. & W.   25, *C. compta* J. & W.   26, *C. cassa* Salt.   27, *C. cassoides* J. & W.   28, *C.? longicauda* Sharpe.   29, *C. decora* (Phillips).   30, *C.? lata* Salt.   31, *C.? insperata* Salt.   32, *C.?*   33, *C. perornata* Salt.   *Xiphocaris ensis* (Salt.)   *Physocaris vesica* Salt·   *Emmelezoe elliptica* (McCoy).   2, *E. crassistriata* J. & W.   3, *E. tenuistriata* J. & W.   4, *E. maccoyiana* J. & W.

—— On some Scandinavian Phyllocaridæ.
In Geol. Mag., new series, decade 3, vol. 5, London, 1888, p. 97, pl. 5.
Ceratiocaris.

—— Sixth report of the committee, consisting of Mr. E. Etheridge, Dr. H. Woodward, and Prof. T. Rupert Jones (Secretary), on the fossil Phyllopoda of the Palæozoic Rocks, 1888.
In Rep. 58th Meeting Brit. Assoc. Adv. Sci., 1888.
Ceratiocaris tyrannus. Aristozoe solitaria. Note on *Cryptocaris.* Note on *Ptychocaris? jaschei. Bactropus. Saccocaris major* Salter. *Lingulocaris salteriana* n. sp., *L. siliquiformis* J. & W. *Hymenocaris vermicauda* Salter. *Estheria membranacea* Pacht. Devonian Phyllocarids of New York State, etc., etc.

—— Notes on the Palæozoic bivalved Entomostraca, No. 27. On some North American (Canadian) species.
In Annals Mag. Nat. Hist., 6th series, vol. 3, 1889, plates 16–17.
Aparchites n. g. *Primitia nundula*, varieties *P. simplex*, *P. scaphoides* n. sp., *P. æqualis* J. & H. *Beyrichia klœdeni* McCoy, var. *acadica* Jones, *B. arcuata* Bean. *Isochilina labrosa* n. sp. *Aparchites whiteavesii* n. sp.

—— and **Kirkby** (J. W.) On some Ostracoda from the Mabou coal fields, Inverness Co., Cape Breton (Nova Scotia).
Geol. Mag. London, decade 3, vol. 6, 1889, p. 300.
Carbonia fabulina var. *altilis*, figs. 1–4.

—— Notes on the Palæozoic bivalved Entomostraca, No. 28. On some Scandinavian species.
In Annals Mag. Nat. Hist., 6th series, vol. 4, 1889, p. 267, plate 15.
Macrocypris? pusella n. sp. *Pontocypris mawii* Jones var. *breviata* n. var., also vars. *proxima* and *divergens. Bythocypris holli* Jones var. *oblonga*

n. var., *B. caudalis* n. sp., *B. symmetrica* Jones var. *obesa, B. phaseolus* Jones var. *elongata* n. var., *B. concinna, B. philliana* J. & H. var. *gotlandica. Aparchites decoratus* n. sp., *A. simplex* n. sp., *A. lindströmi* n. sp.

—— and **Woodward** (H.) On some Palæozoic Ostracoda from Pennsylvania.
In Geol. Mag., decade 3, vol. 6, 1889, p. 385, plate 11.
Echinocaris whidbornei n. sp. *Beyrichia punctata* Hall, *B. devonica* n. sp.

—— On some Palæozoic Ostracoda from Pennsylvania.
In American Geologist, vol. 4, 1889, p. 337, plate.
Devonian: *Primitia mundula* Jones. *Klœdenia simplex* n. sp. *Bythocypris favulosa* n. sp. *Bollia ungula* Claypole. *Primitia pennsylvanica* n. sp.
Silurian: *Bythocypris oriformis* n. sp. *Leperditia subquadrata* n. sp. *Klœdenia pennsylvanica* n. sp.

—— On some Palæozoic Ostracoda from North America, Wales and Ireland.
In Quart. Jour. Geol. Soc., London, vol. 46, 1890, p. 1, plates 1-4.
Primitia mundula Jones var. *cambrica* nov., *P. humilis* J. & H. var. *humilior* nov., *P. seminulum* Jones, *P. morgani* n. sp., *P. ulrichi* n. sp., *P. unicornis* Ulrich, *P. minuta* Eichw., *P. whitfieldi* n. sp. *Primiopsis punctulifera* Hall. *Entomis rhomboidea* n. sp. *Strepula sigmoidalis* n. sp. *Æchmina spinosa* Hall, *.E. byrnesi* Miller. *Bollia symmetrica* Hall, *B. lata* Conrad. *Klœdenia notata* Hall, *K. notata* var. *ventricosa* Hall. *Beyrichia trisulcata* Hall, *B. halli* n. sp. *B. granulata* Hall, *B. oculina* Hall, *B. buehiana?* Jones, *B. parasitica* Hall, *B. clarkei* n. sp., *B. æquilatera* Hall, *B. tuberculata* Boll. var. *pustulosa* Hall, *B. hamiltonensis* n. sp., *B. ciliata* Emm., *B. oculifer* Hall. *Isochilina lineala* n. sp., *I.? fabacea* n. sp., *I. seelyi* Whitf., *I. gregaria* Whitf., *I. cristata* Whitf. *Leperditia? seneca* Hall, *L.? hunsonica* Hall, *L. hudsonica?* *L. claypolei* n. sp., *L. alta* Conrad and *L. jonesi* Hall (*L. alta* var.) *L. nana?* Jones. *Xestoleberis wrightii* n. sp.

—— On some Devonian and Silurian Ostracoda from North America, France and the Bosphorus.
In Quart. Jour. Geol. Soc., vol. 46, 1890, p. 534, plates 20-21.
Devonian species: *Primitia clarkei* n. sp. *Beyrichia devonica* Jones, *B. subquadrata* n. sp., *B. klœdeni* McCoy variety *B. kolmodini* n. sp. *Eurychilina reticulata* Ulrich. *Bollia bilobata* n. sp., *B. hindei* n. sp. *Bollia? Strepula plantaris* n. sp. *Octonaria linnarssoni* n. sp. *Moorea kirkbyi* n. sp. *Primitia? walcotti* n. sp. *Ulrichia conradi* n. sp.

—— Seventh report of the committee, consisting of Mr. R. Etheridge, Dr. H. Woodward, and Prof. T.

Rupert Jones (Secretary), on Fossil Phyllopoda of the Palæozoic Rocks.
In Rep. 59th Meeting Brit. Assoc. Adv. Sci., 1889, p. 63.
1, *Aristozoe* and *Callizoe*. 2, *Bactropus*. 3, *Tropidocaris*. 4, *Echinocaris*. 5, *Nothozoe*. 6, *Protocaris*. 7, *Ceratiocaris patula*, *C. pusilla*, *C. cenomanensis*, *C. grandis*. 8, *Caridolites (Ceratiocaris ?)* 9, *Dithyrocaris*. 10, *Lingulocaris*. 11, *Discinocaris*. 12, *Estheria* and *Estheriella*, *Estheriella lineata* and *costata* Weiss. 13, *Ribeiria*. 14, *Proricaris*.

The authors give a brief description of the fossils *Estheria* and *Estheriella* described and figured by the following authorities: Wenjukoff, Ludwig, Kratow, Weiss and Jones.

Ribeiria pholadiformis Sharpe. *Myocaris lutraria* Salt. *Ribeiria calcifera* and *R.? longiuscula* Billings.

Other species of this genus are also cited. *Proricaris machenrici* Baily is referred to the swimming feet of *Eurypterus* or *Pterygotus*. *Leperditia megalops* Kolm. is referred to *Callizoe*. *Echinocaris whidbornei* J. & W. *Estheria coglani* Etheridge, Jr.

———— Eighth report of the committee, consisting of Mr. R. Etheridge, Dr. H. Woodward and Prof. T. Rupert Jones (Secretary), on the Fossil Phyllopoda of the Palæozoic Rocks.
In Rep. Brit. Assoc. Adv. Sci., 1890.
Saccocaris minor J. & W. *Aristozoe memorauda* Barr., *A. regina* Barr. *Estheria youngii* n. sp., *E. tessellata* n. sp., *E. pulex*, *E. tegulata*.
Estheriæ punctatella is referred to the genus *Posidonomya*.

———— On some *Estheria* and *Estheria*-like shells from the Carboniferous Shales of Western Scotland.
In Trans. Geol. Soc. of Glasgow, 1890, 1 plate.
Estheria youngii n. sp., *E. tessellata* n. sp., *E. tegulata* n. sp.

———— On some more fossil *Estheriæ*.
In Geol. Mag., decade 3, vol. 8, 1891, p. 49, plate.
Estheria membranacea Pacht., *E. andrewsii* n. sp. *E. hindei* n. sp., *E. minuta* Alberti also var. *brodieana*. *Estheriella costata* Weiss, *E. nodocostata* Giebel.

———— and **Kirkby** (J. W.) On the Ostracoda found in the Shales of the Upper Coal Measures at Slade Lane, near Manchester.
In Trans. Manchester Geol. Soc., vol. 21, 1890, part 3, 1 plate.
Carbonia pungens J. & K., *C. secans* J. & K., *C. roederiana* n. sp., *C. fabulina* J. & K., *C. salteriana* Jones.

―― Contributions to Canadian Micro-Palæontology, Part 3.
Geol. and Nat. His. Survey of Canada, Moutreal, 1891, pp. 59-99, plates 10-13.
Cambro-Silurian species: *Aparchites mundulus* n. sp., *A. tyrrelli* n. sp. *Primitia logani* Jones, *P. mundula* Jones var. *effossa* n. var., *P. mundula* Jones var. *incisa* n. var. *Beyrichia clavigera* n. sp., also var. *clavifracta* n. var., *B. quadrifida* n. sp. *Isochilina ottawa* Jones, var. *intermedia* n. var., *I. whiteavesii* n. sp., *I. amii* n. sp., *I. labellosa* n. sp. *Leperditia balthica* His. var. *primæva* n. var., *L.* sp.? (cf. *L. hisingeri*), *L. obscura* n. sp.
From the Silurian rocks: *Primitia mundula* Jones. *Beyrichia æquilatera* Hall, *B. tuberculata* Kloeden, *B. tuberculata* var. *pustulosa* Hall, also var. *strictispiralis* n. var. and *noetlingi* Reuter. *Isochilina grandis* var. *latimarginata* n. var., *I. winnepeyosis* n. sp. *Leperditia balthica* His. var. *guelphica* n. var., *L. hisingeri* Schmidt, *L. hisingeri* var. *fabulina* n. var., also var. *gibbera* n. var. and *egena* n. var., *L. alta* Conrad, *L. phaseolus* His. and var. *guelphica* n. var., *L. marginata* Schmidt, *L. whiteavesii* n. sp. *L. cocca* n. sp., *L. selwynii* n. sp.
From the Devonian rocks: *Aparchites mitis* n. sp. *Primitia scitula* n. sp. *Isochilina bellula* n. sp., *I. dawsoni* n. sp. *Elpe tyrrellii* n. sp. *Leperditia? exigua* n. sp. *Ulrichia conradi* Jones. *Primitiopsis punctulifera* Hall. *Kirkbya? walcotti* Jones.
The appendix contains notes on the species of Ostracoda described and figured in decade 3 of the Geol. Survey of Canada.

―― and **Woodward** (Henry). See **Woodward** (Henry) and **Jones** (T. Rupert).

―― See **Schmidt** (Fr.) and **Jones** (T. Rupert).

Jordan (H.) Entdeckung fossiler Crustaceen im Saarbrücken schen' Steinkohlengebirge.
In Vörhandl. natur. Vereins preussischen Rheinlande, Jahrg. 1847, vol. 4, p. 89, pl. 2, figs. 1, 2.
Gampsonyx fimbriatus n. sp.

―― and **Bronn** (H. G.) On *Posidonomya tenella*.
In Neues Jahrb., 1850, p. 577.
Estheria tenella.

―― and **Meyer** (H. von). Ueber die Crustaceen der Steinkohlenformation von Saarbrücken. Cassel, 1854, 2 pls.
Gampsonyx fimbriatus Jordan. *Adelophthalmus (Eurypterus) granosus* Jordan. *Chonionotus lithanthracis*. *Arthropleura armata* Jordan.

Jukes (F.) and **Sowerby** (J. D. C.) An account of a new species of Trilobite found in the Barr limestone in the neighborhood of Birmingham by Frederick Jukes, with a note by J. D. C. Sowerby.
In Mag. Nat. Hist. (London), vol. 2, London, 1829, p. 41, 5 figs.; Am. Jour. Sci., 1st series, vol. 23, 1833, p. 203; Neues Jahrbuch für Mineral, 1833, p. 624.
(Bumastes barriensis). Asaphus caudatus. Calymene blumenbachii.

Karsten (Gustav). Die Versteinerungen des Uebergangs-Gebirges in den Geröllen der Herzogthümer Schleswig und Holstein. Kiel, 1869, 25 pls.
Ostracoden: *Cythere baltica* His. *Cypridina*. *Beyrichia tuberculata* Boll., *B. proturberans* Boll., *B. buchiana* Jones, *B. nodulosa* Boll., *B. wilkensiana* Jones.
Trilobiten: *Olenus truncatus* Brunn., *O. scarabæoides* Wahl. *Proetus* sp.? *Cyphaspis* sp.? *Phacops stockesii* Milne-Edwards, *P.* sp. *(caudata), P. conicophthalma* Boeck. *Calymene blumenbachii* Brong., *C. duplicata* Murch. *Lichas angusta* Beyr., *L.* sp. *(a), L.* sp. *(b), L.* sp. *(c), L. (scabra* Beyr.*), L. arenswaldii* Boll. *Trinucleus concentricus* Eaton., *T.* sp. (or *Ampyx?) Ampyx longirostris (Lonchodomas longirostris), A. nasustus* Dalm. *Asaphus expansus* Linné, *A. expansus* var. *cornutus* Murch., *A.* sp. (cf. *Ptychoppge globifrons* Eichw.), *A. robustus (Isotelus robustus* Römer), *A* sp. *(tyrannus* Murch.), *A.* sp. *(a), A.* sp. *(b), A.* sp. *(c), A.* sp. *(d), A.* sp. *(e)*. *Illænus* sp.*(a), I.* sp. *(b). Encrinurus punctatus* Brünn. *Agnostus pisiformis* Linné. *Illænus grandis* Romer, *I. crassicauda* Wahl., *I.* sp. *(a), I. (Nileus) armadillo* Dalm. *Ceraurus insignis* Beyr., *C.* sp. *(a), C. exsul* Beyr. *Sphærexochus* sp. *(a), S.* sp. *(b), mirus* Beyr. (Fragment of *Asaphus* or *Ogygia?.)*

Kayser (Emanuel). Beiträge zur Geologie und Palæontologie der Argentinischen Republik. Cassel, 1876, 2 pls.
Agnostus tilcuyensis n. sp. *Olenus argentinus* n. sp. *Arionellus lorentzi* n. sp., *A. hyeronimi* n. sp. *Bathyurus? lajensis* n. sp., *B.? darwinii* n. sp., *orbignyanus* n. sp. *Ogygia* sp.? *Arethusina argentina* n. sp. *Leperditia. Asaphus* sp. *Ogygia cordensis* Murch.? *Ampyx* sp.?

——— Die Fauna der ältesten Devonablagerungen des Harzes.
In Abhandlungen zur geologischen Specialkarte von Preussen und den thüringischen Staaten. Vol. 2, pt. 4. Berlin, 1878. Atlas, 36 pls.
Primitia sp.? *Harpes biochofi* A. Roemer. *Proetus unguloides* Barr., *P. complanatus* Barr., *P.* sp.?, *P. richteri* n. sp., *P. eremita* Barr., *P.*

wiedensis n. sp., *P.* cf. *orbitatus* Barr. *Cyphaspis hydrocephala* A. Roem. *Phacops fecundus* Barr. var. *P. zinkeni* A. Roemer, *P.* sp., *P. fugitivus* Barr., *P. zorgensis* n. sp. *Dalmanites tuberculatus* Emm., *D.* sp., *D. beyrichi* n. sp. *Cryphæus calliteles* Green, *C.* cf. *stellifer* Burm. *Lichas sexlobata* A. Roem. *Acidaspis seleana* A. Roem., *A. glabrata* A. Roem., *A.* sp. *Cheirurus sternbergi* Boeck.? var. *interrupta* n. sp. *Bronteus bischofi* A. Roem., *B.* sp., *B. roemeri* Kayser, *B. elongatus* Barr., *B.* cf. *billingsi* Barr.

——— Ueber *Dalmanites rhenanus*, eine Art der *Hausmanni*-Gruppe, und einige andere Trilobiten aus den älteren rheinischen Dachschiefern.
In Zeitschr. Deutsch. geol. Gesell., vol. 32, 1880, p. 19, pl. 3.
Dalmanites (Odontochile) rhenanus n. sp. *Phacops ferdinandi* n. sp.

——— *Dechenella*, eine devonische Gruppe der Gattung *Phillipsia*.
In Zeitschr. Deutsch. geol. Gesell., vol. 32, 1880, p. 703, pl. 27.
Dechenella n. g., *D. verneuili* Barr., *D. verticalis* Burm., *D. elegans* Münst., *D. haldemani* Hall.

——— Beiträge zur Kenntniss von Oberdevon und Culm am Nordrande des rheinischen Schiefergebirges.
In Jahrbuch Preuss. geol. Landesanst. und Bergakad., für 1881, Berlin, 1882, pp. 51-91, pls. 1-3.
Phacops granulatus Münst. *Phillipsia æqualis* v. Meyer, *P. longicornis* n. sp., *P.* cf. *eichwaldi* Fischer, *P.* sp.?, *P. emarginata* n. sp. *Cypridina subglobulosa* Sandb.

——— Die Orthocerasschiefer zwischen Balduinstein und Laurenberg an der Lahn. Palæontologischer Anhang.
In Jahrbuch Preuss. geol. Landesanst. und Bergakad., für 1883, Berlin, 1884, pp. 56, pls. 1-6.
Phacops fecundus Barr. *Cryphæus rotundifrons* Barr. *C. kochi* n. sp.

——— Mittel- und obersilurische Versteinerungen aus dem Gebirgsland von Tshau Tiens.
In China (Richtofen), vol. 4, Berlin, 1883, p. 37.
Asaphus sp.? *Calymene* sp.? *Trinucleus richthofeni* n. sp. *Encrinurus* sp.

——— Obercarbonische Fauna von Loping.
In China (Richthofen), vol. 4, Berlin, 1883, p. 161.
Phillipsia obtusicauda n. sp.

—— Ueber einige neue order wenig genannte Versteinerungen des rheinischen Devon.

In Zeitschr. Deutsch. geol. Gesell., vol. 41, 1889, p. 288.

Phacops (Trimerocephalus) acuticeps n. sp.

Ketley (Charles).*

In Trans. Dudley & Midland Geol. Soc., 1865, p. 105.

Acidaspis barrandei n. sp. *Proetus fletcheri* n. sp.

Keyes (Chas. R.) On the fauna of the Lower Coal Measures of central Iowa.

In Proc. Acad. Nat. Sci. Phila., 1883, p. 222.

Cythere nebrascensis Geinitz.

Keyserling (Alex. G.) Palaeontologische Bemerkungen 1. Reste der Transitionszeit.—IV. Crustacea, p. 288.

Wissen. Beobachtungen auf einer Reise in das Petschora-land, Jahre, 1843.

Cypridina marginata n. sp. *Illænus (Bumastus) barriensis* Murch. *Phacops odini* Eichw. *Encrinurus punctatus* Brünn. *Phillipsia eichwaldi* Fischer., *P. truncatula* Phill.

Kiesow (J. von). Ueber silurische und devonische Geschiebe West-Preussens.

In Schrift. nat. Gesell. Danzig, new series, vol. 6, 1884, p. 205, 3 pls.

Leperditia hisingeri Schm., *L. baltica* His., *L. eichwaldi* Schm., *L. phaseolus* His. *Primitia oblonga* J. & H., *P. obsoleta* J. & H., *P. ovata* J. & H., *P. semicircularis* J. & H., *P. mundula* Jones. *Cytherellina siliqua* Jones. *Beyrichia tuberculata* Boll., also var. *gedanensis* n. sp., *B. kochii* Boll., *B. salteriana* Jones, *B. buchiana* Jones, *B. maccoyana* Jones, *B. wilckensiana* Jones, *B. klödeni* McCoy, *B. spinigera* Boll.

Trilobites: *Phacops downingiæ* Murch., *P. caudata* Brünn, *P. lævigata* Schm., *P. nieszkowskii* Schm., *P. odini* Eichw., also var. *itferensis* Schm., *P. conicophthalma* Sars. & Boeck., *P. bucculenta* Sjögren, *P. marginata* Schm., *P. maxima* Schm., *P. wesenbergensis* Schm., *P. eichwaldi* Schm., *P. macroura* Sjören. *Cheirurus variolaris* Linrs., *C. (Cyrtometopus) plautini* Schm., *C. exsul* Beyr., *C. conformis* Ang. *Encrinurus* cf. *obtusus* Ang., *E. punctatus* Brünn., *E.* cf. *seebachi* Schm. *Acidaspis. Lichas deflexa* Sjögren, *L. tricuspidata* Beyr., *L.* cf. *augusta* Beyr. *Calymene senaria* Conrad, *C. blumenbachii* Brong., *C. spectabilis* Ang. *Proetus* sp. *P. pulcher* Nieszk. *Ampyx rostratus* Sars., *A. setrostris* Angelin. *Homalonotus* cf. *rhinotropis* Ang. *Asaphus latisegmentatus* Nieszk., *A. tecticaudatus* Steinh., *A. weissii* Eichw. *Ptychopyge rimulosa* Ang. *Illænus linnarssoni* Holm., *I. chiron* Holm. *Proetus concinnus* Dalm.

―――― Ueber gotländische *Beyrichien*.
In Zeitschr. Deutsch. geol. Gesell, vol. 40, 1888, p. 1, 2 pls.
Beyrichia tuberculata Boll., var. *gotlandica* n. sp., *B. lindströmi* n. sp., also var. *expansa*, *B. buchiana* Jones, also var. *nutans*, *B. lauensis* n. sp., *B. klödeni* McCoy, also var. *protuberans* and *bicuspis*, *B. klödeni* var. *nodulosa* Boll., *B. tuberculata* Salter, also cf. var. *granulata* Jones, *B. maccoyana* Jones, *B. jonesii* Boll., also var. *clavata* Kolm.

Kinnear (W. T.) Note on the occurrence of a new Carboniferous Crustacean at Ardross Castle, Fife.
In Trans. Edingburgh Geol. Soc., vol. 5, pt. 3, 1887, p. 467.
Dithyrocaris glabra W. & E. *Palæocrangon sociale* Salt. *Ceratiocaris* n. sp., *Rostrocaris* n. g. (no description given).

Kinsky (Graf von). Schreiben des H. Grafen von Kinsky an von Born über einige mineralogische Merkwürdigkeiten.
In Abhandlungen einer Privat-Gesell. in Böhmen, vol. 1, Prag. 1775, p. 243.
The Comt. Kinsky described and figured under the term of *Entomolithus paradoxus* many fragments of *Conocephalites sulzeri* from Ginetz with those of *Paradoxides*, without distinguishing them from each other.

Kirkby (J. W.) On some Permian fossils from Durham.
In Ann. Mag. Nat. Hist., 3d series, vol. 2, 1858, pp. 317 and 434, plate.
Leperditia? *permiana* Jones. *Bairdia plebeia* Reuss, *B. mucronata?* Reuss, *B. plebeia* Reuss var. *elongata*, also var. *neptuni*,'*B. reussiana* n. sp., *B. ventricosa* n. sp., *B. schaurothiana* n. sp., *B. kingii* Reuss, *B. plebeia* var. *compressa* n. var., *B. berniciensis* n. sp., *B. rhomboidea* n. sp., *B. jonesiana* n. sp., *B. reniformis* n. sp., *B. truncata* n. sp.

―――― On Permian Entomostraca from the shell limestone of Durham. With notes on the species by T. Rupert Jones.
In Trans. Tyne Nat. Field Club, vol. 4, 1859, p. 122, 4 plates; author's edition, 51 pp., 4 plates.
Kirkbya n. g., *K. permiana* Jones. *Cythere* subgenus *Bairdia plebeia* Reuss, also variations *caudata*, *amygdalina*, *elongata*, *neptuni*, *reussiava* and *ventricosa*. *Bairdia schaurothiana* Kirkby, *B. kingii* Reuss, also var. *compressa*, *B.? beniciensis* Kirkby, *B. rhomboides* Kirkby. *Cythere (Cytherideis) jonesiana* Kirkby, *C. sub-reniformis* Kirkby, *C. amputata* Kirkby, *C. morrisiana* Jones, *C. elongata* Münst., *C. kutorginiana* Jones, *C. geintziana* Jones, *C. biplicata* Jones, *C. inornata* McCoy, *C. nuciformis* Jones, *C. (Bairdia) plebeia* Reuss, *C. (Bairdia) plebeia* var. *brevicauda*, *C. (Bairdia)*

curta Jones, *C. (Bairdia plebeia* var. *rhombica,* also var. *grandis, C. (Bairdia) ampla* Reuss, *C. (Bairdia) gracilis* McCoy, *C. (Bairdia) acuta, C. tyronica* n. sp., *C. richteriana* n. sp., *C. ampulata* Kirkby, *C. jonesiana* Kirkby.

———— On the Permian rocks of South Yorkshire and their Palæontological relations.
In Quart. Jour. Geol. Soc. London, vol. 17, 1861, p. 287.
Cythere (Bairdia) plebeia Reuss, *C. (Bairdia) ampla* Reuss, *C. cytherideis jonesiana* Kirkby. *Kirkbya permiana* Jones. *Cythere (Bairdia) schaurothiana* Kirkby.

———— On some additional species that are common to the Carboniferous and Permian strata.
In Annals Mag. Nat. Hist., 3d series, London, vol. 10, 1862, p. 202, plate iv.
Cythere (Bairdia) plebeia Reuss, *C. (Bairdia) schaurothiana* Kirkby.
See Jones (T. Rupert) and Kirkby (J. W.), also Jones (T. Rupert), Kirkby (J. W.) and Brady (M. P.)

Kjerulf (Theo.) Veiviser ved geologiske Excursioner i Christiania Omegn. Christiania, 1865.
The author herein gives a list of Norwegian Silurian fossils, with some notes and figures.
Ceratopyge forficula Sars. *Olenus (?) pusillus* Sars. *Cyrtometopus sarsi* Ang. *Illænus glaber. Phacops elegans* Boeck & Sars. *Illænus barriensis* Murch.

———— Om skuringsmærker, Glacialformationen Terrasser og Strandlinier samt on grundfjeldets og sparagmitfjeldets mægtighed i Norge. Christiania, 1873.
Paradoxides kjerulfi Linrs. *Arionellus* sp. ? *Liostracus aculeatus* Ang. *Agnostus* sp. ? *A. fallax* Ang., *A. planicauda* Ang., *A. exsculptus* Ang.

Klein (J. T.) Specimen descriptionis petrefactorum Gedaunensium order Oryctographia Gedannensis, etc.
In Nuremberg, 1770, folio.
Plate xv, figs. 3–4, *Concha trilobus.* *(Calymene blumenbachii* Brong. Fig. 5, *(Phacops latifrons* Bronn.) Figs. 6–7, (Doubtful species).

Kloden (K. F.) Die Versteinerungen des Mark Brandenburg, etc. Berlin, 1834, 10 pls.
Cytherina phaseolus His. *Calymene blumenbachii* Brong., also var. *(a) tuberculata* Dalm., var. *(b) pulchella* Dalm., var. *(c) tuberculosa, C. bellatula* Dalm., *C. sclerops* Dalm., *C. punctata* Brünn, *C. concinna* Dalm., *C'.*

polytoma? Dalm. *Asaphus extenuatus*? Dalm., *A. angustifrons* Dalm., *A. expansus* Linné, *A. auriculatus* Sternb., *A. dilatatus* Dalm., *A. (Illænus) crassicauda* Wahl., *A. (Ampyx) nasutus* Dalm., *A. caudatus* Brong. *Battus pisiformis, B. tuberculutus* n. sp., pl. 1, figs. 16-23. McCoy, "Brit. Pal. Rocks," p. 135, remarks on this species that the name *Beyrichia tuberculatus* is generally applied to fig. 22 of Klöden's plate, which is a common form in the Gothland beds. *Battus gigas* n. sp., pl. 2, fig. 1.

Koch (C.) Monographie der Homalonotus-Arten des rheinischen Unterdevon.

In Preuss. geol. Landesanstalt, Abhandl., Berlin, 1883, vol. 4, pt. 2.

Homalonotus armatus Burm., *H. subarmatus* n. sp., *H. ornatus* n. sp., *H. aculeatus* n. sp., *H. römeri* DeKon, *H. rhenanus* n. sp., *H. crassicauda* Sandb., *H. scabrosus* Koch, *H. obtusus* Sandb., *H. multicostatus* n. sp., *H. lævicauda* Quenst. *H. planus* Sandb.

Harz: *Homalonotus ahrendi* F. A. Römer, *H. punctatus* F. A. Römer, *H. gigas* F. A. Römer, *H. obtusus* Sandb., *H. minor* A. Römer, *H. barrandei* A. Römer, *H. latifrons* A. Römer, *H. schusteri* A. Römer, *H. latifrons* A. Römer, *H. granulosus* Trenkner, *H. crassicauda* Sandb.

III. Französisch-belgische Ardenen: *Homalonotus roemeri* DeKon.
IV. England: *Homalonotus elongatus* Salt., *H. champernownei* Woodw.. *H. goniopygaeus* Woodw.
V) Westliches Frankreich: *Homalonotus gervillei* Vern., *H. hausmanni* Rouault, *H. legraverendi* Rouault.
VI. Spanien: *Homalonotus pradoanus* Vern.
VII. Turkei: (Bosporus) *Homalonotus gervillei* Vern., *H. salteri* Vern.
VIII. Afrika, Cape Colonie: *Homalonotus herscheli* Murch., *H. crassicauda* Sandb.
IX. Nordamerika: *Homalonotus dekayi* Green, *H. vanuxemi* Hall.
X. Sudamerika: *Homalonotus oiara* H. & R.

—— Ueber das vorkommen der Gattung *Homalonotus* im rheinischen Unterdevon.

In Verhandl. Naturhist. Vereins für Rheinland und Westfalen, 1880.
Homalonotus roemeri DeKon., *H. scabrosus* n. sp., *H. mutabilis*.

Koenen (A. von). Die Kulm-Fauna von Herborn.

In Neues Jahrbuch für Mineral., 1879, p. 309, pls. 6, 7.
Phillipsia æqualis H. v. Mey., *P. latispinosa* Sandb.

The author places *Proetus? posthumus* Richter under *Phillipsia æqualis* and observes: "As von Meyer expressly says that the glabella is reduced in front there is no doubt that Burmeister was in error in illustrating as this species, a form with a club-shaped glabella:"

—— Ueber die Unterseite der Trilobiten.

In Neues Jahb. für Mineral., 1880, vol. l, p. 429, pl. 8, figs. 9, 10.

————— Ueber *Bronteus thysanopeltis* Barr. von Wildungen.
In Neues Jahb. für Mineral, 1882, p. 108.
In this paper the author refers a Middle Devonic species of *Bronteus* cited by Waldschmidt under *B. thysanopeltis* Barr. to *B. waldschmidti* n. sp.

————— Ueber Clymenienkalk und Mittel devon resp. Hercynkalk? bei Montpellier.
In Neues Jahrbuch für Mineral., 1886, p. 163.
Dechenella escoti n. sp. *Phacops cryptophthalmus* Emm.

Kolmodin (Lars). Bidrag till kännedomem on Severiges Siluriska Ostracoder.
In Akad. Afhandling som med tillständ af Tidtberömda Filosofiska Fakultetens i Upsala, etc., 1869, 1 pl.
Leperditia baltica His., *L. megalops* n. sp., *L. nitens* n. sp. *Beyrichia lunata* n. sp., *B. clavata* n. sp., *B. scanensis* n. sp., *B. verrucosa* n. sp., *B. bilobata* n. sp. *Cytheropsis bisulcata* n. sp., *C. concinna* Jones.

————— Ostracoda Silurica Gotlandiæ enumerat.
In Ofversigt kongl. Vet. Akad. Förhandlingar. 1879, No. 9, p. 133, pl. 19.
Leperditia schmidti Kolm., *L. baltica* His., ex parte *L. phaseolus* His., *L. grandis* Schm., *L. nitens* Kolm., *L. tuberculata* n. sp. *Elpe reniformis* n. sp. *Beyrichia tuberculata* Boll., *B. klödeni* McCoy, *B. klödeni* var. *antiquata* Jones, *B. jonesi* Boll., *B. buchiana* Jones, *B. maccoyana* Jones, *B. clavata* Kohm., *B. grandis* n. sp. *Cytheropsis concinna* Jones, *C. bisulcata* Kolm.

Koenig (C. E.) Icones fossilium sectiles.
Centuria prima, text, 4 pp., pls. 1–8; Centura seconda, pls. 9–19.; n. d.
The common edition of this work has only four pages of text and eight plates (London, 1825, 4to).
Calymene decipiens, p. 2, pl. 32 *(Ellipsocephalus hoffi.)* *Asaphus myops*, p. 3. pl. 4, fig. 53. *Homalonotus* n. g., *H. knighti*, p. 4, pl. 7, fig. 85. *Agnostus*, pl. 19, figs. 119, 120. *Isotelus gigas*, pl. 10, fig. 121. *Asaphus extenuatus*, pl. 10, fig. 122, *A. crassicauda*, pl. 13, fig. 150. *Belinurus* n. g., *B. bellulus*, pl. 18, fig. 230.
For a description of the genus *Belinurus* see W. H. Baily's article, Annals Mag. Nat. Hist., 3d series, London, vol. 11, 1863.

Koninck (L.-G. de). Mémoire sur les Crustacés fossiles de Belgique.
In Mém. Acad. Sci. Bruxelles, vol. 14, 1841, 1 pl.
Goldius n. g. *Cyclus* n. g. *Goldius flabellifer* Goldf. *Asaphus gemmuliferus* Phill., *A. brongniarti* Fischer. *Cyclus radialis* Phill. Famile des

Cyproides: *Cytherina phillipsiana* n. sp. *Cypridina edwardsiana* n. sp.
C. concentrica n. sp., *C. annulata* n. sp. *Cyprella chrysalidea* n. sp.
Cypridella cruciata n. sp.

—— Description des animaux fossils qui se trouvent dans le terrain carbonifère de Belgique. Liège, 1842–44.

Cythere phillipsiana DeKon. *Cypridina edwardsiana* DeKon, *C. concentrica* DeKon, *C. annulata* DeKon. *Cyprella chrysalidea* DeKon. *Cypridella cruciata* DeKon. *Cyclus brongniartianus* n. sp., *C. radialis* Phill. *Phillipsia brongniarti* Fisch, *P. globiceps* Phill., *P. derbyensis* Martin, *P. gemmulifera* Phill., *P. pustulata* Schloth., *P. jonesii* Portl.

—— Notice sur quelque fossiles recuellis par G. Dewalque dans le système Gédinien de A. Dumont.

In Annales Soc. Geol. Belgique, vol. 3, 1876, p. 25, plate.

—— Recherches sur les fossiles palæozoïques de la Nouvelle-Galles du Sud Australie.

In Mém. Soc. Sci. Liège, 2d series, vol. 6, 1876, p. 45.

—— Recherches sur les fossiles palæozoïques de la Nouvelle-Galles du Sud Australie.

In Mém. Sci. Liège, 2d series, vol. 7, 1878.
This work was published separately under the above title in two volumes, with an atlas of 24 plates. Bruxelles, 1876–77.

Entomis pelagica Barr. *Illænus wahlenbergi* Barr.? *Staurocephalus clarkei* n. sp. *Cheiruris insignis* Beyr. *Encrinurus punctatus* Brünn., *E. barrandei* n. sp. *Cromus bohemicus* ? Barr., *C. murchisoni* n. sp. *Calymene blumenbachii* Broug. *Proetus stokesii* Murch. *Lichas* sp.? cf. *L. palmata. Bronteus partschii* Barr., *B. goniopeltis* n. sp. *Harpes ungula* Sternb.
Fossiles Carbonifères: *Polycope simplex* J. & K. *Entomis jonesi* n. sp. *Phillipsia seminifera* Phill. *Griffithides eichwaldi* Fischer. *Brachymetopus strzeleckii* McCoy.

—— Sur une nouvelle espèce de crustacé du terrain houiller de la Belgique.

In Bull. Acad. Sci. Bruxelles, 2d series, vol. 45, 1878, p. 409, 1 pl.
Brachypyge carbonis n. sp.
The principal part of this article is a letter from Dr. Henry Woodward to L. G. de Koninck.

—— Notice sur le *Prestwichia rotundata* J. Prestwich découvert dans le schiste houiller de Horme, près Mons.

In Bull. Acad. Sci. Bruxelles, 3d series, vol. 1, 1881. p. 91, 1 pl.

Krotow (P.) Geologische Forschungen am Westlichen Uralabhange in dem Gebieten von Tscherdyn und Ssolikamsk. Zweite Lieferung.
In Mém. Comité géol. St. Petersburg, vol. 6, 1888, plate 2, p. 329.
Bronteus granulatus Goldf. *Leperditia* n. sp. *Isochilina* n. sp.

Krause (A.) Die Fauna der sogen. Beyrichien oder Choneten-Kalke des norddeutschen Diluviums.
In Zeitschr. Deutsch. geol. Gesell., vol. 29, Berlin, 1877, p. 1, 1 pl.
Calymene blumenbachii Brong. *Phacops downingiæ* Murch. *Proetus concinus* Dalm. *Leperditia angelini* Schm. *Beyrichia tuberculata* Boll., *B. buchiana* Jones, *B. kochii* Boll., *B. maccoyana* Jones, *B. salteriana* Jones, *B. wilckensiana* Jones. *Cytherellina siliqua* Jones. *Primitia oblonga* J. & H., *P. obsoleta* J. & H., *P. ovata* J. & H., *P. semicircularis* J. & H., *P. römeriana* J. & H., *P. beyrichiana* J. & H., *P. minuta* n. sp., *P. mundula* Jones.

—— Ueber *Beyrichien* und verwandte Ostracoden in untersilurischen Geschieben.
In Zeitschr. Deutsch. Gesell., vol. 41, 1889, plates 1 and 2.
Primitia simplex J. & H., *P. plana* n. sp., *P. sulcata* n. sp., *P. distans* n. sp., *P. cincta* n. sp., *P. jonesii* n. sp., *P. bursa* n. sp., *P. schmidtii* n. sp., *P. schmidtii* var. *P. intermedia* n. sp. *Entomis sigma* n. sp., *E. sigma* var. *Bollia, v-scripta* n. sp., *B. granulosa* n. sp. *Strepula lineata* n. sp., *S. linnarssoni* n. sp. *Beyrichia erratica* n. sp., *B. marchica* n. sp., *B. digitata* n. sp., *B. palmata* n. sp. *Klœdenia? globosa* n. sp.

—— Ueber *Beyrichien* und verwandte Schalenkrebse in märkischen Silurgeschieben.
In Gesell. naturforschender Freunde No. 1, 1889, p. 11.

—— Beitrag zur Kenntniss der Ostrakoden-Fauna in silurischen Diluvia geschieben.
In Zeitchr. Deutsch. geol. Gesell, vol, 43, 1891, p. 488, plates 29–33.
Leperditia eichwaldi Schm.. *L. hisingeri* Schm., *L. (Isochilina?)* aff. *conspersa* Kiesow. *Isochilina? erratica* n. sp. *Aparchites simplex* Jones, *A. ovatus* J. & H., *A. obsoletus* J. & H., *A. oblongus* J. & H. *Primitia jonesii* Krause, *P. maccoyii* J. & H., *P. elongata* n. sp., *P. mundula* Jones, *P. cristata* J. & H., *P. reticristata* Jones, *P. beyrichia* Jones, *P.? striata* n. sp. *Bollia semicircularis* n. sp., *B. rotundata* n. sp., *B.? sinuata* n. sp. *Strepula limbata* n. sp., *S. simplex* n. sp., *S.* cf. *costata* Linrs. *Beyrichia marchica* Krause, *B. erratica* Krause, var. *acuta* n. v., *B. digitata* Krause, *B. nodulosa* Boll., *B. spinigera* Boll., *B. damesii* n. sp., *B. scanensis* Kolm., *B. reuteri* n. sp., *B. steusloffi* n. sp., *B. hieroglyphica* n. sp. *Klœdenia*

kiesowi n. sp. *Octinaria elliptica* n. sp. *Thlipsura tetragona* n. sp., *T. simplex* n. sp., *T. personata* n. sp. *Entomis sigma* Krause, var. *ornata* n. v. *Bythocypris semicircularis* J. & H., *B. cornuta* n. sp., *B. phillipsiana* J. & H., *B. hollii* Jones, *B. symmetrica* Jones, *B.* aff. *reniformis* Jones. *Pontocypris mawii* Jones. *Xestoleberis ?* aff. *wrightii* Jones. *Bursulella ? rostrata* n. sp.

——— Die Ostrakoden der silurischen Diluvialgeschiebe. Berlin, 1891.

Leperditia baltica His., *L. phaseolus* His., also var. *(a) ornata* Eichw. and var. *(b) subpentagona* Kiesow, *L. gregaria* Kiesow, *L. eichwaldi* Schm., *L. hisingeri* Schm., *L. conspersa* Kiesow, *L. gigantea* F. Römer. *Isochilina? erratica* n. sp. *Aparchites simplex* Jones, *A. ovata* J. & H., *A.? obsoleta* J. & H., *A.? oblonga* J. & H. *Primita plana* Krause, *P. sulcana* Krause, *P. distans* Krause, *P. cincta* Krause, *P. jonesii* Krause, *P. bursa* Krause, *P. schmidti* Krause, *P. intermedia* Krause, *P. maccoyi* J. & H., *P. elongata* n. sp., *P. mundula* Jones, *P. cristata* J. & H., *P. reticristata* Jones, *P. beyrichiana* J. & H., *P. striata* n. sp. *Bollia v-scripta* Krause, *B. granulosa* Krause, *B. semicircularis* n. sp., *B. roundata* n. sp., *B.? sinuata* n.sp., *Strepula lineata* Krause, *S. 'simplex* n. sp., *S. linnarssoni* Krause, *S.* aff. *costata* Linrs. *Beyrichia marchica* Krause, *B. erratica* Krause, *B. digitata* Krause, *B. palmata* Krause, *B. protuberans* Boll., *B. bolliana umbonata* Reuter, *B. gotlandica* Kiesow, *B. nodulosa* Boll., *B. spinigera* Boll., *B. damesii* n. sp., *B. buchiana* Jones, also var. *lata* Reuter, *B. tuberculata* Klöden, also var. *nuda* Jones, var. *antiquata* Jones, var. *gibbosa* Reuter, var. *bigibbosa* Reuter, var. *tuberculato-buchiana* Reuter, var. *tuberculato-kochiana* Reuter, *B. bronni* Reuter, *B. gedanensis* Kiesow, also var. *pustulosa* Hall, *B. baueri* Reuter, *B. jonesii* Boll., also var. *clavata* Kolm., *B. maccoyiana* Jones, *B. kochii* Boll., *B. scanensis* Kolm., *B. reuteri* n. sp., *B. steuslofti* n. sp.? *B.? hieroglyphica* n. sp., *B.? primitiva* Verwon. *Klœdenia ? globosa* Krause, *K. kiesowi* n. sp., *K. wilckensiana* Jones, also var. *plicata* Jones. *Octonaria elliptica* n. sp. *Thlipsura tetragona* n. sp., *T. simolex* n. sp., *T. personata* n. sp., *T. v-scripta*, var. *discreta* Jones. *Entomis sigma* Krause. *Cytherellina siliqua* Jones. *Echmina bovina* Jones. *Bythocypris semicircularis* J. & H., *B. phillipsiana* J. & H., *B. hollii* Jones, *B. symmetrica* Jones, *B.* aff. *reniformis* Jones. *Pontocypris mawii* Jones, var. *proxima* Jones. *Xestoleberis ?* aff. *wrightii* Jones. *Bursulella ? rostrata* n. sp.

——— Neue Ostrakoden aus märkischen Silurgeschieben.

In Zeitschr. Deutsch. geol. Gesell. Jahrg. 1892, p. 383, plates 21-22.

Isochilina canaliculata n. sp. *Primitia plana* var. *tuberculata*, n. var., *P. distans* Krause, *P. elongata* Krause, *P. corrugata* n. sp., *P. plicata* n. sp., *P. (Halliella) seminulum* Jones, *P.* aff. *obliquipunctata* Jones. *P. apil lata* n. sp. *Entomis obliqua* n. sp., *E. (Primitia) flabellifera* n. sp. *Primitia excavata* n. sp., *P. (Ulrichia?) umbonata* n. sp., *P. (Ctenobolbina?) globifera* n. sp., *P. labrosa* n. sp. *Entomis simplex* n. sp., *E. auricularis* n. sp., *E. plicata* n. sp., *E. trilobata* n. sp., *E. (Bursulella ?) quadrispina* n. sp. *Bollia minor* n. sp., *B. major* n. sp., *B. duplex* n. sp. *Beyrichia dissecta* n. sp., *B. mammillosa* n. sp., *B. radians* n. sp., *B. plicatula* n. sp., *B. (Tetradella) harpa* n. sp., *B. (Tetradella) carinata* n. sp., *B. (Tetradella) signata* n. sp., *B. (Ctenobolbina) rostrata* n. sp., *B. (Ulrichia?) bidens* n. sp. *Octonaria bifasciata* n. sp. *Thlipsura v-scripta* var. *discreta* Jones. *Æchmina bovina* Jones, var. *punctata* n. var. *Crustaceum* sp. ?

Kuntgen (K.) Die Trilobiten.
In Mus. Luxemburg, 1877.*

Kutorga (S.) Beiträge zur Kenntniss der organischen Ueberreste des Kupfersandst. am. westl. Abhange des Urals. St. Petersburg, 1837, p. 22, pl. 4, fig. 1-3.
Limulus oculatus Kutorga.
Édouard d'Eichwald ("Lethæa Rossica," vol. 1, p. 1360), uses this species for a type of his new genus, *Compylocephalus*.
For similar fossils Page (Advance Text Book of Geology, 1856), used the generic term of *Stylonurus*.

———— Ueber einige baltische silurische Trilobiten.
In Verhandl. russ. k. mineral. Gesell. zu St. Petersburg, Jahr., 1847, p. 287, plate. St. Petersburg, 1848.
Illænus tauricornus n. sp. *Asaphus expansus (hypostoma)*. *Encrinurus punctatus* Brünn.
The author also illustrates the eyes of *Illænus tauricornis, I. crassicauda, Proetus concinnus, Cheirurus, Phacops, Asaphus expansus*.

———— Einige *Sphærexochus* und *Cheirurus* aus den silurischen Kalksteinschichten des Gov. St. Petersburg.
In Verhandl. russ. k. mineral. Gesell. zu St. Petersburg, Jahr., 1854, p. 105, 3 pls.
Cheirurus zembnitzkii (Calymene zembnitzkii Eichw.), *C. macrophthalmus* n. sp. *(C. exsul* Beyr. (Schm.) *Sphærexochus cranium* n. sp., *S. hemicranium* n. sp., *S. platycranium* n. sp. *(Cheirurus (Pseudosphærexochus) platycranium* Kut. (Schm.), *S. enurus* n. sp.

Lapworth (Charles). On the discovery of the *Olenellus* fauna in the Lower Cambrian rocks of Britain.
In Nature, vol. 39, 1888, p. 212.
Olenellus callavei n. sp.

—— The *Olenellus* zone in North West Europe.
In Geological Mag., new series, decade 3, vol. 6, 1889, p. 190.

—— On *Olenellus callavei* and its Geological relations.
In Geological Mag., December number, 1890.

Laspeyres (H.) Das fossile Phyllopoden-Genus *Leaia*.
In Zeitschr. Deutsch. geol. Gesell., Berlin, 1870, p. 733, 1 pl.
A. Subrectangular type: *Leaia leidyi* Jones.
B. Suboval types: *Leaia salteriana* Jones, *L. baentschiana* Beyr., *L. wettinensis* n. sp.

Latreille (P. A.) Affinités des Trilobites.
In Mém. Mus. Hist. Nat., Bruxelles, vol. 7, 1821, p. 22; Annals Sci. Phys., Bruxelles, vol. 6.
The author endeavors to prove by the absence of feet and other characteristics, that the Trilobites are nearly related to the genus Chiton.

La Touche (J. D.) Guide Book to the Geology of Shropshire. London, 1884.*
Ceratiocaris pardoensis u sp.

Lawrow (N.) Zwei Neue Asaphus-Arten aus dem silurischen Kalksteine.
In Verhandl. russ. k. mineral. Gesell. zu St. Petersburg, 1855-56, p. 237, pls. 4, 5. St. Petersburg, 1856.
Asaphus delphinus n. sp., *A. kowalewskii* n. sp.

—— *Ptychopyge* und *Megalaspis*, Trilobiten der unter-silurischen Kalksteine.
In Verhandl. russ. k. mineral. zu St. Petersburg, 1857-58, p. 146, plate. St. Petersburg, 1858.
The author gives typical illustrations of these genera.

Lebesconte (M. P.) Constitution générale du massif breton comparée à celle du Finistère.
In Bull. Soc. Géol. France, 3d series, vol. 14, 1886, p. 776, pls. 34-36.
Homalonotus barroisi n. sp., *H. heberti* n. sp. *Ogygites armoriana* Trom. and Lebesc.
See also Tromelin (Gaston de) and Lebesconte (M. P.)

Lehmann (J. G.) . Versuch einer Geschichte von Flötzgebirgen betreffend deren Entstehung, Lage, darin befindlichen Metalle, Mineralien und Fossilien. Berlin, 1756. (*)

———— De Entrochis et Asteriis. (*)
In Novi Comm. Sci. Petropol., vol. 10, 1764, p. 429, pl. 12, figs. 8-10.

Lesley (P. J.) A Dictionary of the Fossils of Pennsylvania and neighboring States named in the Reports and Catalogues of the Survey. Harrisburg, 1889-90.
This work contains brief descriptions of the Palæozoic Crustacea, with woodcut illustrations.
The work gives an excellent figure of *Olenoides curticei* Walcott.

Leuchtenberg (M. H. von). Beschreibung einiger neuer Thierreste der Urwelt von Zarskoje-Selo. St. Petersburg, 1843, 2 pls.
Asaphus centron n. sp., *A. longicauda* n. sp., *A. hyorrhinus* n. sp., *A. buchii* Brong. *Metopias huebneri* Eichw., *M. verrucosus* Eichw., *M. coniceps* n. sp., *M. aries* Eichw. *Nileus nanus* n. sp.

Lhwyd (Edward). Part of a letter from Mr. Edward Lhwyd to Doctor Martin Lister.
In Philos. Trans. Royal Soc. London, vol. 20, No. 243, 1698, p. 299, plate, figs. 8, 9 and 10.
This paper contains the earliest account of Trilobites. The author discovered two fragments and one entire specimen of *Ogygia buchii* near Llandeilo, in Carmarthenshire. Lhwyd says in his letter that he did not know what to make of these fossils. The *Ogygia* (fig. 10) he refers to the skeleton of an unknown fish. Fig. 8 represents the head of a species of the genus *Trinucleus*.

———— Lithophylacii Britannici Ichnographia. Londini, 1699, 23 pls.; 2d Edition, Oxonii, 1760.
This work gives a catalogue of English fossils contained in the Ashmolean Museum, and also several essays on fossils. In the author's article "Epistola I. de lapidibus quibusdamé Germania acceptis," pp. 95-100, he alludes to an *Ogygia*, which he names *Buglossam curtam strigosam*, plate 22, fig. 2. He also gives a description and figure of *Trinucleus fimbriatum vulgare*, plates 22 and 23, on which R. I. Murchison afterwards founded the genus *Trinucleus*.
The second edition contains figures of *Entomolithus paradoxus*, the *Dudley fossil* or *Dudley locust*, p. 101, fig. a.

Lindaker (T. J.) Beschreibung einer noch nicht bekannten Käfermuschel. (*)
In Mayer's Sammlung phys. Aufsätze, Dresden, vol. 1, 1791, p. 37, pl. 1.
The author describes under the name of "gegitterte Käfermuschel" a *Trinucleus* from Prague.

Lindaker's description can easily be recognized as that of a *Trinucleus*, approaching *T. ornatus* Sternb. He indicates the limb with the genial angles extending into spines, also the three protuberances of the head, and the trilobation of the body, with its divisions into segments, The author also states that the fossil is often found extended or rolled into a ball.

Lindstrom (G.) Normina fossilium Siluriensium Gotlandiæ, Visby, 1867.

This pamphlet contains a catalogue of the Silurian fossils of Gothland.

—— List of fossils of the Upper Silurian formation of Gothland. Stockholm, 1885.

—— Förteckning pa Götland Siluriska Crustacéer.

In Ofversigt. Kongl. Vet. Akad. Förhandlingar, 1885. No. 6, p. 37, pls. 12-16.

Phacops vulgaris Salt., *P. imbricatula* Ang., *P. obtusa* n. sp., *P. downingiæ* Murch., *P. quadrilineata* Ang., *P. musheni* Salt. *Cheirurus speciosus* His., *C. conformis* Aug., *C. bimucronatus* Murch., *C. gotlandicus* n. sp. *Sphærexochus scabridus* Ang., *S. latifrons* Ang., *S. laciniatus* n. sp., *S. beyrichi* n. sp. *Youngia* n. g., *Y. trispinosa* E. & N., *Y. globiceps* n. sp., *Y. inermis* n. sp. *Deiphon forbesi* Barr. *Encrinurus punctatus* Brünn, *E. lævis* Aug., *E. obtusus* Aug. *Acidaspis crenata* Emm., *A. barrandei* Aug., *A. marklini* Ang., *A. pectinata* Ang., *A. (Trapelocera) bicuspis* Ang. *Lichas concinnus* Aug., *L. latifrons* Ang., *L. palifer* n. sp., *L. araneus* n. sp., *L. ornatus* Ang., *L. marginatus* n. sp., *L. visbyensis* n. sp., *L. plicatus* n. sp., *L. triquetrus* n. sp., *L. rotundifrons* Ang., *L. gotlandicus* Ang. *Trochurus salteri* Fletcher, *T. pusillus* Aug. *Harpes acuminatus* n. sp. *Calymene tuberculata* Brünn, *C. tuberculosa* Dalm., *C. spectabilis* Ang., *C. lævis* n. sp., *C. frontosa* n. sp., *C. intermedia* n. sp., *C. excavata* n. sp., *C. papillata* n. sp. *Homalonotus knighti* König. *Phaëtonides stokesi* Murch., *P. rugolosus* n. sp., *P. longifrons* n. sp. *Cyphaspis elegantula* Ang., *C. punctillosa* n. sp. *Proetus concinnus* Dalm., *P. obconicus* n. sp., *P. distans* n. sp., *P. acutus* n. sp., *P. conspersus* Ang., *P. signatus* n. sp., *P. granulatus* n. sp., *P. verrucosus* n. sp. *Illænus barriensis* Murch., *I. sulcatus* n. sp., *I. holmi* n. sp. *Bronteus platyactin* Aug., *B. marklini* Ang., *B. polyactin* Ang., *B. irradians* n. sp., *B. umbonatus* n. sp., *B. crebristratus* n. sp. *Eurypterus fischeri* Eichw. *Pterygotus osiliensis* Schm.

Linnarsson (J. G. O.) Om de Siluriska bildningarne i Mellersta Westergötland, 1.

In Akad. Afhandling som med tillstond af Vidtberömda Filosofiska Fakulteten i Upsala, etc., 1866, 2 pls.

The Trilobites are grouped as follows: Phacopidæ: *Phacops*. Cheiruridæ: *Pliomera*, *Cyrtometopus*, *Sphærexochus*. Encrinuridæ: *Cybele*.

Acidaspidæ: *Acidaspis.* Lichidæ: *Lichas.* Harpedidæ: *Arraphus.* Calymenidæ: *Calymene, Homalonotus;* Aulacopleuridæ: *Aulacopleura.* Liostracidæ: *Liostracus.* Olenidæ: *Paradoxides, Olenus, Peltura, Parabolina, Eurycare, Sphærophthalmus.* Remopleuridæ: *Remopleurides.* Proetidæ: *Forbesia.* Asaphidæ: *Asaphus, Megalaspis, Ptychopyge, Ogygia?, Niobe.* Illænidæ: *Illænus, Rhodope, Nileus, Symphysurus, Æglina.* Trinucleidæ: *Trinucleus, Ampyx, Lonchodomas, Raphiophorus, Dionide.* Agnostidæ: *Agnostus.* Incertæ sedis: *Holometopus.*
The author also gives descriptions and figures of the following species: *Phacops pulchellus* n. sp., *Cyrtometopus latilobus* n. sp., *Sphærexochus laticeps* n. sp., *Lichas segmentatus* n. sp., *Ogygia ? apiculata* n. sp., *Illænus limbatus* n. sp., *Agnostus affinis* n. sp.

—— Diagnoses specierum novarum e classe Crustaceorum in depositis Cambricis et Siluricis Vestrogotiæ Sueciæ repertarum.

In Ofversight. Kongl. Svenska Vet.-Akad. Förhandl., 1869, No. 1, p. 191.

Phacops recurvus n. sp. *(Ogygia apiculata* Linrs.) *Cheirurus subulatus* n. sp., *C. variolaris* n. sp. *Cybele aspera* n. sp., *C. loveni* n. sp. *Dindymene ornata* n. sp. *Acidaspis furcata* n. sp. *Lichas validus* n. sp. *Remmopleurides dubius* h. sp. *Triarthrus angelini* n. sp. *Symphysurus socialis* n. sp. *Niobe obsoleta* n. sp., *N. insignis* n. sp. *Ogygia? concentrica* n. sp. *Panderia megalophthalma* n. sp. *Trinucleus latilimbus* n. sp. *Agnostus gibbus* n. sp., *A. fallax* n, sp., *A. parvifrons* n. sp., *A. sidenbladhi* n. sp. *Trilobites ænigma* n. sp. *Leperditia (Isochilina) primordialis* n. sp. *Beyrichia costata* n. sp. *Primitia tenera* n. sp.

The new species are illustrated in Om Vestergötlands Cambriska och Siluriska Aflagringar, 1869, plates 1 and 2.

—— Om vestergötlands Cambriska och Siluriska Aflagringar.

In Kongl. Svenska Vet.-Akad. Handl., vol. 8, No. 2, 1869, 2 pls.

Phacops macrourus Siögren, *P. conicophthalmus* S. & B., *P. sclerops* Dalm., *P. recurvus* n. sp., *P. granulosus* Ang., *P. mucronatus* Brong., *P. pulchellus* Linrs. *Cheirurus subulatus* n. sp., *C.* sp.?, *C. variolaris* n. sp., *C. (Cyrtometopus) clavifrons* Dalm., *C. (Cyrtometopus) latilobus* Linrs., *C. (Cyrtometopus) octacanthus* Ang., *C. (Cyrtometopus) foveolatus* Ang., *C. (Cyrtometopus) longispinus* Ang., *C. (Cyrtometopus?) decacanthus* Ang.

Sphærexochus laticeps Linrs. *Staurocephalus clavifrons* Ang., *S. (Sphærocoryphe) dentatus* Ang., *S. (Sphærocoryphe) granulatus* Ang. *Pliomera fischeri* Eichw., *P. mathesii* Ang., *P. primigena* Ang. *Encrinurus* sp.? *Cybele verrucosa* Dalm., *C. aspera* n. sp., *C. loveni* n. sp. *Dindymene ornata* n. sp. *Acidaspis centrina* Dalm., *A.* sp.?, *A. cornuta* Beyr., *A. furcata* n. sp. *Lichas laciniatus* Wahl., *L. polytomus* Ang., *L. laxatus*

McCoy, *L. validus* n. sp. *Harpes (Arraphus) corniculatus* Aug. *Harpides rugosus* S. & B. *Remopleurides radians* Barr., *R. dorsospinifer* Portl., *R. sexlineatus* Aug., *R. dubius* n. sp. *Paradoxides tessini* Brong., *P. forchhammeri* Aug.? *Olenus (Parabolina) spinulosus* Wahl., *O. gibbosus* Wahl., *O. (Eurycare) latus* Boeck, *O. (Sphærophthalmus) alatus* Boeck, *O.* sp.?, *O. (Peltura) scrabæoides* Wahl. *Triarthrus becki* Green, *T. angelini* n. sp. *Dikelocephalus angusticauda* Aug., *D. dicræurus* Aug. *Ceratopyge forficula* Sars. *Liostracus costatus* Aug., *L. aculeatus* Aug. *Arionellus (Anomocare) difformis* Ang. *Anomocare* sp.? *Conocoryphe* sp.?, *C. (Solenopleura) stenometopa* Aug. *Euloma ornatum* Aug. *Calymene tuberculata* Brünn., *C.* sp.? *Homalonotus platynotus* Dalm. *Proetus (Forbesia) brevifrons* Ang., *P.* sp.? *Phillipsia parabola* Barr. *Megalaspis heros* Dalm., *M. rotundata* Aug., *M. explanata* Aug., *M. limbata* Boeck., *M. planilimbata* Ang., *M. planilimbata* aff. *Asaphus (Ptychopyge) applanatus* Ang., *A. (Ptychopyge) glabratus* Ang., *A. acuminatus* Boeck, *A. platyurus* Ang., *A. lævigatus* Aug. *Nileus armadillo* Dalm., *N. lineatus* Aug. *Symphysurus palpebrosus* Dalm., *S. breviceps* Ang., *S. socialis* n. sp. *Niobe obsoleta* n. sp., *N. insignis* n. sp., *N. emarginula* Ang., *N. lata* Aug. *Ogygia* sp.? *O. concentrica* n. sp. *Stygina latifrons* Portl. *Illænus crassicauda* Wahl., *I. limbatus* Linrs., *I.* sp.? *Panderia megalophthalma* n. sp., *P. (Rhodope) lata* Aug., *Trinucleus wahlenbergi* Rouault, *T. ceriodes* Aug., *T. carinatus* Ang., *T. seticornis* His., *T. latilimbus* n. sp. *Ampyx nasutus* Dalm., *A. costatus* Boeck, *A. (Lonchodomas) rostratus* Sars., *A. (Lonchodomas) carinatus* Aug., *A. (Lonchodomas) domatus* Ang., *A. (Lonchodomas) tetragonus* Aug., *A. (Raphiophorus) tumidus* Aug., *A. (Raphiophorus) culminatus* Ang., *A. (Raphiophorus)* sp.? *Dionide euglypta* Aug. *Agnostus pisiformis* Linné, *A. gibbus* n. sp.? *A. fallax* n. sp., *A. parvifrons* n. sp., *A. lævigatus* Dalm., *A. sidenbladhi* n. sp., *A. trinodus* Salt. *Ægilina oblongata* Ang. *Holometopus aciculatus* Ang., *H. ornatus* Ang.? *H. elatifrons* Ang. *Trilobites ænigma* n. sp. *Leperditia primordialis* n. sp., *L.* sp.? *Beyrichia costata* n. sp. *Primitia strangulata* Salt., *P. tenera* n. sp.

——— Om nogra försteningar fron Sveriges och Norges "Primordialzon."

In Ofversigt Kongl. Svenska Vet.-Akad. Förhandl., 1871, No. 6.

Paradoxides kjerulfi n. sp., pl. 16, figs. 1–3.

——— Ofversigt af Nerikes öfvergongsbildningar.

In Ofversigt Kongl. Svenska Vet.-Akad. Förhandl., 1875, No. 5, pl. 5; Sveriges Geologiska Undersökning Ser. C, No. 21, 1875.

Leptoplastus n. g. *Paradoxides tessini* Brong., *P. forchhammeri* Ang., *P.* sp.? *Ellipsocephalus muticus* Aug. *Olenus gibbosus* Wahl. *Parabolina spinulosa* Wahl. *Leptoplastus stenotus* Ang. *Sphærophthalmus alatus* Boeck. *Peltura scarabæoides*. *Liostracus costatus* Aug., *L. aculeatus* Ang. *Megalaspis planilimbata* Ang., *M. extenuata* Wahl., *M.* sp.? *Asaphus expansus*

Linné, A. raniceps Dalm.   Nileus armadillo Dalm.   Symphysurus palpebrosus Dalm., S. breviceps Ang., S. socialis Linrs.   Niobe frontalis Dalm., N. læviceps.   Ptychopyge angustifrons Dalm.   Illænus crassicauda Wahl. Dysplanus centrotus Dalm.   Ampyx nasutus Dalm.   Agnostus pisiformis Linné, A. gibbus Linrs., A. fallax Linrs., A. lævigatus Dalm., A. recticulatus Ang. Leperditia primordialis Linrs.   Beyrichia angelini Barr.

The author illustrates *Harpes excavatus* n. sp., *Ellipsocephalus muticus* Ang., *Leptoplastus stenotus* Aug. and *Beyrichia angelini* Barr.

────── On the Brachiopoda of the Paradoxides beds of Sweden.
In Kongl. Svenska Vetenskaps Akad. Band 3, No. 12, 1876.

Linnarsson divides the Swedish Paradoxides beds into six divisions as follows:*

6 Strata with *Agnostus lævigatus*
5 Strata with *Paradoxides forchhammeri*.
4 Strata with *Paradoxides ölandicus*.
3 Strata with *Paradoxides davidis*.
2 Strata with *Paradoxides tessini*.
1 Strata with *Paradoxides kjerulfi* (Olenellus zone).

The strata containing *Agnostus lævigatus* is wanting in the Island of Öland, and the beds containing *Paradoxides forchhammeri* are immediately overlain by the Olenus beds.

────── Trilobiter fron Vestergötlands Andrarum kalk.
In Geol. Föreningens Stockholm Förhandl., vol. 2, 1875, p. 491.

Paradoxides forchhammeri Ang. *Arionellus difformis* Aug., *A. aculeatus* Ang. *Liostracus microphthalmus* Ang. *Conocoryphe (Solenopleura) brachymetopa* Ang. *Dolichometopus specicus* Ang. *Agnostus* sp.? *Trilobites ænigma* Linrs.

────── En egendomlig Trilobitfauna fron Jemtland.
In Geol. Föreningens Stockholm Förhandl., vol. 2, No. 12, 1875, p. 491, pl. 22, figs. 1–5.

Dicellocephalus billingsi n. sp.   *Triarthrus jemtlandicus* n. sp.   *Remopleurides microphthalmus* n. sp.   *Bohemilla (?) denticulata* n. sp.

────── Tvo nya Trilobiter fron Skones alumskiffer.
In Geol. Föreningens Stockholm Förhandl., vol. 2, No. 12, 1875, p. 498.
Liostracus (?) superstes n. sp.   *Cyclognathus* n. g., *C. micropygus* n. sp.

────── Om faunan i lagren med *Paradoxides ölandicus*.
In Sveriges Geologiska Undersökning, ser. C, No. 22, 1877, 2 pls.

Paradoxides ölandicus Sjögren, *P. sjogreni* n. sp., *P. aculeatus* n. sp.? *Ellipsocephalus polytomus* n. sp.?   *Conocoryphe emarginata* n. sp.   *Solenopleura cristata* n. sp.   *Agnostus fallax* Linrs., *A. regius* Sjögr., *A. gibbus* Linrs.

―――― On the Trilobites of the Shineton shales.
In Geol. Mag., new series, decade 2, vol. 5, 1878, p. 188.
The author makes the following remarks on Callaway's paper, Quart. Jour. Geol. Soc., vol. 33. p. 652: "I do not think that *Conocoryphe monile* is nearly related to such species as *C. striata*. By the strongly lobed glabella and the dotted marginal furrow it approaches to *Euloma*."
Conophrys can hardly be separated from the *Shumardia* of Billings.
As to *Lichapyge*, it cannot have any affinity to *Paradoxides* and hardly to *Lichas*, it is most nearly allied to *Remopleurides;* the subgenus *Platypeltis* to *Niobe*.

―――― Om faunan i kalken med *Conocoryphe exsulans*.
In Sveriges Geologiska Undersökning, series C, No. 35, 1879, 3 pls.
Paradoxides tessini Brong., *P. hicksii* Salt. var. *palpebrosus*. *Liostracus aculeatus* Ang., *L. linnarssoni* n. sp. *Solenopleura parva* n. sp. *Conocoryphe exsulans* n. sp., *C. tenuicincta* n. sp., *C. dalmani* Ang., *C. impressa* n. sp. *Agnostus gibbus* Linrs., *A. fallax* Linrs., *A. fissus* Lundgren MS.

―――― Om försteningarne i de Svenska lagren med *Peltura* och *Sphærophthalmus*.
In Geol. Föreningens Stockholm Förhandl., vol. 5, 1880, No. 4; Sveriges Geologiska Undersökning, Ser. C, No. 43, 1880.
Ctenopyge n. g. *Peltura scarabæoides* Wahl. *Sphærophthalmus alatus* Boeck., *S. majusculus* n. sp., *S. flagellifer* Ang. *Ctenopyge pecten* Salt., *C. concava* n. sp., *C. teretifrons* Ang., *C. bisulcata* Phill., *C.* sp.? *Agnostus trisectus* Salt.

―――― Promemoria lemnad af A. G. Nathorst för resa po Öland.
In Geol. Föreningens Stockholm Förhandl., vol. 5, No. 13, 1881.

―――― De undre Paradoxides lagren vid Andrarum.
In Sveriges Geologiska Undersökning, Ser. C, No. 54, 1883, 4 pls.
Paradoxides tessini Brong.,, *P. davidis* Salt., *P. hicksii* Salt., *P. brachychachis* n. sp., *P.* sp.? *Olenellus kjerulfi* Linrs. *Ellipsocephalus nordenskiöldi* n. sp. *Arionellus primævus* Brögger. *Liostracus aculeatus* Ang., *L. linnarssoni* Brögger. *Solenopleura parva* Linrs. *Conocoryphe exsulan* Linrs., *C. tenuicincta* Linrs., *C. æqualis* n. sp., *C. dalmani* Ang., *C. impressa* Linrs. *Harpides breviceps* Ang. *Microdiscus scanicus* n. sp., *M. eucentrus* n. sp. *Agnostus atavus* Tullb., *A. gibbus* Linrs., *A. fissus* Lundg. MS., *A. intermedius* Tullb., *A. punctuosus* Ang., *A. incertus* Brögger, *A. elegans* Tullb., *A. lundgreni* Tullb., *A. cicer* Tullb., *A. nudus* Beyr. var. *scanicus* Tullb., *A. fallax* Linrs., *A. parcifrons* Linrs., *A. pusillus* Tullb., *A. rex* Barr.

Linné (C.) Olandska och Gotländska resa. Stockholm och Upsal, 1745.

On p. 147, under the name of *Entomolithus paradoxus*, there is in this work a rough woodcut of the pygidium of *Asaphus expansus* Linné.

—— Wästgöta resa förratad or 1746. Stockholm, 1747.

On p. 88 of this work there are some rough woodcuts of Trilobites, probably an *Olenus* and the head of a *Paradoxides*.

—— Museum Tessinianum. Holmiæ, 1753, p. 123, pl. 3.

Entomolithus, pl. 3, fig. 2 *(Calymene tuberculatus)*; E. paradoxus, pl. 3, fig. 1 *(Paradoxides tessini.)*
An excellent figure of *Paradoxides tessini* Linné is given by Dr. Lindström in his revised edition of Angelin's Pal. Scand., plate 1.
The *Paradoxides tessini* figured by Wahlenberg is therein referred to *P. tessini* var. *wahlenbergii*.

—— Skanska resa. Stockholm, 1757, p. 121.

The author remarks that the slates at Andrarum contain two species so common and numerous that the whole slate seems to be composed of them.
He cites Bromell's figures: *(Peltura scarabæoides, Olenus truncatus* and *Agnostus pisiformis* of other authors).

—— Petrificatet Entomolithus paradoxus beskrifed.

In Acta Regiæ Acad. Sci. Holmiens, vol. 21, 1759, p. 19, pl. 1, figs. 1–4.
Entomolithus paradoxus, pl. 1, fig. 1 *(Olenus spinulosus)*; Entom. paradoxus β cantharidum, pl. 1, fig. 4, head *(Olenus gibbosus)*; Entom. No. 3, pl. 1, fig. 3 *(Calymene blumenbachii;* Entom. No. 2, pl. 1, fig. 2 *(Encrinurus punctatus).*

—— Systema naturæ. Ed. 12. Vol. 3. Holmiæ, 1768, p. 161.

Entomolithus paradoxus α expansus *(Asaphus expansus* Linné), E. paradoxus β cantharidum *(Olenus gibbosus* Wahl.), E. paradoxus γ pisiformis *(Agnostus pisiformis* Linné).

Locke (John). On *Isotelus maximus*, found near Trebers, in Adams County, Ohio.

In Second Ann. Rept. Geol. Survey Ohio (W. W. Mather), 1838, p. 247, fig. 8.

—— On a new species of Trilobite of very large size.

In Trans. Assoc. Am. Geol. and Naturalists, vol. 1, 1843, p. 221, plate.
Isotelus megistos n. sp.
Dr. John Locke has herein given a new name to the species which he called *Isotelus maximus* in the Second Ohio Report.
See, also, Am. Jour. Sci., 1st series, vol. 42, 1843, p. 366, 1 pl.

――― Notice of a new Trilobite.
In Am. Jour. Sci., 1st series, vol. 44, 1843, p. 346.
Ceraurus crosotus n. sp.

――― Supplementary notice of the *Ceraurus crosotus* Locke.
In Am. Jour. Sci., 1st series, vol. 45, 1843, p. 223.

Logan (W. E.) *Bronteus ? canadensis.*
In Rep. Geol. Survey Canada, 1844, p. 54, 2 figs.

――― On the tracks of an animal lately found in the Potsdam sandstone.
In Canadian Naturalist, vol. 5, 1860, p. 279.

Lossen (K. A.) Ueber *Cryphæus rotundifrons* aus dem zoryer Schiefer des südlichen Unterharzes.
In Zeitschr. Deutsch. Gesell., vol. 31, 1879, p. 215.

Lovén (S. L.) Svenska Trilobiter.
In Ofversigt Kongl. Svenska Vet.-Akad. Förhandl., 1845, p. 62.
Calymene clarifrons Dalm., *C. ornata* Dalm.

――― Svenska Trilobiter.
In Ofversigt Kongl. Svenska Vet.-Akad. Förhandl., 1845, pp. 46 and 104, pls. 1, 2.
Cybele n. g. *Ceraurus crenatus* Emm. *Proetus (Calymene) concinnus* Dalm., *P. (Asaphus) stokesii* Murch., *P. (Calymene) elegantulus* Ang. *Trilobites (Calymene) verrucosus* Dalm. *Metopias (Ampyx ?) pachyrrhinus* Dalm. *Lichas (Entomostr.) laciniatus* Wahl., *L. cicatricosus* n. sp. *Trinucleus (Asaphus) seticornis* His.. *T. (Entomostr.) granulatus* Wahl. *Cybele bellatula* Dalm., *C. (Calymene) verrucosa* Dalm., *C. (Trilobites) relata* Schloth. The author refers species of genus *Odontopleura* Emm. to Dr. Green's genus *Ceraurus*.

Lundgren (B.) Om den vid Ramsoca och Ofvedskloster i Skåne förekommande sandstenens alder.
In Acta Univ. Lundensis, Ars-skrift för or 1874, vol. 10.
Beyrichia salteriana Jones, *B. buchiana* Jones. *Leperditia angelini* Schmidt. *Cytheropsis concinna* Jones.

Lyttelton (Charles.) A letter from the Rev. Charles Lyttelton, read Dec. 20th, 1750.
In Philos. Trans. Royal Soc. London, vol. 46, No. 296, 1750, p. 598, pl. 1, figs. 3–12. pl. 2.
(Calymene blumenbachii.)

MacLeay (W. S.) Observations on Trilobites, founded on a comparison of their structure and that of living Crustacea.
In Murchison's "Silurian system," London, 1839, p. 666; Annals Mag. Nat. Hist., 1st series, London, vol. 4, 1839, p. 16.

The author compares them with *Apus* and other *Aspidophora*, animals which, in his opinion, of all the *Entomostraca*, appear to come nearest to the *Trilobita*.

Mantell (G.) Medals of creation, or first lessons in geology and the study of organic remains. 2 vols. London, 1854.
The author gives figures of *Illænus perovalis* Murch. *Trinucleus lloydii*. *Calymene blumenbachii*. *Homalonotus delphinocephalus*. *Phacops caudatus*. *Paradoxides bohemicus*. *Limulus rotandatus*, *L. trilobitoides*.

Marcou (Jules). On the Primordial fauna and the Taconic system, by Joachim Barrande; with additional notes by Jules Marcou.
In Proc. Boston Soc. Nat. Hist., vol. 7, 1861, p. 369.

The author remarks that "it is evident from the new specimen of *Elliptocephalus asaphoides* from the shales of Washington county, N. Y., figured in Emmon's American Geol., pl. 1, fig. 18, that this trilobite is a *Paradoxides* related to the group of *P. spinosus*, perhaps identical with the *P. harlani* of Braintree. Besides, the trilobite figured pl. 1, fig. 16, and called *Atops trilineatus*, is a true *Sao*, which genus is the most characteristic of the primordial fauna of Bohemia and Scandinavia."

—— Notice sur les gisements des lentilles trilobitifères taconique de la Pointe Lévis on Canada.
In Bull. Soc. Géol. France, 2d series, vol. 21, 1864, p. 236.

Marr (J. E.) On some well-defined life zones in the lower part of the Silurian (Sedgwick) of the Lake District.
In Quart. Jour. Geol. Soc. London, vol. 34, 1878, p. 871.
Phacops (Chasmops) macroura Sjögren, *P. (Odontochile) mucronatus* Brong. *(Phacops appendiculatus* Salt.) *Phacops* sp.?, *P. (Odontochile) obtusicaudatus* Salt.

—— and **Nicholson** (H. A.) The Stockdale shales.
In Quart. Jour. Geol. Soc. London, vol. 44, 1888, p. 654, pl. 16.
Phacops elegans var. *glaber* n. var., *P. (Dalmanites) micronatus* Brong. *Cheirurus bimucronatus* Murch. var. *acanthoides* n. var., *C. (Pseudosphœrexochus) moroides* n. sp. *Acidaspis erinaceus* n. sp. *Harpes judex* n. sp., *H. angustus* n. sp. *Ampyx (Rhaphiophorus) aloniensis* n. sp. *Proetus brachypygus* n. sp.

Martin (D. S.) A new Eurypterid from the Catskill group.
In Trans. New York Acad. Sci., vol. 2, 1892, p. 8.
This notice was based upon a cast of *Stylonurus excelsior* Hall, obtained from the Museum of Natural History, Albany, New York.

Martin (W.) Petrificata Derbiensia: a description of petrifications collected in Derbyshire. Wigan, 1809, 4to, 53 pls.
Part of this work was published in 1794, in detached numbers.
Entomolithus onicites (derbyensis), pl. 45, figs. 1-2. *Entomolithus monoculites ? (lunatus)*, pl. 45, fig. 4; *Entom. paradoxus*, from Dudley, pl. 45, fig. 3.
These fossils have been referred to the following modern genera: *Phillipsia derbyensis* Martin, pl. 45, figs. 1-2; *Belinurus*, pl. 45, fig. 4; *Calymene blumenbachii* Brong., pl. 45, fig. 3; *Griffithides seminifera* Phill., pl. 45*, fig. 1.

Matthew (G. F.) Illustrations of the fauna of the St. John group. No. 1. The Paradoxides; their history.
In Trans. Royal Soc. Canada, vol. 1, 1882, sec. 4, p. 87, pl. 8.
Paradoxides eteminicus n. sp., also vars. *breviatus, suricoides, malicitus, quacoensis* and *pontificalis*. *Paradoxides acadicus* n. sp., *P. lamellatus* Hartt. var. *loricatus*.

—— Illustrations of the fauna of the St. John group. No. 1. The Paradoxides (supplementary section describing the parts).
In Trans. Royal Soc. Canada, vol. 1, 1882, sec. 4, p. 271, pl. 10.
Paradoxides eteminicus Matt., *P. acadicus* Types A, B, C, *(Paradoxides pontificalis* and a large undescribed species).

—— Illustrations of the fauna of the St. John group continued: On the *Conocoryphea*, with further remarks on *Paradoxides*.
In Trans. Royal Soc. Canada, vol. 2, 1884, sec. 4, p. 99, pl. 1.
On page 103 and in his explanation of the plate the author uses the new

subgenus *Ctecocephalus (Hartella) matthewi*, and *Conocoryphe (Bailiella) baileyi*.
Paradoxides acadicus Matt., *P. lamellatus* Hartt., *P. micmac* Hartt. *Ctenocephalus matthewi* Hartt. (also vars. *geminispinosus, hispidus* and *perhispidus*) *Conocoryphe baileyi* Hartt. (also var. *arcuata*), *C. elegans* Hartt. also var. *granulatus, C. walcotti* n. sp.

—— An outline of recent discoveries in the St. John group.
In Bull. New Brunswick Nat. Hist. Soc., No. 4, 1884.

—— Illustrations of the fauna of the St. John group. No. 3. Descriptions of new genera and species (including a description of a new species of Solenopleura, by J. F. Whiteaves).
In Trans. Royal Soc. Canada, vol. 3, 1885, sec. 4, p. 29, 3 pls.
Lepiditta n. g., *L. alata* n. sp., *L. curta* n. sp. *Lepidilla* n. g., *L. anomala* n. sp.
Dr. Matthew in a letter to the author says: "I found *Lepidilla* to be made up of parallel somewhat spreading tubes? when I obtained better specimens, hence I do not now consider it a crustacean."
Beyrichona n. g., *B. papilio* n. sp., *B. tinea* n. sp. *Hipponicharion* n. g., *H. eos* n. sp. *Primitia acadica* n. sp. *Agnostus regulus* n. sp., *A. partitus* n. sp., *A. vir* n. sp., also var. *concinnus* n. var., *A. acadicus* Hartt., also var. *declivis* n. var., *A. tessella* n. sp., *A. umbo* n. sp., *A. obtusilobus* n. sp., *A. acutilobus* n. sp. *Microdiscus punctatus* Salter var. *pulchellus* Hartt., also var. *precursor* n. var., *M. dawsoni* Hartt. *Agraulos? articephalus* n. sp. *Solenopleura acadica* Whiteaves MS. *Paradoxides acadicus* var. *suricus, P. abenacus* n. sp., *P. micmac* Hartt.

—— On the probable occurrence of the great Welsh *Paradoxides, P. davidis*, in America.
In Am. Jour. Sci., 3d series, vol. 30, 1885, p. 72; Nature, vol. 32, 1885, p. 358.

—— Abstract of a paper on the Cambrian fauna of Cape Breton and Newfoundland.
In Canadian Record Sci., vol. 2, 1886, p. 235.

—— Note on the occurrence of *Olenellus? kjerulfi* in America.
In Am. Jour. Sci., 3d series, vol. 31, 1886, p. 472.

————— On the Cambrian fauna of Cape Breton and Newfoundland.
In Trans. Royal Soc. Canada, vol. 2, 1886, sec. 4, p. 147.
Agraulos socialis Billings, *A. affinis* Billings, *A. strenuus* Billings, *A. (Strenuella* n. subgenus) *strenuus. Solenopleura communis* Billings, *S. bombifrons* n. sp.

————— Illustrations of the fauna of the St. John group. No. 4. On the smaller eyed Trilobites of Division 1, with a few remarks on the species of the higher division of the group.
In Canadian Record Sci., vol. 2, No. 6, 1887, p. 357.

————— On the kin of *Paradoxides (Olenellus?) kjerulfi*.
In Am. Jour. Sci., 3d series, vol. 33, 1887, p. 390.

————— Illustrations of the fauna of the St. John group. No. 4. Part 1. Description of a new species of *Paradoxides (P. regina)*. Part 2. The smaller Trilobites with eyes *(Ptychoparidæ* and *Ellipsocephalidæ)*.
In Trans. Royal Soc. Canada, vol. 5, 1887, sec, 4, p. 115, 3 pls.
Paradoxides regina n. sp. *Ellipsocephalus* sp.? *Agraulos ? whitfieldianus* n. sp., also var. *compressa. Strenuella ? halliana* n. sp. *Liostracus tener* Hartt., *L. ouangondianus* Hartt. (also vars. *immarginata, aurora, gibba, plana). Ptychoparia linnarssoni* Brögger var. *alata* (narrow form). *Solenopleura robbii* Hartt., *S. acadica* Whiteaves, also var. *elongata* n. var. *S. robbii* var. *orestes. Ellipsocephalus* sp. cf. *polytomus. Agraulos hallianus* n. sp.

————— The great Acadian *Paradoxides*.
In Am. Jour. Sci., 3d series, vol. 33, 1887, p. 388.
Paradoxides regina Matt.

————— On the occurrence of *Leptoplastus* in Acadian Cambrian Rocks.
In Canadian Record of Science, October, 1889.
Leptoplastus stenotoides n. sp., *L. spiniger* n. sp.
These species are referred to *Anomocare* in Illustrations of the St. John fauna No. 6, 1891, p. 61.

————— How is the Cambrian divided? A plea for the classification of Salter and Hicks.
In American Geologist, vol. 4, 1889, p. 139.

―――― On some remarkable Organisms of the Silurian and Devonian rocks in Southern New Brunswick.
In Trans. Royal Soc. Canada, vol. 6, 1888, sec. 4, p. 49, plate 4.
Ceratiocaris pusillus n. sp. *Bunodella* n. g., *B. horrida* n. sp. *Eurypterella* n. g.. *E. ornata* n. sp.

―――― Illustrations of the fauna of the St. John group. No. 5.
In Trans. Royal Soc. Canada, vol. 3, 1890, sec. 4, p. 166, plates 11-16.
Conocoryphe walcotti Matt., *C. baylei* Hartt. *Paradoxides lamellatus* Hartt., *P. mimac* Hartt. var. *pontificalis*. *Agraulos whitfieldi* Matt., *A. socialis* Billings, *A.? holocephalus* n. sp.

―――― Notes on Cambrian Faunas.
In American Geologist, vol. 8, 1891, p. 287.
Conocoryphe trilineatus Emmons. *Olenellus asaphoides* Emmons. *Microdiscus connexus*, and *Leperditia dermatoides*.

―――― Note on *Leptopalstus*.
In the Canadian Record of Science, December, 1891.
Leptoplastus latus n. sp.

List of fossils found in the Cambrian rocks in or near St. John, New Brunswick. St. John, 1892.
Reprinted from Natural History Society Bulletin, No 10, 12 pp.

―――― Illustrations of the fauna of the St. John group. No. 6.
In Trans. Royal Soc. Canada, vol. 9, 1891, sec. 4, p. 33, plates 12 and 13.
Agnostus bisectus n. sp. *Parabolina spinulosa* Wahl., *P. heres* Brög., also var. *lata* n. var. and var. *grandis* n. var. *Protopeltura acanthura* Ang. *Peltura scarabæoides* Wahl. *Leptoplastus latus* Matt. *Ctenopyge* (variety) *flagillifer* Ang., *C. spectabilis* Brög., *C. pecten* Salter? *Agnostus pisiformis*. *Anomocare stenotoides* Matt., *A. spiniger* Matt. *Conocephalites? contiguus* n. sp. *Protopeltura acanthura* var. *tetracanthura* n. var.

―――― Protolenus: A new genus of Cambrian Trilobites.
In Bull. Nat. Hist. Soc. New Brunswick, No. 10, Sept., 1892.
Protolenus elegans (W. D. Matthew sp. MS.) *P. paradoxides* n. sp.

―――― Notes on Cambrian Faunas.
In Canadian Record of Science, October, 1892.
Protolenus elegans Matt., *P. paradoxides* Matt., *P. roppii* Meneghini. *Ellipsocephalus galeatus* n. sp., fig. 4a-e, *E. articephalus* Matt. fig. 5a-b, *E. hoffi* Schloth.

Maurer (F.) Paläontologische Studien im Gebiet des rheinischen Devon. Die Fauna des Rotheisenstein der Grube Haina.

In Neues Jahrbuch für Mineral, 1875, p. 596, pl. 14.
Phacops latifrons Bronn. *Cyphaspis ceratophthalmus* Goldf.

―――― Paläontologische Studien im Gebiet des rheinischen Devon. Die Thonschiefer des Ruppbachthales bei Diez.

In Neues Jahrbuch für Mineral., 1876, p. 808, pl. 14.
Phacops latifrons Bronn. *Acidaspis* sp.? *Bronteus cameratus* n. sp.

―――― Der Kalk bei Greifenstein.

In Neues Jahrb., 1881.
Proetus informis n. sp., *P. kaneni* n. sp., *P.* cf. *curtus* Barr., *P.* cf. *neglectus* Barr., *P. glaber* n. sp., *P. saturni* n. sp., *P.* cf. *erenita* Barr., *P.* cf. *natator* n. sp., *P.* cf. *complanatus* Barr., *P.* cf. *myops* Barr., *P. orbitatus* Barr., *P. strengi, P. urani. Lichas haueri* Barr. *Bronteus thysanopeltis* Barr.

―――― Die Fauna der Kalke von Waldgirmes bei Giessen.

In Abhandl. grossh.-hess. geol. Landesanstalt zu Darmstadt, vol. 1, pt. 2, 1885, atlas of 11 pls.
Primitia pila n. sp., *P. contusa* n. sp., *P. leviter* n. sp., *P. fabula* n. sp. *Orozoe marginata* n. sp. *Harpes macrocephalus* Goldf. *Proetus suplantus* n. sp., *P. informis* Maur., *P. consonus* n. sp., *P. lævigatus* Goldf., *P. gracilis* n. sp., *P. quadratus* n. sp. *Cyphaspis hydrocephala* A. Römer, *C. strengi* n. sp. *Phacops latifrons* Bronn. *Cheirurus gibbus* Beyr.? *Bronteus* cf. *umbellifer* Goldf., *B. fædus* n. sp., *B. geminatus* n. sp.

McCoy (Frederick.) On *Entomoconchus scouleri* n. sp.

In Jour. Geol. Soc. Dublin, vol. 2, 1844, p. 91, pl. 5, figs a, c.
Entomoconchus n. g.

―――― A synopsis of the characters of the Carboniferous limestone fossils of Ireland. Dublin, 1844, 29 pls.

The title-page bears no author's name. The book shows that it was prepared by Frederick McCoy, under the direction of Richard Griffith, the collector, in whose cabinet all the species were contained.
Phillipsia coela n. sp., *P. colei* n. sp., *P. discors* McCoy, *P. gemmulifera* Phill., *P. jonesii* Portl., *P. kellii* Portl., *P. marcoyii* Portl., *P. mucronata* n. sp., *P. quadriserialis* n. sp., *P. truncatula* Phill. *Griffithides calcaratus* n. sp., *G. globiceps* Phill., *G. granuliferus* Phill., *G. longiceps* Portl., *G.*

longispinus Portl., *G. obsoletus* Phill. *Calymene ? granulata* Münst., *C. lævis* Münst., *C. latreillii* Stein. *Daphinia primæra* McCoy. *Bairdia curtus* n. sp., *B. gracilis* n. sp. *Cythere amygdalina* n. sp., *C. arcuata* n. sp., *C. bituberculata* n. sp., *C. costata* n. sp., *C. cornuta* n. sp., *C. elongata* n. sp., *C. excavata* n. sp., *C. gibberula* n. sp., *C. impressa* n. sp., *C. hibbertii* n. sp., *C. inflata* n. sp., *C. inornata* n. sp., *C. oblonga* n. sp., *C. orbicularis* n. sp., *C. pusilla* n. sp., *C. scutulum* n. sp., *C. spinigera* n. sp., *C. trituberculata* n. sp. *Dithyrocaris colei* Portl., *D. orbicularis* n. sp., *D. scouleri* n. sp., *D. tenuistriatus* McCoy. *Entomoconchus scouleri* McCoy. The genera *Bairdia* and *Entomoconchus*. *Astacus? phillipsi*, an undeterminable figure of a crustacean probably not an *Astacus* but probably Entomostracan.

The localities of these fossils was not published in the Synopsis and did not appear until 1861. See The Dublin Quart. Jour. Science, No. 1, Jan., 1861, p. 20.

This work was reissued in 1860 by Sir Richard Griffith with a new title page and an Appendix of the Localities of the Irish Carboniferous Limestone Fossils, pp. 209-271, which was not given in the first edition.

―――― A synopsis of the Silurian fossils of Ireland, collected from the several districts by Richard Griffith, F. G. S. The whole being named, and the new species drawn and described, by Frederick McCoy. Dublin, 1846, 5 pls.

Forbesia n. g. *Portlockia* n. g. *Trinodus* n. g. *Tiresias* n. g. *Beyrichia* n. g.

A notice of the new genera described by Frederick McCoy is given in the Ray Society's edition of Burmeister's "Organization of Trilobites. Supplementary appendix," p. 123.

The generic name *Tiresias* was used in 1845 for a genus of the Coleoptera, and that of *Forbesia* for one of the Polypi in 1845. *Trinodus* was used in 1846 for a genus of the Coleoptera.

Tiresias insculptus n. sp. *Remopleurides colbii* Portl., *R. dorsospinifer* Portl., *R. laterspinifer* Portl., *R. platyceps* n. sp. *Cheirurus (a) bimucronatus* Murch., *C. brevimucronatus* Münst., *C. gelasinosus* Portl. *Sphærexochus (b) clavus* n. sp. *Acidaspis bispinosus* n. sp. *Encrinurus (b) multisegmentatus* Portl., *E. stokesii* n. sp. *Calymene arenosa* n. sp., *C. blumenbachii, (b) pulchella* Dalm., *C. brevicapitata* Portl., *C. forcipata (Lichas laxatus* (McCoy) Linrs.), *C. variolaris* Murch. (Brong. pars.) *Forbesia latifrons* n. sp. *Phacops brongniarti* Portl., *B. dalmani* Portl., *P. jamesii* Portl., *P. murchisonii* Portl., *P. truncato-caudatus* Portl. *Portlockia sublævis* n. sp.

Lichas hibernica Portl., *L. laxata* n. sp., *L. pumila*. *Brontes hibernicus* Portl., *B. signatus* Phill. *Ogygia murchisoniæ* Murch. *Asaphus buchii* Brong., *A. latifrons* Portl., *A. marginatus* Portl. *Homalonotus ophiocephalus*

n. sp. *Isotelus gigas* Dekay, *I. powisii* Murch.? var. *I. rectifrons* Portl. *Illænus centrotus* Dalm., *I. perovalis* Murch. *Otarion obtusum* n. sp. *Harpes flanagani* Portl., *H.? megalops* n. sp. *Trinucleus caractaci* Murch., *T. longatus* Portl., *T. fimbriatus* Murch., *T. radiatus* Murch., *T. seticornis* His. *Ampyx nasutus* Dalm., *A. rostratus* Sars. *Trinodus agnostiformis* n. sp. *Beyrichia klödeni* n. sp., *B. tuberculatus* Klöden, *B. latus*. *Cythere phaseolus* His. (*Primitia maccoyi* Salt.)

—— On the fossil botany and zoology of the rocks associated with the coal of Australia.
In Annals Mag. Nat. Hist., 1st series, London, vol. 20, 1847, p. 226, pl.
Brachymetopus n. g.
The author establishes the genus *Brachymetopus* to contain *Phillipsia ? discors* McCoy, *P. maccoyi* Portl. and *Brachymetopus strzeleckii* described and figured in this article.

—— On the classification of some British fossil Crustacea, with notices of new forms in the University collection at Cambridge.
In Annals Mag. Nat. Hist., 2d series, London, vol. 4, 1849.
Chasmops n. g. *Barrandia* n. g. *Ceratiocaris* n. g. *Cytheropsis* n. g. *Harpidella* n. g.
Asaphinæ: 1, *Phacops*. 2, *Calymene*. 3, *Trimerocephalus* n. g. 4, *Asaphus* (subgenera *Isoletus*, *Basilicus*). 5, *Illænus* (subgenera *Illænus*, *Bumastus*, *Dysplanus*). 6, *Forbesia*. 7, *Phillipsia*.
Paradoxinæ: 1, *Paradoxides* (subgenus *Olenus*). 2, *Ceraurus*. 3, *Cryphæus*. 4, *Sphærexochus*. 5, *Acidaspis*. 6, *Staurocephalus*. 7, *Remopleurides*. 8, *Zethus*.
Ogyginæ: 1, *Trinucleus* (subgenus *Tetrapsellium*). 2, *Tretaspis* n. g. 3, *Ampyx*. 4, *Ogygia*. 5, *Bronteus*. 6, *Lichas* (subgenera *Trochurus*, *Acanthopyge*).
Harpedinæ: 1, *Harpes*. 2, *Harpidella* n. g. 3, *Amphion*.
Agnostinæ: 1, *Trinodus*. 2, *Agnostus*. Subgen. *Diplorhima*.
Partly republished in the author's "Contributions to British Palæontology, etc., London, 1854.
Pterygotus leptodactylus McCoy (subg. *Leptocheles* for *P. anglicus*). *Daphnia ? primæva*. *Chasmops odini* Eichw. *Trimerocephalus lævis* Münst. *Illænus latus* McCoy. *Isotelus affinis* McCoy. *Griffithides meso-tuberculatus* McCoy. *Cryphæus sedgwickii* McCoy. *Ceraurus octo-lobatus* McCoy, *C. williamsii* McCoy. *Ogygia radians* McCoy. *Barrandia cordai* McCoy. *Ampyx latus* McCoy. *Tretaspis gibbifrons* McCoy. *Harpidella megalops* McCoy. *Ceratocaris solenoides* McCoy, *C. ellipticus* McCoy. *Cytheropsis*.

—— On some new Cambro-Silurian fossils.
In Annals and Mag. of Natural History for Nov. 1851.
Cytheropsis aldensis McCoy. *Harpes parvulus* McCoy.

―――― On the supposed fish remains figured on plate 4 of "The Silurian System."
In Quart. Jour. Geol. Soc. London, vol. 9, 1853, p. 13.

The author divides the genus *Pterygotus* into two subgenera: 1, *Pterygotus* proper, in which the didactyle claws are very thick and armed with powerful teeth. 2, *Leptocheles* McCoy, in which the pincers are very slender and unarmed. He refers *Onchus murchisoni* to *Leptocheles murchisoni*.

―――― A synopsis of the classification of the British Palæozoic rocks, by Adam Sedgwick; with a systematic description of the British Palæozoic fossils in the Geological Museum of Cambridge, by Frederick McCoy, with figures of the new and imperfectly known species. London and Cambridge, 1855, 25 pls. Pt. 2, Palæontology.

This work was published in three parts.

Fasciculus 1. Radiata and Articulata. London, 1851.
Beyrichia klödeni McCoy, *B. complicata* Salt., *B. strangulata* Salt. *Ceratiocaris ellipticus* McCoy, *C. inornatus* n. sp., *C. solenoides* McCoy, *C.? umbonatus* Salt. *Trinodus agnostiformis* McCoy, *T. tardus ?* Barr. *Diplorhina triplicata* H. & C. *Harpidella megalops* McCoy. *Trinucleus caractaci* Murch., *T. gibbifrons* McCoy, *T. latus* Portl., *T. ? radiatus* Murch., *Tretaspis fimbriatus* Murch., *T. seticornis* His. *T. latus* McCoy. *Ampyx nudus* Murch. *Ogygia buchi* Brong, *O. radians* McCoy. *Barrandia cordia* McCoy. *Lichas propinqua?* Barr. *Acanthopyge angilica* Beyr. *Trochurus nodulosus* Salt. *Acidaspis brightii* Murch. *Staurocephalus murchisoni* Barr. *Ceraurus clavifrons* Dalm., *C. octolobatus* McCoy, *C. williamsii* McCoy. *Eccoptochile sedgwicki* McCoy. *Zethus atractopyge* McCoy, *Z. sexcostatus* Salt., *Z. variolaris* Brong. *Harpes parvulus*, pls. 1, 2. *Encrinurus punctatus* Brünn. *Phacops alifrons* Salt., *P. downingiæ* Murch. *Odontochile caudata* Brünn, *O. longicaudata* Murch., *O. obtusi-caudata* Salt., *O. truncata-caudata* Portl. *Portlockia ? apiculata* Salt., *P. stokesii* M. Edw. *Chasmops odini ?* Eichw. *Calymene baylei* Barr., *C. blumenbachii* Brong., *C. brevicapitata* Portl., *C. sub-diademiata* McCoy, *C. parvifrons* Salt., *C. tuberculosa* Salt. *Homalonotus bisulcatus* Salt., *H. knightii* König., *H. rudis* Salt. *Isotelus affinis* McCoy, *I. (Basilicus?) laticostatus* Green, *I. (Basilicus) powisii* Murch., *I. (Basilicus) tyrannus* Murch. *Illænus davisii* Salt., *I. latus* McCoy, *I. rosenbergi* Eichw. *Dysplanus centrotus* Dalm. *Forbesia latifrons* McCoy, *F. stokesii* Murch. *Eurypterus cephalaspis* Salt. *Pterygotus lepto-dactylus* McCoy. *Portlockia granulata* Münst., *P. latifrons* Broun. *Trimerocephalus lævis* Münst. *Bronteus alutaceus* Goldf. *Dithyrocaris scouleri* and *D. lateralis* n. sp. *Griffithides meso-tuberculatus* McCoy. *Phillipsia gemmulifera* Phill., *P. jonesii* Portl., *P. seminifera* Phill.

For the species of Crustacea described in this volume, Appendix A, see Salter (J.W.), British Palæozoic fossils in the Geol. Mus. Univ. of Cambridge, 1852.

———— Contributions to British palæontology, or first description of 360 species and several genera of fossil Radiata, Articulata, Mollusca and Pisces from the Tertiary, Cretaceous, Oolitic and Palæozoic strata of great Britain. Republished from the Annals Mag. Nat. Hist. Cambridge, 1854.

For a list of the genera of fossil Crustacea, see entries under McCoy (Frederick), Annals Mag. Nat. Hist., 2d series, London, vol. 4, 1849; also Annals for November, 1851.

———— Geological Survey of Victoria. Prodromus of the palæontology of Victoria, or figures and descriptions of Victorian organic remains, decade 3. Melbourne and London, 1875, pls. 22, 23.

Phacops (Odontochile) caudatus, *P. (Portlockia) fecundus*. *Forbesia euryseps* McCoy. *Lichas australis* McCoy. *Homalonotus harrisoni* McCoy.

Meek (F. B.) Note on *Belinurus danæ*, from the Illinois Coal Measures.

In Am. Jour. Sci., 2d series, vol. 43, 1867, p. 257; Geol. Survey Illinois, vol. 2, p. 395.

———— Note on a new genus of fossil Crustacea.

In Am. Jour. Sci., 2d series, vol. 43, 1867, p. 394; Geol. Survey Illinois, vol. 3, p. 547.

Euproops n. g., *E. danæ* n. sp.

———— Descriptions of fossils collected by the U. S. Geological Survey, under charge of Clarence King.

In Proc. Acad. Nat. Sci. Phila., 1870, p. 56.

Republished and illustrated in Rep. U. S. Geol. Expl., 40th Parallel, vol. 4, pt. 1, pp. 1-197, pls. 1-17.

Conocoryphe (Ptychoparia) kingii Meek. *Paradoxides? nevadensis* Meek.

———— Descriptions of new species of invertebrate fossils from the Carboniferous and Devonian rocks of Ohio.

In Proc. Acad. Nat. Sci. Phila., 1871, p. 57.

Republished and figured in Palæontology of Ohio, vol. 1, 1873, p. 233, 23 pls.

Proetus planimarginatus n. sp. *Dalmanites ohioensis* n. sp.

—— Descriptions of new Western Palæozoic fossils, mainly from the Cincinnati group of the Lower Silurian series of Ohio.
In Proc. Acad. Nat. Sci. Phila., for 1871, p. 308, Phila., 1872.
Redescribed and illustrated in Palæontology of Ohio, vol. 2, 1875.
Solenocaris n. subgen. (Not *Solenocaris* Young, 1868.) *Cythere cincinnatiensis* n. sp. *Ceratiocaris (Colpocaris* n. subgen.) *bradleyi* n. sp., *C. (Colpocaris) elytroides* n. sp., *C. (Solenocaris* n. subgen.) *strigata* n. sp. *Archæocaris* n. g., *A. vermiformis* n. sp.

—— List of Carboniferous fossils from West Virginia, with descriptions of new species.
In Appendix B, Rep. of Regents of West Virginia Univ., 1870, Wheeling, 1871, p. 68.
Phillipsia stevensoni n. sp.

—— Final report of the U. S. Geological Survey of Nebraska and portions of the adjacent Territories, by F. V. Hayden, etc. Report on the palæontology of eastern Nebraska, with some remarks on the Carboniferous rock of that district by F. B. Meek. Washington, 1871, p. 83, 11 pls. (Two editions; first in 8vo, second in 4to.)
Cythere nebrascensis Geinitz, *C.* sp.? *Phillipsia* sp.?, *P. scitula* M. & W., *P. major* Shumard.

—— Description of a new species of fossils from Cincinnati group, Ohio.
In Am. Jour. Sci, 3d series, vol. 3, 1873, p. 423.
Dalmanites carleyi n. sp. *Proetus spurlocki* n. sp.

—— Preliminary palæontology report, consisting of list and descriptions of fossils, with remarks on the age of the rocks in which they are found, etc.
In Sixth Ann. Rep. U. S. Geol. Survey Territories, embracing portions of Montana, Idaho, Wyoming and Utah, etc. F. V. Hayden, Washington, 1873, p. 431.
Agnostus maladensis n. sp. *Bathyurus serratus* n. sp. *Asaphus (Megalaspis?) goniocercus* n. sp. *Bathyurus? haydeni* n. sp. *Bathyuriscus* n. g., *B. (Asaphiscus) bradleyi* n. sp. *Asaphiscus* n. g. *Conocoryphe (Ptychoparia) gallatensis* n. sp. *Asaphiscus wheeleri* n. sp. *Bathyurus saffordi* Billings. *Bathyurellus (Dikelocephalus?) truncatus* n. sp. This species is given in the list of fossils collected, but not described in the report. *Bathyurus serratus* is also referred to *Corynexochus serratus*.

―――― Report of the Geological Survey of Ohio.
Vol. 1. Geology and palæontology. Pt. 2. Palæontology. Descriptions of invertebrate fossils of the Silurian and Devonian systems, by F. B. Meek.
Columbus, 1873, pp. 1-243, pls. 1-23.

Cythere cincinnatiensis Meek. *Asaphus (Isotelus) megistos* Locke? *Proetus spurlocki* Meek. *Ceraurus icarus* Billings. *Acidaspis crosotus* Locke? *A. cincinnatiensis* n. sp. *A. ceralepta* Anthony. *Dalmanites carleyi* Meek. *Calymene senaria* Conrad. *Leperditia alta* Conrad. *Illænus (Bumustus) insignis* Hall. *Proetus planimarginatus* Meek. *Dalmanites ohioensis* Meek.

―――― Descriptions of *Olenus (Olenellus) gilberti* and *Olenus (Olenellus) howelli*.
In Rep. Geog. and Geol. Expl. Surveys West 100th Mer., vol. 3, Washington, 1875, p. 182.

―――― A report on some of the invertebrate fossils of the Waverly group and Coal Measures of Ohio.
In Rep. Geol. Survey Ohio. Vol. 2. Geology and Palæontology. Pt. 2. 1875, pp. 269-347, pls. 10-20.

Ceratiocaris (Colpocaris) bradleyi Meek, *C. (Colpocaris) elytroides* Meek., *C. (Solenocaris) strigata* Meek. *Archæocaris vermiformis* Meek. *Phillipsia (Griffithides?) lodiensis* n. sp., also *Proetus (Phæton) lodiensis*.

―――― Palæontology.
In U. S. Geol. Expl. 40th Parallel, vol. 4, pt. 1, 1877, pp. 1-197, pls. 1-17.
Conocoryphe (Ptychoparia) kingii Meek. *Paradoxides nevadensis* Meek. The author suggests the new generic name of *Olenoides* for this species.

―――― and **Hayden** (F. V.) Descriptions of new Lower Silurian, Jurassic, Cretaceous, and Tertiary fossils collected in Nebraska by the exploring expedition under the command of Capt. W. F. Reynolds, U. S. Top. Eng., with remarks on the rocks from which they were obtained.
In Proc. Acad. Nat. Sci. Phila., for 1861. Phila., 1862, p. 415.
Arionellus (Crepicephalus) oweni n. sp.

―――― Palæontology of the Upper Missouri. A report upon collections made principally by the expeditions under command of Lieut. G. K. Warren, U. S. Top. Eng., in 1855–56. Invertebrates. Pt. 1.
In Smithsonian Contrib., No. 172, Washington, 1864, 5 pls.
Agraulos oweni M. & H.

―――― and **Worthen** (A. H.) Notice of some new types of organic remains from the Coal Measures of Illinois.

In Proc. Acad. Nat. Sci. Phila., 1865, p. 41.
Acanthotelson n. g. *Palæocaris* n. g. *Anthrapalæmon gracilis* n. sp. *Acanthotelson inæqualis* n. sp., *A. stimpsoni* n. sp. The species described as *Euproops danæ* is referred to as *Belinurus danæ, Palæocaris typus* n. sp.

―――― Contributions to the palæontology of Illinois and other Western States.

In Proc. Acad. Nat. Sci. Phila., 1865, p. 245.
Dalmania danæ n. sp. *Lichas cucullus* n. sp. *Proetus ellipticus* n. sp. *Phillipsia (Griffithides) portlocki* n. sp., *P. (Griffithides) scitula* n. sp., *P. (Griffithides?) sangomonensis* n. sp.

―――― Descriptions of invertebrates from the Carboniferous system.

In Geol. Survey Illinois, vol. 2, 1866, p. 145, pls. 14-20, 23-32.
Belinurus danæ M. & W. *Acanthotelson stimpsoni* M. & W., *A. inæqualis* M. & W. *Palæocaris typus* M. & W. *Anthrapalæmon gracilis* M. & W.

―――― Preliminary notice of a scorpion, a *Eurypterus*, and other fossils from the Coal Measures of Illinois.

In Am. Jour. Sci., 2d series, vol. 46, 1868, p. 19. Republished and illustrated in Geol. Survey Ill., vol. 3, p. 560.
Eurypterus (Anthraconectes n. subgen.) *mazonensis* n. sp. It is also referred to *Adelopthalmus mazonensis*. *Ceratiocaris ? sinuatus* n. sp. *Acanthotelson eveni* n. sp. *Palæocaris*, of the Illinois Report, plate 31, figs. 5 and 5a, is referred to the same species.

―――― Palæontology.

In Geol. Survey Illinois, vol. 3, 1868, pp. 291-565, pls. 1-20.
Trenton group: *Lichas cucullus* M. & W.
Galena beds: *Illænus taurus* Hall, *I. crassicauda* Wahl.?
Niagara group: *Dalmanites danæ* M. & W.
Lower Helderberg group: *Dalmanites tridentiferus* Shumard. *Acidaspis hamala* Conrad.
Corniferous group: *Dalmanites (Odontocephalus) ægera* Hall.
Hamilton group: *Phacops rana* Green.
Kinderhook group: *Proetus ellipticus* M. & W.
Coal Measures: *Ceratiocaris ? sinuatus* M. & W. *Leaia tricarinata* M. & W. *Eurypterus (Anthraconectes) mazonensis* M. & W. *Euproops danæ* M. & W. *Acanthotelson stimpsoni* M. & W., *A. eveni* M. & W. *Anthrapalæmon gracilis* M. & W. *Palæocaris typus* M. & W. *Gampsonyx fambriatus* Jord.

—— Descriptions of a new species and genera of fossils from the Palæozoic rocks of the Western States.
In Proc. Acad. Nat. Sci. Phila., 2d series, 1870, p. 22.
Phillipsia tuberculata n. sp. *Phillipsia (Griffithides) bufo* n. sp. *Asaphus (Isotelus) vigilans* n. sp. *Illænus (Bumastus) graftonensis* n. sp. *Dithyrocaris carbonarius* n. sp.

—— Descriptions of invertebrates from the Carboniferous system.
In Geol. Survey Illinois, vol. 5, 1873, pp. 323-619, pls. 1-32.
Keokuk group: *Phillipsia (Griffithides) portlockii* M. & W., *P. (Griff.) bufo* M. & W.
Coal Measures: *Phillipsia (Griffithides) scitula* M. & W., *P. (Griffithides) sangamonensis* M. & W. *Dithyrocaris carbonarius* M. & W.

—— Description of the invertebrates.
In Geol. Survey Illinois, vol. 6, 1875, pp. 489-532, pls. 23-32.
Lichas boltoni Bigsby. *Illænus (Bumastus) graftonensis* M. & W. *Sphærexochus romingeri* Hall. *Asaphus (Isotelus) vigilans* M. & W.

Meneghini (Giuseppe). Nuovi fossili siluriani di Sardegna.
In Atti reale Accad. Lincei, Roma, 1880, 1 pl.
Dalmanites lamarmora n. sp.

—— Posizione relativa dei varii piani siluriani dell' Iglesiente in Sardegna.
In Proc. verb. Soc. Toscana Sci. Nat., 1881, p. 258.
Paradoxides gennarii Mgh., *P. bornemanni* Mgh., *P. armatus* Mgh. *Conocephalites bornemanni* Mgh. *Trinucleus ornatus* and *Dalmanites lamarmoræ*.

—— Fauna Cambriana dell' Iglesiente in Sardegna.
In Proc. verb. Soc. Toscana Sci. Nat., 1881, p. 158.
Paradoxides gennarii Mgh., *P. bornemanni* Mgh. *Olenus zoppii* n. sp., *O. armatus* Mgh. *Conocephalus bornemanni* Mgh.

—— Nuovi Trilobiti di Sardegna.
In Proc. verb. Soc. Toscana Sci. Nat., 1881, p. 200.
Paradoxides gennarii Mgh. *Conocephalites bornemanni* Mgh.

—— Note alla fauna Cambriana dell' Iglesiente.
In Proc. verb. Soc. Toscana Sci. Nat., 1882, p. 7.
Olenus zoppii Mgh., *O. armatus* Mgh., *O.* sp.? *Conocephalites* sp.? *Anomocare arenivagum* n. sp. *Platypeltis meneghinii* Born.

——— Nuovi fossili Cambriani di Sardegna.
In Proc. verb. Soc. Toscana Sci. Nat., 1884, p. 56.
Conocephalites lamberti n. sp., *C. inops* n. sp.?, *C.* sp.? *Anomocare pusillum* n. sp.

——— Palæontologia dell Iglesiente in Sardegna. Fauna Cambriana trilobiti.
In Mem. R. Com. Geol. Italia, vol. 3, 1888, pt. 2, 49 pp., 7 pls.
Olenus sp.?, *O. armatus* Mgh. *Paradoxides gennarii* Mgh., *P. bornemanni* Mgh., *P. torosus* Mgh. *Conocephalites bornemanni* Mgh., *C. phialare* n. sp., *C. lamberti*, *C. frontosus* n. sp., *C. inops* Mgh., *C.* sp.? *Anomocare* sp.?, *A. arenivagum* Mgh., *A. pusillum* Mgh. *Conocoryphe* sp.? *Asaphus* (*Platypeltis ?*) *meneghinii* Born., *A.* (*Psilocephalus ?*) *gibber* Mgh. *Encrinurus* sp.?

Meyer (H. von). Beiträge zur Petrefacten-Kunde.
In Nova Acta Acad. Leop. Carol., vol. 15, 1831, p. 57, pl. 56, fig. 13.
See, also, Neues Jahrbuch für Mineral., 1833, p. 481.
Calymene æqualis n. sp.

——— *Jonotus reflexus* ein Trilobit aus der.Grauwacke der Eifel.
In Dunker & von Meyer's Palæontographica, vol. 1, 1846-51, p. 182, pl. 26.

——— Squaliden-Reste aus dem Posidonomyen-Schiefer des Oberharzes bei Ober-Schulenburg.
In Dunker & von Meyer's Palæontographica, vol. 3, Cassel, 1854, p. 53.
Cytherina intermedia n. sp. *Harpes*. *Phacops latifrons* Brong., *P. bronni* Barr. *Cheirurus jaschei* n. sp. *Phacops pectinatus* n. sp., *P. stellifer* Burm. *Lichas crassirhachis* n. sp. *Cheirurus myops* n. sp. *Proetus crassimargo* n. sp.
See, also, Jordon (H.) and Meyer (H. von).

Mickleborough (John). Locomotory appendages of Trilobites.
In Jour. Cincinnati Soc. Nat. Hist., vol. 6, 1883, p. 200, figs. 1-3.
Noticed: Am. Jour. Sci., 3d series, vol. 27, 1884, p. 409; Geol. Mag., new series, decade 3, vol. 1, London, 1884, p. 80.
Asaphus megistos Locke.

Miller (S. A.) Monograph of the Crustacea of the Cincinnati group.
In Cincinnati Quart. Jour. Sci., vol. 1, 1874, p. 115.
The author, in his classification and generic descriptions, has generally followed McCoy (" Brit. Palæozoic Rocks," 1855). He gives a description

of the different fossil Crustacea occurring in the Cincinnati group, and figures *Leperditia byrnesi* Miller, and a woodcut representing the tracks of an *Asaphus ?*.

Beyrichia oculifer Hall, B. tumifrons Hall. Cythere cincinnatiensis Meek. Leperditia cylindrica Hall, L. minutissima Hall, L. byrnesi Miller, fig. Trinucleus concentricus Eaton. Lichas trentonensis Conrad. Acidaspis crosotus Locke, A. ceralepta Anthony, A. cincinnatiensis Meek. Ceraurus pleurexanthemus Green, C. icarus Billings. Asaphus (Isotelus) megistos Locke, A. (Isotelus) gigas Dekay. Calymene senaria Conrad, C. christyi Hall. Dalmanites carleyi Meek, D. breviceps Hall. Proetus parvisculus Hall, P. spurlocki Meek. Triarthrus becki Green.

—— [No title.—Description of new species of Palæozoic Entomostraca.]

In Cincinnati Quart. Jour. Sci., vol. 1, 1874, p. 232.
Beyrichia duryi n. sp., B. striato-marginatus, B. chambersi n. sp.

—— Some new species of fossils from the Cincinnati group, and remarks upon some described forms.

In Cincinnati Quart. Jour. Sci., vol. 2, 1875, p. 349, figures.
Acidaspis anchoralis n. sp. Beyrichia cincinnatiensis n. sp., B. ciliata Emm., B. regularis Emm. Leperditia cylindrica Hall.

—— The American Palæozoic fossils. A catalogue of the genera and species, with names of authors, dates, places of publication, group of rocks in which found, and the etymology and signification of the words, etc. Cincinnati, 1877. Also, a 2d ed., with supplement.

—— Description of a new genus and eleven new species of fossils, with remarks upon others well known, from the Cincinnati group.

In Jour. Cincinnati Soc. Nat. His., vol. 1, 1878-79, p. 100, pl. 3.
Lichas harrisi n. sp. Cythere irregularis n. sp.

—— Description of two new species from the Niagara group, and five from the Keokuk group.

In Jour. Cincinnati Soc. Nat. Hist., vol. 2, 1879, p. 254, pl. 15.
Encrinurus egani n. sp.

—— Description of new species of fossils.

In Jour. Cincinnati Soc. Nat. Hist., vol. 4, 1881, p. 251, pl. 6.
Leperditia cæcigena n. sp.

―――― Description of three new species, and remarks upon others.
In Jour. Cincinnati Soc. Nat. Hist., vol. 5, 1882, p. 116, pl. 5.
Calymene callicephala Green.

Milne-Edwards (H.) Sur les affinités des Trilobites.
In Institute de France, Acad. Royale Sci., vol. 1, 1827, p. 254.

―――― Trilobites dans Lamarck's "Histoire naturelle des animaux sans vertèbres." Vol. 5. Paris, 1838, p. 220.
Eurypterus remipes Dekay, *E. lacustris* Harlan, *E. souleri* Hibbert. *Cytherina*.
(A) Trilobites oculés: *Calymene blumenbachii* Brong. var. 1, *tuberculata* Dalm. var. 2, *tuberculosa* Dalm. 2, *C. tristani* Brong. 3, *C. bellatula* Dalm. 4, *C. polytoma* Dalm. 5, *C. actinura* Dalm. 6, *C. latifrons* Bronn. 7, *C. schlotheimii* Bronn. 8, *C. sclerops* Dalm. 9, *C. macrophthalma* Brong. 10, *C. punctata* Brünn. 11, *C. concinna* Dalm. 12, *C. arachnoides* Hoeninghaus.
Espèces ayant les angles posterieurs de la téte allongés et amincis: 13, *Calymene variolaris* Brong. 14, *C. speciosa* Dalm. 15, *C. clavifrons* Dalm. *Trimerus delphinocephalus* Green.
Première section Asaphus articulés: *Asaphus caudatus* Brünn. 2, *A. mucronatus* Brong.
aa Extrêmité de l'abdomen arronde: 3, *Asaphus buchii* Brong. 4, *A. hausmanni* Brong. 5, *A. frontalis* Dalm. 6, *A. brongniartii* DeLouch. 7, *A. fischeri* Eichw.
The author also doubtfully places under this section the following species: *Asaphus gemmuliferus* Phill., *A. seminiferus* Phill., *A. globiceps* Phill., *A. quadrilimbatus* Phill., *A. ? grypturus* Green.
Deuxième section, Asaphus anchyloures: 8, *Asaphus dilatatus* Brünn. 9, *A. angustifrons* Dalm. 10, *A. cornigerus* Schloth. 11, *A. lichtenstenii* Eichw. 12, *A. weissii* Eichw. 13, *A. schlotheimii* Eichw. 14. *A. læviceps* Dalm. 15, *A. gigas* Dekay. 16, *A. palpebrosus* Dalm. 17, *A. extenuatus* Wahl. 18, *A. grandis* Sars. 19, *A. brevicaudatus* DeLouch. 20, *A. crassicauda* Wahl. 21, *A. centrotus* Dalm. 22, *A. laticauda* Wahl.
Troisième section, Asaphus oniscoides: 13, *Asaphus (Nilleus) armadillo* Dalm. *Depleura dekayi* Green.
Trilobites typhlien: *Ampyx nasutus* Dalm. 2, *A. rostratus* Sars. 3, *A. mammillatus* Sars. 4, *A. incertus* DeLouch. *Conocephalus costatus* Zenk., *C. sulzeri* Schloth. *Ogygia guettardi* Brong. 2, *O. desmarestii* Brong. *Otarion diffractum* Zenk., *O. squarosum* Zenk. *Cryptolithus tessellatus* Green. *Paradoxides tessini* Brong. 2, *P. spinulosus* Wahl. 3, *P. longicaudatus* Zenk. 4, *P. pyramidalis* Zenk. 5, *P. latus* Zenk. 6, *P. bucephalus*

Wahl. 7, *P. forficula* Sars. 8, *P. scrabæoides* Wahl. 9, *P. gibbosus* Wahl. 10, *P. triarthrus* Harlan. *Triarthrus beckii* Green.
Trilobites anormaux ou Battoides: *Agnostus pisiformis* Linné.

———— Histoire naturelle des Crustacés, comprenant l'anatomie la physiologie et la classification de ces animaux. Three vols., with atlas of plates. Paris, 1834–40.
The Trilobites are described in vol. 3, pp. 285–346. *Pleuracanthe* n. g. (referred by author to *Dalmanites*). *Peltura* n. g. (the first species cited by the author under this new genus was: *Peltura scarabæoides* Wahlenburg; the second, *Peltura bucklandi* Milne-Edwards, is a species of the genus *Lichas*). Classifies the species under 13 genera.
I. Division Trilobites proprement dits.
Famille des Isoteliens: 1, *Nileus armadillo* Dalm.; 2, *N. chiton* Pand. 1, *Ampyx nasutus* Dalm., 2, *A. rostratus* Sars., 3, *A. mammillatus* Sars. 1, *Isotelus gigas* Dekay, 2, *I. megalops* Green, 3, *I. cyclops* Green, 4, *I. palpebrosus* Dalm., 5, *I. læviceps* Dalm.
a, Thorax composé de plus de 8 anneaux—6, *Isotelus crassicauda* Wahl., 7, *I. centrotus* Dalm., 8, *I. extenuatus* Wahl.
Especes dont bouclier abdominal est bien distinctement trilobé dans la plus grande partie de sa longueur et présente des sillons transversaux sur le lobe médian ou sur toute sa largeur. 9, *Isotelus dilatus* Brünn, 10, *I. angustifrons* Dalm.
bb. Angles latéraux de la téte arrondis—11, *Isotelus lichtensteinii* Eichw., 12, *I. expansus* Linné, 13, *I. weissii* Eichw.
Famille des Calyméniens. Asaphus I. Espèces dont l'extrémité de l'abdomen est prolongée en pointe ou garnie d'un appendice caudal. 1, *Asaphus caudatus* Brünn, 2, *A. tuberculato-caudatus* Murch., 3, *A. longicaudatus* Murch., 4, *A. mucronatus* Dalm., 5, *A. buchii* Brong., 6, *A. corndensis* Murch., 7, *A. tyranus* Murch., 8, *A. frontalis* Dalm., 9, *A. grandis* Sars.
Homalonotus delphinocephalus Murch. (*Trimerus delphinocephalus* Green.)
Calymene 1. Espèce dont le lobe médian de la téte est divisé en plusieurs lobules. (a) Lobe médian de la téte au moins aussi large en arrière qu' en avant. 1, *Calymene blumenbachii* Brong., 2, *C. callicephala* Green, 3, *C. selenocephala* Green, 4, *C. platys* Green, 5, *C. tristani* Brong., 6, *C. polytoma* Dalm. (aa) Lobe médian de la téte plus large en avant qu'en arrière. 7, *Calymene bellatula* Dalm., 8, *C. odontocephala* Green, 9, *C. sclerops* Dalm., 10, *C. diops* Green, 11, *C. macrophthalma* Brong., 12, *C. downingiæ* Murch., 13, *C. stokesii* n. sp., 14, *C. tuberculata* Murch., 15, *C. anchiops* Green, 16, *C. concinna* Dalm., 17, *C. microps* Green, 18, *C. variolaris* Brong., 19, *C. bufo* Green.
Pleuracanthe n. g., *P. arachnoides* Hoeningh. 1, *Trinucleus caractaci* Murch., 2, *T. fimbriatus* Murch., 3, *T. radiatus* Murch., 4, *T. lloydii* Murch. *Otarion. Ogygia guettardii* Brong., 2, *O. desmarestii* Brong.,

3, *O. murchisonii* Murch. 1, *Paradoxides tessini* Brong., 2, *P. longicauda* Zenk., 3, *P. latus* Zenk., 4, *P. pyramidalis* Zenk., 5, *P. spinulosus* Wahl. *Peltura* n. g., 1, *P. scarabæoides* Wahl., 2, *P. bucklandii* n. sp.
2. Division Trilobites anormaux ou Battoides: *Agnostus pisiformis* Linné.

—— Report on the palæontological researches of M. Marie Rouault in Britanny and Anjou.
In Comptes Rendus Acad. Sci. Paris, 1847, vol. 24, p. 593; Bull. Soc. Géol. France, vol. 4, 1847, p. 309; Quart. Jour. Geol. Soc. London, vol. 4, 1848, pt. 2, Miscellaneous, p. 35.
The Quart. Jour. contains an extract from Bull. Geol. Soc. France, vol. 4, 1847, with brief descriptions of *Trinucleus pongerardi*, etc.

—— Structure des Trilobites.
In Annales Sci. Nat., vol. 12, 1881, 33 pp., pls. 1-3.
A review of C. D. Walcott's "Organization of Trilobites."

Mitchell (John). On some new Trilobites from Bowning, N. S. W.
In Proc. Linhean Soc. New South Wales, 2d series, vol. 2, 1887, p. 435, pl. 16.
Proetus bowningensis n. sp. *Cyphåspis bowningensis* n. sp. *Bronteus longispinifer* n. sp.

—— On a new Trilobite from Bowning.
In Proc. Linnean Soc. New South Wales, 2d series, vol. 3, 1888, p. 397, pl. 16 of vol. 2.
Acidaspis longispinis Mitchell.

Mitchell (Samuel L.) An account of the impressions of a fish in the rocks of Oneida County, New York.
In Am. Monthly Mag., vol. 3, 1818, p. 291.
(*Eurypterus remipes* Dekay.)

Moberg (J. C.) Till frogan om pygidiets byggnad hos *Ctenopyge pecten* Salter.
In Geol. Föreningens Stockholm Förhandl., vol. 14, part 4, 1892, p. 351.

Modeer (A.) Anmerkungen über einige nerikische Versteinerungen. (*)
In Berlin. Gesell. naturf. Freunde, vol. 6, 1775, p. 247, pl. 2, figs. 1-8-12.
(*Agnostus pisiformis*), pl. 2, figs. 1, 2; (*Olenus gibbosus*), pl. 2, figs. 3-5, (*Olenus scarabæoides*), pl. 2, fig. 7.

Moller (Valerian von). Ueber die Trilobiten der Steinkohlenformation des Ural, nebst einer Uebersicht und einigen Ergänzungen der bisherigen Beobachtungen über Kohlen-Trilobiten im Allgemeinen. Moscou, 1867. 1 pl.

In Bull. Soc. Imp. des Naturalistes Moscou, No. 1, 1867.

Phillipsia romeri n. sp., *P. gruenewaldti* n. sp.

The author gives a historical sketch with bibliographical notes of the Carboniferous Trilobites. In addition to the new species described he illustrates *Phillipsia mesotuberculata* McCoy, *P. mucronata* McCoy, *P. eichwaldi* Fischer, *P. pustulata* Schloth, and *Brachymetopus ouralicus* DeVern.

Morris (J.) Catalogue of British fossils comprising all the genera and species hitherto described with references to their Geological distribution and the localities in which they have been found. London, 1843, 222 pp. Second Edition, London, 1854, 372 pp.

Mortimer (C.) Some further account of the before mentioned Dudley fossil by the editor of these transactions (Charles Mortimer).

In Philos. Trans. Roy. Soc. London, vol. 46, 1760, p. 602, pl. 1, 2.

The author states that he has consulted all the books which give figures of insects and crustaceous animals in their natural and petrified state and finds none resembling this Dudley fossil so near as Mr. Klein's insect, " therefore I shall till we get more information call it *Scolopendræ aquaticæ scutata affine animal petrificatum.*"

Müller (O. F.) Entomostraca seu insecta testacea quæ in aquis Daniæ et Norwegiæ reperit, etc. Lipsiæ et Haniæ, 1785. 21 pls.

Cythere n. g.

The generic term *Cytherina* was substituted by Lamarck for Müller's *Cythere*, and has been used by many authors. Prof. J. D. Dana uses the term *Cytherina* to represent a subgenus.

The author describes twenty-two species, only a few of them belong to the Palæozoic Crustacea.

Münster (G. G.) Ueber einige fossile Arten *Cypris* und *Cythere*.

In Jarbuch für Mineral., 1830, p. 60.

Cythere hisingeri n. sp.

The author describes and illustrates *Cythere okeni* n. sp. and *C. scrobiculata* n. sp. now classed under the genus *Leperditia*.

────── Verzeichniss der Versteinungen welche in der Kreis-Naturalien Sammlung zu Baireuth.
In Jahrb., 1836, p. 121.
Asaphus nilssoni n. sp. *Battus tricostatus* n. sp. *Calymene obscura* n. sp.

────── In Golduss' Petrefacta Germaniæ, vol. 2, p. 159, pl. 59, fig. 19.
Sanguinolaria striata (*Esthere stridta*).

────── Die Versteinerungen des Uebergangskalkes mit Clymenien und Orthoceratiten, von ober franken.
In Beiträge zur Petrefacten-Kunde Bayreuth, 1840, pp. 33-121, pl. 5.
Trilobiten: *Calymene* sub. *variolaris* Brong. 2, *C. intermedia* n. sp. 3, *C. granulata* n. sp. 4, *C. lævis* n. sp. 5, *C. sternbergii* Boeck. 6, *C. propinqua* n. sp. 7, *C. articulata* n. sp.
Asaphus ? cawdori Murch, 2, *A. pusillus* n. sp. 3, *A. ? brevis* n. sp. 4, *A. grandis* n. sp. Subgenus *Illænus perovalis ?* Murch. *Paradoxides brevimucronatus* n. sp.
Brontes radiatus n. sp. 2, *B. costatus* n. sp. 3, *B. subradiatus* n. sp. 4, *B. neptuni* n. sp. *Bumastus frankonicus* n. sp. 2, *B.? planus* n. sp. *Harpes speciosus* n. sp. *Trinucleus gracilis* n. sp. 2, *T. wilkensii* n. sp. 3, *T. ellipticus* n. sp. 4, *T. lævis* n. sp. 5, *T. nilsoni* n. sp. 6, *T.? otarion* n. sp. 7, *T. ? gibbosus* n. sp. *Agnostus pisiformis* Linné.

────── Nachtrag zu den Versteinerungen des Uebergangs-Kalkes mit Clymenien von Oberfranken vom Herausgeber.
In Beitrage zu Petrefacten-Kunde. Bayreuth, 1842, pt. 5, pp. 112-128, pl. 10.
Calymene marginata n. sp. 2, *C. furcata* Braum. *Asaphus dubius* n. sp. *Ellipsocephalus hoffii ?* var. *pygmæus* n. var. *Otarion elegans* n. sp. 2, *O. pygmæum* n. sp. *Harpes. Trinucleus? lævis* n. sp., *T. intermedius* n. sp., *T. nilssoni* n. sp.

Murchison (R. I.) The Silurian system, founded on geological researches in the counties of Salop, Hereford, Radnor, etc. London, 2 vols., 1839, 40 pls., and 3 large maps in folio.
Bumastus n. g. *Acidaspis* n. g. *Trinucleus* n. g.
Trilobites of the Upper Silurian (Ludlow & Wenlock): *Homalonotus knightii* König, *H. ludensis* n. sp., *H. delphinocephalus* (*Trimerus delphino-*

cephalus Green), *H. herschelii* n. sp. *Calymene blumenbachii* Brong. *Asaphus caudatus* Brong., *A. tuberculato-caudatus* n. sp., *A.* (*Olenus*) *flabellifer?* Steininger, *A. cawdori* n. sp., *A. subcaudatus* n. sp. *Calymene variolaris* Brong., *C. macrophthalma* Brong., *C.? downingiæ* n. sp., *C. tuberculata* n. sp. *Asaphus stokesii* n. sp., *A. longicaudatus* n. sp. *Bumastus barriensis* n. sp. *Paradoxides bimucronatus* n. sp., *P. quadrimucronatus* n. sp. *Acidaspis brightii* n. sp.

Trilobites of the Lower Silurian (Caradoc & Llandeilo): *Trinucleus caractaci* n. sp., *T. fimbriatus* n. sp., *T. radiatus* n. sp., *T. lloydii* n. sp., *T. nudus* n. sp., *T.? asaphoides* n. sp. *Calymene? punctata* Brünn. *Asaphus powisii* n. sp. *Illænus ? perovalis* n. sp. *Asaphus duplicatus* n. sp.

Trilobites from the Llandeilo: *Asaphus buchii* Brong., *A. tyrannus* n. sp., also var. *ornata*, *A. corndensis* n. sp., *A. ? vulcani* n. sp. *Ogygia murchisoni* n. sp. *Agnostus pisiformis?* Linné.

Cypris inflata, p. 84, has been referred to *Leperditia* by Jones.

—— Metamorphosis of certain Trilobites as recently discovered by M. Barrande.

In Rep. 19th Meeting Brit. Assoc. Adv. Sci., 1849, Trans. Sec., p. 58.

Sir R. Murchison brought before the Section the important discovery of Barrande of the metamorphosis of Trilobites as exhibited in *Sao hirsuta*.

—— On the Silurian rocks of the south of Scotland, with a list and description of the Silurian fossils of Ayrshire, by J. W. Salter.

In Quart. Jour. Geol. Soc. London, vol. 7, 1851, p. 137, pls. 8–10.

Cheirurus gelasinosus Portl. *Asaphus* sp.? *Calymene blumenbachii* Brong, *Phacops stokesii* Milne-Edwards. *Illænus bowmanni?* *Lichas laxatus* McCoy. *Encrinurus punctatus* Brünn.

—— Siluria. A history of the oldest rocks in the British Isles and other countries. London, 1854, geol. map., 41 pls. 4th ed., London, 1867, geol. map., 42 pls. 5th ed., 1872, with atlas of 42 plates and geological map.

The figures are all transferred from the original plates in the "Silurian System."

The appendix contains a table showing the vertical range of the Silurian fossils of Britian.

—— and **Verneuil** (E. de) and **Keyserling** (Alex. von). Geology and palæontology of Russia and the Ural Mountains. London, 1845, 2 vols., 50 pls.

Palæontology, vol. 2; Crustacea, p. 376, pl. 27.

Asaphus expansus Linné *Philipsia eichwaldi* Fisch., *P. ouralica* n. sp.
Calymene odeni Eichw., *C. fischeri* Eichw.

M'Murtrie (H.) Sketches of Louisville and the Fall of the Ohio. Louisville, 1819.

Somatrikelon megalomaton, p. 74, n. g. and n. sp.

This new genus under the law of priority should replace the genus *Phacops* if it were not for the objection to *Somatrikelon*, which means the same as the word Trilobite used for the class. The name given to the species "having very large eyes" corresponds to that given by Dr. Emmrich to the genus *Phacops*.

The author refers "this very curious animal to the vermes class," and remarks: "At first view it appears to have three distinct bodies and one head, which is owing to the former being very broad (nearly one-half an inch) and divided throughout its length by two deep clefts, one on each side of the central part or back, which is elevated a fourth of an inch above the sides, is convex, and terminated at one end by a large head, furnished with unusually prominent eyes and a mouth formed like that of a snapping turtle. From the Falls of the Ohio." The following diagnosis is given: "Corpus tripartitum, rugosum, sinuatum; extremitate altera acuminata curvataque dorso, arcuato, serie tuberculorum utro, bique, capite subrotundo, oculis peramplis."

(*Phacops rana* Green.)

Nathorst (A. G.) Om lager-följdem inom Cambriska formationen vid Andrarum i Skone.

In Ofversigt kongl. Svenska Vet.-Akad. Förhandl., 1869, No. 1, p. 61.
Contains list of fossils and observations.

―――― Om de Kambriska och Siluriska lagren vid Kiviksesperid i Skone jemte anmärkningar om primordial faunans lager vid Andrarum.

In Geol. Föreningens Stockholm Förhandl., No. 37, vol. 3, 1876.

―――― Om det inbördes förhollandet af lagren mit *Paradoxides ölandicus* och *P. tessini* po Öland.

In Geol. Föreningens Stockholm Förhandl., No. 69, vol. 5, 1881.

―――― Om det inbördes aldersförhollandet mellan zonerna med *Olenellus kjerulfi* och *Paradoxides ölandicus*.

In Geol. Föreningens Stockholm Förhandl., No. 71, vol. 6, 1882.

―――― The position of the Olenellus beds.

In Am. Geol., vol. 2, 1888, p. 356.

Newberry (J. S.) China, by Ferdinand F. von Richtofen. Vol. 4. Berlin, 1883.
In Am. Jour. Sci., 3d series, vol. 26, 1883, p. 152.
This is a notice of the palæontological volume.

Nicholson (H. A.) Report upon the Palæontology of the Province of Ontario. Toronto, 1874.
Proetus crassimarginatus Hall. *Dalmanites boothi* Green. *Phacops rana* Green and *Cythere? punctulifera* Hall.

—— A manual of palæontology for the use of students, with a general introduction on the principles of palæontology, in two volumes. Vol. 1. Edinburgh and London, 1879. Crustacea, Chapter xviii.
Harpes, Remopleurides, Paradoxides, Anopolenus, Dicellocephalus, Hydrocephalus, Olenus, Parabolina, Triarthrus, Bohemilla, Conocephalites, Angelina, Ellipsocephalus, Sao, Proetus, Arethusia, Phillipsia, Griffithides, Bathymethopus, Phacops, Dalmanites, Calymene, Homalonotus, Lichas, Trinucleus, Ampyx, Dionide, Æglina, Asaphus, Ogygia, Barrandia, Niobe, Nileus, Stygina, Illænus, Acidaspis, Cheirurus, Sphærexochus, Areia, Placoparia, Cromus, Encrinurus, Cybele, Dindymene, Bronteus, Agnostus, Microdiscus, Protichnites, Aristozoe, Beyrichia, Leperditia, Primitia, Entomis, Cythere, Eurypterus, Pterygotus, Isochilina, Cypris, Candona, Orozoe, Callizoe, Kirkbya, Moorea, Crypidina, Cytherella, Cytherellina, Æchmina, Bairdia, Hymenocaris, Caryocaris, Peltocaris, Discinocaris, Aptychopsis, Pterocaris, Cryptocaris, Ceratiocaris, Dithyrocaris, Aspidocaris, Estheria, Psilocephalus, Hemiaspis, Exapinurus, Pseudoniscus, Stylonurus, Neolimulus, Prestwichia, Belinurus.

—— and **Etheridge, Jr.** (R.) A monograph of the Silurian fossils of the Givan District, in Ayrshire, etc. Vol. 1, 1878–80, 24 pls.
Fasciculus I. Rhizopoda, Actinozoa, and Trilobites (1878).
Acidaspis callipareos Wyv. Thom., *A. grayæ* R. Eth., *A. hystyix* Wyv. Thom., *A. lalarge* Wyv. Thom. *Cheirurus bimucronatus* Murch., *C. clavifrons* Dalm.? *C. gelasinosus* Portl., *C. trispinosus* Young. *Cybele bellatula* Dalm. *Dindymene cordai* n. sp. *Encrinurus punctatus* Brünn., *E. punctatus* var. *arenaceus* Salt., also var. *calcareus*, *E. sexcostatus*, *E. variolaris* Brong. *Lichas barrandei* Fletch., *L. hibernicus* Portl., *L. laxatus* McCoy. *Phacops downingiæ* Murch., *P. stokesi* Milne-Edw., *P. brongniarti* Portl. *Sphærexochus mirus* Beyr.
Phacops truncato-caudatus n. sp. is referred to *Dionide* sp. ind. on p. 293.
Fasciculus II. Trilobita, Phyllopoda, Cirripedia, and Ostracoda (1879).
Bronteopsis n. g. *Lichas geikiei* E: & N. *Cyphaspis megalops* McCoy.

Calymene blumenbachii Brong. *Remopleurides colbii* Portl., *R. laterispinifer* Portl., *R.* sp.? *R. (Caphyra) barrandii* n. sp. *Asaphus radiatus* Salt., *A. gigas* Dekay ? *Illænus bowmani* Salt., *I. thomsoni* Salt., *I. læmulus* Salt.? *I. nexilis* Salt., *I. davisii* Salt., *I. rosenbergii* Eichw.? *I. crassicauda* Wahl.? *I. murchisoni* Salt.?, *I. macallumi* Salt. *Bronteus andersoni* n. sp., *B.* sp.? *Bronteopsis* n. g., *B. scotica* Salt. MS. *Proetus givanensis* n. sp., *P. procerus* n. sp. *Ampyx rostratus* Sars., *A. macallumi* Salt. MS., *A. macconochiei* n. sp., *A. hornei* n. sp. *Trinucleus seticornis* His. var. *bucklandi* Barr., *T. concentricus* Eaton, *T.* sp.? *Salteria primæva* Wyv. Thom. *Agnostus agnostiformis* McCoy. *Phacops brongniarti* Portl. *Cheirurus bimucronatus* Murch., *C. clavifrons* Dalm., *C.* sp.? *Encrinurus punctatus* var. *calcareus* Salt. *Cybele rugosa* Portl. *Acidaspis* sp.? *Ampyx* sp.? *Solenocaris solenoides* Young. *Pinnocaris lapworthi* Eth. jr. *Peltocaris* sp.? *Dictyocaris* sp.? *Turrilepas scotica* Eth. jr. *Cythere aldensis* McCoy, also. var. *major* Jones, *C. grayana* Jones, *C. wrightiana* J. & H. *Beyrichia klödeni* McCoy, *B. impendens* Jones, *B. comma* Jones. *Primitia barrandiana* Jones. *Entomis globulosa* Jones.

The authors follow Salter's division of *Remopleurides* in two sections:
1. *Remopleurides.* The glabella devoid of furrows.
2. Section *Caphgra.* Glabella with three pairs of furrows, not inflated.

The genus *Illænus* is subdivided into the sections *Dysplanus* and *Illænus.* Under the genus *Ampyx* the authors use *Raphiophorus, Lonchodomus* and *Ampyx* in the sectional sense. On p. 288 *Ampyx macconochiei* n. sp. is referred to the genus *Trinucleus.*

Fasciculus III. The Annelida and Echinodermata, with supplements on the Protozoa, Cœlenterata, and Crustacea (1880).

Staurocephalus sp.? *Cyclopyge rediviva* Barr., *C. armata* Barr. *Trinucleus macconochiei* E. & N., *T.* sp.? *Dionide lapworthi* n. sp., *D.* sp.? *Agnostus perrugatus* Barr.? *Turrilepas peachii* n. sp.

Nieszkowski (J.) Versuch einer Monographie der in den silurischen Schichten der Ostseeprovinzen vorkommenden Trilobiten.

In Archiv Naturk. Liv-Ehst. u. Kurl., vol. 1, 1857, p. 517, 3 pls.

Phacops stokesi Milne-Edw., *P. downingiæ* Murch., *P. dubius* n. sp., *P. conophthalmus* Boeck., *P. truncato-caudatus* Portl., *P. caudatus* Brünn. *Calymene blumenbachii* Brong., *C. brevicapitata* Portl. *Asaphus expansus* Linné, *A. tyrannus* Murch, *A. raniceps* Dalm., *A. platycephalus* Stokes, *A. acuminatus* Boeck, *A. latisegmentatus* n. sp. *Proetus concinnus* Dalm., *P. latifrons* McCoy, *P. pulcher* n. sp., *P. ramisulcatus* n. sp. *Cyphaspis megalops* McCoy. *Lichas marginitifer* n. sp., *L. deflexa* Ang. *L. eichwaldi* n. sp., *L. verrucosa* Eichw., *L. ornata* Aug., *L. darlecarlica* Ang., *L. laticeps* Ang., *L. platyura* n. sp., *L* sp.? *Illænus schmidti* n. sp., *I. centrotus* Dalm., *I. crassicauda* Wahl., *I. barriensis* Murch. *Bronteus signatus* Phill. *Cheirurus spinulosus* n. sp., *C. exul* Beyr., *C. octolobatus* McCoy.

Sphœrexochus mirus Beyr., *S. deflexus* Aug., *S. conformis* Aug., *S. cephaloceras* n. sp., *S. minutus* n. sp., *S. hexadactylus* n. sp. *Encrinurus punctatus* Brünn., *E. multisegmentatus* Portl., *E. sexcostatus* Salt. *Zethus bellatulus* Dalm., *Z. rex* n. sp., *Z. atractopyge* McCoy, *Z. brevicauda* Aug. *Amphion fischeri* Eichw., *A. actinurus* Dalm. (*A. fischeri* not *Calymene actinurus* Dalm.) *Platymetopus illænoides* n. sp.

—— Der *Eurypterus remipes* aus den ober-silurischen Schichten der Insel Oesel.

In Archiv. Naturk. Liv-Ehst. u. Kurl., vol. 2, 1858, p. 299, 2 plates.

—— Zusätze zur Monographïe der Trilobiten der Trilobiten der Ostseeprovinzen, nebst der Beschreibung einiger neuen ober-silurischen Crustaceen.

In Archiv. Naturk. Liv-Ehst. u. Kurl., vol. 1, 1859, p. 345, 2 pls.

Exapinurus n. g. *Pseudoniscus* n. g. *Asaphus truncatus* n. sp., *A. lepidurus* n. sp., *A. acuminatus* Boeck. *Cyphaspis elegantulus* Aug. *Lichas conico-tuberculata* n. sp., *L. angusta* Beyr., *L. gothlandica* Aug. *Bronteus laticauda* Wahl. *Cheirurus spinulosus* Nieszk., *C. ornatus* Dalm., *C.* sp.? *Sphœrexochus cephalocerus* Nieszk., *S. cranimum* Kut., *S. pseudohemicranium* n. sp. *Encrinurus obtusus* Aug. *Trilobites* sp.? *Bunodes lunula* Eichw., *B. rugosus* n. sp. *Exapinurus schrenkii* n. sp. *Pseudoniscus aculeatus* n. sp.

Novák (Ottomar). Studien an Hypostomen böhmischer Trilobiten. No. 1.

In Sitzungsberichte k. böhm. Gesell. Wiss., Jahrg., 1879, p. 475.

The author illustrates the hypostoma of *Paradoxides sacheri* Barr. *Harpes venulosus* Cord. *Dalmanites rugosa* Cord. *Calymene declinata* Cord. *Lichas palmata* Barr. *Cromus intercostatus* Barr. *Bronteus palifer* Beyr.

—— Studien an Hypostomen böhmischer Trilobiten. No. 2.

In Sitzungsberichte k. böhm. Gesell. Wiss., Jahrg., 1884, 1 pl.

The author figures the hypostoma of *Harpes venulosa* Cord., *H. l'orbignyanus* Barr. *Harpina prima* Barr., *H. benignensis* Barr. *Asaphus ingens*, *A. tyrannus* Murch., *A. striatus* Boeck. *Ogygia desiderata* Barr., *O. cordensis* Murch., *O. selwynii* Salt., *O. homfrayi* Salt., *O. scutatrix* Salt. *Niobe discreta* Barr., *N. peltata* Salt., *N. insignis* Linrs.

—— Zur Kenntniss der Böhmischen Trilobiten.*

In Beiträge zur Palæontolographie von Oesterrich, Wien, 1884, 5 pls.

Harpes, *Remopleurides*, *Phillipsia*, *Dalmanites*, *Calymene*, *Homalonotus*, *Æglina*, *Trinucleus*, *Ampyx*, *Dionide*, *Asaphus*, *Barrandia*, *Ptychocheilus*, *Illænus*, *Acidaspis*, *Cheirurus*, *Placoparia*, *Cromus*, *Bronteus*, *Agnostus*.

―――― Studien an Hypostomen böhmischer Trilobiten. No. 3.
In Sitzungsberichte k. böhm. Gesell. Wiss., Jahrg., 1885, p. 581, 1 pl.
Phillipsinella n. g.
The author illustrates *Phillipsinella parabola* Barr. *Phillipsia eichwaldi*, P. sp.?

―――― Remarques sur le genre *Aristozoe* Barrande.
In Sitzungsberichte k. böhm. Gesell. Wiss., Jahrg., 1885. p. 239, 1 pl.
Aristozoe regina Barr.

―――― Nouveau Crustacé phyllocaride de l'étage F-f_2 en Bohême.
In Sitzungsberichte k. böhm. Gesell. Wiss., Jahrg., 1885, p. 343, 1 pl.
Ptychocaris n. g., *P. parvula* Nov., *P. simplex* Nov.

―――― Studien an Hypostomen böhmischer Trilobiten. No. 4.
In Sitzungsberichte k. böhm. Wiss., Jahrg., 1886, p. 429, plate.
The author illustrates *Encrinurus punctatus* Brünn., *E. (Cromus) beaumonti* Barr., *E. (Cromus) transiens* Barr., *E. (Cromus) bohemicus* Barr., *E. (Cromus) intercostatus* Barr.

―――― Zur Kenntniss der Fauna der Étage F–f_1 in der palæozoischen Schichtengruppe Böhmens.
In Sitzungsberichte k. böhm. Gesell. Wiss., Jahrg., 1886, p. 660, 2 pls.
Aristozoe solitaria Nov.

―――― Note sur Phasganocaris, nouveau Phyllocaride de l'étage F–f_2 en Bohême.
In Sitzungsberichte k. böhm. Gesell. Wiss., Jahrg., 1886, p. 498, 1 pl.
Phasganocaris pugio Barr.

―――― Vergleichende studien an einigen Trilobiten aus dem Hercyn von Bicken, Wildungen Greifenstein und Böhmen.
In Dames & Kayser's Palæont. Abhandl. Neue Folge, vol. 1, part 3, 1890, 46 pp., 5 plates.
The following genera are discussed: *Proetus*, *Arethusina*, *Tropidocoryphe* n. g. *Phaëtonellus* n. g., *Acidaspis*, *Cheirurus* and *Bronteus*.
Proetus eremita Barr., *P. filicostata* n. sp., *P. holzapfeli* n. sp., *P. köneni* Maur., *P. myops* Barr., *P. orbitatus* Barr., *P. planicauda* Beyr., also var. *rhenana* Nov., referred to subgenus *Phaëtonellus*. *Proetus unguloides* Barr., *P. waldschmidti* n. sp. *Arethusina beyrichi* n. sp., *A. peltata* n. sp. *Cyphaspis hydrocephalus* A. Römer. *Phacops breviceps* Barr., also var.

minuscula Nov. and *rhenana* Nov., *P. fecundis* Barr., *P.* cf. *ferdinandi* Kayser, *P.* cf. *zorgensis* Kayser, *P.* sp.? *Harpes kayseri* n. sp., *H. reticulatus* Cord. *Lichas (Arges) haueri* Barr., *L. (Arges) maureri* n. sp. *Acidaspis pigra* Barr., *A. (Trapelocera) vesiculosa* Beyr. *Cheirurus (Crotalocephalus) cordai* Barr. *Bronteus speciosus* Corda., *B. dormitzeri* Barr. var. *applanata*. *Harpes montagnei* Cord., *H. reticulatus* Cord. *Cyphaspis scuticauda* n. sp. *Harpes fornicatus* n. sp. *Bronteus brevifrons* Barr. *Proetus crassirhachis* A. Römer, *P. crassimargo* A. Römer, *P. frechi* n. sp., *P. orbitatus* Barr.

The author considers *Phacops filicostatus* of such distinctive importance that he has proposed to designate the group which it represents by the new generic name of *Tropidocoryphe*.

This species is remarkable for its broad flabellate pygidium and rapidly-tapering glabella.

Oehlert (D.) Sur les fossiles dévoniens du département de la Mayenne.

In Bull. Soc. Géol. France, 3d series, vol. 5, 1877, p. 578, pl. 9.

Cryphæus michelini Marie Rouault, *C. jonesi* n. sp., *C. munieri* n. sp. *Homalonotus gervillei* De Vern. *Leperditia britannica* Marie Rouault. *Primitia fischeri* n. sp.

—— Description de *Goldius gervillei* Barrande.

In Bull. Soc. d'Études Sci., Angers, 1885, 1 pl.

—— Etudes sur quelques Trilobites du groupe des Proetidæ.

In Bull. Soc. d'Études Sci., Angers, 1885, 2 pls.

The author divides the Proetidæ into two divisions on the form of the glabella.

Sec. A. Glabella conic, genera *Proetus, Phæton, Dechenella* and *Brachymetopus*.

Sec. B. Glabella pyriform, genera *Phillipsella, Phillipsia* and *Griffithides*. *Brachymetopus ouralicus* De Vern, *B. maccoyi* Portl. *Dechenella haldemani* Hall. *Proetus micropygus* Cord. *Dechenella verneuili* Barr. *Phillipsia truncaluta* Phill., *P. colei* McCoy. *Proetus astyanax* Cord., *P. unguloides* Barr. *Phillipsia eichwaldi* var. *mucronata* McCoy, *P. gemmulifer* Phill. *Proetus intermedius* Barr., *P. ryckholti* Barr. *Griffithides acanthiceps* Woodw., *G. seminiferus* Phill., *G. globiceps* Phill. *Proetus (Cyphaspis) depressa* Barr. *Phillipsella (Phillipsia) parabola* Barr. *Proetus (Cyphaspis) depressa*. *Cyphaspis halli* Barr. *Arethusina konincki* Barr. *Harpes ungula*. *Dechenella verticalis* Burm. *Calymene baylei* Barr. *Conocephalites emmrichi* Barr., *C. striatus* Emm. *Phillipsia derbyensis* Martin. *Dalmanites auriculata* Dalm. *Ogygia buchii* Bront., *O. desiderata* Barr. *Asaphus alienus* Barr. *Griffithides acanthiceps* Wood. *Phacops fecundus* var. *dejener* Barr. *Asaphus expansus* Linné. *Bronteus brongniarti* Barr., *B. tenellus* Barr. *Phillipsella (Phillipsia) parabola* Barr., *P. colei* McCoy.

―――― Études sur quelques fossiles dévoniens de l'ouest de la France.
In Bibliothèque de l'École des Hautes Études, Sec. Sci. Nat., vol. 33, 1887, 5 pls.
Proetus oehlerti Bayle. *Phacops portieri* Bayle.

Oehlert and Davoust. Sur le Devonien du Department de la Sarthe.
In Bull. Soc. Géol. France, 3d series, vol. 7, 1880-81, p. 693, pl. 13.
Proetus guerangeri n. sp. *Bronteus verneuili* n. sp.

Oehlert (D. P.) Sur la Dévonien des environs d'Angers.
In Bull. Soc. Géol. France, 3d series, vol. 17, 1889, p. 742, 3 pls.
Cryphina n. g. *Dalmanites (Probolin)* n. subgen. *Proetus gosseleti* Barrois, *P. rondeaui* n. sp. *Dechenella ? incerta* n. sp. *Acidaspis* sp.? *Phacops potieri* Bayle. *Cryphina* n. g., *C. andegavensis* n. sp. *Cryphœus* cf. *barrandei* Cailliaud. *Dalmanites stillifer* var. Burm., *D. (Probolium) galloisi* n. sp. *Goldius gervillei* Barr., *G. galloisi* n. sp. *Goldius* sp.? *Calymene reperta* n. sp. *Aristozoe aff. memoranda* Barr.
The genus *Cryphina* is distinguished from *Cryphœus* Green, which has five spines on the pygidium, in having seven spines.
In *Probolium* the head has on the right of the front a median prolongation, generally well developed, and usually ending in a bifid expansion.

Oldham (J.) On *Griffithides globiceps*.
In Jour. Geol. Soc. Dublin, vol. 3, pt. 3, 1846, p. 188, pl. 2.
Redrawn, pl. 6, fig. 1 a, b, Woodward's "Monograph British Carboniferous Trilobites," London, 1883.

Ordway (Albert). On the supposed identity of *Paradoxides harlani* Green with *P. spinosus* Boeck.
In Proc. Boston Soc. Nat. Hist., vol. 8, 1861, p. 1, 2 figs.

―――― The organization of the Trilobites.
In Geol. Nat. Hist. Repertory, vol. 1, 1865, p. 12.*

Owen (D. D.) Report of a geological exploration of part of Iowa, Wisconsin and Illinois, made under instructions from the Secretary of the Treasury of the United States, in the autumn of the year 1839. Washington, 1840.
In 26th Cong., H. of Rep., Doc. 233, 2d ed., 1844; 28th Cong., Sen. Doc. 407, 18 pls. of fossils.
Pl. 12, fig. 3, *Calymene bufo* Green; pl. 16, fig. 1, *Illænus trentonensis ?*

(referred by Dr. D. D. Owen, in 1852, to *Illænus ovalis*); pl. 17, fig. 9, *Thaleops* ? ; pl. 18, fig. 2, *Asaphus:* fig. 10 (*Asaphus platycephalus* Stokes); fig. 11, *Ceraurus*.

——— Report of a geological reconnoissance of the Chippewa Land District of Wisconsin and incidentally of a portion of the Kickapoo country, and a part of Iowa and of the Minnesota Territory, etc. Washington, 1848.

This report contains figures of the following fossils, which were afterwards described by Dr. Owen under the names of *Dikelocephalus minnesotensis*, pl. 7, figs. 2, 3; *Lonchocephalus hamulus*, pl. 7, fig. 5; *Dikelocephalus ? iowensis*, pl. 7, fig. 1.

——— Report of a geological survey of Wisconsin, Iowa and Minnesota, and incidentally a portion of Nebraska Territory. Philadelphia, 1852, 638 pp., 27 pls., 16 sections and maps.

Dr. Owen described four new genera of trilobites under the names of *Dikelocephalus*, *Lonchocephalus*, *Crepicephalus* and *Menocephalus*. The second genus is now classed as a synonym to *Anomocare*, with *Anomocare chippewaensis* and *A. hamulus* described in the report. There is a third species mentioned by Owen on p. 624, and also pl. 1a, fig. 15. Pygidium of *Lonchocephalus ?*, with long, slender, divergent caudal spines; also a similar pygidium, figured pl. 1a, fig. 13, as *Dikelocephalus ? iowensis*. Walcott, Bull. U. S. Geol. Sur., No. 10, p. 36, places the latter species under a subgenus to *Ptychoparia*, using Dr. Owen's *Crepicephalus* for it, on account of the peculiar pygidium, overlooking, however, the fact that Owen has referred a similar pygidium directly to his genus Lonchocephalus on p. 624 of his report. Under the law of priority the genus *Ptychoparia* replaces *Crepicephalus*, and the latter name should be cancelled *in toto*, and not be retained in a modified sense. *Lonchocephalus* forms an exception to this rule, if we divide the genus into two sections, retaining *Anomocare* for the two species described by Owen, and *Lonchocephalus*, in a restricted sense, for "the pygidium with long, slender, divergent caudal spines," including the *Dikelocephalus ? iowensis* Owen.

The author also describes and illustrates the following species: 1, *Dikelocephalus minnesotensis* n. sp. 2, *D. miniscænsis* n. sp. 3, *D. pepinensis* n. sp. 4, *D. ? iowensis* n. sp. 5, *D. granulosus* n. sp. 6, *Lonchocephalus chippewaensis* n. sp. 7, *L. hamulus* n. sp. 8, *L.* sp.? 9, *Menocephalus minnesotensis* n. sp. 10, *Crepicephalus miniscænsis* n. sp. 11, *C. wisconensis* n. sp.

The following species are mentioned or illustrated: 12, *Phillipsia* sp. undet. 13, *P. granulifera ?* 14, *P. pustulata*. 15, *Calymene senaria* Conrad, pl. 2a, fig. 12. 16, *C. crassimarginata*, pl. 3a, fig. 6. 17, *C.* n. sp., pl. 2a, fig. 6. 13, *Asaphus gigas.* 19, *A.* n. sp. 20, *A. iowensis,* pl. 2a, figs. 1-7. 21, *Illænus crassicauda* Wahl. 22, *I.* sp. undet. 23, *I. ovalis* n. sp. 24, *I.* n. sp., pl. 2a, fig. 10. 25, *Ceraurus pleurexanthemus* Green 26, *Phacops macrophthalma* Brong. 27-28, *P.* n. sp. 29, *P. callicephalus.* 30, *Lichas* sp.? 31-32, *Cytherina* sp. undet. 33, *Bumastus barriensis* Murch.

Owen (Richard). Description of the impressions and foot-prints of the *Protichnites* from the Potsdam sandstone of Canada.

In Quart. Jour. Geol. Soc. London, vol. 8, 1852, p. 214, pls. 9-14a.

——— Palæontology, or a systematic summary of extinct animals and their geological relation. Edinburgh, 1861.

Class 3. Crustacea, pp. 45-51.

The author illustrates *Leperditia baltica, Entomoconchus scouleri, Beyrichia complicata, Dithyrocaris scouleri, Pterygotus anglicus, Belinurus bellulus, Illænus davisii, Phacops caudata, Calymene blumenbachii, Trinucleus ornata, Agnostus trinodus.*

Pacht (R.) Der Devonische kalk in Livland, 1849.

Asmusia membranacea n. sp.

——— Geognostische Untersuchungen zwischen Orel Woronasch und Simbirsk in Jahre, 1853.

In Beiträge zur Kenntniss des russischen Reiches und der angränzenden Länder Asiens, St. Petersburg, 1858.

Packard (A. S., Jr.) On the embryology of *Limulus polyphemus.*

In Proc. Am. Assoc. Adv. Sci., 1870, 19th Meeting, p. 247.

Notes on the larva of Trilobites, etc.

——— The structure of the eye of Trilobites.

In Am. Naturalist, vol. 14, 1880, p. 503.

——— A monograph on the Phyllopod Crustacea of North America, with remarks on the order Phyllocarida.

In Twelfth Rep. U. S. Geol. Survey Territories (Hayden), Washington, 1883, pt. 1, p. 295, pls. 1-38.

The author illustrates *Echinocaris punctatus* Hall, *E. multinodosus* Whitf., *E. sublevis* Whitf. *Discinocaris browniana* H. W. *Dithyrocaris neptuni* Hall. *Estheria ovata* Jones. *Leaia leidyi* Jones. *Estheria dawsoni* Pack.

―――― Types of Carboniferous *Xiphosura* new to North America.

In Am. Naturalist, vol. 19, 1885, p. 291.
Dipeltis n. g. *Belinurus lacoei* n. sp. *Euroops longispina* n. sp. *Cyclus americanus* n. sp. *Dipeltis diplodiscus* n. sp.

―――― The *Syncarida*, a group of Carboniferous Crustacea.

In Am. Naturalist, vol. 19, 1885, p. 700.
Acanthotelson.

―――― On the *Gampsonychidæ*, an undescribed family of fossil Schizopod Crustacea.

In Am. Naturalist, vol. 19, 1885, p. 790; Mem. Natl. Acad. Sci., vol. 3, 1886, p. 129, pl. 3.
Palæocaris typus M. & W. *Gampsonyx fimbriatus* J. & von M.

―――― On the *Anthracoridæ*, a family of Carboniferous Macrurous Decapod Crustacea allied to the *Eryonidæ*.

In Am. Naturalist, vol. 19, 1885, p. 880; Mem. Natl. Acad. Sci., vol. 3, 1886, p. 135, pl. 4.
Anthrapalæon gracilis M. & W., *A. grossarti* Salt.

―――― Discovery of the thoracic feet in a Carboniferous *Phyllocaridan*.

In Am. Philos. Soc., vol. 23, 1886, p. 380, pl.
Cryptozoe n. g., *C. problematicus* n. sp.

―――― On the class *Podostomata*, a group embracing the Merostomata and Trilobites.

In Am. Naturalist, vol. 20, 1886, p. 1060.
Merostomata: *Xiphosura, Synziphosura, Eurypteridæ*.

―――― On the Syncarida, a hitherto undescribed synthetic group of extinct Malacostracous Crustacea.

In Mem. Natl. Acad. Sci., vol. 3, pt. 2, 1886, p. 123, pls. 1, 2.
Acanthotelson stimpsoni M. & W., *A. magister* Pack.

―――― On the *Gampsonychidæ*, an undescribed family of fossil Schizopod Crustacea.

In Mem. Natl. Acad. Sci., vol. 3, pt. 2, 1886, p. 129, 1 pl.
Gampsonyx fimbriatus J. & von M. *Palæocaris typus* M. & W.

―――― On the *Anthracaridæ*, a family of Carboniferous Macrurous Decapod Crustacea.
In Mem. Natl. Acad. Sci., vol. 3, pt. 2, 1886, p. 135, pl.
Anthrapalæmon gracilis M. & W.

―――― Note on the Palæozoic Shrimps (Carididæ).
Archicaris salteri. Crangopsis soliates.

―――― On the Carboniferous Xiphosurous fauna of North America.
In Mem. Natl. Acad. Sci., vol. 3, pt. 2, 1886, p. 143, 8 figs., 3 pls.
Protolimulus n. g. *Cyclus americana* Pack. *Dipeltis diplodiscus* Pack.
Belinuridæ: *Prestwithia danæ* Meek, *P. longispina* Pack. *Belinurus lacoei* Pack. *Prestwichia rotundatus* Woodw. *Protolimulus eriensis* Williams. *Palæocaris typus* M. & W. *Anthrapalæmon gracilis* M. & W.

Page (David). On the *Pterygotus* and *Pterygotus* beds of Great Britain.
In Rep. 25th Meeting Brit. Assoc. Adv. Sci., for 1855, Trans. Sec., p. 89. London, 1856.
The genus *Stylonurus* was proposed by Mr. David Page in his paper read before the Association, and the name published in his " Advance text-book of geology," 1856, where he also figured and named the then only known species *Stylonurus powriei*, but without giving any description.

―――― Further contributions to the palæontology of the Tilestones or Silurio-Devonian strata of Scotland.
In Rep. 28th Meeting Brit. Assoc. Adv. Sci., for 1858, Trans. Sec., p. 104. London, 1859.
Two new species are provisionally named *Stylonurus spinipes* and *S. clavipes* in allusion to their pointed style-shaped caudal termination, and the characteristic form of their swimming paddles or third pair of organs, which spring from the under side of their cephalo-thorax.
The author also mentions the genera *Beyrichia, Ceratiocaris, Himanthopterus, Pterygotus* and *Kampecaris*.

―――― Advance text-book of geology. London, 1856.
Slimonia n. g. *Stylonurus* n. g. *Kampecaris*.
In the second edition, 1859, Mr. David Page gives restorations of *Pterygotus anglicus. Slimonia acuminatus. Stylonurus powriei, S. spinipes. Eurypterus clavipes, E. scouleri, and Limuloides (Belinurus) rotundatus*.

―――― Note on the genus *Stylonurus* from the old red sandstone of Forfarshire.
In Proc. Royal Phys. Soc. Edinburgh, session 1864-65, p. 230.

Pander (C. H.) Beiträge zur Geognosie des russischen Reiches. St. Petersburg, 1830; Leipsig, 1839, 31 pls.
    Amphion n. g. *Zethus* n. g. *Nileus armadillo* Dalm., *N. chiton* n. sp. *Asaphus cornigerus* Broug. *Illænus crassicauda* Wahl. *Calymene macrophthalma* Broug., *C. sclerops* Dalm. *Amphion frontiloba (Asaphus fischeri* Eichw.) *Zethus uniplicatus* n. sp. This species is referred by Schmidt to *Cheirurus (Pseudosphærexochus) hemicranium* Kut. *Zethus verrucosus* n. sp. This species is referred by Schmidt to *Cybele bellatula* Dalm.

Parkinson (James). Organic remains of a former world, an examination of the mineralized remains of the vegetable and animal antediluvian world generally termed extraneous fossils. London, 1808, 3 vols.
Crustacea, vol. 3, letter 16,* pl. 17.
    The Dudley fossil *(Calymene blumenbachii* Broug.) A trilobite from Llanelly *(Ogygia buchi* Brong), pl. 17, fig. 16. *(Encrinurus variolaris* Broug.) pl. 17, fig. 17. *(Dalmanites caudatus* Brünn.) pl. 17, fig. 18. *(Belinurus lunatus* Martin.)

Payton (—). Fossils found in the Transition limestone at Dudley, from the collection of Mr. Payton, at Dudley. London, 1827, 2 pp., 2 pls.
    Calymene blumenbachii Broug. and *Asaphus caudatus* Brünn.

Peach (B. N.) On some new Crustaceans from the Lower Carboniferous rocks of Eskdale and Liddesdale.
    In Trans. Royal Soc. Edinburgh, vol. 30, pt. 1, 1880, p. 73, 4 pls.
    Ceratiocaris scorpioides n. sp., *C. elongatus* n. sp. *Anthrapalæmon etheridgii* n. sp., *A. parki* n. sp., *A. traquairii* n. sp., *A. macconochii* Etheridge jr., *A. ornatissimus* n. sp., *A. formosus* n. sp. *Palæocrangon eskdalensis* n. sp., *S. typus* M. & W. *Palæocaris scoticus* n. sp.

——— Further researches among the Crustacea and Arachnida of the Carboniferous rocks of the Scottish border.
    In Trans. Royal Soc. Edinburgh, vol. 30, pt. 2, 1880, p. 511, 2 pls.
    Glyptoscorpius n. g. *Acanthocaris* n. g. *Pseudo-Galathea* n. g. *Acanthocaris attenuatus* n. sp. *Anthrapalæmon formosus* Peach., *A. etheridgii* var. *latus* n. var. *Pseudo-Galathea rotunda* n. sp., *P. ornatissima* Peach. *Palæocrangon elegans* n. sp. *Palæocaris scoticus* Peach. *Glyptoscorpius perornatus* n. sp., *G. caledonicus* Salt. *(Cycadites caledonicus* Salt.) *Prestwichia alternata* n. sp. *Cyclus testudo* n. sp.

―――― On the occurrence of *Pterygotus* and *Limuloid* in the Caithness Flagstones, and on the nature and mode of formation of "Adam's plate."
In Proc. Royal Phys. Soc. Edinburgh, 1883, p. 343, pl.
Pterygotus dicki n. sp. Cast of a Limuloid allied to *Belinurus*.

―――― and **Horne** (J.) On the *Olenellus* zone in the Northwest Highlands.
In Quart. Jour. Géol. Soc. London, vol. 48, 1892, p. 227, plate 5.
Olenellus lapworthi n. sp.

Phillips (John). On the occurrence of some minute fossil Crustaceans in Palæozoic rocks.
In Rep. 11th Meeting Brit. Assoc. Adv. Sci., for 1841, Trans. Sec., p. 64. London, 1842.
Cyprides of the Yorkshire coal shales.

―――― Illustrations of the geology of Yorkshire. Part 2. The mountain limestone district. London, 1836, p. 239.
Under the generic term of *Asaphus* the author illustrates and describes the following Trilobites: *Phillipsia quadrilimbus* n. sp. *Griffithides obsoletus* n. sp. *Phillipsia granuliferus* n. sp. *Griffithides seminiferus* n. sp. *Phillipsia gemmuliferus* n. sp., *P. truncatula* n. sp. *Asaphus raniceps* has been referred to *Phillipsia derbyensis*. *Griffithides globiceps* n. sp. *Cypridiform* shell. *Agnostus ? radialis* n. sp.

―――― Figures and descriptions of the Palæozoic fossils of Cornwall, Devon and West Somerset, observed in the course of the Ordnance Geological Survey of that district. London, 1841, 60 pls. Crustacea, p. 127.
Harpes macrocephalus Goldf. *Calymene sternbergii* Münst., *C. granulata* Münst., *C. lateillii* Steininger, *C. lævis* Münst. *Asaphus granuliferus* Phill. *Homalonotus knightii ? Bronteus flabellifer* Goldf., *B. signatus* Murch.

―――― The Malvern Hill compared with the Palæozoic district of Abberley, Woolhope, May Hill, Tortworth and Usk.
In Mem. Geol. Survey Gt. Brit., vol. 2, pt. 1, London, 1848. Palæontological appendix to Prof. John Phillips's memoir on the "Malvern Hills compared with the Palæozoic district of Abberley," etc., by John Phillips and J. W. Salter. *Ibid*, pp. 331–386, 30 pls.

For list of genera, see Salter (J. W.) and Phillips (John). The species therein described have the initials of the respective authors attached. The following Crustacea are accredited to Prof. Phillips: *Olenus bisulcatus* n. sp. and *O. humilis* n. sp.

Pictet (F. J.) Traité de paléontologie, ou histoire naturelle des animaux fossiles considérés dans leurs rapports zoologique et géologique. 4 vols. and atlas. First ed., Paris, 1844–46. Second ed., Paris, 1853–57. Crustacea, vol. 2, 1854, p. 413.

The author arranges the families and genera as follows: Harpidæ: *Harpes*. Paradoxidæ: *Remopleurides, Paradoxides, Hydrocephalus, Sao, Arionellus, Ellipsocephalus, Olenus, Peltura, Triarthrus, Conocephalites*. Calymenidæ, 1st section, Proetidæ: *Proetus, Phillipsia, Cyphaspis, Arethusina, Harpes;* 2d section, Phacopidæ: *Phacops, Dalmania;* 3d section, Calymenidæ: *Calymene, Homalonotus*. Lichasidæ: *Lichas*. Trinucleidæ: *Trinucleus, Ampyx, Dionide*. Asaphidæ: *Asaphus, Symphysurus, Ogygia*. Æglinidæ: *Æglina*. Illænidæ: *Illænus, Nileus*. Odontopleuridæ: *Acidaspis, Cheirurus, Placoparia, Sphærexochus, Staurocephalus, Deiphon, Zethus, Dindymene*. Amphionidæ: *Amphion, Cromus, Encrinurus*. Brontidæ: *Bronteus*. Agnostidæ: *Agnostus*. Phyllopodæ: *Apusiens, Dithyrocaris, Apus, Estheria*. Copepodæ: *Eurypterus*. Cyproidæ: *Cythere, Bairdia, Beyrichia, Leperditia, Cypris*. Cypridinæ: *Cyprella, Cypridella*. Xiphosuræ: *Limulus, Belinurus, Pterygotus*. Amphipodæ: *Gampsonyx*.

The author illustrates Plate 44.—*Harpes ungula* Sternb. *Remopleurides radiana* Barr. *Paradoxides bohemicus* Boeck. *Hydrocephalus saturnoides* Barr. *Sao hirsuta* Barr. *Olenus gibbosus* Wahl. *Peltura scrabæoides*. *Conocephalites sulzeri* Schlot. *Proetus ryckolti* Barr., *P. myops* Barr., *P. striatus* Barr. *Phillipsia derbyensis* Martin. *Cyphaspis burmeisteri* Barr. *Arethusina konincki* Barr. *Phacops latifrons* Broun.

Plate 45. *Dalmania caudata* Brünn., *D. punctata* Barr. *Calymene blumenbachii* Brong. *Homalonotus armatus* Burm. *Lichas palmata* Barr. *Trinucleus ornatus.*, *T. pongerardi* Ronault. *Ampyx rouaulti* Barr. *Dionide formosa* Barr. *Asaphus tyrannus* Murch. *Ogygia buchii* Brong. *Æglina rediviva* Barr. *Illænus davisii* Salt. *Nileus armadillo* Dalm.

Plate 46. *Acidaspis mira* Barr., *A. elliptica* Burm. *Cheirurus quenstedti* Barr. *Sphærexochus mirus* Beyr. *Staurocephalus murchisoni* Barr. *Deiphon forbesi* Barr. *Dindymene frederici-angusti* Cord. *Amphion lindaueri* Barr. *Cromus intercostatus* Barr. *Bronteus partschi* Barr. *Agnostus nudis* Beyr. *Bronteus haidingeri* Barr. *Eurypterus remipes* Dekay. *Cyprella chrysalidæ* DeKon. *Belinurus bellulus* König.

Pitt. See **Grote and Pitt.**

Plant (John). On the discovery of *Paradoxides davidis* at Tyddyngwladis, near Dolgelly, North Wales.
In Trans. Manchester Geol. Soc., vol. 5, 1886, p. 76.

Pohlman (Julius). On certain fossils of the Waterlime group near Buffalo.
In Bull. Buffalo Soc. Nat. Sci., vol. 4, 1881, p. 17, 7 figs.
Pterygotus buffaloensis n. sp. *Ceratiocaris grandis* n. sp.

—— Additional notes on the fauna of the Waterlime group.
In Bull. Buffalo Soc. Nat. Sci., vol. 4, 1882, p. 41, 2 pls.
Eurypterus giganteus n. sp. *Pterygotus globicaudatus* n. sp., *P. acuticaudatus* n. sp., *P. quadraticaudatus* n. sp., *P. macrophthalmus ?* Hall, *P. buffaloensis* Pohlm.

—— Fossils from the Waterlime group near Buffalo, N. Y.
In Bull. Buffalo Soc. Nat. Sci., vol. 5, 1886, p. 23, pl. 3.
Pterygostus buffaloensis Pohlm., *P. bilobus* H. & S. *Ceratiocaris acuminatus* Hall. *Eusarcus, E. grandis* G. & P. *Eurypterus scorpionis* G. & P. (the term *Eurypterus scorpiœdes* was used by Henry Woodward for a species of this genus in 1868. See *Eurypterus pohlmani*.) Vogdes, Bull. U. S. Geol. Sur., No. 63, p. 167.

Pompecki (J. F.) Die Trilobiten Fauna der Ost-und Westpressischen Diluvialgeschiebe.
In Beiträge zur Naturkunde Preussens VII, 1890, 97 pp., 6 pls.
Agnostus pisiformis Linné, also var. *socialis* Tulb. *Harpes spaski* Eichw. *Ampyx foveolatus* Ang., *A. (Lonchodomas) rostratus* Sars., *A. (Raphiophorus) culminatus* Ang., *A. (Raphiophorus) setirostris* Ang. *Phacops prussica* n. sp., *P. (Acaste) downingiœ* Murch., *P. (Dalmania) imbricatula* Ang., *P. (Dalmania) caudata* Brünn, *P. (Pterygometopus) exilis* Eichw., *P.(Pterygometopus) lævigata* Schm., *P. (Pterygometopus) kegelensis* Schm., *P. (Chasmops) præcurrens* Schm., *P. (Chasmops) odini* Eichw., *P. (Chasmops) cf. marginata* Schm., *P. (Chasmops) bucculenta* Sjögr., *P. (Chasmops) wesenbergensis* Schm., *P. (Chasmops) macroura* Sjögr., *P. (Chasmops) maxima* Schm., *P. (Chasmops) eichwaldi* Schm.
Cheirurus exsul Beyr., *C. exsul* forma *gladiator* Eichw., *C. exsul* cf. forma *macrophthalma* Kut., *C. gotlandicus* Lind., *C. speciosus* His., *C. dubius* n. sp., *C. (Cyrtometopus) clavifrons* Dalm., *C. (Cyrtometopus)* cf. *plautini* Schm., *C. (Cyrtometopus) pseudo-hemicranium* Niesz., *C. (Nieszkowskia) variolaris* Linrs., *C. (Nieszkowskia) cephaloceras* Niesz. *Amphion priscus* n. sp. *Cheirurus* sp.? *Sphærexochus* sp.? *Cybele revaliensis* Schm., *C.* cf.

grewingki Schm., *C. rex* Niesz., *C.* sp. *(a) C.* sp. *(b) Encrinurus obtusus* Aug., *E.* cf. *obtusus* Ang., *E.* cf. *seebachi* Schm., *E. punctatus* Brünn. *Encrinurus* sp.? *Calymene tuberculata* Brünn, *C. intermedia* Lind., *C. spectabilis* Aug. *Homalonotus* cf. *rhinotropis* Ang. 1, Arges: *Lichas salteri* Flet. 2, Leilichas: *Lichas illænoides* Niesz. *Lichas gageli* n. sp. 3, Hoplochas: *Lichas tricuspidata* Beyr., *L.* affin. *tricuspidatæ* Beyr., *L. plautini* Schm., *L, proboscidea* Dames, *L. media* n. sp. 4, Conolichas: *Lichas æquiloba* Steinh. 5, Homolichas: *Lichas branconis* n. sp., *L. deflexa* Sjögr., *L. eichwaldi* Niesz., *L.* cf. *pahleni* Schm., *L.* cf. *angusta* Beyr. *Lichas* sp.? *L. lindströmi* n. sp. 6. Oncholichas: *Lichas aranea* Lind. *Acidaspis mutica* Emm. *Proetus verrucosus* Lind., *P. signatus* Lind., *P. concinnus* Dalm., *P.* affin. *concinnus* Dalm., *P. distinctus* n. sp. *Proetus* sp.? *Proetus* sp.? *Cyphaspis parvula* n. sp. *Cyphaspis* sp.? *Phaëtonides.*

Illænus jevensis Holm., *I. chiron* Holm., *I. tauriconis* Kut., *I. revaliensis* Holm., *I. schmidti* Niesz., *I. oblongatus* Aug., *I. angustifrons* Holm., *I. roemeri* Volb., *I. comes* n. sp., *I. vanhoeffeni* n. sp., *I. bisulcatus* n. sp., *I.* sp. cf. *dalmani* Volb. (Holm.) *I. masckei* Holm., *I. linnarssoni* Holm., *I. nuculus* n. sp., *I. caecus* Holm., *I. (Bumastus) barriensis* Murch., *I. (Bumastus) sulcatus* Lind., *I. (Bumastus) holmi* Lind.

1, *Asaphus (Ptychopyge) rimulosus* Aug., *A. (Ptychopyge) multicostatus* Aug., *A. (Ptychopyge)* sp.? *A. (Ptychopyge) teticaudatus* Stein., *A. (Ptychopyge) undulatus* Stein., *A. (Ptychopyge)* cfr. *aciculatus* Aug., *A. oculosus* n. sp., *A.* sp.? *A. jevensis* Schm. MS., *A. branconis* n. sp., *A. steinhardti* n. sp., *A.* cf. *raniceps* Dalm., *A. ornatus* n. sp., *A.* sp.? *A. obtusus* n. sp., *A.* sp.? *A. devexus* Eichw., *A.* cf. *platyrus* Aug., *A. (Isotelus)platyrhachis* Steinh. 2, *Megalaspis. Megalaspis limbata* Boeck, form *typica* Brögger, *M. limbata* Boeck, var. *elongata, M. planilimbata* Aug., *M. gigas* Aug. 3, *Niobe.* 4, *Nileus armadillo* Dalm., n. sp.? *Holometopus? gracilis* n. sp., *H.? lævis* n. sp., *H.? radiatus* n. sp. *Remopleurides jentzscki* n. sp. *Olenus truncatus* Brünn. *Sphærophthalmus alatus* Boeck. *Peltura scarabæoides* Wahl.

—— Ueber das Einrollungsvermögen der Trilobiten. In Schrift K. phys.-ökon. Gesell. in Königsberg, Jahrg. 31, 1890.

—— Bemerkungen über das Einrollungsvermögen der Trilobiten. In Jahreshefte des Vereins für vaterl. Naturkunde in Württ, 1892, p. 93.

Portlock (J. E.) Report on the geology of the county of Londonderry, and of parts of Tyrone and Fermanagh. Dublin, 1843, 28 pls., also maps and sections.

Remopleurides n. g. *Phillipsia* n. g. *Griffithides* n. g. *Dithyrocaris* (Scouler) Portlock.

The author gives a brief abstract of the various opinions which have been published on fossil crustacea in regard to their nature, analogies and affinities. The following species are described and illustrated:

Remopleurides colbii n. sp., *R. lateri-spinifer* n. sp., *R. dorso-spinifer* n. sp., *R. longi-capitatus* n. sp., *R. longi-costatus* n. sp. *Ceraurus globiceps* n. sp. *Paradoxides ? bucephali* (var.)

Ampyx rostratus Sars., name changed to *A. sarsii* Portl.; *A. nasutus* Dalm., name changed to *A. dalmani* Portl. *Ampyx sarsii* (var.), *A. rostratus* Sars. (var.), *A. austinii* n. sp., *A. mammillatus ?* Sars., *A.?* *baccatus* n. sp. *Trinucleus caractaci* Murch., *T. elongatus* n. sp., *T. (Asaphus* His.) *seticornis* His:, *T. radiatus ?* Murch., *T. fimbriatus* Murch., *T. latus* Portl. *Harpes doranni* n. sp., *H. flanaganni* n. sp. *Brontes hibernicus* n. sp. *Arges* (Goldfuss), *A. plano-spinosus (Paradoxides bimucronatus* Murch.)

Nuttainia hibernica n. sp., *N.?* *obscura* n. sp. Genus *Asaphus*. *Calymene*. *Phacops truncato-caudatus* n. sp., *P. brongniartii* n. sp., *P. dalmani* n. sp., *P. murchisonii* n. sp., *P. jamesii* n. sp., *P.?* *tuberculatus* Murch. *Calymene blumenbachii* Brong., *C. pulchella* Dalm., *C. brevi-capitus* n. sp. *Amphion*, *Zethus*, *A. gelasionosus* n. sp., *A. frontilobus* Pand., *A. multisegmentatus* n. sp., *A. pseudo-articulatus* n. sp. *Asaphus latifrons* n. sp., *A. marginatus* n. sp., *A. dilatatus* Dalm. *Isotelus gigas* Dekay, also *Isotelus gigas (planus* Dekay), *I. ovatus* n. sp., *I. powisii* Murch., *I. rectifrons* n. sp., *I. arcuatus* n. sp., *I. læviceps* Dalm., *I. intermedius* n. sp., *I. scleropa?* Green. *Illænus centrotus* Dalm., *I. crassicauda* Wahl., *I. quadratocaudatus* n. sp. *Ogygia ?* *rugosa* n. sp. *Proetus. Phillipsia kellii* n. sp., *P. ornata (Asaphus truncatulus* Phill.), *P. jonesii* n. sp., *P. jonesii* var. *seminiferus* Phill., *P. maccoyi* n. sp. *Griffithides longiceps* n. sp., *G. globiceps* Phill., *G. longispinus* n. sp.

Dithyrocaris (Scouler). In the Records of Science, 1835, Dr. Scouler first noticed this crustacean, and, depending as a principal character on the lateral appendages, he pointed out its analogy to the genera *Cyclops* and *Apus*. He also noted two new species in a paper read before the British Association at Glasgow. Dr. Scouler having subsequently altered his views of the affinities of the genus, he replaced the original name of *Argas* by that of *Dithyrocaris*. The authority for this change is generally quoted from Portlock.

Dithyrocaris colei n. sp., *D. orbicularis* n. sp. *Limulus trilobitoides* Buckland ? *Cypris scoto-burdigalensis* Hibbert, *C. subrecta* n. sp.

Portlock figures the teeth of *Dithyrocaris* on plate xii, fig. 6.

Postlethwaite (J.) and **Goodchild** (J. G.) On some Trilobites from the Skiddaw slates.

In Proc. Geologist Assoc., vol. 9, No. 7, 1886, p. 455, pls. 6-9.

In this contribution the initials of each author designate the paragraphs written by him.

The following list is given of the Trilobite fauna of the Skiddaw slates: *Phacops nicholsoni* Salt. *Cybele ovata* Eth. *Calymene brevicapitata* ? *Remopleurides* sp.? *Olenus* sp.? *Trinucleus gibbsii* ? Salt. *Ogygia selwyni* Salt. *Ogygia* sp.? *Barrandia falcata* P. & G. *Niobe doveri* Eth. *Eurymetopus* n. g., *E. cambrianus* n. sp. P. & G., *E. harrisonii* P. & G. *Æglina* ? *rediviva* Barr., *A. binodosa* Salt., *A. obtusicauda* Hicks, *A.? caliginosa* Salt. *Agnostus morei* Salt.

Prestwich (S.) On the geology of Coalbrook Dale.

In Trans. Geol. Soc. London, vol. 5, 1840, p. 413, map, 2 sections, 4 pls.

The author mentions *Asaphus caudatus* Brünn, *A. tuberculato-caudatus* Murch., *A. longicaudatus* Murch. *Calymene blumenbachii* Brong.

On plate 41 he illustrates *Limulus anthrax* n. sp., *L. rotundatus* n. sp., *L. trilobitoides* Buckland. *Apus dubius* n. sp. *Anthrapalæmon (Palæocarabus) dubius* Salt.

Dr. Woodward uses *Limulus anthrax* and *L. rotundata* for the types of his new genus *Prestwichia*.

After a careful study of the Limuli belonging to Prestwich's collection, and also other specimens, this author remarks, that they are divisible into two well-marked genera:

(a) Those having movable thoracic segments and anchylosed abdomina ones; to be included in the genus *Belinurus*, namely *Belinurus trilobitoides* 1 Buckland, etc.

(b) Those in which the thoracic and abdominal segments are not divided, and which the former appear to be anchylosed. *Prestwichia anthrax* and *P. rotundata*.

Quenstedt (F. A.) Beiträge zur Kenntniss der Trilobiten, mit besonderer. Rücksicht auf ihre bestimmte Gliederzahl.

In Archiv für Naturgesch. (Wiegmann), vol. 1, Berlin, 1837, p. 337.

This work divides the Trilobites into two groups:
1. Trilobites with faceted eyes.
2. Trilobites with non-faceted eyes.

The genera are classified according to the number of segments in the thorax, as follows:

With 13 segments in the thorax: *Trilobites blumenbachii* Brong. *(Calymene callicephala* Green, *C. selenocephala* Green?, *C. platys* Green.

Trimerus delphinocephalus Green. *Calymene polytoma* Dalm., *C. bellatula* Dalm. *Asaphus fischeri* Eichw. These Trilobites form, perhaps, a second section among the Trilobites with 13 segments. Their segments are smooth and without grooves.

12 Segments: *Tril. hoffii* Schloth. *Ent. scarabæoides* Wahl.

11 Segments: After all the observations made, the Trilobites which have 11 thorax segments are generally provided with large eyes composed of a reticulated cornæ, and form, without any known exception, a very natural group. *Tril. macrophthalmus* Brong. *Calymene bufo* Green, *C. microps* Green, *C. diops'* Green, *C. variolaris* Brong. *Tril. caudatus* Brünn. *Asaphus wetherilli* Green. *Calymene scleops* Dalm. *Asaphus mucronatus* Brong., *A. selenurus* Eaton. *Calymene arachnoides* Hön., *C. anchiops* Green.

10 Segments: *Tril. esmarkii* Schloth. *Calymene concinna* Dalm. *Asaphus dalmani* Goldf. *Otarion diffractum* Zenk.

9 Segments: *Asaphus centrotus* Dalm.

8 Segments: The Trilobites with this number of segments form a very natural group. Their glabella is smooth, distinctly separated from the cheeks. Segments deeply grooved. *Tril. conigerus* Schloth. *Asaphus angustifrons* Dalm., *A. frontalis* Dalm. *Cryptonymus schlotheimi* Eichw., *C. weissi* Eichw., *C. panderi* Eichw., *C. lichtensteinii* Eichw. *Isotelus gigas* Dekay, *I. megalops* Locke, *I. stegops* Green, *I. planus* Green. *Asaphus læviceps* Dalm., *A. palpebrosus* Dalm., *A. armadillo* Dalm., etc.

7 Segments: *Asaphus buchii* Brong. *Ogygia guettardi* Brong., *O. desmarestii* Brong.

6 Segments: *Asaphus granulatus* Wahl. *Tril. dalmani* Boeck. *Asaphus nasutus* Dalm. *Cryptolithus tessellatus* Green.

4 Segments: *Triarthrus beckii* Green.

Without Segments: *Agnostus*.

14 Segments: *Tril. sulzeri* Schloth. *Calymene tristani* Brong. *Entom. gibbosus* Wahl. *Asaphus tetragonocephalus* Green. *Dipleura*.

Number not determined: *Entom. paradoxissimus* Linné. *Paradoxides longicaudatus* Zenk., *P. pyramidalis* Zenk., *P. latus* Zenk. *Tril. gracilis* Boeck. *Entom. spinulosus* Wahl. *Tril. problematicus* Schloth., *T. bituminosus* Schloth.

Rafinesque (C. S.) Prodrome de 70 nouveaux generes découverts dans Etats Unis.
In Jour. de Physique de Chemie His. Nat., Paris, vol. 88, 1819.
Dr. Green gives the following list of Trilobites from this article: *Trilobites cephaleurya, T. simia, T. granulata. Bilobites lunulata, B. lobata*.

―――― On the genera of fossil Trilobites or Glomerites of North America.
In Atlantic Jour. and Friend to Knowledge, vol. 1, 1832, p. 71.
The author states that he was among the first to call attention to the Trilobites of North America. In 1817 Dr. Shæffer presented to the Lyceum of New York a fossil from the Catskill mountains, which the author described as a new genus of fossil Entomostraca, under the name of *Glomerites eurycephala*, in a paper read before the society.

Classifies the Trilobites into 4 sections and 15 genera.

I—*Polyopsites*, more than two eyes: 1, *Alloctops* Raf. 1821. 2, *Diplopsites* Raf. 1821. 3, *Tomoligus* Raf. 1821.

II—*Diopsites*, with two eyes: 4, *Isoctomesa* Raf. 1821 (*Isotelus* Dekay 1824, *Nileus* Dalm. 1826), subgen. *Illænus* Dalm. 5, *Dipleura* Green. 6, *Asaphus* Brong. 7, *Trilobites* Park. 1812 (*Calymene* Brong., *Glomerites* Raf. 1817). 8, *Telesiops* Raf. 1832 (*Phacops*). 9, *Promenites* Raf. 1832. 10, *Ogygia* Brong.

III—*Monopsites*, only one eye: 11, *Metapteles* Green, or *Monopsites* Green, or *Cryptolithus* Green (*Trinucleus*).

IV—*Anopsites*, no eyes: 12, *Agnostus* Brong. 13, *Paradoxides* Brong. 14, *Ampyx* Dalm. 15. *Retusites* Raf. 1821.

Ramsay (A. C.) The Geology of North Wales, with an appendix on the fossils; with plates by J. W. Salter.
In Mem. Geol. Survey Gt. Brit., vol. 3, 1866.
Hymenocaris vermicauda Salt. *Lingulocaris lingulæcomes* n. sp. *Ceratiocaris? latus* n. sp., *C. insperatus* n. sp. *Beyrichia complicata* Salt.
Trilobita: *Agnostus princeps* n. sp., *A. pisiformis* Linné, *A. maccoyi* Salt., *A. trinodus* Salt., *A. limbatus* Salt., *A.* sp.? (allied to *A. limbatus*). *Paradoxides hicksii* n. sp. *Olenus micrurus* Salt., *O. cataractes* n. sp., *O. scarabæides* Wahl., *O. (Parabolina) serratus* n. sp., *O. (Sphærophthalmus) flagellifer* Ang.? *O.* sp.? *O. (Sphærophth.) alatus* Boeck, *O. (Sphærophth.) humilis* Phill., *O. impar* n. sp. *Dikelocephalus (Centropleura?) furca* n. sp., *D. (Centropleura?) celticus* n. sp., *D.? (Centropleura?) discoidalis* n. sp., *D.? (Centropleura?)* sp.? *Conocorphype invita* Salt., *C. abdita* n. sp., *C.* sp.? *C. simplex* n. sp., *C. vexata* n. sp., *C. (Solenopleura) depressa* Salt., *C.? verisimilis* n. sp., *C.? olenoides* n. sp. *Angelina sedgwickii* Salt. *Asaphus (Isotelus) affinis* McCoy, *A. homfrayi* n. sp., *A. radiatus* Salt., *A. powisii* Murch., *A. (Basilicus) tyrannus* Murch. *Ogygia scutatrix* Salt., *O. peltata* n. sp., *O. selwynii* Salt. *Niobe homfrayi* n. sp. *Psilocephalus* n. g., *P. innatatus* n. sp., *P. inflatus* n. sp. *Illænopsis* n. g., *I. thomsoni* n. sp. *Illænus bowmanni* Salt., *I. davisii* Salt. *Æglina grandis* Salt., *Æ.* sp.? *Æ. binodosa* Salt., *Æ.? caliginosa* n. sp. *Trinucleus murchisoni* Salt., *T. sedgwicki* n. sp., *T. gibbsii* Salt., *T. concentricus* Eaton, *T. favus* n. sp. *Ampyx tumidus* Forbes, *A. prænuntius* n. sp. *Dionide atra* n. sp. *Cheirurus frederici* n. sp., *C. juvenis* Salt., *C. octolobatus* McCoy, *C. bimucronatus* Murch. *Encrinurus sexcostatus* Salt. *Cybele verrucosa* Dalm. *Lichas laxatus* McCoy. *Calymene parvifrons* n. sp., *C. blumenbachii* also var. *brevicapitata* (or *caractaci*), *C. blumenbachi* var. *cambrensis*, *C. duplicata* Murch. *Homalonotus bisulcatus* Salt.? *H. rudis* Salt.

Rathbun (Richard). See **Hartt** (C. F.) and **Rathbun** (Richard).

Ratte (F.) Note on some Trilobites new to Australia.
In Proc. Linn. Soc. New South Wales, 2d series, vol. 1, 1886, p. 1967, 3 pls.
Lichas sinuata n. sp. *Proetus ascanius* Cord. *Acidaspis verneuili* Barr., or *A. vesiculosa* Beyr. *Acidaspis* affin. *A. prevosti, A.* affin. *A. mira.*

——— Note on some Trilobites new to Australia.
In Proc. Linn. Soc. New South Wales, 2d series, vol. 2, 1887, p. 95, 2 pls.
Lichas palmata var. *sinuata* emend. from *L. sinuata* Ratte. *Acidaspis* affin. *A. dormitzeri* Cord., *A.* affin. *A. leonharti* Barr. *Staurocephalus* affin. *S. murchisoni* Barr. *Lichas hirsutus* Fletcher, *L. palmata* Barr.

Rasoumowsky (Comte G. de). Quelques observations sur les Tribolites et leur gisemens.
In Annales Sci. Nat. Paris, vol. 8, 1826, p. 186, pl. 28.
Trilobites de Tzarsko-sélo.

Remelé (A.) Katalog der von Dr. Ad. Remelé beim international Geologen Congress zu Berlin im September und October, 1885, ausgestellten Geschiebesammlung Berlin, 1855.
The author illustrates *Paradoxides glandicus* Sjögren and *Homalops* n. sp., *H. altumii* n. sp.

Reuss (A. E.) Ueber Entomostraceen und Foraminifera im Zechstein der Wetterau.
In Jahres Ber. d. Wetterauer Gesell. für 1851–53, p. 59, pl. 1.
Bairdia gracilis McCoy, *B. geintziana* Jones, *B. kingi* n. sp., *B. mucronata* n. sp., *B. ampla* n. sp., *B. frumentum* n. sp. *Cytherella nuciformis* Jones. *Cythere bituberculata* n. sp., *C. roessleri* n. sp.

——— Plæontologische Miscellen.
In Denkschriften k. Akad. Wiss. Wien, vol. 10, 1856, 7 pls,
Lepidoderma n. g., *L. imhoffi* n. sp.
This is referred by H. Woodward to the genus *Eurypterus.*

Reuss (F.) Geognostische Bemerkungen auf einer Reise durch einen Theil des Pilsner Kreises.
In Sammlung naturhist. Aufsätze, vol. 4, 1794, p. 47.*

Reuter (G.) Die *Beyrichien* der obersilurischen Diluvialgeschiebe Ostpreussens.
In Zeitschr. Deutsch Geol. Gesell., Berlin, vol. 37, 1885, p. 621, pls. 25, 26.
Beyrichia tuberculata Klöden, also variations *nuda* Jones, *antiquata* Jones, *gibbosa,* also *bigibbosa, B. noetlingi-conjuncta* Reuter, *B. noetlingi* n. sp.,

B. bronni n. sp., *B. baueri tripartita* n. sp., *B. baueri* n. sp., *B. tuberculato-buchiana* n. sp., *B. buchiano-tuberculata*, *B. buchiana* var. *lata*, *B. buchiana* var. *angusta*, also var. *incisa*, *B. buchiana* Jones, *B. tuberculo-kochiana* n. sp., *B. kochii* Boll., *B. maccoyana* Jones, also var. *sulcata* and *lata*, *B. salteriana* Jones, *B. bolliana* n. sp., *B. bolliana umbonata* n. sp., *B. wilkensiana* Jones, *B. dubia* Reuter.

Richter (R.) Beiträge zur Palæontologie Thüringens. Waldes. Die Grauwacke des Bohlens und des Pfaffenberges bei Saalfeld. Dresden and Leipsig, 1848, 6 pls.
    Cytherina striatula n. sp., *C. hemisphærica* n. sp. *Phacops cryptophthalmus* Emm., *P. limbatus* n. sp *Asaphus? læviceps* Dalm.

———— Aus dem thüringischen Zechstein.
In Zeitschr. Deutsch. Geol. Gesell., Berlin, vol. 7, 1855, p. 526.*
Cythere, Cytherella, Cytheris, Bairdia.

———— Aus dem thüringischen Schiefergebirge.
In Zeitschr. Deutsch. Geol. Gesell., Berlin, vol. 15, 1863, p. 659, 2 pls.*
Harpes. Proetus. Arethusina. Phyllaspis n. g. *Phacops. Acidaspis. Beyrichia.*

———— Aus dem thüringischen Schiefergebirge.
In Zeitschr. Deutsch. Geol. Gesell., Berlin, vol. 17, 1865, p. 361, 2 pls.
Proetus expansus n. sp., *P. playiophthalmus* n.sp. *Cheirurus. Beyrichia klödeni* McCoy, *B. subcylindria* Münst.

———— Das thüringischen Schiefergebirge.
In Zeitschr. Deutsch. Geol. Gesell., Berlin, vol. 21, 1869, p. 341.
Contains a catalogue.

———— Devonische Entomostraceen in Thüringen.
In Zeitschr. Deutsch. Geol. Gesell., Berlin, vol. 21, 1869, p. 757, pls. 20, 21.
Cypridina ava n. sp., *C. scrobiculata* n. sp., *C. serratostriata* Sandb., *C. tenella* n. sp., *C. labyrinthica* n. sp., *C. gyrata* Richt., *C. costata* n. sp., *C. sandbergeri* n. sp., *C. barrandei*, *C. tæniata* Richt., *C. calcarata* Richt. *Cytherina striatula*, *C. costata* n. sp. *Beyrichia dorsalis* n. sp., *B. nitidula* n. sp., *B. aurita* n. sp.

Roberg (L.) Dissertatio academica de fluviatite Astaco, etc.' Upsaliæ, 1715.
This work contains a plate with the figure of the pygidia of Trilobites and a fossil Crustacean. Fig. H is of a large pygidium of some species of *Asaphus*. Fig. J is a cast in stone of a smaller species of the same genus. The author remarks "that he found the specimens during an excursion along the shores of the Baltic Sea, in a red limestone." He has the merit of referring these fossils to the class Crustacea.

Roemer (C. F.) Das rheinische Uebergangsgebirge, eine palæontologisch-geognostiche Darstellung. Hannover, 1844, pl. 2.

Phacops latifrons Bronn. *Pleuracanthus lacimatus* n. sp., *P. punctatus* Stein. *Gerastos lævigatus* Goldf.

Roemer (F. A.) Die Versteinerungen der Hartzgebirge. Hannover, 1843, 12 pls. Crustacea, p. 37.

Bronteus flabellifer Goldf., *B. signatus* Phill., *B.? glabratus* n. sp. *Calymene? jordoni* n. sp., *C. schusteri* n. sp., *C. subornatus* n. sp., *C. hydrocephala* n. sp. *Asaphus zinckeni* n. sp. *Paradoxides grotei* n. sp. *Homalonotus ahrendi* n. sp., *H. punctatus* n. sp., *H. gigas* n. sp.

―――― Beiträge zur geologischen Kenntniss der nordwestlichen Harzgebirges.

In Palæontographica, vols. 3, 5, and 13, 5 pts., Cassel, 1854-66, 38 pls. Vol. 3, 1854: *Acidaspis horrida* n. sp., *A. selcana* n. sp. *Bronteus intumescens* n. sp., *B. minor* n. sp. *Cheirurus jaschei* n. sp., *C. myops* n. sp. *Conocephalus longecornutus* n. sp. *Cyphaspis spinulosa* n. sp. *Harpes bischoffi* n. sp. *Homalonotus barrandei* n. sp., *H. minor* n. sp. *Lichas crassirhachis* n. sp., *L. granulosus* n. sp. *Phacops bronni* Barr., *P. cryptophthalmus* Emm., *P. latifrons* Br., *P.>micromma* n. sp., *P. pectinatus* n. sp., *P. stellifer* Burm., *P. tuberculatus* n. sp. *Phillipsia cassimargo* n. sp., *P. alternans* n. sp. *Proetus barrandei* n. sp., *P. crassimargo* n. sp., *P. latispinosus* Sandb., *P. orbicularis* n. sp.

Vol. 5, 1855: *Acidaspis selcana* A. Römer. *Lichas sexlobatus* n. sp. *Cyphaspis hydrocephala* A. Römer. *Phacops tuberculatus* A. Römer. *Dithyrocaris jaschei* n. sp. *Homalonotus schusteri* n. sp., *H. latifrons* n. sp. *Cheirurus myops* Römer. *Bronteus alternans* n. sp., *B. minor* Römer. *Lichas granulosus* Römer. *Cyphaspis truncata* n. sp. *Crypidina oculata* n. sp., *C. elliptica* n. sp., *Phacops lævis* Phill.

Vol. 13, 1866: *Acidaspis horrida* F. A. Römer. *Phacops granulatus* Münst. *Bronteus minor* Römer. *Phacops (Trimerocephalus) incisus* n. sp.

Roemer (Ferdinand). Texas, mit besonderer Rücksicht auf deutsche Auswanderung und die physischen Verhältnisse des Landes, nach eigener Beobachtung geschildert. Mit einem naturwissenschaftlichen Anhange und einer topographisch-geognostischen Karte von Texas. Bonn, 1849.

In the list of fossils on p. 421 *Pterocephalia sancti sabæ* Römer is mentioned.

―――― Ueber ein bisher nicht beschriebenes Exemplar von *Eurypterus* aus devonischen Schichten des Staates New York in Nord Amerika.

In Palæontographica, vol. 4, 1848, pp. 190-193, pl. 27.
Describes and figures a specimen of *Eurypterus remipes* Dekay.

―――― Die Kreidebildung von Texas. Bonn, 1852, 11 pls.

Pterocephalia n. g. *P. sancti sabæ* n. g.

―――― Notiz über eine riesenhafte neue Art der Gattung *Leperditia* in silurischen Diluvial-Geschieben Ost-Preussens.

In Zeitschr. Deutsch. Geol. Gesell., Berlin, vol. 10, 1858, p. 356, 2 figs.
Leperditia baltica His., *L. gigantea* n. sp.

―――― Die Versteinerungen der silurischen Diluvial-Geschiebe von Gröningen in Holland.

In Neues Jahrbuch für Mineral., 1858, p. 257.
Encrinurus punctatus Brünn. *Proetus concinnus* Dalm. *Cytherina baltica* His., *C. phaseolus* His. *Beyrichia tuberculata* Boll.

―――― Bericht über eine geologische Reise nach Norwegen im Sommer 1859.

In Zeitschr. Deutsch. Geol. Gesell., Berlin, vol. 11, 1860.*
Dalmania. Cryphæus. Asaphus. Illænus. Phacops.

―――― Die fossile Fauna der Silurischen Diluvial-Geschiebe von Sadewitz bei Oels in Niederschlesien. Breslau, 1861, 8 pls.

Asaphus expansus Linné. *Calymene pediloba* n. sp. *Ceraurus ornatus* Ang. *Chasmops conicophthalmus* Emm. *Encrinurus multisegmentatus* Portl. *Illænus crassicauda* Wahl., *I. grandis* n. sp. *Isotelus robustus* n. sp. *Lichas augusta* Beyr. *Proetus concinna* Dalm. *Remopleurides nanus* Leucht.

―――― Die Silurische Fauna des westlichen Tennessee. Breslau, 1860, 5 pls.

Calymene blumenbachii Brong., *(C. niagarensis* Hall). *Cheirurus bimucronatus* Murch., *(C. insignis* Hall). *Sphærexochus mirus* Beyr. *Dalmania caudata* Brünn, *(Phacops limulurus* Green). *Bumastus harriensis* Murch. *Illænus* sp.?

——— Lethæa geognostica oder Beschreibung und Abbildung der für die Gebirgs-formationen bezeichnendstein Versteinerungen Herausgegeben von einer Vereinigung von Paläontologen. 1 Theil Lethæa palæozoica. Stuttgart, 1876, 1880, 1883.

The Atlas, 1876, contains illustrations of the following Crustacea:
Cambrisch: *Paradoxides bohemicus* Barr. *Agnostus pisiformis* Linné. *Arionellus ceticephalus* Barr. *Hydrocephalus carens* Barr. *Olenus (Peltura) scarabæoides* Wahl., *O. truncatus* Ang. *Conocephalus sulzeri* Schloth. *Sao hirsuta* Barr. *Ellipsocephalus hoffi* Schloth.
Unte-Silur.: *Asaphus (Basilicus) tyrannus* Murch., *A. expansus* Linné. *Illænus crassicauda* Wahl. *Chasmops conicophthalmus* Boeck. *Amphion fischeri* Eichw. *Harpides hospes* Beyr. *Dionide formosa* Barr. *Trinucleus goldfussi* Barr. *Nileus frontalis* Dalm. *Ampyx tetragonus* Ang. *Ogygia buchii* Brong. *Dindymene haidingeri* Barr. *Remopleurides radians* Barr. *Æglina rediviva* Barr. *Placoparia zippei* Barr.
Ober-Silur.: *Acidaspis prevosti* Barr. *Harpes ungula* Burm., *H. venulosus* Cord. *Bronteus haidingeri* Barr. *Dalmania caudata* Brünn. *Lichas scabra* Beyr. *Cromus intercostatus* Barr. *Encrinurus punctatus* Brünn. *Sphærexochus mirus* Beyr. *Ceraurus insignis* Beyr. *Deiphon forbesi* Barr. *Arethusina konincki* Barr. *Cyphaspis burmeisteri* Barr. *Staurocephalus murchisoni* Barr. *Calymene blumenbachii* Brong. *Pterygotus anglicus* Ag. *Neolimulus falcatus* H. W. *Eurypterus remipes* Dekay. *Slimonia acuminata* Salt. *Hemiaspis limuloides* H. W. *Antifopsis prima* Barr. *Ceratiocaris papilio* Salt., *C. scharyi* Barr. *Leperditia baltica* His., *L. (Isochilina) gigantea* F. Römer. *Beyrichia tuberculata* Boll. *Cryptocaris pulchra* Barr. *Turrilepas wrightii* H. W.
Unter-Devon.: *Dipleura dekayi* Green. *Homalonotus crassicauda* Sandb. *Cryphæus laciniatus* F. Römer.
Mittel-Devon.: *Harpes macrocephalus* Goldf. *Proetus cuvieri* Stein. *Bronteus flabellifer* Goldf. *Cyphaspis burmeisteri* Barr.
Ober-Devon.: *Phacops cryptophthalmus* Emm.
Culm.: *Bostrichopus antiquus* Goldf.
Kohlengebirge: *Anthrapalæmon gracilis* M. & W. *Limulus rotundatus* Prest. *Eurypterus scouleri* Hibbert. *Phillipsia derbyensis* Mart., *P. globiceps* Phill., *P. gemmulifera* Phill., *P. acuminata* n. sp. *Cyclus radialis* DeKon. *Entomononchus scouleri* McCoy. *Carbonia agnes* Jones. *Leaia leidyi* Jones.

——— Ueber die Diluvial-Geschiebe von nordischen Sedimentär-Gesteinen in den norddeutschen Ebene, etc.
In Zeitschr. Deutsch. Geol. Gesell., Berlin, vol. 14, 1862, p. 575.
Beyrichia. Gives list of fossils; not descriptive.

——— Ueber eine marine Conchylien-Fauna im productiven Steinkohlengebirge Oberschlesiens.
In Zeitschr. Deutsch. Geol. Gesell., Berlin, vol. 15, 1863, p. 567, pls. 14–16.
Griffithides mesotuberculata McCoy. *Phillipsia*.

——— Geologie und Palæontol. von Oberschlesien Breslau, 1870. Atlas of 50 plates, maps and sections.
In this work are mentioned or described *Acidaspis* sp.?; *Homalonotus crassicauda* Saudb.; *Phillipsia macronata* F. Römer, *P. margintifera* F. Römer.

——— Notiz über das Vorkommen von *Eurypterus scouleri* im niederschlesischen Steinkohlengebirge.
In Zeitschr. Deutsch. Geol. Gesell., Berlin, vol. 25, 1873, p. 562, 1 fig.

——— Ueber die Gattung *Arthropleura* aus dem schlesischen Steinkohlengebirge.
In Sitzungsber. schles. Gesell. für vaterland. Cultur., Breslau, Jahrg. 1880, p. 128.

——— Ueber eine Kohlenkalk-Fauna der Westküste von Sumatra. Cassel, 1880, 3 pls.
Phillipsia sumatrensis n. sp.

——— Ueber eine Art der Limuliden-Gattung *Belinurus* aus dem Steinkohlengebirge Oberschlesiens.
In Zeitschr. Deutsch, Geol. Gesell., Berlin, vol. 35, 1883, p. 429, 2 figs.
Belinurus silesiacus n. sp.

——— Lethæa erratica, oder Aufzählung und Beschreibung der in den norddeutschen Ebene vorkommenden Diluvial Geschiebe nordischen Sedimentär-Gesteine.
In Palæont. Abhandl. Herausgegeben von Dames und Kayser, vol. 2, pt. 5. Berlin, 1885, 11 pls.
The author illustrates *Paradoxides ölandicus* Sjorgren, *P. tessini* Broug. *Agnostus lœvigatus* Dalm., *A. pisiformis* Linné. *Peltura scaraboeoides* Wahl. *Parabolina spinulosa* Wahl. *Asaphus expansus* Linné. *Illænus crassicauda* Wahl. *Ceraurus exsul* Beyr. *Chasmops macrura* Ang. *Cybele bellatula ?* Dalm. *Chasmops conicophthalmus* S. & B. *Proetus concinnus* Dalm. *Encrinurus multisegmentatus* Portl. *Remopleurides nanus* Leucht. *Ceraurus (Pseudosphærexochus) roemeri* Schm. *Calymene blumenbachii* Broug. *Beyrichia tuberculata* Boll. *Leperditia angelina* Schm. *Phacops downingiæ* Murch. *Encrinurus punctatus* Brünn. *Beyrichia maccoyiana*

Jones, B. wilkensiana Jones, B. buchiana Jones. Dalmania caudata Brünn. Ampyx parvulus Forbes. Homalonotus delpninocephalus Green. Odontopleura ovata Emm., O. mutica Emm., O. barrandei Ang. Beyrichia jonesii Boll.
For other references, see Bronn and Roemer, p. 31.

Rogers (Henry D.) The geology of Pennsylvania. A Government survey with a general view of the geology of the United States, essays on the coal formation and its fossils, and a description of the coal fields of North America and Great Britain. 2 vols. Edinburgh, London, and Philadelphia, 1858. Atlas of plates and maps.

Organic remains of the Palæozoic strata of Pennsylvania, vol. 2, pp. 815-834.

Notes on the *Beyrichia* and *Leperditia* by T Rupert Jones, vol. 2, p. 834, figs. 595-699.

Paradoxides spinosus Boeck *(P. harlani* Green*)* from the quarry in Quincy, Mass. *Isotelus gigas* Dekay. *Trinucleus concentricus* Eaton. *Calymene beckii* Green. *Agnostus lobatus* Hall. *Calymene clintoni* Vanuxem. *Beyrichia maccoyiana* Jones, *B. pennsylvanica* Jones. *Leperditia ovata* Jones, *L. gibbera* Jones, *L. pennsylvanica* Jones.

Rogers (W. B.) Discovery of *Paradoxides* in the altered rocks of Massachusetts.
In Edinb. New Philos. Jour., vol. 4, 1856.

—— On the discovery of Cambrian fossils at Braintree, Mass.
In Proc. Boston Soc. Nat. Hist., vol. 6, 1856,. p. 27 (continued on p. 40).
Paradoxides harlani Green.

—— *Paradoxides* near Boston.
In Edinb. New Philos. Jour., vol. 6, 1857, p. 314.

Rominger (C.) Description of Primordial fossils from Mount Stephens, N. W. Territory of Canada.
In Proc. Acad. Nat. Sci. Phil., 1887, p. 12, 1 pl.
Embolimus n. g. *Ogygia klotzi* n. sp., *O. serrata* n. sp. *Embolimus spinosa* n. sp., *E. rotundata* n. sp. *Monocephalus salteri* Billings. *Conocephalites cordilleræ* n. sp. *Bathyurus* sp.? *Agnostus* sp.?

—— Rejoinder to Mr. C. D. Walcott.
In Am. Geologist, vol. 2, 1888, p. 356.

Rouault (M.) Extrait du mémoire sur les Trilobites du Départment d'Ille et Vilaine.

This paper was read Dec., 1846, and published in the Bull. Soc. Géol. France, vol. 4, 1847, pp. 309-328.

The author gives a general discussion of organization and structure of the Trilobite especially of the eye.

The text gives a list of 20 species with figures of the following: *Trinucleus pongerardi* n. sp. *Prinocheilus* n. sp., *P. verneuili* n. sp. *(Calymene) Nileus beaumonti* n. sp. *Calymene tournemini* n. sp. *Phacops dujardini* n. sp., *P. downingiæ* Murch., *P. macrophthalmus* Brong., *P. longicaudatus* Zenker.

——— Mémoire sur l'organization des Trilobites.

In Institute de France Acad. Sci., vol. 27, 1848.

The author remarks on the fossil Crustacea described by Guettard from Angers in the year 1757: that Burmeister correctly indicated the genus of the species described by Bronguiart as *Ogygia desmaresti*, but it is to be regretted that he gave it the new name of *Illænus giganteus* inasmuch as Bronguiart had already named it.

——— Mémoire. 1. Sur la composition du test des Trilobites. 2. Sur les changement de formes dus à des causes accidentelles, ce qui a pu permettre de confondre des espèces différentes.

In Bull. Soc. Géol. France, vol. 6, 1848, p. 67, pls. 1-2.

Ogygia brongniarti n. sp., *O. edwardsi* n. sp. *Illænus desmaresti* n. sp. *Calymene arago* n. sp. *Trinucleus.*

——— Note sur des nouvelles espèces de fossiles découvertes en Bretagne.

In Bull. Soc. Géol. France, vol. 6, 1849, pp. 377-381.

Describes *Lichas heberti* n. sp. and gives figure.

——— Sur les Trilobites et les terrains de transition de la Bretagne.

In Bull. Soc. Géol. France, vol. 8, 1851, pp. 167-68.

——— Mémoire sur le terrains paléozoques des environs de Rennes.

In Bull. Soc. Géol. France, vol. 8, 1851, pp. 358-399.

Calymene salteri n. sp., *C. verneuili* n. sp. *Dalmannia vetillarti* n. sp., *D. incerta* n. sp. *Placoparia tournemini* n. sp. *Ogygia desmaresti* Broug. *Homalonotus brongniarti* n. sp., *H. barrandei* n. sp., *H. hausmanni* n. sp., *H. legraverendi* n. sp. *Plæsiacomia kieneria* n. sp. *Proetus huhayi* n. sp., *P. michelini* n. sp. *Cyphaspis. Beyrichia. Leperditia* n. g., *L. brittannica* n. sp. *Phacops.*

Safford (J. M.) Geology of Tennessee. Nashville, 1867, 7 pls. and map.

The author, in his catalogue of Trenton and Nashville fossils occurring in the Central Basin, p. 285, gives the names of *Dalmanites troosti* n. sp. and *Encrinurus excedrinus* n. sp., but without specific descriptions.

―――― and **Vogdes** (A. W.) Description of new species of fossils. Crustacea from the Lower Silurian of Tennessee.

In Proc. Acad. Nat. Sci., Philadelphia, 1889, p. 166, 3 wood cuts.
Ampyx americanus n. sp. *Encrinurus varicostatus* Wal. *Chasmops troosti* Saff.

Sandberger (F.) Bemerkungen über einige Arten der Gattung *Bronteus*.

In Jahrbuch Nassauischer Verein für Naturkunde, Jahrg. 44, 1891, p. 1, pl. 1.
Bronteus acanthopeltis Schnur., *B. laciniatus* Sandb.

Sandberger (G.) and Sandberger (F.) Die Versteinerungen des rheinischen Schichten-System in Nassau. Wiesbaden, 1850–56, atlas in folio, 39 pls.

Trigonaspis n. g. *Cylindraspis* n. g. *Bostrichopus antiquus* Goldf. *Cypridina subglobularis* Sandb., *C. serratostriata* Sandb., *C. subfusiformis* Sandb. *Phacops laciniatus* C. F. Römer, *P. brevicauda* Sandb., *P. cryptophthalmus* Emm., *P. latifrons* Bronn. *Cheirurus gibbus* Beyr. *Bronteus alutaceus* Goldf., *B. lacinatus*. *Cyphaspis ceratophthalmus* Goldf. *Odontopleura*. *Homalonotus obtusus* Sandb., *H. crassicauda* Sandb. *Harpes gracilis* Sandb. *Trigonaspis lævigata* Goldf., *T.?* cornuta Goldf. *Cylindaspis latispinosa* Sandb., *C.?* macrophthalmus Sandb.

Salter (J. W.) Descriptions of *Typhloniscus* n. g., *T. bainii* n. sp., *Phacops africanus* n. sp., *P. caffer* n. sp.

In Geo. Trans. 2d Series, vol. 7, 1846, pl. 25.

―――― On the structure of *Trinucleus*, with remarks on other species.

In Quart. Jour. Geol. Soc. London, vol. 3, 1847, p. 251, 4 figs.

The author remarks: "The puncta are almost always arranged in radiating rows; three, four, or more holes in each row, and these being at equal distances they form concentric lines. In *T. granulatus*, two of the rows are separated by a furrow from the rest; in *T. seticornis*, three are distinct from the remaining two or three, by the front rows being sunk in a deep concentric furrow. Other modifications take place. In *T. fimbriatus*, the two front rows are turned downwards; lastly in *T. ornatus*, the dots occur most frequently in quincunx order."

—— Palæontological appendix.
In Mem. Geol. Survey Gt. Britain, vol. 2, pt. 1, 1848, pp. 331-352, figure, pls. 5-8.
Phacops stokesii Milne-Edw., *P. downingiæ* Murch. *Dalmania affinis* Salt. *Proetus latifrons* McCoy.
Sec. *Asaphus* divided into genera *Asaphi* and *Ogygia*.
1 Sec. *Illænus* and *Asaphus. Illænus rosenbergii* Eichw., *I bowmanni* Salt. *Ogygia. Lichas laxatus* McCoy, *L. verrucosus* Eichw., *L.* sp.? *Calymene brevicapitata* Portl., *C. tuberculata* Salt. *Cybele sexcostata* Salt., *C. verrucosa* Dalm., *C. variolaris* Broug. *Sphærexochus juvenis* Salt. *Cheirurus speciosus* Dalm. *Olenus bisulcatus* Phill., *O. humilis* Phill., *O. spinulosus? Acidaspis brightii* Murch., *A. bispinosus* McCoy. *Trinucleus ornatus* Sternb., *T. ornatus* var. *favus* Salt. *Ampyx parvulus* Forbes. *Agnostus trinodus* Salt., *A. convexus* Salt. *Beyrichia tuberculata* Klöden, *B. complicata* Salt., *B. gibba* Salt.

—— Appendix A to British Palæozoic fossils in the Geological Museum, University of Cambridge. London, 1852 (Sedgwick and McCoy).
Description of a few species from Wales and Westmoreland, referred to in the foregoing work by John William Salter, p. 645.
Cythere? umbonata n. sp. *Beyrichia complicata* Salt., *B. strangulata* n. sp. *Phacops (Odontochile) obtusicaudatus* n. sp., *P. (Phacops) alifrons* n. sp., *P. (Phacops) apiculatus* n. sp. *Calymene parvifrons* n. sp. *Cheirurus clavifrons* Dalm. *Cybele rugosa* Portl. *Encrinurus sexcostatus* Salt. *Lichas nodulosus* n. sp. *Illænus davisii* Salt., *I. murchisoni* Salt., *I. bowmanni* Salt. *Homalonotus bisulcatus* n. sp., *H. bisulcatus* var. *minor, H. rudis* n. sp., *H.* sp.? *Eurypterus cephalaspis* n. sp.

—— On the lowest fossiliferous beds of North Wales.
In Rep. 22d Meeting Brit. Assoc. Adv. Sci., 1852, Trans. of Sec., p. 56.
Hymenocaris n. g., *H. vermicauda* n. sp. *Asaphus selwynii* n. sp.

—— On a few genera of Irish Silurian fossils.
In Rep. 22d Meeting Brit. Assoc. Adv. Sci., 1852, Trans. of Sec., p. 59.
Stygina n. g. *Cyphoniscus* n. g.
1. With head spines short: *Stygina latifrons* Portl., *A. marginatum* Portl.
2. With head spines long: *Stygina murchisoni* Murch.
Cyphoniscus socialis n. sp. *Acidaspis jamesii* n. sp. *Æglina mirabilis* n. sp. *Remopleurides dorsospinifer* Portl. *Cyphaspis megalops* McCoy.

—— and **Forbes** (Edward). Figures and descriptions illustrative of British organic remains. Decade 2.
In Mem. Geol. Survey United Kingdom London, 1849, 10 pls.
Parabolina n. g. *Brachampyx* (Forbes) n. g.

The description of all the species, with the exception of *Ampyx nudus*, was made by J. W. Salter.
Phacops caudatus Brünn. var. *(a) vulgaris*, var. *(b) longicaudatus*.
Other British species of *Phacops* of the section *Dalmania*.
Phacops sp.? *P. obtusi-caudatus* Salt., *P. truncato-caudatus* Portl., var. *(b) affinis* Salt.
Subgenera: 1, *Portlockia*. 2, *Phacops*. 3, *Dalmania*. 4, *Cryphæus* Green *(Pleuracanthus* Milne-Edw.)
Illænus davisii Salt., *I. bowmanni* Salt., *I. portlocki* Salt., *I. perovalis* Murch., *I. ocularis* n. sp., *I. murchisoni* Salt., *I. barriensis* (subgenus *Bumastus)*. *Asaphus tyrannus* Murch. *Ogygia buchii* Brong., *O. portlockii* Salt. *Calymene tuberculosa* Salt. *Olenus micrurus* Salt. *Ampyx nudus* (Murch.) Forbes, *A. tumidus* Forbes, also mentions *Ampyx austini* Portl., *A. mammillatus* Sars., *A. sarsii* Portl., *A. rostratus* Sars., *A. parvulus* Forbes, *A. bruckneri* Boll., *A. portlockii* Barr.

——— Description of *Pterygotus problematicus* Agass,
In Quart. Jour. Geol. Soc. London, vol. 8, 1852, p. 386, pl. 21.

——— Notes on the Trilobites.
Appendix C to an article by Señor Carlos Ribeiro on the Carboniferous and Silurian formations of the neighborhood of Bussaco, in Portugal.
In Quart. Jour. Geol. Soc. London, vol. 9, 1853, p. 158, pl. 7.
Calymene tristani Brong., *C. arago* Rouault. *Asaphus guettardi* Brong., *A. desmaresti* Brong. *Asaphus* sp.? *Illænus giganteus* Burm. *Placoparia zippei* Boeck ? *Trinucleus pongerardi* Rouault. *Ogygia ? glabrata* n. sp.

——— On Arctic Silurian fossils.
In Quart. Jour. Geol. Soc. London, vol. 9, 853, p. 312.
Encrinurus lævis Ang.? *Proetus* sp.? *Leperditia balthica* His.

——— Figures and descriptions illustrative of British organic remains. Decade 7.
In Mem. Geol. Survey United Kingdom, London, 1853, 10 pls.
The genera are arranged in the following order: *Phacops downingiæ* Murch. Sec. *Acaste:* 1, *P. apiculatus* Salt. 2, *P. brongniarti* Portl. 3, *P. dalmani* Portl. 4, *P. jamesi* Portl. 5, *P. alifrons* Salt. 6, *P. jukesii* n. sp. 7, *P. conophthalmus* Boeck. Sec. *Odontochile: P. mucronatus* Brong., *P. amphora* n. sp. *Cheirurus bimucronatus* Murch. Sec. *Crotalocephalus:* 1, *C. articulatus* Münst. Sec. *Cheirurus* proper: 2, *C. speciosus* Dalm. 3, *C. gelasinosus* Portl. 4, *C. cancurus* n. sp. 5, *C. octolobatus* McCoy. Sec. *Actinopeltis:* 6, *C. clavifrons* Dalm.

Sphærexochus mirus Beyr. 1, *Encrinurus sexicostatus* Salt. 2, *E. punctatus* Brünn. 3, *E. variolaris* Brong. 4, *E. multisegmentatus* Portl. *Cyphaspis megalops* McCoy, *C. pygmæus* n. sp. *Acidaspis jamesii* Salt., *A. bispinosus* McCoy. 3, *A.* sp.? 4, *A. biserialis* n. sp. 5, *A.* sp.? 6, *A. caractaci* n. sp., *A. coronatus* n. sp. *(A. brightii,* Mem. Geol. Sur., pl. 9, figs. 8–9).

Trinucleus lloydii Murch.
Sec. 1. 1, *T. lloydii* Murch. 2, *T. concentricus* Eaton, vars. *junior, caractaci, portlocki, elongatus, favus, goldfussii* Barr. 3, *T. thersites* n. sp.
Sec. 2. *Tretaspis.* 4, *T. seticornis* His. 5, *T. fimbriatus* Murch. 6, *T. radiatus* Murch. *Remopleurides colbii* Portl., *R. laterispinifer* Portl., *R. dorsospinifer* Portl.
Sec. 1. Remopleurides, Glabella furrows quite obsolete. *R. platyceps* McCoy, *R. obtusus* n. sp.
Sec. 2. Caphyra, Glabella moderate, not inflated, with three pairs of furrows. *R. (Caphyra) radians* Barr.
Cyphoniscus socialis Salt. *Æglina mirabilis* Forbes, *Æ. major* n. sp.

———— On some new crustacea from the uppermost Silurian rocks, by J. W. Salter. With a note on the structure and affinities of Himantopterus, by T. H. Huxley.

In Quart. Jour. Geol. Soc. London, vol. 12, 1856, pl. 26, figs. 1–7.
Himantopterus n. g., *H. bilobus* n. sp., *H. maximus* n. sp., *H. lanceolatus* n. sp., *H. perornatus* n. sp., *H. acuminatus* n. sp., *H. banksii* n. sp. *Ceratiocaris.*

———— Description of the Crustaceans from the uppermost Silurian rocks near Lesmahago. (*)

In London, Edinb., Dublin Philos. Mag., vol. 11, 1856, p. 83.
Himantopterus.

———— On the great *Pterygotus* (Seraphim) of Scotland and other species.

In Rep. 26th Meeting Brit. Assoc. Adv. Sci., 1856, Trans. of Sec., p. 75.

———— On fossil remains in the Cambrian rocks of the Longmynd and North Wales.

In Quart. Jour. Geol. Soc. London, vol. 12, 1856, p. 246.
Palæopyge n. g., *P. ramsayi* n. sp.

———— On two Silurian species of *Acidaspis* from Shropshire.

In Quart. Jour. Geol. Soc. London, vol. 13, 1857, p. 210, pl. 6.
Acidaspis coronata n. sp., *A. caractaci* Salt.

———— On some new species of *Eurypterus*. With note on the distribution of the species.

In Quart. Jour. Geol. Soc. London, vol. 15, 1859, p. 229, pl. 10.
Eurypterus symondsii n. sp., *E. scouleri* ?, *E. pygmæus* Salt., *E. megalops* n. sp., *E. acuminatus* n. sp., *E. linearis* n. sp., *E. abbreviatus* n. sp., *E. chararius* n. sp.

─── On the fossils of the Lingula flags or zone Primordial *Paradoxides* and *Conocephalus* from North America.
In Quart. Jour. Geol. Soc. London, vol. 15, 1859, p. 551, 4 figs.
Paradoxides bennetti n. sp.   *Conocephalus antiquatus* n. sp.   *Asaphus* or *Olenus* sp.?

─── New Silurian Crustacea.
In Annals Mag. Nat. Hist., 3d series, London, vol. 5, 1860, p. 153, figs.
Dictyocaris n. g. *Physocaris* n. g. *Ceratiocaris papilio* n. sp., *C. stygius* n. sp., *C. inornatus* Salt., *C. murchisoni* McCoy, *C. leptodactylus* McCoy, *C. robustus* n. sp., *C. decorus* Phill., *C.? ensis* n. sp., *C. cassia* n. sp. *Cultellus? (Ceratiosolen?) rectus* n. sp. *Ceratocaris (Physocaris) vesica* n. sp., *C. ellipticus* McCoy, *C. aptychoides* was referred by Salter in 1863 to the genus *Peltocaris*.

─── On the fossils from the High Andes, collected by David Forbes.
In Quart. Jour. Geol. Soc. London, vol. 17, 1861, p. 62, pls. 4, 5.
Proetus (Cryphæus) pentlandi n. sp. *Homalonotus linares* n. sp., *H.* sp.? *Beyrichia forbesii* (Jones) n. sp. *Phacops latifrons* Bronn.

─── On some of the higher Crustacea from the British Coal Measures.
In Quart. Jour. Geol. Soc. London, vol. 17, 1861, p. 528, figs. 1-8.
Anthrapalæmon n. g., *Palæocrangon* n. g.
Name changed to that of *Crangopsis*. See same journal, vol. 19, p. 80.
Subgenus *Palæocarbus*, type *Anthrapalæmon dubius* Prestw. *Anthrapalæmon grossarti* n. sp. *Palæocrangon socialis* Salt.

─── Crustacea in the Coal Measures.
In London, Edinb., Dublin Philos. Mag., vol. 22, 1861, p. 325.
Anthrapalæmon.
This is merely a notice of the article referred to in the Quart. Jour. Geol. Soc., vol. 17, 1861, p. 528.

─── Description of *Uronectes (Gamponyx) socialis* in the Rev. T. Brown's paper, "Mountain limestone and Lower Carboniferous rocks of the Fifeshire coast."
In Trans. Royal Soc. Edinb., vol. 22, pt. 2, 1861, p. 394.

─── A monograph of British Trilobites.
In Palæont. Soc. London, vol. 16, 1862, pp. 1-83, pls. 1-6, pt. 1.
This work forms different parts of the Palæont. Soc. London, volumes for 1862-68, as separately itemized in this bibliography.

Subgenus 1, *Trimerocephalus* McCoy. *Phacops (Trimerocephalus) lævis* Münst., *P. cryphthophthalmus* Emm.
Subgenus 2, *Phacops* Emm. *P. (Phacops) granulatus* Münst., *P. (Phacops) latifrons* Bronn, also var. *P. (Phacops) stokesii* Milne-Edw., *P. (Phacops) nudus* n. sp., *P. (Phacops) mushini* n. sp.
Subgenus 3, *Acaste* Goldf. *P. (Acaste) downingiæ* Murch., also variations *(a) vulgaris, (b) macrops, (c) inflatus, (d) spinosus.* Variety or subspecies *P. constrictus, P. (Acaste) apiculatus* Salt., *P. (Acaste) minus* n. sp., *P. (Acaste) incertus* Deslong., *P. (Acaste) jamesii* Portl., *P. (Acaste) alifrons* Salt., *P. (Acaste) brongniarti* Portl., also var. *dalmani.*
Subgenus 4, *Chasmops* McCoy. *P. (Chasmops?) jukesii* Salt., *P. (Chasmops) macroura* Sjögr., *P. (Chasmops) conophthalmus* Boeck? *P. (Chasmops) amphora* Salter, *P. (Chasmops) truncato-caudatus* Portl., *P. (Chasmops) bailyi* n. sp.
Subgenus 5, *Odontochile* Cord. *P. (Odontochile) obtusicaudatus* Salt., *P. (Odontochile) mucronatus* Brong., *P. (Odontochile) imbricatulus* Ang., *P. (Odontochile) caudatus* Brünn, var. *(a) vulgaris, (b) tuberculato-caudatus* Murch., *(c) nexilis* Salt., *(d) aculeatus* Salt., *P. (Odontochile) longicaudatus* Murch. var. *(a) armiger, (b) grindrodianus, P. weaveri* Salt.
Subgenus 6, *Cryphæus* Green. *P. (Cryphæus) punctatus* Stein.
Fam. Cheirurus subgenus *Crotalocephalus* Salt. *Cheirurus (Crotalocephalus) articulatus* Münst.? *C. bimucronatus* Murch. var. *(a) bimucronatus, (b) centralis.*
Subgenus *Cheirurus* Beyr. *C. (Cheirurus) gelasinosus* Portl., *C. (Cheirurus) cancrurus* Salt.
Subgenus *Eccoptochile* Cord. *C. (Eccoptochile) sedgwicki* McCoy, *C. (Eccoptochile) frederici* n. sp.
Subgenus *Actinopeltis* Cord., *C. (Actinopeltis) juvenis* Salt., *C. (Actinopeltis) octolobatus* McCoy.
Sphærexochus mirus Beyr., *S.? boops* n. sp. *Amphion pseudo-articulatus* Portl., *A. benevolens* n. sp., *A. pauper* n. sp.

—— On some fossil Crustacea from the Coal Measures and Devonian rocks of New Brunswick, Nova Scotia, and Cape Breton.
In London, Edinb., Dublin Philos. Mag., 4th series, vol. 24, 1862. p. 166.

—— A monograph of British Trilobites. Pt. 2. Pp. 81–128, pls. 7–19.
In Palæont. Soc. London, vol. 17, 1863.
Staurocephalus murchisoni Barr., *S. globiceps* Portl., *S.? unicus* Wyv. Thom. *Deiphon forbesi* Barr. *Calymene tuberculosa* Salt., *C. blumenbachii* Brong., also vars. *allportiana* and *caractaci.* Subspecies 1, *Calymene senaria* Conrad. 2, *C. cambrensis* Salt. *Calymene tristani* Brong., *C. duplicata* Murch., *C. parvifrons* Salt. also var. *murchisoni, C.? daviesii* n. sp.

Homalonotus Sec. 1. *Brongniartia* Salt., *H. bisulcatus* Salt., *H. sedgwicki*
n. sp., *H. edgelli* n. sp., *H. rudis* Salt., *H. brongniarti* Deslong., *H. vicary*
n. sp., *H. sp.*? *H. vulcani* Murch.
Sec. 2. *Trimurus* Green. *Homalonotus delphinocephalus* Green, *H. cylindricus* n. sp., *H. johannis* n. sp.
Sec. 3. *Koenigia* Salt. *Homalonatus knightii* König., *H. lundensis* n. sp.
Sec. 4. *Dipleura* Green. *Homalonotus dekayii* Green.
Sec. 5. *Burmeisteria* Salt. *Homalanotus elongatus* n. sp.
Ogygia buchii Brong., *O. angustissima* Salt. *Staurocephalus nodulosus* MS. Salt., pl. 7, fig. 25.

————— On some fossil Crustacea from the Coal Measures and Devonian rocks of British North America.
In Quart. Jour. Geol. Soc. London, vol. 19, 1863, p. 75, figs. 1-12.
Amphipeltis n. g. *Diplostylus* n. g.
The author herein changes the name *Palæocrangon* Salter to that of *Crangopsis*.
Amphipeltis paradoxus n. sp. *Diplostylus dawsoni* n. sp. *Eurypterus pulicaris* n. sp. *Eurypterus* allied to *E. scouleri* Hibb., *E.* sp.?

————— On some species of *Eurypterus* and allied forms.
In Quart. Jour. Geol. Soc. London, vol. 19, 1863, p. 81, 8 wood cuts.
Eurypterus scouleri Hibb., *E. (Arthropleura?) mammatus* (Woodward says this is an undoubted plant), *E.? (Arthropleura) ferox* (this species has been referred by Woodward to the Myriopodous genus *Euphoberia*).

————— On Peltocaris, a new genus of Silurian Crustacea.
In Quart. Jour. Geol. Soc. London, vol. 19, 1863, p. 87.
Peltocaris apychoides Salt., *P. harknessi* n. sp.
The author gives a series of wood cuts illustrating the genera *Peltocaris* n. g., *Hymenocaris*, *Ceratiocaris*, *Dithyocaris*, *Arges*, *Apus*, *Nebalia*, *Estheria*, *Leperditia*, *Beyrichia*.

————— On some tracks of Lower Silurian Crustacea.
In Quart. Jour. Geol. Soc. London, vol. 19, 1863, p. 92, figs.

————— Note on the Skiddaw slate fossils.
In Quart. Jour. Geol. Soc. London, vol. 19, 1863, p. 135, figs.
Caryocaris n. g., *C. wrightii* n. sp.

————— On the discovery of Paradoxides in Britain.
In Quart. Jour. Geol. Soc. London, vol. 19, 1883, p. 274.
Paradoxides davidis n. sp.

—— On a new Crustacean from Glasgow Coal Field.
In Quart. Jour. Geol. Soc. London, vol. 19, 1863, p. 519, figs. 1, 2,
Palæocarabus russellianus n. sp.

—— A monograph of British Trilobites. Pt. 3, pp. 129–179, pls. 15–25.
In Palæot. Soc. London, vol. 18, for 1864 (London, 1866).

Ogygia angustissima n. sp., *O.?* (vel. *Phacops) subduplicata*, see *Phacops* pl. xv, figs. 7 and 8), *O. (Ptychopyge) corridensis* Murch., *O. scutatrix* Salt., *O. peltata* Salt., *O. selwyni* Salt.

Barrandia—Subgenus *Homalopteon*. *Barrandia (Homalopteon) portlocki* Salt., *B. (Homalopteon) radians* McCoy.

Subgenus *Barrandia*. *B. (Barrandia) cordai* McCoy. *Niobe homfrayi* Salt.

Asaphus—Subgenus 1, *Ptychopyge* Ang. *Asaphus angustifrons* Dalm., *A. latus* Ang.

Subgenus 2, *Basilicus*. *Asaphus (Basilicus) tyrannus* Murch., *A. (Basilicus) peltastes* n. sp., *A.? (Basilicus) hybridus* n. sp., *A. (Basilicus) powisii* Murch., *A. (Basilicus) marstoni* n. sp., *A. (Basilicus?) radiatus* Salt., *A. (Basilicus) laticostatus* McCoy. *Ogygia corndensis* Murch.

Subgenus 3, *Megalaspis*. *M. extenuatus* Wahl.

Subgenus 4, *Isotelus*. *Asaphus (Isotelus) gigas* Dekay, *A. (Isotelus) affinis* McCoy, *A. (Isotelus) homfrayi* Salt.

Subgenus 5, *Cryptonymus*. *Asaphus expansus* Linné, *A. (Cryptonymus) scutalis* Salt., *A. (Cryptonymus) kowalewskii*.

Subgenus 6, *Symphysurus* Goldf., type *S. palpebrosus* Dalm.

Subgenus 8, *Nileus* Dalman. *Nileus armadillo* Dalm.

Stygina latifrons Portl., *S. murchisoni* Murch., *S. musheni* n. sp. *Psilocephalus innotatus* Salt., *P. inflatus* Salt. *Phacops subduplicata* Salt., pl. 15, figs. 7-8. *Ogygia (Ptychopyge) cordensis* Murch. *Illænopsis thomsoni* Salt., pl. 20, figs. 1, 2.

—— Figures and descriptions illustrative of British organic remains. Decade XI. Trilobites (chiefly Silurian)
In Mem. Geol. Survey United Kingdom, London, 1864, 10 pls.

The article on *Salteria primæva* n. g. et sp., pl. 6, was edited by P. Wyville Thomson.

The genera are arranged in this work in the following order:

Pl. 1. *Agnostus princeps* Salt., also var. (a) *ornatus*, (b) *rudis*.

Sec. 1. *Condylopyge*. *Agnostus maccoyi* Salt., *A. rex* Barr., *A. morea* n. sp.

Sec. 2. *Agnostus*, *A. pisiformis* Linné, *A. princeps* Salt., *A. trisectus* n. sp.

Sec. 3. *Trinodus*. *A. tardus* Barr., *A. trinodus* Salt., *A. limbatus* n. sp.

Sec. 4. *Phalacroma. A. integer* Beyr.
Pl. 2. *Stygina latifrons* Portl.
Pl. 3. *Asaphus gigas* Dekay. *Isotelus rectifrons* Portl., *I.* sp.?
Pl. 4. *Æglina binodosa* Salt., *A. grandis* Salt.
Pl. 5. *Staurocephalus murchisoni* Barr., *S.? unicus*, Wyv. Thom., *S. globiceps* Portl.
Pl. 6. *Salteria primæva* Wyv. Thom., *S. involuta* n. sp.
Pl. 7. *Angelina sedgwicki* Salt. *Conocoryphe invita* Salt., with list of British species.
Pl. 8. *Olenus cataractes* Salt., *O. (Spærophthalmus) flagellifer* Ang., *O. (Parabolina) serratus* Salt., *O. scarabæoides* Wahl., also var. *obesus*, *O. (Sphæroph.) humilis* Phill., *O. bisulcatus* Phill., *O. (Sphæroph.) pecten* Salt., *O. cataractes* Salt., *O. (Sphæroph.) alatus* Boeck, *O. (Leptoplastus)* or *Sphærophthalmus raphidophorus* Ang., *O. (Parabolina) spinulosa* Wahl.
Pl. 9. *Phacops (Trimerocephalus) lævis* Münst., *P. (Trimer.?) cryptophthalmus* Emm.
Pl. 10. *Paradoxides davidis* Salt., *P. forchhammeri* Ang.

———— On some fossils from the Lingula flags of Wales.
In Quart. Jour. Geol. Soc. London, vol. 20, 1864, p. 233, pl. 13.
Anopolenus n. g. *Holocephalina* n. g. *Anopolenus henrici* n. sp. *Conocoryphe? variolaris* n. sp. *Holocephalina primordialis* n. sp. *Agnostus princeps* Salt. *Microdiscus punctatus* n. sp. *Leperditia solvensis* Jones.

———— Note on the fossils from the Budleigh sandstone pebble bed.
In Quart. Jour. Geol. Soc. London, vol. 20, 1864, p. 286, pl. 15.
Phacops incertus Deslong. *Calymene tristani* Brong. *Homalonotus brongniarti* Deslong. *Myocaris* n. g., *M. lutraria* n. sp.
For detailed description of Myocaris, see Geol. Mag., vol. 1, London, 1864, p. 11, fig. 4.

———— On the new genus *Anopolenus*.
In Geol. Mag., vol. 1, London, 1864, p. 6.
Anopolenus henrici Salt. *Myocaris lutraria* Salt.

———— Catalogue of the Museum of Practical Geology of London. London, 1865.

———— and **Woodward** (Henry). Chart of fossil Crustacea, accompanied by a descriptive catalogue of all the genera and species figured. London, 1865, pls. 3 and 4.
See, also, Rep. 35th Meeting Brit. Assoc. Adv. Sci., 1865, Trans. of Sec., p. 79; Geol. Mag., vol. 2, London, 1865, p. 468.

Order Decapoda, Sub-Orders 1 and 2. Brachyura and Anomura. No Palæozoic genera.

Sub-Order 3: Macrura. 1, *Palæocrangon socialis* Salt. 2, *Palæocarabus dubius* Prestw. 3, *P. russellianus* Salt.

Orders Stomapoda, Amphipoda, Isopoda. 1, *Kampecaris forfariensis* Page. 2, *Gitocrangon granulatus* Richter. 3, *Amphipeltis paradoxus* Salt. 4, *Pygocephalus cooperi* Huxley. 5, *Uronectes (Gampsonyx) fimbriatus* Jordan. 6, *Diplostylus dawsoni* Salt. 7, *Prosoponiscus problematicus* Geintz. 8, *Dalmanites kablikiæ* Geintz. 9, *Bostrichopus antiquus* Goldf.

Order Trilobita. *Agnostus nudus* Beyr., *A. rex* Barr., *A. granulatus* Barr., *A. princeps* Salt., *A. maccoyi* Salt., *A. trinodus* Salt. *Sao hirsuta* Barr. *Anomocaris difformis* Aug. *Dolichometopus suecicus* Aug. *Anopocaris pusillus* Aug. *Arionellus ceticephalus* Barr. *Conocoryphe sulzeri* Schloth., *C. breviceps* Aug., *C.? (Harpides) rugosa* Aug. *Solenopleura canaliculata* Aug. *Acerocaris ecornis* Aug. *Ellipsocephalus hoffi* Schloth. *Olenus gibbosus* Wahl., *O. (Parabolina) spinulosus* Wahl., *O. (Peltura) scarabæoides* Wahl., *O. (Sphærophthalmus) alatus* Boeck, *O. micrurus* Salt., *O.? holopyge* Hall. *Paradoxides bohemicus* Boeck, *P. bennetti* Salt., *P. inflatus* Cord. *Hydrocephalus carens* Barr. *Dikelocephalus minnesotensis* Owen, *D. magnificus* Bill., *D. granulosus* Owen. *Lonchocephalus chippewaensis* Owen. *Palæopyge ramsayi* Salt. *Menocephalus globosus* Bill. *Centropleura dicræura* Ang., *C. angusticauda* Aug. *Corynexochus umbonatus* Aug. *Atops trilineatus* Emm. *Triarthus beckii* Green. *Cyphoniscus socialis* Salt. *Remopleurides colbii* Portl., *R. radians* Barr. *Harpes doranni* Portl., *H. ungula* Sternb., *H. macrocephalus* Goldf. *Calymene declinata* Barr., *C. tuberculosa* Salt., *C. bluembachii* Brong. *Cybele verrucosa* Dalm. *Staurocephalus murchisoni* Barr. *Acidaspis buchii* Barr., *A. mira* Barr. *Encrinurus punctatus* Brünn, *E. subvariolaris* Münst. *Deiphon forbesii* Barr. *Eccoptochile sedgwickii* McCoy. *Cheirurus quenstedti* Barr., *C. sternbergii* Münst., *C. myops* Roemer. *Sphærexochus mirus* Beyr. *Placopharia zippei* Boeck. *Lichas palmata* Beyr. *Typhloniscus bainii* Salt. *Arges speciosus* Goldf. *Trinucleus ornatus* Sternb. *Ampyx nudis* Murch. *Dionide formosa* Barr. *Æglina speciosa* Barr. *Ogygia buchii* Brong. *Asaphus expansus* Linné, *A. platycephalus* Stokes, *A. (Megalaspis) extenuatus* Wahl., *A. gigas* Dekay. *(Isotelus megistos* Locke.) *Niobe frontalis* Dalm. *Nileus armadillo* Dalm. *Stygina latifrons* Portl. *Illænus davisii* Salt., *I. barriensis* Murch. *Phacops proetus* Emm., *P. phillipsia* Barr., *P. truncato-caudatus* Portl., *P. downingiæ* Murch., *P. cephalotes* Cord., *P. (Dalmania) auriculatus* Dalm., *P. caudatus* Brünn, *P. (Dalmania) nasutus* Hall, *P. laciniatus* Roemer, *P. latifrons* Bronn, *P. granulatus* Münst., *P. (Trimerocephalus) lævis* Münst., *P. cryptophthalmus* Emm., *P. microma* Roemer, *P. africanus* Salt., *P. caffer* Salt. *Homalonotus delphinocephalus* Green, *H. armatus* Burm., *H. pradoanus* De Vern. *Aulacopleura koninckii* Barr. *Proetus striatus* Barr., *P. lepidus* Barr., *P. curieri* Goldf., *P. lævigatus* Goldf., *P.? (Cylindraspis) latispinosus* Sandb. *Cyp-*

haspis burmeisteri Barr., *C. megalops* McCoy. *C. ceratophthalmus* Goldf. *Bronteus dormitzeri* Barr., *B. flabellifer* Goldf., *B.* (*Bumastus*) *planus* Münst. *Phillipsia pustulata* Schloth., *B. seminifera* Phill., *P. derbyensis* Mart., *P. brongniarti* Fischer. *Griffithides* or *Phillipsia, G. longiceps* Portl., *G. eichwaldi* Fischer, *G. globiceps* Portl. *Brachymetopus ouralicus* DeVern., *B.* (*Trinucleus*) *issedon* Eichw., *B. discors* McCoy.

Orders Phyllopoda and Ostracoda. *Hymenocaris vermicauda* Salt. *Peltocaris harknessi* Salt., *P. aptychoides* Salt. *Ceratiocaris ? latus* Salt., *C.* (*Leptocheles*) *bohemicus* Barr., *C. inequalis* Barr., *C.* (*Physocaris*) *vesica* Salt., *C. papilio* Salt. *Dictyocaris ramsayi* Salt. *Dithyrocaris scouleri* McCoy, *D. testudineus* Scouler, *D.* (*Argas*) *tricornis* Scouler, *D. tenuistriatus* McCoy. *Cyclus radialis* Phill. *Estheria membrancea* Pacht., *E. striata* Münst., *E. tenella* Jones. *Leaia leidyi* Lea. *Leperditia solvensis* Jones, *L. canadensis* Jones, *L. arctica* Jones, *L. gibbera* Jones, *L. alta* Jones, *L. balthica* His., *L.* (*Cythere*) *elongata* McCoy, *L.* (*Cythereopsis*) *subrecta* Portl., *L. scoto-burdigalensis* Hibb. *Beyrichia strangulata* Salt., *B. logani* Jones, *B. complicata* Salt., *B. plagosa* Jones, *B. kloedeni* McCoy, also var. *torosa, B. seminulum* Jones, *B. clathrata* Jones, *B.* (*Cythere*) *bituberculata* McCoy, *B. gibberosa* Eichw., *B. arcuata* Beau., *B. subarcuata* Jones. *Kirkbya striolata* Eichw., *K. umbonata* Eichw., *K. permina* Jones, *K. biplicata* Jones. *Cypridella cruciata* DeKon. *Entomis serratostriata* Sandb., *E. concentrica* DeKon. *Cypridina? striatula* Richter, *C.? subfusiformis* Sandb., *C.? hemispharica* Richter, *Ç. primæva* McCoy. *Cyprella chrysalidea* DeKon. *C. annulata* DeKon. *Entomoconchus scouleri* McCoy. *Cytheropsis umbonata* Salt., *C. aldensis* McCoy. *Cythere* (*Leperditia*) *pusilla* McCoy, *C. orbicularis* McCoy, *C. impressa* McCoy, *C. costata* McCoy, *C. trituberculata* McCoy, *C. excavata* McCoy, *C. subreniformis* Kirkby, *C. nuciformis* Jones, *C. kutorgiana* Jones, *C. subelongata* Münst., *C. amputata* Kirk. *Candona salteriana* Jones. *Bairdia distracta* Eichw., *B. curta* McCoy, *B. qualeni* Eichw., *B. plebeia* var. *caudata* and *elongata* Jones, *B. brevicauda* Jones, *B. ampla* Reuss., *B. rhomboidea* Kirkby.

Order Eurypterida. *Protichnites octonotatus* (foot prints). *Hemiaspis limuloides* H. W. *Bunodes lunula* Eichw. *Pterygotus problematicus* Agass., *P. anglicus* Agass., *P. bilotus* Agass. *Slimonia acuminata* Salt. *Stylonurus logani* H. W. *Eurypterus pygmæus* Salt., *E.* (*Stylonurus?*) *symondsii* Salt., *E.* (*Stylonurus?*) *megalops* Salt., *E. tetragonophthalmus* Fischer, *E. lacustris* Hall, *E. granosus* (*Adelophthalmus*) Jordan. *Arthropleura armata* Jord., *A. ferox* Salt. *Hemiaspis selweyni* Salt. MS.

Order Poecilopoda Xiphosura. *Belinurus* (*Limulus*) *trilobitoides* König, *B.* (*Limulus*) *anthrax* Prestw., *B.* (*Limulus*) *rotundatus* Prestw., *B. reginæ* Baily. *Limulus? oculatus* Kut.

―――― and **Blanford** (H. F.) Palæontology of Niti in the Northern Himalaya, being descriptions and figures of the Palæozoic and Secondary Fossils collected by

Col. Richard Stachey, R. E. Descriptions by J. W. Salter and H. F. Blanford. Calcutta, 1865.
Reprinted with slight corrections for private circulation from Col. R. Strachey's forthcoming work on the Physical Geography of the Northern Himalaya.
The following new Crustacea are described therein by J. W. Salter:
Asaphus emodi n. sp., also var. *brevier*. *Illænus brachyoniscus* n. sp., *I. punctulosus* n. sp. *Cheirurus mitis* n. sp. *Prosopiscus* n. g., *P. mimis* n. sp. *Sphærexochus idiotes* n. sp. *Lichas tibestanus* n. sp. *Calymene nivalis* n. sp.
Diagonis of *Prosopiscus:* "Head transverse with deep furrows, a short glabella with strong side lobes, broad convex, punctata cheeks without eyes or facial suture. Thorax not known, Pygidium probably as in *Cheirurus* spinose."

Salter (J. W.) Unpublished plates on the genus *Acidaspis* probably for a Decade. Geol. Sur. United Kingdom.
Acidaspis brightii Murch., *A. dumetosus* n. sp., *A. quinquespinosus* n. sp., *A. barrandii* n. sp., *A. crematus* Emm.

—— In Brit. Assoc. Report, 1865.
Agnostus davidis n. sp. *A. scutalis* n. sp., *A. barrandei* n. sp. *Erinnys* n. g., *E. venulosa* n. sp. *Leperditia bupestis* n. sp. *Conocoryphe bufo* Hicks, *C. applanata* Salt., *C. humerosa* Salt. *Paradoxides aurora* Salt., *P. hicksii* Salt.

—— On some additional fossils from the Lingula flags, with a note on the genus *Anopolenus*, by Henry Hicks.
In Quart. Jour. Geol. Soc. London, vol. 21, 1865, p. 746.
Anopolenus salteri (Hicks) n. sp., *A. henrici* Salt. *Olenus (Sphærophthalmus) pecten* Salt.
Erinnys n. g. This genus was proposed in 1865, Brit. Assoc. Report, p. 136. The same was used in 1810 for a genus of Lepidoptera.

—— On the fossils of North Wales.
In Mem. Geol. Survey Gr. Brit., vol. 3, London, 1866, p. 239, 37 pls.
For a list of the genera and species see Ramsay (A. C.), p. 182.

—— A monograph of British Trilobites. Pt. 4, pp. 177–216, pls. 25–30.
In Palæont. Soc., vol. 21, 1867.
Ogygia peltata Salt., *O. butlina* n. sp. *Barrandia (Homalopteon) longifrons* Edgell.

Illænus; subgenus *Octillænus* Salt. *Illænus hisingeri* Barr. 2, *Panderia* Volb. *Illænus (Panderia) lewisii* n. sp., *I. triquetra* Volb. 3, *Dysplanus* Burm. *Illænus (Dysplanus) centrotus* Dalm., *I. (D.) bowmanni* Salt., *I. (D.) æmulus* n. sp., *I. (D.) thomsoni* Salt., *I. nexilis* n. sp. 4, *Illænus* Dalm. *Illænus (Illænus) crassicauda* Wahl., *I. (I.) baylyi* n. sp., *I. (I.) davisii* Salt., also var. *involutus, I. (I.) ocularis* Salt., *I. portlockii* Salt., *I. rosenbergii* Eichw., *I. (I.) murchisoni* Salt., *I. dalmani* Volb. 5, *Ectillænus* Salt. *Illænus (Ectillænus) perovalis* Murch. 6, *Hydrolænus* Salt. *Illænus conifrons* Billings. 7, *Illænopsis* Salt. *Illænopsis thomsoni* Salt. 8, *Bumastus* Murch. *Illænus (Bumastus) barriensis* Murch., *I. (B.) insignis* Hall, *I. (B.) carinatus* n. sp., *I. (B.) maccallumi* n. sp. *Stygina musheni* Salt., pl. 29, fig. 1.

——— On *Saccocaris major*.
In Rep. Proc. Geol. Polytech. Soc. West. Riding of Yorkshire (for 1867), vol. 4, 1868, p. 588.

——— and **Hicks** (Henry). On some fossils from the Menevian group.
In Quart. Jour. Geol. Soc. London, vol. 25, 1869, p. 51, pls. 2, 3.
In a note on p. 52, Mr. J. W. Salter suggests the new generic name of *Plutonia*, but leaves the description for another article.
A species of this genus was afterwards described and figured in the same journal, vol. 27, 1871, pl. 15, figs. 1-8, under the name of *Plutonia sedgwicki* Hicks. This large species has affinities with the genera *Paradoxides* and *Anopolenus* in the form of its glabella, but differs, according to Hicks, "in having the whole surface covered with coarse tubercles or spines." The term *Plutonia* is objectionable, inasmuch as it was used in 1864 for a genus of the Mollusca.
Conocoryphe bufo Hicks, *C. applanata* Salt., *C.? humerosa* Salt. *Paradoxides aurora* Salt., *P. hicksii* Salt. *Conocoryphe perdita* n. sp.

———Catalogue of the collection of Cambrian and Silurian fossils contained in the Geological Museum of the University of Cambridge, 1873.

——— and **Etheridge** (R.) On the fossils of North Wales.
In Geology of North Wales, with an appendix on the fossils by J. W. Salter, revised and added to by R. Etheridge; Mem. Geol. Survey Gt. Brit., vol. 3, London, 1881, 2d ed., 623 pp., 113 illustrations, 2 geol. maps, 26 plates of fossils.

——— Monograph of British Trilobites from the Cambrian, Silurian and Devonian formations. Pt. 5.
Palæont. Soc. London, vol. 37, 1883, p. 215 and index, 3 figs.
Illænus bowmanni ? Salt., *I. crassicauda* Wahl., *I. (Bumastus)* sp. ?

Sars (M.) Ueber einige neue oder unvollständig bekannte Trilobiten.

In Isis (oder Encycl. Zeitung), Oken, 1835, p. 333, pls. 8-9.

Olenus forficula n. sp. *Ampyx rostratus* n. sp., *A. mammillatus* n. sp. *Asaphus dilatatus* Dalm., *A. grandis* n. sp. *Calymene speciosa* Dalm., *C. clavifrons* Dalm.

Angelin describes and illustrates *Ampyx costatus* Boeck, taking Sars' figure of *Ampyx mammillatus*, pl. 8, figs. 4 a–b and d, for the type of the species. On comparing Angelin's illustrations of *Ampyx costatus*, pl. 40, fig. 1, we observe that they resemble Sars' figure, pl. 8, fig. 4 b, in having a produced glabella and spine. Sars' figures given on pl. 8, fig. 4 a, represent a species with a blunt glabella terminating in a tubercle; fig. 4 b of the same plate has an extended glabella prolonged into a spine. Both these figures do not appear to be the same, therefore Angelin's description of *Ampyx costatus* should be confined to Sars' fig. 4 b, and *Ampyx mammillatus* should be restricted to Sars' figures of this species given on pl. 8, figs. 4 a–c.

These illustrations do not quite coincide, but they agree much better than fig. 4 b. That both Boeck and Angelin were correct in splitting up *Ampyx mammillatus* Sars there can be no doubt; it is a question of the correct separation. *Ampyx costatus* should not include the spined and non-spined glabella represented on Sars' pl. 8, figs. 4 a–b, as Angelin has reclassified it.

———— Bemerkungen über die untere seite von einigen Trilobiten.

In Isis (oder Encycl. Zeitung, Oken), 1835, p. 340, pl. 9.

Asaphus expansus Linné, pl. 9, fig. 9., *A. dilatatus* Brünn, pl. 9, fig. 11. *Illænus crassicauda* Wahl., pl. 9, fig. 10.

Sars describes the hypostomæ of these species. The author remarks that the hypostomæ may have surrounded the mouth and served as a support for the mandibles.

Scheuchzer (J. J.) Specimen lithologiæ Helvetia. Zürich, 8vo, 1702.

———— Meteorologia et Oryctographia Helvetica, oder Beschreibung der Lufftgeschichten, Steinen, Metallen, etc. Zürich, 1718.

The author classes the Trilobites with the Mollusca; being only acquainted with fragments or parts of the pygidium, he compares them with Patella.

Schlotheim (E. Fr. von.) Beschreibung einer seltenen Trilobiten-Art.

In Taschenbuch für gesam. Mineral., Leonhard, 1810, pl. 1, figs. 1-6. *Trilobiten cornigerus* n. sp.

———— Die Petrefactenkunde auf ihrem jetzigen Standpunkte durch die Beschreibung seiner Sammlung versteinerter und fossiler Überreste des Thier- und Pflanzenreichs der Vorwelt erläutert. Mit XV Kupfertafeln. Gotha, 1820.

Entomolithen. B. Trilobiten.

1, *Trilobites cornigerus*. 2, *T. paradoxus*. 3, *T. bituminosus*. 4, *T. tentaculatus*. 5, *T. problematicus*.

———— Nachträge zur Petrefactenkunde. II. Gotha, 1823, pl. 22, p. 1.

1, *Calymene blumenbachii* Brong. 2, *C. tristani* Brong. 3, *C. variolaris* Brong. 4, *C. macrophthalma* Brong.

1, *Asaphus cornigerus* Schloth. (*A. expansus* Linné). 2, *A. buchii* Brong. 3, *A. hausmanni* Brong. 4, *A. caudatus* Brünn. 5, *A. laticauda* Wahl.

1, *Ogygia guettardi* Brong. 2, *O. desmarestii* Brong.

1, *Paradoxides tessini* Brong. 2, *P. spinulosus* Brong., *P. scarabæoides* Wahl. 4, *P. gibbosus* Wahl. 5, *P. laciniatus* Wahl.

Agnostus pisiformis Linné.

Pl. 20. Fig. 1, *Trilobites sulzeri* (*Conocoryphe sulzeri* Schloth.) Fig. 2, *Tril. hoffi* (*Ellipsocephalus hoffi* Schloth.) Fig. 3, *Tril. schroet* (*Illænus chiron* Holm.) Fig. 4, *Tril. spærocephalus* (*Hypostoma Paradoxides ?*). Fig. 5, *Tril. velatus*. Fig. 6, *Tril. pustulosus* (*Phacops latifrons* Bronn). Fig. 7, *Tril. hausmanni* (*Dalmanites hausmanni* Brong.) Fig. 8, *Tril. problematicus*. Fig. 9, *Tril. bituminosus* (*Ianassa bituminosa* Schloth.)

———— Beschreibung einiger abgebildeten Arten von Echinosphæriten und Trilobiten.

In Isis (oder Encycl. Zeitung), Oken, 1826, p. 309, pl. 1, figs. 8, 9*; also Féruss., Bull., 1827, vol. 12, p. 406.

Tril. esmarkii (*Illænus esmarki* n. sp.), *T. granum* n. sp. (*Agnostus granum* n. sp.)

———— Systematisches verzeichniss der Petrefacten Sammulung des verstorbenen wirklichen. Gotha, 1832, 80 pp.

1, *Calymene blumenbachii* Brong. 2, *C. tristani* Brong. 3, *C. variolaris* Brong.

2. *Asaphus*. 1, *A. cornigerus* Schloth. 2, *A. hausmanni* Brong. 3, *Tril, expansus* Linné.

3. *Ogygia*. 1, *O. guettardi* Brong.

4. *Paradoxides.* 1, *P. tessini* Brong. 2, *P. spinulosus* Brong. 3, *P. gibbosus* Wahl. 4, *P. bucephalus* Wahl.
5. *Agnostus.* 1, *A. pisiformis* Linné.
1, *Tril. (Calymene) sulzeri* Schloth. 2, *Tril. (Calymene) hoffi* Schloth. 3, *Tril. problematicus* Schloth. 4, *Tril. cassicauda* Wahl. 5, *Tril. bituminosus* Schloth. 6, *Tril. gracilis* Boeck. 7, *Tril. conophthalmus* S. & B. 8, *Tril. schlotheimii* Bronn. 9, *Tril. catenatus* n. sp. 10, *Tril. esmarkii* Schloth. 11, *Tril. velatus* Schloth. 12, *Tril. borealis* n. sp. 13, *Tril.* bruschst. von. Tril. aus Reval. 14, *Tril.* sp.? 15, *Tril. sternbergii* Boeck.

Schmid (E. E.) Die kleineren organischen Formen des Zechsteinkalkes von Selters in der Wetterau.
In Neues Jahrbuch für Mineral., 1867, p. 576, pl.
The author gives a brief notice of the following species:
Cythere (Cytherella) nuciformis Jones, *C. pyrrhœ* Eichw., *C. elongata* Geinitz, *C. morrisiana* Jones, *C. (Bairdia) mucronata* Reuss, *C. (Bairdia) rhomboidea* Kirkby, *C. (Cytherella) tyronica* Jones, *C. richteriana* Jones, *C. kutorgana* and *C. (Bairdia) subreniformis* Kirkby, *C. (Bairdia) frumentum* Reuss, *C. (Bairdia) plebeja*, pl. 6, figs. 1–45, Reuss, *C. (Bairdia) brevicauda* Jones, *C. (Cytherella) inornata*, *C. (Bairdia) kingi* Reuss, *C. (Bairdia) acuta* Jones, *C. (Bairdia) ampla* Reuss.

Schmidt (Fr.) Untersuchungen über die silurische Formation von Ehstland, Nord-Livland und Oesel.
In Archiv Naturk. Liv-, Ehst- u. Kurl., Dorpat, vol. 2, 1858–59, pp. 443, 445, 448, 453, 461, 463.*

—— Miscellanæ Silurica. I. Ueber die russischen silurischen Leperditien mit Hinzuziehung einiger Arten aus den Nachbarländern.
In Mém. Acad. Imp. Sci. St. Petersburg, vol. 21, No. 2, 1873, 1 pl.
Leperditia grandis Schrenk, *L. barbotana* n. sp., *L. tyraica* n. sp., *L. angelini* n. sp., *L. balthica* His., *L. hisingeri* n. sp., *L. eichwaldi* n. sp., *L. wiluiensis* n. sp., *L. parallela* n. sp., *L. marginata* Keys., *L. keyserlingi* n. sp. *Isochilina biensis* Grünew, *I. punctata* Eichw., *I. maakii* n. sp.

—— Einige Bemerkungen über die podolisch-galizische Silurformation und deren Petrefacten. St. Petersburg, 1875, 1 pl.
Eurypterus fischeri Eichw. (*E. tetragonophthalmus* Fischer). *Encrinurus punctatus* Brünn, *E. obtusus* Aug. *Illœnus (Bumastus) barriensis* Murch. *Calymene blumenbachii* Broug. *Phacops caudatus* Brünn, *P. downingio* Murch. *Proetus concinna* Dalm. *Cyphaspis elegantulus* Lovén. *Pterygotus anglicus* Agas. The author illustrates *Phacops downingiœ* and *Illœnus (Bumastus) barriensis*.

—— Revision der ost-baltischen silurischen Trilobiten, nebst geognostischer Uebersicht des ost-baltischen Silurgebiets.

In Mém. Acad. Imp. Sci. St. Petersburg, vol. 30, No. 1, 1881, 16 pls.

Subgenus *Phacops* Emm., *P. elegans* S. & B. *Acaste* Goldf. *Phacops downingiæ* Murch. *Pterygometopus* Schmidt. *Phacops sclerops* Dalm., *P. trigonocephala* n. sp., *P. panderi* n. sp., *P. exilis* Eichw., *P. lævigata* n. sp., *P. kuckersiana* n. sp., *P. kegelensis* n. sp., *P. nieszkowskii* n. sp.

Subgenus *Chasmops* McCoy, *P. ingrica* n. sp., *P. nasuta* n. sp., *P. præcurrens* n. sp., *P. odini* Eichw., *P. marginata* n. sp., *P. bucculenta* Sjögr., *P. wrangeli* n. sp., *P. brevispina* n. sp., *P. mutica* n. sp., *P. wenjukowi* n. sp., *P. maxima* n. sp., *P. wesenbergensis* n. sp., *P. eichwaldi* n. sp. Cheiruridæ.

Subgenus *Cheirurus*, *C. ornatus* Dalm., *C. ingricus* n. sp., *C. exsul* Beyr., *C.* subsp. *macrophthalmas* Kut., *C.* subsp. *gladiator* Eichw., *C. spinulosus* Nieszk., *C.* cf. *glaber* Ang.

Subgenus *Cyrtometopus* Ang., *C. clavifrons* Dalm., *C. affinis* Ang., *C. plautini* n. sp., *C. aries* Leuchtbg., *C.* cf. *pseudohemicranium* Nieszk.

Subgenus *Sphærexocoryphe* Ang., *C. cranium* Kut., *C. hübneri* n. sp., *C.* cf. *granulatus* Ang.

Subgenus *Pseudosphærexochus* Schmidt, *C. hemicranium* Kut., *C. conformis* Ang., *C. pahnschi* n. sp., *C. roemeri* n. sp.

Subgenus *Nieszkowskia* Schmidt, *C.*°*tumidus* Ang., *C. variolaris* Linrs., *C. cephaloceros* Nieszk.

Sphærexochus angustifrons Ang. *Amphion fischeri* Eichw. *Diaphanometopus* n. g., *D. volbothi* n. sp. *Cybele bellatula* Dalm., *C. revaliensis* n. sp., *C. rex* Nieszk., *C. grewingki* n. sp., *C. coronata* n. sp., *C. wörthi* Eichw., *C. affinis* n. sp., *C. kutorgæ* n. sp., *C. brevicauda* Ang. *Encrinurus obtusus* Ang., *E. punctatus* Brünn, *E.* cf. *multisegmentatus* Portl., *E. seebachi* n. sp. *Phacops lævigata* n. sp., *P. kegelensis, marginata, bucculenta, P. wenjukowi* n. sp., *P. maxima* n. sp. *Cheirurus rosenthali* n. sp., *C. plautini* n. sp.

—— and Jones (T. Rupert). On some Silurian Leperditia.

In Annals Nat. Hist., 4th ser., vol. 9, London, 1882, p. 168.

—— Miscellanæ Silurica. III. 1. Nachtrag zur Monographie der russischen silurischen Leperditien. 2. Die Crustaceenfauna der Eurypterschichten von Rootziküll auf Oesel.

In Mém. Acad. Imp. Sci. St. Petersburg, vol. 31, No. 5, 1883, 9 pls.

Leperditia grandis Schrenk, *L. phaseolus* His., *L. eichwaldi* Schm., *L. balthica* His., *L. keyserlingi* Schm., *L. hisingeri* Schm., also var. *abbreviata*, *L. marginata* Key., *L. wiluiensis* Schm., *L. barbotana* Schm., *L. mölleri*

n. sp., *L. grandis* Schrenk var. *uralensis*, *L. nordenskjöldi* n. sp., *I. waigatschensis* n. sp. *Bunodes lunula* Eichw., *B. schrenki* Nieszk., *B. rugosus* Nieszk. *Pseudoniscus aculeatus* Nieszk. *Eurypterus fischeri* Eichw., also var. *rectangularis*, *E. laticeps* n. sp. *Pterygotus osiliensis* n. sp. *Ceratiocaris nötlingi* n. sp. *Leperditia lindströmi* n. sp., also var. *mutica*.

—————— Revision der Ostbaltischen Silurischen Trilobiten.

In Mém. Acad. Imp. Sci. St. Petersburg, 7th ser., vol. 33, No. 1, 1885.
Acidaspis emarginata n. sp., *A. kuckersiana* n. sp.
Lichidæ. 1, Group *Arges:* *Lichas wesenbergensis* n. sp.
2. Group *Leiolichas:* *Lichas illænoides* Nieszk.
3. Group *Platymetopus:* *Lichas lævis* Eichw., *L. dalecarlica* Ang., *L. holmi* n. sp.
4. Group *Metopias:* *Lichas celorrhin* Ang., *L. pachyrhina* Dalm., *L. verrucosa* Eichw., *L. hübneri* Eichw., *L. kuckersiana* n. sp.
5. Group *Hoplichas:* *Lichas tricuspidata* Beyr., *L. longispina* n. sp., *L. plautini* n. sp., *L. furcifer* n. sp., *L. conicotuberculata* Nieszk.
6. Group *Conolichas:* *Lichas triconica* Dames, *L. aquiloba* Steinh., *L. schmidti* Dames.
7. Group *Homolichas:* *Lichas depressa* Ang., *L. pahleni* n. sp., *L. deflexa* Sjogr., *L. eichwaldi* Nieszk., *L. angusta* Beyr.
8. Group *Oncholichas:* *Lichas ornata* Ang., *L. gotlandica* Ang.
Miscellaneous: *Lichas st. mathiæ* n. sp., *L. margaritifer* Nieszk., *L. docens* n. sp., *L. cicatricosa* Lov., *L. hamata* n. sp., *L. laxata* McCoy, *Lichas* sp. ?

—————— Ueber einige neue ostsibirische Trilobiten und verwandte Thierformen.

In Bull. Acad. Sci. St. Petersburg, vol. 30, No. 4, 1886, p. 502, pl.
Anomocare pawlowskii n. sp. *Liostracus ?* *maydeli* n. sp. *Agnostus czekanowski* n. sp.
Phacops (Monorakos n. subgen.) *lopatini* n. sp., pl. 30, figs. 6-9. *Proctus (Phæton) slatkowskii* n. sp. *Cyphaspis sibirica* n. sp. *Eurypterus ? czekanowskii* n. sp., *E. ? punctatus* n. sp. *Phacops (Monorakos) sibiricus* n. sp., pl., fig. 10.

—————— and **Jones** (T. Rupert). See Jones (T. Rupert) and Schmidt (Fr.)

—————— Ueber eine neuentdeckte untercambrische Fauna in Estland.

In Mem. Acad. Imp. Sci. St. Petersburg, series 7, vol. 36, 1888, No. 2.
Olenellus mickwitzi n. sp.

Schrenk (G. A.) Uebersicht des obersilurischen Schichten-Systems Liv- und Ehstlands, etc.

In Archiv Naturk. Liv-, Ehst- u. Kurl., Dorpat, vol., 1854, pp. 35 and 86.
Leperditia.
The author describes and figures *Cypridina grandis* from the Silurian of the Island of Oesel. This species is identical with *Leperditia gigantea* Römer, 1858 (Barrande).

Scouler (John). Description of a fossil crustaceous animal.

In Edinb. Jour. Nat. Geogr. Sci., vol. 3, 1831, p. 352, pl. 10.*
The author describes in this article, under the new generic name of *Eidothea*, certain fragments of an *Eurypterus*.

—— Account of some fossil Crustacea which occur in the coal formation *(Argas)*.

In Record of Geol. Sci. (Thomson's), vol. 1, 1835, p. 136.
According to Capt. J. E. Portlock (Rep. Geol. Londonderry, etc., 1843, p. 313), "Dr. John Scouler first noticed this remarkable Crustacean found by him in the Carboniferous strata of Paisley; and, depending as a principal character on the lateral caudal appendages, he pointed out its analogy to the genera *Cyclops* and *Apus*. And again, at the meeting of the British Association at Glasgow, he described two species, one of which, *Argas tricornis*, in all probability depends on an illusive appearance produced by the overlapping of the crust of different individuals. * * * Dr. John Scouler having subsequently altered his view of the affinities of the genus, he now considers it a bivalved Crustacean, distinguished by not being able to retract the tail within the shell, and hence has named it *Dithyrocaris*, or bivalved shrimp, abandoning the original name as one already preoccupied."

The Rep. 10th Meeting Brit. Assoc. Adv. Sci., 1840, held at Glasgow, does not contain in the index Dr. Scouler's name or that of the genus *Argas* or *Dithyrocaris*. In the 5th Rep. of the "Committee on the Fossil *Phyllopoda* of the Palæozoic Rocks," 1887, the genus *Dithyrocaris* Scouler is quoted on the authority of Captain Portlock as follows: "*Dithyrocaris* Scouler in Portlock's Geol. Rep., Londonderry, etc., 1843," p. 313.

Scudder (Samuel H.) *Rhachura*, a new genus of fossil Crustacea.

In Proc. Boston Soc. Nat. Hist., vol. 19, 1878, p. 296, pl. 9, figs. 3, 3 *a*.
In the 5th Rep. to the Brit. Assoc. Adv. Sci., 1887, on the fossil *Phyllopoda* by the committee, consisting of Mr. R. Etheridge, Dr. H. Woodward and Prof. T. Rupert Jones, this new genus of S. H. Scudder is referred to as probably that of *Dithyrocaris* Scouler, 1843.

Shaler (N. S.) and **Foerste** (A. F.) Preliminary description of North Attleborough fossils.
In Bull. Mus. Comp. Zoöl. Harvard Coll., vol. 16, No. 2, 1888, Geol. Series, vol. 2, 1888, p. 27, 2 pls.
Aristozoe sp.? *Microdiscus belli-marginatus* n. sp., *M. lobatus* Hall. *Paradoxides walcotti* n. sp. *(Olenellus walcotti* Wal.) *Ptychoparia mucronatus* n. sp., *P. attleborensis* n. sp.

Sharpe (Daniel). On the geology of the neighborhood of Oporto, including the Silurian coal and slates of Valongo.
In Quart. Jour. Geol. Soc., London, vol. 5, 1849, p. 142, pl. 6.
Isotelus powisii Portl. *Illænus lusitanicus* Sharpe.

────── Description of the new species of Zoophyta and Mollusca. Appendix B to an article on the Carboniferous and Silurian formations of the neighborhood of Bussaco, in Portugal, by C. Ribeiro.
In Quart. Jour. Geol. Soc. London, vol. 9, 1853, p. 146.
Dithyrocaris ? longicauda Sharpe.

Shumard (B. F.) Palæontology. Crustacea.
In 1st and 2d Ann. Reps. of the Geol. Survey of Missouri (Swallow), 1855, pt. 2, p. 195, pl. B.
Proetus swallowi n. sp., *P. missouriensis* n. sp. *Cyphaspis girardeauensis* n. sp. *Encrinurus deltoideus* n. sp. *Phillipsia meramecensis* n. sp. *Dalmania tridentifera* n. sp. *Calymene rugosa* n. sp. *Acidaspis halli* n. sp.

────── Notice of new fossils from the Permian strata of New Mexico and Texas, collected by Dr. George G. Shumard, geologist of the U. S. Government expedition for obtaining water by means of artesian wells along the 32d parallel, under the direction of Capt. John Pope, U. S. Corps of Top. Eng.
Trans. Acad. Sci. St. Louis, vol. 1, 1858, p. 290.
Phillipsia perannulata n. sp.

────── The Primordial zone of Texas, with descriptions of new fossils.
In Am. Jour. Sci., 2d series, vol. 32, 1861, p. 213.
Agnostus coloradoensis n. sp. *Arionellus (Bathyurus) texanus* n. sp., *A. (Bathyurus) planus* n. sp. *Conocephalites depressa* n. sp., *C. billingsi* n. sp. *Dikelocephalus römeri* n. sp.

——— Notice of some new and imperfectly known fossils from the Primordial zone (Potsdam sandstone and calciferous sand group) of Wisconsin and Missouri.
In Trans. Acad. Sci. St. Louis, vol. 2, 1863, p. 101.
Dikelocephalus latifrons n sp. *Arionellus bipunctatus* n. sp. *Conocephalites iowensis* Owen, *C. wisconsensis* Owen, *C. chippewaensis* Owen, *C. hamulus* Owen, *C. minor* n. sp. *Agnostus orion ?* Billings.

——— Descriptions of a new Palæozoic fossil.
In Trans. Acad. Sci. St. Louis, vol. 2, 1863, p. 108.
Proetus proutii n. sp. This species belong to the section *Dechenella*.

Simpson (G. B.) New species from Pennsylvania formations.
In Trans. Amer. Philos. Soc., 1889, p. 435.
Homalonotus trentonensis n. sp.

Sjogren (A.) Anteckningar om Öland, ett bidrag till Sveriges geologi.
In Öfversigit K. Vet.-Akad. Förhandl. Stockholm, vol. 8, 1851, No. 2, p. 36.
Enumerated species of the following genera: *Remopleurides, Paradoxides, Ellipsocephalus, Olenus, Phacops, Calymene, Homalonotus, Lichas, Ampyx, Cheirurus, Asaphus, Illænus, Nileus, Battus, Cytherina*.

——— In Angelin's Palæontologia Scandinavica.
Phacops bucculenta n. sp., *P. macroura* n. sp. *Lichas deflexus* n. sp.

——— Om nogra försteningar i Ölands Kambriska lager.
In Geol. Föreningens Stockholm Förhandl., vol. 1, No. 5, 1871, 1 pl.
Paradoxides tessini Brong., *P. ölandicus* n. sp., *P.* sp. indet. *Ellipsocephalus hoffi* Schloth, *E.* sp. indet. *Conocoryphe dalmani ?* Ang. *Agnostus regius* n. sp., *A.* sp. indet.

——— Bidrag till Ölands Geologi.
In Geol. Föreningens Stockholm Förhandl., vol. 1, 1871.
The Primordial rocks of the Island of Öland have been made known chiefly by the labors of M. Sjogren. The *Paradoxides* beds are divided by him into two divisions, of which the lowest consists of arenaceous flagstones—this is characterized by *Paradoxides tessini;* the upper consisting of greenish shale with intercalated calcareous and arenaceous beds with *Paradoxides ölandicus*. The strata with *P. forchhammeri* (that there occurs) making a third division. They are immediately overlain by the *Olenus* beds.

Smith (J.) Notes on a collection of Bivalved Entomostraca and other microzoa from the Upper Silurian strata of the Shropshire district.
In Geol. Mag., London, Dec. 2, vol. 8, 1881, p. 70.
The author gives the method of collecting these small Entomostraca. List of species by Prof. T. Rupert Jones.

Sowerby (J. D. C.) Description of *Homalonotus knightii* in Adam Sedgwick and R. 1. Murchison's article on the older formation on the right bank of the Rhine, etc.
In Trans. Geol. Soc. London, 2d series, vol. 6, 1842, p. 275, pl. 38, fig. 17.

Stainier (X.) Note sur un nouveau Trilobite,
In Annales Soc. Géol. Belgique; Mémoires, vol. 14, 1887.
Dechenella.

Steinhart (E. T. G.) Die bis jetzt in preuss. Geschieben gefundenen Trilobiten.
In Beiträge zur Natur. Preuss., herausgegeben von der phys. ökonomischen Gesell., Königsberg, 1874.
Calymene blumenbachii Broug. *Phasops latifrons* Broun., *P. dubius* Nieszk., *P.* sp.? *Chasmops conicophthalmus* Boeck. *Dalmania caudata* Brünn, *D. sclerops* Dalm. *Asaphus expansus* Linné, *A. raniceps* Dalm., *A. cornutus* Pand.,* *A. platyrhachis* n. sp., *A. (Basilicus) tyrannus* Murch., *A. tecticaudatus* n. sp., *A. undulatus* n. sp. *Lichas eichwaldi* Nieszk., *L. conico-tuberculata* Nieszk., *L. dissidens* Beyr., *L. velata* n. sp., *L. aquiloba* n. sp., *L. gibba* Aug., *L. tricuspidata* Beyr., *L. quadricornis* n. sp., *L. convexa?* Aug. *Harpes spaskii* Eichw. *Proetus ramisulcatus* Nieszk., *P.* sp.? *Cyphaspis megalops* McCoy. *Ampyx (Lonchodomas) rostratus* Sars, *A. (Raphiophorus) culminatus* Aug. *Illanus crassicauda* Wahl. (*a*) forma typica *dalmani*, (*b*) varietas *I. wahlenbergi* Eichw., *I. schmidti* Nieszk., *I. centaurus* Dalm., *I. eichwaldi* Volb., *I. tauricornis* Kut. *Nileus armadillo* Dalm. *Bumastus* sp. a.? *barriensis* Murch., *B.* sp. b.? *Ceraurus (Cheirurus) exsul* Beyr., *C. speciosus* His., *C. spinulosus* Nieszk. *Encrinurus punctatus* Brünn. *Zethus* sp.? *Sphærexochus clavifrons* Dalm., *S. pseudohemicranium* Nieszk. *Agnostus pisiformis* Linné. *Acidaspis* sp.?

---

* A variety of *Asaphus expansus* remarkable for its projecting eyes. Pander has suggested the new name of *Asaphus cornutus* for this variety in his excellent work, Beiträge zur Geognosie, p. 137, pl. 7, figs. 5 and 6. The name has been revived by De Verneuil in Murchison's Geol. Russia and the Ural Mts., vol. 1, p. 37*, figs. 1-4, in which he exhibits the transitions in the length of the eye and the variations between *Asaphus expansus* and *A. cornutus*.

The author gives a short sketch of the "Panderian organs" illustrating those of *Asaphus expansus*, p. 20, pl. 2.

Steininger (Jean). Observations sur les fossiles du calcaire intermédiaire de l'Eifel.
In Mém. Soc. Géol. France, vol. 1, 1831, p. 331, pl. 21.
Calymene brongniarti n. sp., *C. latreilii* n. sp., *C. schlotheimii* Bronn, *C. tristani* Brong. *Proetus* n. g., *P. cuvieri* n. sp. *Olenus punctatus* n. sp. *Asaphus mucronatus* Brong., *A. laticauda* Wahl.

Sternberg (Kasper Graf von). Uebersicht der in Böhmen dermalen bekannten Trilobiten.
In Verhandl. minèral. Gesell. vaterländisch. Mus. Böhm., 1825, p. 69, pls. 2, 3; Isis (oder Encycl. Zeitung), Oken, vol. 7. 1827, p. 627.
Trilobites macrophthalmus Brong., *T. hausmanni* Brong., *T. sulzeri* Schloth, *T. hoffi* Schloth, *T. tessini* Brong.
The author illustrates plate 1, fig. 1 a-b, *Phacops sternbergii* Cord.; figs. 2 a-b, *P. protuberans* Dalm.; fig. c-d, *P. cephalotes* Cord.; fig. 3, *Cheirurus sternbergii* Boeck; fig. 4, *Paradoxides bohemicus* Boeck; fig. 5, *Placoparia zippei* Boeck.
Plate 2, fig. 1 a, *Ptychoparia striatus* Emm.; fig. 1 b, *Conocoryphe sulzeri* var. *kinskyanus;* fig. 3, *Dalmanites auriculata* Dalm.; fig. 3 a, c, sp.? fig. 3 b, *D. rugosa* Cord.; fig. 3 b, *D. hausmanni* Brong.; fig. 4, *Ellipsocephalus hoffi* Schloth.
These references are mostly taken from Barrande's Sys. Sil. Bohème.

——— Ueber die Gliederung und die Füsse der Trilobiten.
In Isis (oder Encycl. Zeitung.), Oken, 1830, pp. 516, 563, pl. 5, figs. 1-3.

——— Ueber böhmische Trilobiten.
In Verhandl. mineral. Gesell. vaterländisch. Mus. Böhm., 1833, p. 17, plate.
Olenus longicaudatus Zenk. *Trilobites spinosus* Zenk. *Olenus pyramidalis* Zenk., *O. latus* Zenk. *Conoceph. costatus* Zenk. *Trilobites sulzeri* Schloth, *T. zippei* Boeck. *Ellipsocephalus ambiguus* Zenk. *Trilobites hoffi* Schloth. *Otarion diffractum* Zenk. *Trilobites sternbergii* Boeck, *T. ungula* n. sp., *T. ornatus* n. sp.

——— Ueber die böhmischen Trilobiten mit Beziehung auf die Arbeiten von Boeck und Zenker darüber.
In Neues Jahrbuch für Mineral., 1835, p. 727.
Olenus longicaudatus Zenk., *O latus* Zenk. *Trilobites minor* Boeck., *T. gracilus* Boeck, *T. sulzeri* Schloth, *T. hoffi* Schloth, *T. sternbergii* Boeck, *T. ungula* Sternb., *T. ornatus* Sternb., *T. gibbosus* Wahl., *T. zippei* Boeck. *Otarion diffractum* Zenk.

Stock (Thomas). Note on the occurrence of *Anthrapalæmon etheridgi* B. N. Peach, in the ironstones above the sandstones at Craigleith Quarry, near Edinburgh.
In Trans. Edinb. Geol. Soc., vol. 4, 1881, p. 97.

—— Note on the occurrence of the remains of Decapod Crustaceans in the Wardie shales.
In Trans. Edinb. Geol. Soc., vol. 4, 1882, p. 219.
Not descriptive: *Anthrapalæmon traquairii* Peach., *A. etheridgii* Peach. *Palæocrangon* sp.?

Stoddard (W. W.) On Trilobites.
In Proc. Bristol. Nat. Sci., vol. 1, 1866, p. 82.*

Stokes (Charles). On a Trilobite from Lake Huron.
In Trans. Geol. Soc. London, 2d series, vol. 1, 1823, p. 208, pl. 27.
Asaphus platycephalus n. sp.

Strauss-Durckheim (H.) Ueber *Estheria dahalacensis* Rüppell, eine neue Gattung aus der Familie der Daphniden.
In Mus. Senckenberg'anum, vol. 2, 1837, pp. 119-128, pls. 7, 7 b.
The name *Estheria* was originally proposed by Dr. Rüppell for a species brought from Abyssinia; its description, with figures of typical species, appeared in the above cited work. In the same year Mr. Audoin (Annales Soc. Entom., vol. 6), proposed that of *Cyzicus* for similar species, without a generic description. The simultaneous publication of these two generic names for the same genus lead Mr. Joly (Annales Sci. Nat., 2d series, vol. 17, 1842, p. 293), to propose that of *Isaura*.
Estheria is now generally used by authors.

Symonds (W. S.) On a Phyllopod Crustacean in the Upper Ludlow rock of Ludlow.
In Rep. 25th Meeting Brit. Assoc. Adv. Sci., 1858, Trans. of Sec., p. 98.
Hymenocaris vermicauda Salt.

—— On a new species of *Eurypterus* from the Old Red Sandstone of Herefordshire.
In Rep. 27th Meeting Brit. Assoc. Adv. Sci., 1857, Trans. of Sec., p. 93.
The name of this species is not given. It was described by Salter in 1859 as *Eurypterus symondsii*.

—— On a new species of *Eurypterus* from the Old Red Sandstone of Herefordshire.
In Edinburgh New Philos. Jour., vol. 6, 1857, p. 257.

The species described as *Eurypterus symondsii* Salter is noted in this paper.

Tate (George). The Farne Islands, with an account of their Geology, Botany, Zoology and Ancient History.
In Proc. Berwickshire Nat. Club, vol. 3, 1857, p. 222, pl. 1, fig. 4.
Griffithides farnensis n. sp.

Thomson (P. Wyville). On some species of *Acidaspis* from the Lower Silurian beds of the south of Scotland.
In Quart. Jour. Geol. Soc. London, vol. 13, 1857, p. 206, pl. 6.
Acidaspis lalage n. sp., *A. hystrix* n. sp., *A. callipareos* n. sp., *A. unica* n. sp.

—— The genus *Bronteopsis*.
In the author's investigations of the fossils of the Girvan District in 1856 he used this MS. generic name. Salter refers to it in his Mon. Brit. Tril., pp. 143 and 216.
The first published figure and description is given in the Mon. Sil. Foss. Girvan District, 1879, p. 157, plates 11 and 12.

—— Figures and descriptions illustrative of organic remains.
In Geol. Survey United Kingdom, decade 11, London, 1864, pl. 6.
Salteria n. g., *S. primæva* n. sp.

Tietze (E.) Ueber die devonischen Schichten von Ebersdorf unweit Neurode in der Grafschaft Glatz.
Geognostisch-palæont. Monographie, Cassel, 1870, 56 pp., 2 pls.
Phacops sp. indet., *P. cryptophthalmus* Emm. *Proetus* sp. indet. *Harpes* sp. indet. *Leperdia* sp. indet. *Cypridina serrato-striata* Sandb.

Tilesius (A. v.) Naturhistor. Abhandlung zur Petrefacten-kunde (Trilobiten), Kassel, 1820, fol. 8, pls. (*)

Todd (J. E.) A description of some fossil tracks from the Potsdam sandstone.
In Trans. Wisconsin Acad. Sci., vol. 5, 1882, p. 276, 1 fig.

Toll (Eduard von). Die paläozoischen Versteinerungen der Neusibirishen Insel Kotelny.
In Mém. Acad. Imp. Sci. St. Petersburg, vol. 37, 1889, No. 3, 5 pls.
Phacops quadrilineata Ang. *Monorakos schmidti* n. sp. *Proetus* sp. *Bronteus andersoni* E. & N. *Leperditia kotelnyensis* n. sp., *L. arctica* Jones, *L. czerskii* n. sp., *L. sannikowi* n. sp., *L. keyserlingi* Schm. var. *Leperditia ?*

Törnquist (S. L.) Undersökningar öfver Siljansomrodets Trilobitfauna.

In Svenska Geol. Undersökning, Stockholm, series C, No. 66, 1884, 2 pls.
Phacops elliptifrons Esm., *P. (Pterygometopus) trigonocephalus* Schm., *P. (Chasmops) odini* Eichw., *P. (Pterygometopus) panderi* Schm., *P. (Chasmops) maximus* Schm. *Cheirurus speciosus* His., *C. insignis* Beyr., *C. exsul* Beyr., *C. ingricus* Schm., *C. glaber* Ang., *C. punctatus* Ang., *C. ? tenuispinus* n. sp. *Cyrtometopus affinis* Ang., *C. clavifrons* Dalm. *Sphærocoryphe granulata* Ang. *Pseudosphærexochus conformis* Ang., *P. wegelini* Ang. *Nieszkowskia variolaris* Linrs. *Sphærocohus mirus* Beyr. *Deiphon lævis* Ang., *D. forbesi* Barr. *Pliomera fischeri* Eichw., *P. törnquisti* Holm. *Encrinurus schisticola* n. sp., *E. multisegmentatus* Portl., *E. striatus* Ang. *Cybele adornata* n. sp., *C. brevicauda* Ang. *Acidaspis dalecarlica* n. sp., *A. evoluta* n. sp., *A. breviloba* Ang. *Lichas elegans* n. sp., *L. palmatus* Barr., *L. laxatus* McCoy, *L. æqualis* n. sp., *L. conformis* Ang., *L. cicatricosus* Lovén, *L. affinis* Ang., *L. dalecarlicus* Ang., *E. brevilobatus* n. sp., *L. planifrons* Ang., *L. lineatus* Ang.
Harpes wegelini Ang., *H. costatus* Ang. *Remopleurides radians* Barr., also var. *angustata*, *R. emarginatus* n. sp., *R. dorsospinifer* Portl. *Triarthrus pygmæus* n. sp. *Calymene trinucleina* Linrs. MS., *C. blumenbachii* Brong., *C. leptænarum* n. sp., *C. forcolata* n. sp. *Homalonotus punctillosus* n. sp. *Proetus modestus* n. sp., *P. brevifrons* Ang., *P. papyraceus* n. sp. *Cyphaspis rastritum* n. sp., *C. burmeisteri* Barr. *Arethusina konincki* Barr. *Bronteus laticauda* Wahl., *B. ? nudus* Ang. *Illænus esmarkii* Schloth, *I. sphæricus* Holm., *I. gigas* Holm., *I. scrobiculatus* Holm., *I. vivax* Holm., *I. fallax* Holm., *I. chiron* Holm., *I. crassicauda* Wahl., *I. linnarssoni* Holm., *I. parvulus* Holm., *I. centrotus* Dalm. *Nileus armadillo* Dalm.
Asaphus platyurus Ang., *A. rusticus* n. sp., *A. demissus* n. sp., *A. lepidus* n. sp., *A. raniceps* Dalm., *A. expansus* Linné, *A. vicarius* n. sp., *A. tecticaudatus* Steinh., *A. brachyrachis* Remelé MS., *A. angustiforns* Dalm., *A. applanatus* Ang., *A. ptychopyge-rimulosus* Ang., *A.* cf. *undulatus* Steinh., *A. densistrius* n. sp., *A. ludibundus* n. sp., *A. prætextus* n. sp., *A. plicicostis* n. sp.
Megalaspis acuticauda Ang., *M. gigas* Ang., *M. grandis* Sars., also vars. *lata* and *rudis*, *M. lambens* n. sp., *M. formosa* n. sp., *M. patagiata* n. sp., *M. dalecarlica* Holm. *Niobe læviceps* Dalm., *N. frontalis* Dalm., *N. emarginula* Ang. *Trinucleus seticornis* His. *Ampyx pater* Holm., *A. nasutus* Dalm., *A. foveolatus* Ang., *A. portlocki* Barr., *A. setirostris* Ang. *A. crassirostris* Ang. *Isocolus sjögreni* Ang. *Telephus fractus* Barr. *Agnostus törnquisti* Holm., *A. glabratus* Ang., *A. trinodus* Salt. *Trilobites brevifrons* Holm., *T. triradiatus* n. sp. *Megalaspis limbata* Boeck forma *lata*.

Torrubia (J.) Aparato para la historia natural española. Vol. 1. Madrid, 1754, p. 83, par. 13, No. 96.

The Rev. Father Joseph Torrubia has the merit of referring the Trilobites to the Crustacea.

Traquair (R. H.) Occurrence of Trilobites in the Carboniferous Limestones of Fifeshire.
In Proc. Royal Phys. Soc. Edinburgh, vol. 2, 1859-62, p. 253.
Griffithides mucronatus McCoy.

―――― On the *Griffithides mucronatus* McCoy.
In Jour. Royal Geol. Soc. Ireland, vol. 2, 1869, p. 213, pl.

Trautschold (H.) Die Trilobiten als Erstgeborene.
In Bull. Soc. Imp. des Naturalistes, Moscou, 1871, p. 297.

―――― Die Kalkbrüche von Mjatschkowa. Eine Monographie des oberen Bergkalkes.
In Mém. Soc. Imp. des Naturalistes, Moscou, vol. 13, 1874, p. 277.
Phillipsia globiceps Phill., *P. grünewaldti* Möller, *P. pustulata* Schloth.

Trenkner (M. W.) Paläontologische Novitäten vom nordwestlichen Harze Iberger kalk und kohlengebirge von Grund.
In Abhandl. d. Naturf. Gesellich. zu Halle Bd. 10, 1867, pl. 1.
Harpes convexus n. sp. *Bronteus alutaceus* Gold. *Cyphaspis ellipsocephalus* n. sp. *Cypridina gigantea* n. sp.

―――― Paläontologische Novitäten vom Nordwestlichen Harze.
In Abhandl. d. Naturf. Gesellsch. zu Halle Bd. 10, 1867, pl. 5.
Homalonotus granulosus n. sp. *Cylindrocephalus angustus* n. sp. *Homalonotus trigonalus* n. sp.

Tristan (J.) and **Bigot** (P. M. S.) Sur un crustacé renfermé dans quelques schistes, notamment dans ceux des environs de Nantes (dépt. de la Loire-Inférieure) et d'Angers (dépt. de Maine-et-Loire).
In Jour. Mines, vol. 23, 1808, p. 2.
(Calymene tristani Broug.)

Tromelin (G) and **Lebesconte** (P.) Observations sur les terrains primaires du nord du dépt. d'Ille-et-Vilaine et de quelques autres parties du massif breton.
In Bull. Soc. Géol. France, 3d series, vol. 4, 1875-76, p. 583.
Dalmanites rouaulti n. sp. *Homalonotus gahardensis*.

Troost (Gerhard). Description d'un nouveau genre de fossiles.
In Mem. Soc. Géol. France, vol. 3, 1834, p. 87, pls. 9, 10, 11.

The author describes a new species of *Asaphus* from the Upper Silurian of Perry County, Tennessee, as *Asaphus megalophthalmus*, p. 94, pl. 11, figs. 1-5, and an undescribed trilobite, pl. 11, figs. 6-7. The first is a *Phacops* approaching *P. hudsoni* Hall, the second a species of the genus *Dalmanites*.

Tschernyschew (Th.) Die Fauna des mittleren und oberen Devon am West-abhange des Urals.
In Mém. du Comité Géol., vol. 3, No. 3, 1887, p. 13.
Dechenella romanovski n. sp., *D. haldemani* Hall., *D.* aff. *verneuili* Barr. *Goldius (Bronteus) granulatus* Goldf. *Leperditia mölleri* Schm. *Isochilina biensis* Grün.

Tullberg (S. A.) Om *Agnostus*-arterna i de Kambriska aflagringarne vid Andrarum.
In Svenska Geol. Undersök., series C, No. 42, 1882, 2 pls.
The author divides the Agnosti into four groups, as follows:
Longifrontes: This section is distinguished by its strongly projecting glabella and axis, which is usually moderately long. Crust smooth. The cheeks grooved. The shell on both the cheeks and tail is provided with elevated granulations. Limb is generally narrow. The checks in front of the glabella, and side lobes of the tail behind the axis, are divided by a groove.
Agnostus atavus n. sp., *A. gibbus* Linrs., *A. fissus* Lundgr., *A. intermedius* n. sp., *A. punctuosus* Ang., *A. elegans* n. sp., *A. hybridus* Brög., *A. incertus* Brög., *A. lundgreni* n. sp., *A. nathorsti* Brög., *A. exsculptus* Ang., *A. aculeatus* Ang., *A. reticulatus* Ang., *A. trisectus* Salt., *A. pisiformis* Linné, *A. cyclopyge* n. sp., *A. pusillus* n. sp.
2. Lævigati: The dorsal groove outlining the glabella and axis of the thorax are obsolete. Crust smooth and shining, sometimes with slight indications of striæ. Limb on the head disappearing; on the pygidium it becomes broad.
Agnostus cicer n. sp., *A. nudus* Beyr., also var. *scanicus* n. var. and var. *marginatus* Brög., *A. glandiformis* Ang., *A. lævigatus* Dalm., *A. bitubercu-latus* Ang.
Limbati: General form subquadrate. Head has a broad limb. Basal lobes are large. The cheeks in front of the glabella are divided by a central groove, striated on the sides. Pygidium is usually provided with lateral spines.
(a) Regii: *Agnostus regius* Sjogr., *A. rex* Barr. This subdivision is distinguished by its broad limb diminishing cheeks on the head and side lobes of the tail (especially by the front lobe of the glabella).
(b) Fallaces: This series has a smaller head, moderately broad limb. Cheeks large. Basal lobe rather large, also by the third joint of the tail which is enlarged.

Agnostus fallax Linrs., forma *typica*, forma *ferox*, *A. kjerulfi* Brög., *A. planicauda* Ang., *A. quadratus* n. sp.
Parvifrontes: The glabella is only partly developed in this section, and without lobes.
Agnostus parvifrons Linrs., forma 1, 2, 3; *A. truncatus* Brög., *A. brevifrons* Ang.

Ulrich (E. O.) Description of some species of fossils from the Cincinnati group.
In Jour. Cincinnati Soc. Nat. Hist., vol. 1, 1878-79, p. 92.
Trinucleus bellulus n. sp.

—— Description of a Trilobite from the Niagara group of Indiana.
In Jour. Cincinnati Soc. Nat. Hist., vol. 2, 1879, p. 131, 3 figs.
Calymene nasuta n. sp.

—— Descriptions of new genera and species from the Lower Silurian about Cincinnati.
In Jour. Cincinnati Soc. Nat. Hist., vol. 2, 1879, p. 8, pl. 9.
Leperditia radiata n. sp., *L. crepliformis* n. sp., *L. unicornis* n. sp., *L. bivertex* n. sp. *Beyrichia persulcata* n. sp.

—— New and little known American Palæozoic Ostracoda.
In Jour. Cincinnati Soc. Nat. Hist., vol. 13, 1890, p. 104, 18 pls.
Lower Silurian Species: *Entomis madisonensis* n. sp. *Pontocypris ? illinoisensis* n. sp. *Ctenobolbina* n. g., type *Beyrichia ciliata*, *C. ciliata* var. *curta*, also var. *emaciata*, *C. alata* n. sp., *C. bispinosa* n. sp., *C. tumida* n. sp. *Tetradella* n. g., type *Beyrichia quadrilirata* H. & W., *T. oculifera* Hall., *T. subquadrata* n. sp. or var. *Bollia persulcata* Ul., *B. pumila* n. sp. *Depranella* n. g., *D. crassinoda* n. sp., *D. nitida* n. sp. or var. *D. macra* n. sp., *D. ampla* n. sp., *D. elongata* n. var. *Jonesella* n. g., type *Leperditia crepiformis* Ul., *J. pedigera* n. sp., *J. digitata* n. sp., *J. crassa* n. sp. *Placentula marginata* n. sp., *P. inornatus* n. sp. *Beyrichia ? (Primitia) parallela* Ul. *Eurychilina subradiata* n. sp., *E. reticulata* n. sp., *E. longula* n. sp., *E. granosa* n. sp., *E. æqualis* n. sp., *E. obesa* n. sp., *E. striatomarginata* S. A. Miller. *Primitia centralis* n. sp., *P. perminima* n. sp., *P. impressa* n. sp., *P. cincinnatiensis* S. A. Miller, *P. medialis* n. sp., *P. milleri* n. sp., *P. glabra* n. sp., *P. nodosa* n. sp., *P. nitida* n. sp., *P. rudis* n. sp., *P.? sculptilis* n. sp. *Aparchites oblongus* n. sp. *Leperditia fabulites* Con., *L. linneyi* n. sp., *L. tumidula* n. sp., *L. appressa* n. sp., *L. cæcigena* S. A. Miller var. *frankfortensis*. *Isochilina subnodosa* n. sp., *I. saffordi* n. sp., *I. ampla* n. sp., *I. jonesi* Weth., *I. kentuckyensis* n. sp., *I. amiana* n. sp. var. *insignis*.

Upper Silurian and Devonian Species: *Leperditia ? subrotunda* n. sp. *Isochilina rectangularis* n. sp. *Aparchites inornatus* n. sp. *Entomis waldronensis* n. sp. *Æchmina abnormis* n. sp., *A. marginata* n. sp. *Halliella* n. g., *H. retifera* n. sp. *Ctenobolbina punctata* n. sp., *C. papillosa* n. sp., *C. informis* n. sp., *C.? (Bollia) antespinosa* n. sp. *Bollia ungula* Jones., *B. obesa* n. sp. *Beyrichia tricollina* n. sp., *B. lyoni* n. sp., *B.? (Depranella) kalmodini* Jones. *Moorea bicornuta* n. sp. *Kirkbya subquadrata* n. sp., *K. parallela* n. sp., *K. semimuralis* n. sp. *Octonaria stigmata* n. sp. var. *loculosa* var. *oblonga*, *O. orata* n. sp., *O. clavigera* n. sp., *O. curta* n. sp. *Bythocypris devonica* n. sp., *B. punctulata* n. sp., *B. indianensis* n. sp. *Bairdia leguminoides* n. sp. *Pachydomella* n. g., *P. tumida* n. sp. *Barychilina* n. g., *B. puncto-striata* n. sp., also var. *curta*, *B. pulchella* n. sp. *Ctenobolvina minima* n. sp.

Carboniferous Species: *Leperditia nicklesi* n. sp. *Primitia granimarginata* n. sp., *P. simulans* n. sp., *P. chesterensis* n. sp., also var. *caldwellensis*, *P. subæqualis* n. sp. *Ulrichia emarginata* n. sp., *U.? confluens* n. sp. *Beyrichia radiata* J. & K. var. *chesterensis*, *B. simulatrex* n. sp. *Bollia granifera* n. sp. *Moorea granosa* n. sp. *Kirkbya oblonga* J. & K., *K. lindahli* n. sp., *K. tricollina* J. & K., *K. renosa* n. sp., *K.? (Barychilina) costata* McCoy. *Cypridina herzeri* n. sp. *Cytherella ovatiformis* n. sp. *Bairdia chesterensis* n. sp. *Ponocypris ? acuminata* n. sp.

Ure (David). History of Rutherglen and East Kilbride.

In this work, published in 1793, the Rev. David Ure noticed the existence of certain "microscopic bivalved shells" in the Carboniferous limestone near Glasgow.

Four or five of these fossils were illustrated and described. One of them, pl. 14, fig. 15, is a subreniform *Cythere*; fig. 20 is a somewhat crushed *Bairdia*; fig. 21 represents a *Kirkbya* poorly drawn *(M. urei* Jones); figs. 16 and 17 are *Beyrichian* fossils. Referred to *Beyrichia bituberculata* McCoy by Mr. John Young of Glasgow.

Verneuil (E.) Remarques sur les fossiles Paléozoique communs a l'Amerique et a l'Europe et sur les rapports qu'ils offrent dans leur distribution.

In Bull. Soc. Géol. France, 2d series, vol. 4, 1846–47, p. 688.

Agnostus latus Conrad. *Phillipsia seminifera* Phill. *Calymene blumenbachii* Broug. var. *senaria* Conrad, *C. platys* Green, *C. fischeri* Eichw., *C. punctata* Brünn. *Illænus crassicauda* Dalm. *(Bumastus trentonensis* Emm.) *Lichas laciniata* Wahl., *L. scabra* Beyr. *Ceraurus pleurexanthemus* Green. *Trinucleus caractaci* Murch., *T. tessellatus* Green. *Phacops hausmanni* Broug. *(Asaphus micrurus* Green), *P. dalmani* Portl., *P. limulurus* Green, *P. macrophthalmus* Broug. *(Calymene bufo* Green. *Asaphus megalophthalmus* Troost.) *Cryphæus callitelus* Green. *Bumastus barriensis* Murch.

Homalonotus delphinocephalus Green. *Cheirurus insignis* Beyr. *Sphærexochus mirus* Beyr. *Phillipsia seminifera* Phill., *P.* cf. *ornata* Portl.

Verneuil (Edouard). Sur les fossiles dévoniens des environs de Sabers dans les montagnes de Léon.
In Bull. Soc. Géol. France, 2d series, vol. 7, 1850, p. 155, pl. 3.
Cryphæus callitelus Green. *Phacops latifrons* Bronn. *Homalonotus pradoanus* n. sp. *Cryphæus sublacininatus* n. sp.

―――― Géologie de la Russie d'Europe et des montagnes de l'Oural, par R. I. Murchison, Édouard de Verneuil et le comte Alex. de Keyserling. Vol. 2. 3 part. Palæontologie. Londres et Páris, 1845, 50 pls.
Vol. 1. Geology. *Asaphus cornutus* Pand., *A. expansus* Linné.
Vol. 2. Palæontology. *Calymene odini* Eichw., *C. fischeri* Eichw. *Asaphus expansus* Linné. *Phillipsia eichwaldi* Fischer, *P. ouralica* n. sp.

―――― Note sur les fossiles dévonien du district de Sabero Léon.
In Bull. Soc. Géol. France, 2d series, vol. 7, 1850.
Homalonotus gervillei n. sp.

―――― and **Barrande** (Joachim). Description des fossiles trouvés dans les terrains silurien et dévonien d'Almaden d'une partie de la Sierra Morena et des montagnes de Tobide.
In Bull. Soc. Géol. France, 2d series, vol. 12, 1854-55, p. 904, pls. 23, 24.
Ellipsocephalus pradoanus n. sp. *Calymene pulchra* Barr., *C. tristani* Broug., *C. arago* Rou., *C. transienus* n. sp. *Homalonotus rarus* Cord., *H. brongniarti* Deslongch. *Placoparia tourneminei* n. sp. *Dalmanites socialis* Barr., *D. dujardini* n. sp., *D. vetillari* n. sp., *D. torrubiæ* n. sp., *D. phillipsi* Barr., *D. downingiæ* Murch. *Lichas hispanica* n. sp. *Trinucleus goldfusi* Barr. *Cheirurus marianus* n. sp. *Asaphus nobilis* Barr., *A. cianus* n. sp., *A. glabratus* n. sp., *A. contractus* n. sp. *Ilælnus hispanicus* n. sp., *I. sanchezi* n. sp. *Ogygia*.

―――― Sur l'existence de la faune primordiale dans la chaine Cantabrigue, par M. C. de Prado, suivie de la description des fossiles par MM. de Verneuil et Barrande.
In Bull. Soc. Géol. France, 2d series, vol. 13, 1860, pp. 5-6, pls. 6-8.
Paradoxides pradoanus n. sp. *Arionellus ceticephalus* Barr. *Conocephalites sulzeri* Schloth., also var. *C. coronatus* Barr., *C. ribero* n. sp. *Agnostus* cf. *integer* Barr.

—— and **d'Archiac** (Vicomte). Paléontologie sur la faune dévonienne des rives du Bosphore. Liste des fossiles recueillis par Abdullah-Bey.
In Comptes Rendus Acad. Sci. Paris, vol. 64, 1867, p. 1217.
Cryphœus abdullahi n. sp., *C. callitelus* Green., *C. stellifer* Burm., *C. pectinatus* Römer. *Phacops latifrons* Broun. *Homalonotus gervillei* DeVern.

—— Note sur les fossiles recueillis en 1863, par M. de Tchihatcheff, aux environs de Constantinople.
In Bull. Soc. Géol. France, vol. 21, 1864, p. 147.*
Homalonotus gervillei DeVern. *Phacops longicaudatus* Murch. *Cheirurus* (*Cryphœus asiaticus* DeVern) *Phacops latifrons* Broun.

Verworm (M.) Zur Entwicklungsgeschichte der Beyrichien.
In Zeitschr. Deutsch. Geol. Gesell. Berlin, vol. 39, 1887, p. 27, pl. 3.
Beyrichia primitiva n. sp., *B. salteriana* Jones, *B. tuberculata* Boll.

Vine (G. R.) Notes on the distribution of the Entomostraca in the Wenlock shales.
In Proc. Yorkshire Geol. Polytechnic Soc., vol. 9, pt. 3, 1887–88, p. 393.

Vitry (Abbé de). Mémoire sur les fossiles du Tournaisés et lés pétrifications en général, relativement à leur utilité pour la vie civile.
In Mém. Acad. Royale Sci. Bruxelles, vol. 3, 1780, p. 15, pl. 1, fig. 1 (*Phillipsia*); pl. 3, figs. 8–11.
This species is referred by DeKoninck to *Phillipsia gemmulifera* Phill.

Vogdes (A. W.) Notes on the genera *Acidaspis*, *Odontopleura* and *Ceratocephala*, etc.
In Proc. Acad. Nat. Sci., 1877, p. 138.
On the priority of the genus *Ceratocephala* Warder, 1838, over *Acidaspis* Murch, 1839.
The genus is subdivided in this paper into two groups:
1, *Ceratocephala*; type *C. vesiculosa*.
2, *Odontopleura* Emm.; type *C. ovata* Emm.

—— A Monograph on the genera *Zethus*, *Cybele*, *Encrinurus* and *Cryptonymus*. Charleston, 1878, 35 pp. and 4 pls.
Cryptonymus punctatus Brünn., *C. variolaris* Brong., *C. deltoideus* Shum., *C. lævis* Aug., *C. obtusus* Aug., *C. ornatus* H. & W., *C. nereus* Hall, *C. caudatus* Aug., *C. striatus* Aug., *C. sexcostatus* Salt., *C. raricostatus* Wal.,

C. trentonensis Wal., *C. vigilans* Hall., *C. multisegmentatus* Portl., *C. (Cybele) rex* Nieszk., *C. (Cybele) bellatulus* Dalm., *C. (Cybele) atractopyge* McCoy, *C. (Cybele) brevicauda* Aug., *C. (Cybele) dentata* Esmk., *C. (Cybele) woerthii* Eichw., *C. (Atractopyge) mirus* Bill., *C. (Atractopyge) verrucosus* Dalm.

―――― Short notes on the geology of Catoosa County, Georgia.
In Am. Jour. Sci., 3d series, vol. 23, 1879, p. 475.
Calymene rostrata n. sp.
Similar species have been classed by Bergeron under a new genus (*Calymenella*).

―――― Description of a new Crustacean from the Upper Silurian of Georgia, with remarks on *Calymene clintoni*.
In Proc. Acad. Nat. Sci. Phila., 1880, p. 176, 4 figs.
Calymene rostrata Vogd., *C. clintoni* Vanuxem.
Fig. 3, *Calymene clintoni*, has been referred to A. F. Foerste (Bull. Den. Univ., vol. 2. 1887, pl. 8, figs. 12, 16) to *Calymene vogdesi*.

―――― A new Trilobite.
In 12th Ann. Rep. Geol. Nat. Hist. Survey Minnesota, 1884, p. 8.
Bathyurus stonemanni n. sp.

―――― Description of a new Crustacean from the Clinton group of Georgia, with remarks upon others. New York City, 1886, 5 pp., 4 figs.
Encrinurus americanus n. sp. *Calymene rostrata* Vogd., and *C. clintoni* Vanuxem.

―――― The genera and species of North American Carboniferous Trilobites.
In Annals New York Acad. Sci., vol. 4, 1887, p. 69, 2 pls.
Proetus loganensis H. & W., *P. peroccidens* H. & W., *P. tennesseensis* Winchell, *P. trinucleatus* Herrick (*Phillipsia trinucleatus*), *P. ellipticus* M. & W. *Phillipsia perannulata* Shumard, *P. cliftonensis* Shumard, *P. major* Shumard, *P. missouriensis* Shumard, *P. meramecensis* Shumard, *P. insignis* Winchell, *P. stevensoni* Meek, *P. vindorbonensis* Hartt., *P. doris* Hall, (*Proetus doris*), *P. rockfordensis* Winchell, *P. howi* Bill., *P. tuberculata* M. & W. *Griffithides portlockii* M. & W., *G. buffo* M. & W., *G. scitula* M. & W., *G. sangamonensis* M. & W., *G. granulatus* Wetherby. *Brachymetopus lodiensis* Meek, also a plate of generic illustrations.

―――― Description of two new species of Carboniferous Trilobites.
In Trans. New York Acad. Sci., vol. 7, 1888, p. 247, 1 wood cut.
Phillipsia sampsoni n. sp. *Griffithides? sedaliensis* n. sp.

―――― A catalogue of North American Palæozoic Crustacea, confined to the non-trilobitic genera and species.
In Annals N. Y. Acad. Sci., vol. 5, 1889, p. 1, 2 pls.
Strigocaris n. g., *S. strigata* Meek.
This article also contains generic descriptions and illustrations.

―――― Notes on Palæozoic Crustacea. No. 1. On some new Sedalia Trilobites.
In Trans. St. Louis Acad. Sci., vol. 5, 1891. Author's Ed., 4 pp., 1 pl.
Brachymetopus armatus n. sp.

―――― A Bibliography of Palæozoic Crustacea from 1698 to 1889, including a list of North American species and a systematic arrangement of genera.
In Bull. U. S. Geol. Survey, No. 63, Washington, 1890.

―――― Notes on Palæozoic Crustacea. No. 2. On the North American species of the genus Agnostus.
In the American Geologist, vol. 9, 1892, p. 377, pls. 9–10.
Agnostus acutilobus Matt., *A. obtusilobus* Matt., *A. tesella* Matt., *A. acadicus* Hartt., *A. regulus* Matt., *A. umbo* Matt., *A. vir* Matt., *A. vir* var. *concinus* Matt., *A. americanus* Bill., *A. canadensis* Bill., *A. communis* H. & W., *A. coloradensis* Shumard, *A. orion* Bill., *A. josephi* Hall, *A. richmondensis* Wal., *A. maladensis* Meek, *A. bidens* Meek, *A. tumidosus* H. & W., *A. interstrictus* White, *A. disparilis* Hall, *A. parilis* Hall, *A. prolongus* H. & W., *A. seclusus* Wal., *A. galba* Bill., *A. fabius* Bill.

―――― Notes on Palæozoic Crustacea. No. 3. On the genus *Ampyx*, with descriptions of the American species.
In The American Geologist, vol. 11, 1893, p. 99.
Ampyx nudus Murch., *A. nasutus* Dalm. *Lonchodomas domastus* Ang. *Ampyx americanus* Vogd. & Saff., *A. halli* Bill., *A. normalis* Bill., *A. rutilius* Bill., *A. semicostatus* Bill.
The author arranges this genus into three sections, as follows: Sec. 1, Brevifrontes, type *Ampyx nudus* Murch. Sec. 2, Longifrontes, type *Ampyx nasutus* Dalm. Sec. 3, Lonchodomas, type *L. domastus* Ang.
For other references see Safford (J. M.) and Vogdes (A. W.)

Volborth (A.) Ueber einige russiche Trilobiten.
In Verhandl. russ. k. mineral. Gesell. zu St. Petersburg, 1847, p. 1, pl. 1.
The author redescribed the genus *Zethus* Pander and its type *Zethus verrucosus*, placing *Cybele bellatulus* Eichw. under the same genus.
Zethus verrucosus Pand., *Z. bellatulus* Dalm.
Volborth on p. 15 signifies the particular affinities which exist between *Dindymene* and *Zethus*. He expresses the opinion that these two genera are identical.

——— Ueber das Prioritätsrecht der Trilobiten-Gattung *Zethus* Pander gegen die Gattung *Cryptonymus* Eichwald.
In Bull. Physico-Math. Acad. Sci., St. Petersburg, 1854, p. 251.

——— Ueber die Bewegungs-Organe der Trilobiten. Folgendes mit.
In Verhandl. russ. k. mineral. Gesellsch. zu St. Petersburg, 1857-58, p. 168.

——— Ueber die *Crotaluren* und *Remopleuriden*, ein Beitrag zur Kenntniss der russichen Trilobiten.
In Verhandl. russ. k. mineral. Gesell. zu St. Petersburg, 1857-58, p. 126. pl. 13.
Crotalurus n. g., *C. barrandei* n. sp. *Remopleurides nanus* Leucht.

——— Ueber die mit glatten Rumpfgliedern versehenen russischen Trilobiten, nebst einem Anhange über die Bewegungsorgane und über das Herz derselben.
In Mém. Acad. Imp. Sci. St. Petersburg, vol. 6, No. 2, 1863, 48 pp., 4 pls.
Panderia n. g. *Illænus crassicauda* Wahl., *I. dalmani* n. sp., *I. tauricornis* Kut., *I. triodontusus* n. sp. *Dysplanus centrotus* Dalm., *D. muticus* n. sp. *Panderia triquetra* n. sp., *P. minima* n. sp. *Nileus armadillo* Dalm. *Bumastus barriensis* Murch.
This article also contains an essay on the organs which were found by Dr. Pander on the under side of the pleuræ of *Asaphus*, which resemble very much the feet of Trilobites.
There is also a second paper on the heart of the trilobites. The author states that he found on an *Illænus* from which the shell had been removed in the region of the axis, on the median line, a small, flat, depressed, tubular organ, horizontal furrowed and divided into thirteen chambers. This tube extends from the margin of the pygidium in a curved line to the glabella. The chambers are all of uniform length—1¼ mm. The width enlarges towards the front in such a manner that the first nine chambers from the pygidium are only 1¼ mm., whereas the tenth and eleventh

chambers are 2½ to 3 mm. in width. The anterior chambers are nearly double as wide, and they are divided by a transverse furrow (which is probably a line of fracture) into two equal parts. The thirteenth chamber lies under the glabella.

―――― Ueber einige neue ehstländische Illænen.
In Mém. Acad. Imp. Sci. St. Petersburg, vol. 8, No. 9, 1864, 2 pls.
Illænus schmidtii Nieszk., also var. *I. roemeri* n. sp., *I. tauricornis* Kut.

―――― Ueber Herrn von Eichwald's Beitrag zur näheren Kenntniss der Illænen, etc.
In Bull. Soc. Imp. des Naturalistes, Moscou, 1866, No. 1, p. 77.

Wahlenberg (Geo.) Petrificata telluris Suecana. Upsaliæ, 1818.
This work was published and distributed in 1818 as a separate article, in advance of the Nova Acta Reg. Soc., vol. 8, 1821.

―――― Petrificata telluris Suecana examinata a Georgio Wahlenberg.
In Nova Acta Reg. Soc. Sci., Upsal., vol. 8, 1821, pp. 1-116, pls. 1-4.
1, *Entomostracites expansus* Linné. 2, *E. crassicauda* n. sp., pl. 2, figs. 5 and 6. 3, *E. laticauda* n. sp., pl. 2, figs. 7 and 8. 4, *E. caudatus* Brünn, pl. 2, fig. 3. 5, *E. granulatus* n. sp., pl. 2, fig. 4. 6, *E. tuberculatus* Brünn. 7, *E. punctatus* Brünn, pl. 2, fig. 1.* 8, *E. lacinatus* n. sp., pl. 2, fig. 2.* 9, *E. paradoxissimus* Linné, pl. 1, fig. 1. *(Paradoxides tessini* Brong.) 10, *E. bucephalus* n. sp., pl. 1, fig. 6 (hypostoma of *Paradoxides tessini* Brong.) 11, *E. spinulosus* n. sp., pl. 1, fig. 3. 12, *E. gibbosus* n. sp., pl. 1, fig. 4. 13, *E. scarabæoides* n. sp., pl. 1, fig. 2. 14, *E. pisiformis* Linné, pl. 1, fig. 5.

―――― Additamenta quædam ad petrificata telluris Suecana.
In Nova Acta Reg. Soc. Sci., Upsal., vol. 8, 1821, pp. 293-296, pl. 7.
1, *Entomostracites crassicauda*, pl. 7, fig. 5 *(Illænus esmarki* Schloth.) 2, *E. laticauda* n. sp., pl. 2, fig. 8 *(Bronteus laticauda).* Fig. 7 is referred to *Homalonotus punctill* Törnq. *Entom. expansus* Linné. 4, *E. extenuatus* n. sp., pl. 7, fig. 4. 5, *E. tuberculatus* Brünn. 6, *E. caudatus* Brünn.

Walch (J. E. T.) Des Trilobites dans le règne des pétrifications, ou de la conque ridée à trois lobes (concha triloba rugosa). Recueil des monumens des catastrophes que le globe de la terre a essuiées, contenant des pétrifications. (Knorr, G. W.) Vol. 3. Nuremberg, 1775, p. 105.

See, also, vol. 2, 1768, sec. 1, p. 81; vol. 3, 1775, pp. 184, 185, 186, 193, 222; suppl. pls. 8, 9a, 9b, 9c, 9e, 9f. German and French editions.

This author gives the name of Trilobite to the family, a designation which has been used by all subsequent authors, with the exceptions of J. W. Dalman and C. S. Rafinesque.

Calymene blumenbachii Broug., vol. 3, suppl. p. 222, pl. 9, figs. 1-5.

Walcott (C. D.) Description of a new species of Trilobite from Trenton Fall, New York.
In Cincinnati Quart. Jour. Sci., vol. 2, 1875, p. 273, 2 figs.
Sphærocoryphe robustus n. sp.

—— New species of Trilobite from the Trenton limestone at Trenton Falls, N. Y.
In Cincinnati Quart. Jour. Sci., vol. 2, 1875, p. 347, wood cuts.
Remopleurides striatulus n. sp.

—— Notes on *Ceraurus pleurexanthemus* Green.
In Annals Lyceum Nat. Hist. N. Y., vol. 11, 1876, p. 155, pl. 11.

—— Preliminary notice of the discovery of the remains of the natatory and branchial appendages of Trilobites.
In 28th Rep. New York State Mus. Nat. Hist., 1875, p. 89.

—— Description of new species of fossils from the Trenton limestone.
In 28th Rep. New York State Mus. Nat. Hist., 1875, p. 93.
Bathyurus longispinus n. sp. *Asaphus romingeri* n. sp., *A. wisconsensis* n. sp.

—— Notes on some sections of Trilobites from the Trenton limestone.
In 31st Rep. New York State Mus. Nat. Hist, 1879, p. 61. Advance sheets published Sept. 20, 1877, 1 pl.

—— Notes upon the legs of Trilobites.
In 31st Rep. New York State Mus. Nat. Hist., 1879, p. 64.

—— Notes upon the eggs of the Trilobites.
In 31st Rep. New York State Mus. Nat. Hist., 1879, p. 66. Advanced sheets published September 20, 1877.

—— Descriptions of new species of fossils from the Chazy and Trenton limestones.
In 31st Rep. New York State Mus. Nat. Hist., 1879, p. 69. Advanced sheets published September 20, 1877.

Arionellus pustulatus n. sp. *Ceraurus rarus* n. sp. *Encrinurus trentonensis* n. sp., *E. raricostatus* n. sp. *Acidaspis parvula* n. sp. *Dalmanites intermedias* n. sp. *Illænus indeterminatus* n. sp., *I. milleri* Bill. *Asaphus homalonotides* n. sp.

———— The Utica slate and related formations of the same geological horizon.

In Trans. Albany Inst., vol. 10, 1879, p. 2, pls. 1, 2.
Metamorphoses of *Triarthrus beckii* Green. *Triarthrus beckii* and *Beyrichia cincinnatiensis* S. A. Miller.

———— Description of new species of fossils from the Calciferous formation.

In 32d Rep. New York State Cab. Nat. Hist., 1879, p. 129. Pamphlet published in advance of the report.
Conocephalites calciferus n. sp., *C. hartti* n. sp. *Ptychaspis speciosus* n. sp. *Bathyurus armatus* Bill.

———— The Trilobite, new and old evidence relating to its organization.

In Mus. Comp. Zool. Harvard Coll., vol. 8, 1881, p. 191, 6 pls.
Ceraurus pleurexanthemus Green. *Calymene senaria* Conrad. *Asaphus platycephalus* Stokes. *Eurypterus remipes* Dekay.

———— Description of a new genus of the order *Eurypterus* from the Utica slate.

In Am. Jour. Sci., 3d series, vol. 23, 1882, p. 213, figures.
Echinognathus n. g., *E. clevelandi* n. sp.

———— Description of new species of fossils from the Trenton group, New York.

In 35th Rep. New York State Cab. Nat. Hist., 1883, p. 206, 1 pl. Author's edition, October 15, 1883, 8 pp., 1 pl.
Beyrichia bella n. sp. *Leperditia (Isochilina) armata* n. sp.

———— Injury sustained by the eye of a Trilobite at the time of moulting of the shell.

In Annals Nat. Hist., 5th series, vol. 13, London, 1884, p. 69.

———— On the locomotory appendages of Trilobites.

In Science, vol. 3, 1884, p. 279, 3 figs.
Asaphus megistos Locke. *Calymene senaria* Conrad.

———— Palæontology of the Eureka district.

In Mon. U. S. Geol. Survey, vol. 8, 1884, 24 pls.
Cambrian: *Agnostus bidens* Meek, *A. communis* H. & W., *A. neon* H. &

W., *A. prolongus* H. & W., *A. richmondensis* n. sp., *A. seclusus* n. sp. *Olenellus gilberti* Meek, *O. howelli* Meek, *O. iddingsi* n. sp. *Dicellocephalus ? angustifrons* n. sp., *D. bilobatus* H. & W., *D.? expansus* n. sp., *D. flabellifer* H. & W., *D. iole* n. sp., *D. marica* n. sp., *D. nasutus* n. sp., *D. osceola* Hall, *D.? quadriceps* H. & W., *D. richmondensis* n. sp. *Ptychoparia ? angulatus* H. & W., *P. anytus* H. & W., *P. (Solenepleura?) breviceps* n. sp., *P. granulosus* H. & W., *P. haguei* H. & W., *P. læviceps* n. sp., *P.? linnarssoni* n. sp., *P. maculosus* H. & W., *P. nitidus* H. & W., *P. occidentalis* n. sp., *P. oweni* M. & H., *P. pernasutus* n. sp., *P.? prospectensis* n. sp., *P.? similis* n. sp., *P.? similis* var. *robustus*, *P. unisulcatus* H. & W., *P. (Euloma?) affinis* n. sp., *P. (Euloma?)·dissimilis* n. sp., *P. (Pterocephalus) laticeps* H. & W., *P. (Pterocephalus) occidens* n. sp. *Anomocare? parvum* n. sp. *Ptychaspis minuta* Whitf., *P. pustulosa* H. & W. *Chariocephalus tumifrons* H. & W. *Agraulos? globosus* n. sp. *Arethusina americana* n. sp. *Ogygia? problematica* n. sp., *O.? spinosa* n. sp. *Illænurus* sp.?

Lower Silurian: *Leperditia bivia* White. *Beyrichia* sp.? *Plumulites* sp.? *Agnostus communis* H. & W., *A. tumidosus* H. & W. *Dicellocephalus inexpectans* n. sp., *D. finalis* n. sp., *D. multicinctus* H. & W. *Ptychoparia annectans* n. sp., *P. granulosus* H. & W., *P. haguei* H. & W., *P. maculosus* H. & W., *P. oweni* H. & M., *P. unisulcatus* H. & W., *P. (Euloma) affinis* n. sp. *Arethusina americana* n. sp. *Bathyurus? congeneris* n. sp., *B. pogonipensis* H. & W., *B.? simillimus* n. sp., *B. tuberculatus* n. sp., *B.* sp.? *Cyphaspis? brevimarginatus* n. sp. *Amphion nevadensis* n. sp., *A.* sp.? *Ceraurus* sp.? *Symphysurus? goldfussi* n. sp. *Barrandia? maccoyi* n. sp. *Illænus eurekensis* n. sp., *I. crassicauda* Wahl. *Asaphus caribouensis* n. sp., *A.? curiosus* Bill., *A.* sp.?

Lone Mountain Silurian: *Ceraurus. Dalmanites. Trinucleus concentricus* Eaton. *Illænus. Asaphus platycephalus* Stokes.

Devonian: *Beyrichia occidentalis* n. sp. *Leperditia rotundata* n. sp. *Phacops rana* Green. *Dalmanites meeki* n. sp. *Proetus haldemani* Hall, *P. marginalis* Conrad. *Phillipsia coronata* Hall.?

Carboniferous: *Leperditia. Griffithides portlocki* M. & W.

———— On the Cambrian faunas of North America. Preliminary studies.

In Bull. U. S. Geol. Survey, No. 10, 1884, 10 pls.

Harttia n. g. *Protocaris* n. g.

The St. John formation: *Agnostus acadicus* Hartt. *Microdiscus dawsoni* Hartt., *M. punctatus* Salt. *Paradoxides lamellatus* Hart., *A. acadicus* Matt., *P. eteminicus* Matt. *Conocoryphe* (subgenus?) *matthewi* Hartt., *C. walcotti* Matt., *C. (Salteria) baileyi* Hartt., *C. elegans* Hartt. *Ptychoparia robbi* Hartt., *P. ouangondiana* Hartt., also var. *aurora*, *P. quadrata* Hartt., *P. orestes* Hartt., also var. *thersites* Hartt., *P. tener* Hartt.

Fauna of the Braintree Argillites: *Paradoxides harlani* Green. *Ptychoparia rogersi* n. sp. *Agraulos quandrangularis* Whitf.

Georgian formation: *Protocaris marshi* n. sp.

———— Palæozoic notes: New genus of Cambrian Trilobites, *Mesonacis*.

In Am. Jour. Sci., 3d series, vol. 29, 1885, p. 328, figs.

Describes a new genus, *Mesonacis*, and substitutes the generic name *Bailiella* of Matthew for the subgenus *Salteria*, proposed by him (Bull. U. S. Geol. Survey, No. 10, 1884).

Mesonacis vermontana Hall, also probably *Olenellus howelli*.

———— Second contribution to the studies on the Cambrian fauna of North America.

In Bull. U. S. Geol. Survey, No. 30, 1886.

Oryctocephalus n. g. *Protypus* n. g. *Leperditia troyensis* Ford, *L.? argenta* n. sp. *Protocaris marshi* Wal. *Agnostus interstricus* White, *A. nobilis* Ford. *Microdiscus speciosus* Ford, *M. meeki* Ford, *M. lobatus* Hall., *M. parkeri* n. sp. *Mesonacis vermontana* Hall. *Olenellus thompsoni* Hall, *O. asaphoides* Emmons, *O. iddingsi* Wal., *O. gilberti* Meek. *Olenoides nevadensis* Meek, *O. typicalis* n. sp., *O. spinosus* Wal., *O.? flayricaudus* White, *O.? marconi* Whitf., *O. lævis* n. sp., *O. quadriceps* H. & W., *O. wahsatchensis* H. & W. *Bathynotus holopyga* Hall. *Ptychoparia kingi* Meek, *P. adamsi* Bill., *P. teucer* Bill., *P. vulcanus* Bill., *P. miser* Bill., *P. quadrans* H. & W., *P. housensis* n. sp., *P. piochensis* n. sp., *P.? prospectensis* Wal., *P. trilineata* Emmons, *P. subcoranata* H. & W. *Crepicephalus liliana* n. sp., *C. angusta* n. sp. *Anomocare? parvum* Wal. *Oryctocephalus primus* n. sp. *Protypus hichcocki* Whit., *P. senectus* Bill. *Solenopleura nana* Ford. *Bathyuriscus howelli* n. sp., *B. productus* H. & W. *Asaphiscus wheeleri* Meek. *Dopypyge*.

———— Fauna of the upper Taconic of Emmons, in Washington County, New York, with one plate.

In Am. Jour. Sci., 3d series, vol. 34, 1887, p. 187.

Leperditia dematoides n. sp. *Aristozoc troyensis* Ford, *A. rotundata* n. sp. *Microdiscus connexus* n. sp. *Olenoides fordi* n. sp. *Solenopleura? tumida* n. sp., *S.? nana* Ford. *Conocoryphe trilineata* Emmons. *Ptychoparia fitchi* n. sp., *P.?* (subgenus?) *clavata* n. sp.

———— Cambrian fossils from Mount Stephens, North West Territory of Canada.

Am. Jour. Sci., 3d series, vol. 36, 1888, p. 161.

The author classes the following species under *Olenoides: O. nevadensis, O. marconi, O. quadriceps, O. wasatchensis*.

Under the genus *Zacanthoides* he arranges *Olenoides typicalis, O. spinosus, C. lævis* and *O. flayricaudatus*.

―――― Stratigraphic position of the Olenellus fauna in North America and Europe.

Am. Jour. Sci., 3d series, vol. 37, 1889, p. 374, continued on p. 392 of the same journal.

―――― Descriptive notes on new genera and species, from the Lower Cambrian or Olenellus zone of North America.

In Proc. U. S. Natl. Mus., vol. 12, 1889, p. 34.

Agnostus desideratus n. sp., *A.* sp.? *Microdiscus helena* n. sp. *Olenellus (Mesonacis) asaphoides* Emmons, *O. (M.) brüggeri* Walcott. *Avalonia* n. g., *A. manuelensis* n. sp. *Zacanthoides eatoni* n. sp. *Solenopleura harveyi* n. sp., *S. howleyi* n. sp.

―――― The fauna of the Lower Cambrian or Olenellus Zone.

In 10th Rep. U. S. Geol. Sur. Washington, 1890, pp. 515-638, pls. 43-98.

Isoxys n. g., *I. chilhoweana* n. sp. *Leperditia dematoides* Wal. *Aristozoe rotundata* Wal., *A. troyensis* Ford. *Nortozoe? vermontana* Whitf. *Protocaris marshi* Wal. *Agnostus desideratus* Wal., *A. nobilis* Ford., *A.* sp.? *Microdicus bellimarginatus* S. & F., *M. connexus* Wal., *M. meeki* Ford., *M. parkeri* Wal., *M. lobatus* Hall, *M. speciosus* Ford., *M. helena* Wal., *M.* sp.? *Olenellus thompsoni* Hall, *O. gilberti* Meek, *O. iddings* Wal.

Mesonacis: *Olenellus (Mesonacis) vermontana* Hall, *O. (M.) mickwitzia* Schm. and *O. (M.) asaphoides* Emmons.

Holmia: *Olenellus (Holmia) kjerulfi* Linrs., *O. (H.) brüggeri* Wal., *O. (H.) callavei* Lapw. *Olenoides fordi* Wal., *O. marcoui* Whitf., *O. ellsi* n.sp., *O. (Dorypyge) disiderata* n. sp., *O. quadriceps* H. & W. *Zacanthoides lævis* Wal., *Z. eatoni* Wal. *Bathynotus holopyge* Hall. *Avalonia manuelensis* Wal. *Conocoryphe trilineata* Emmons, *C. reticulata* n. sp. *Ptychoparia adamsi* Bill., *P.? attleborensis* S. & F., *P. fitchi* Wal., *P. miser* Bill., *P. metisensis* n. sp., *P. subcoronata* H. & W., *P. teucer* Bill., *P. vulcanus* Bill. *Crepicephalus angusta* Wal., *C. lilliana* Wal. *Oryctocephalus primus* Wal. *Agraulos strenuus* Bill., also var. *nasutus*, *A. redpathi* n. sp. *Protypus hitchcocki* Whitf., *P. senectus* Bill., *P. clavatus* Wal. *Solenopleura bombifrons* Matt., *S. harveyi* Wal., *S.? howleyi* Wal., *S.? nana* Ford., *S.? tumida* Wal.

In this contribution to the *Olenellus* zone Walcott gives a bibliographic reference to the articles on the subject, also a historical review, with geologic description of the Lower Cambrian or Olenellus zone as known to the geologist, etc.

―――― Notes on the Cambrian rocks of Virginia and the Southern Appalachians.

In Am. Jour. Sci., 3d series, vol. 44, 1892, p. 52.

The author includes in the fauna of the lowest horizon in Wisconsin: *Ptychoparia calymenoides* Whitf., *Agraulos? fecundus* n. sp., *Crepicephalus onustus* n. sp., *Agraulos woosteri* Whitf.
In the next zone above, which is also included in the Middle Cambrian, the following species occur: *Pemphigaspis bullata* Hall. *Agnostus?* *Crepicephalus texanus* Shum., *C. iowensus* Owen. *Ptychoparia connata* n. sp., *P. optatus* Hall, *P. (Lonchocephalus) chippewensis* Owen, *P. (L.) minor* Shum. *Amphion matutina* Hall, and *Agraulos thea* n. sp.

Waldschmidt (E.) Ueber die devonischen Schichten der Gegend von Wildungen.

In Zeitschr. Deutsch. Geol. Gesell., vol. 37, 1885, p. 906, pls. 37-40.
Bronteus thysanopeltis Barr. *Phacops fecundus* Barr. *Acidaspis* n. sp. *Proetus* n. sp.

Warder (John A.) New Trilobites.

In Am. Jour. Sci., 1st series, vol. 34, 1838, p. 377.
Ceratocephala n. g.
This article on the new genus *Ceratocephala* was read before the Western Academy of Science, May 25, 1838.
The name *Ceratocephala* was proposed for a fossil from Springfield, Ohio, well described and illustrated. The original specimen which received the name of *Ceratocephala goniata* was a large head of a trilobite with bases of conspicuous spines upon the occipital ring. The author illustrated the fossil in an inverted position and regarded the occipital spines as antennæ.

Washburn (W. T.) New Trilobites.

In Am. Jour. Sci., 3d series, vol. 6, 1873, p. 186.
Not descriptive; contains important notes on fossil Crustacea.

Weiss (C. E.) Note on *Esthere (Estheriella) costata* and *Esthere (Estheriella) lineata* Weiss.

In Zeitsch Deutsch Geol. Gesell., Berlin, vol. 27, 1875, p. 711.

Weitenweber (W. R.) Systematisches Verzeichniss der böhmischen Trilobiten in der Sammlung des Dr. Zeidler.

In Lotos, vol. 7, 1847, pp. 5-42.

Wetherby (A. G.) Description of a new family and genus of Lower Silurian Crustacea.

In Jour. Cincinnati Soc. Nat. Hist., vol. 1, 1878-79, p. 162, fig.: vol. 2, pl. 7, figs. 1, 1 a-g.
Enoploura balanoides n. sp.
This fossil is probably a Cystidæ and not of the order Crustacea.

―――― Description of new fossils from the Lower Silurian and Subcarboniferous rocks of Ohio and Kentucky.
In Jour. Cincinnati Soc. Nat. Hist., vol. 4, 1881, p. 77.
Proetus granulatus n. sp. *Isochilina jonesi* n. sp.

White (C. A.) and St. John (O. H.) Preliminary notice of new genera and species of fossils. Iowa City, May 8, 1867, 3 pp.
Beyrichia lithofactor n. sp., *B. lithofactor* var. *velata* n. sp. *Cythere simplex* n. sp.
Of this small pamphlet only fifty copies were published by the Iowa State Geological Survey. Republished in Trans. Chicago Acad. Sci., vol. 1, 1867.

―――― Description of new subcarboniferous and Coal-Measure fossils collected upon the geological survey of Iowa, together with a notice of new generic characters observed in two species of Brachiopods.
In Trans. Chicago Acad. Sci., vol. 1, pt. 1, 1867, p. 115, 2 figs.
Beyrichia petrifactor, *B. petrifactor* var. *velata*, *B. fœtoidea* n. sp. *Cythere simplex*.
In this article the authors give the new names of *Beyrichia petrifactor* to the species described in the preliminary notice as *B. lithofactor*, and also change the name of *B. lithofactor* var. *velata* to that of *Beyrichia petrifactor* var. *velata*.

―――― Preliminary report upon invertebrate fossils collected by the expeditions of 1871, 1872 and 1873, with descriptions of new species.
In Geog. and Geol. Expl. Surveys West 100th Meridian, Washington, 1874.
Agnostus interstricta n. sp. *Olenellus gilberti* Meek MS., *O. howelli* Meek MS. *Leperditia bivia* n. sp. *Meglaspis belemurus* n. sp. *Dicellocephalus flagricaudus* n. sp.

―――― Report upon the invertebrate fossils collected in portions of Nevada, Utah, Colorado, New Mexico and Arizona, by parties of the expeditions of 1871, 1872, 1873 and 1874.
In Geographical Surveys West. 100th Meridian, Palæontology, vol. 4, 1877, pt. 1, pp. 1-219, pls. 1-21.
Agnostus interstrictus White. *Conocoryphe (Ptychoparia) kingi* Meek.

Asaphiscus wheeleri Meek. *Olenellus gilberti* Meek, *O. howelli* Meek. *Leperditia bivia* White. *Meglaspis belemurus* White. *Dicellocephalus flagricaudus* White.

——— Palæontology: Fossils of the Indiana rocks.
In 2d Ann. Rep. Dept. Statistics and Geology, Indiana, 1880. pp. 471-522, pls. 1-11.
Calymene senaria Conrad. *Cyphaspis christyi* Hall. *Phillipsia bufo* Meek.

——— The fossils of the Indiana rocks. No. 3.
In 13th Ann. Rep. Dept. Geol. and Nat. Hist., Indiana, 1883, pt. 2, pp. 107-180, pls. 23, 39.
Phillipsia (Griffithides?) scitula Meek, *P. (G.?) sangamonensis* M. & W. *Eurypterus (Anthraconetes) mazonensis* M. & W., *A. eveni* M. & W. *Palæocaris typus* M. & W. *Dithyrocaris carbonarius* M. & W. *Leaia tricarinata* M. & W. *Euproops danæ* Meek, *E. colletti* n. sp.

Whiteaves (J. F.) Palæozoic fossils. Vol. 3, pt. 1, 1884.
In Geol. and Nat. Hist. Survey Canada, 1884, 8 pls.
Eurypterus boylei n. sp.

——— For description of *Solenopleura acadia* Whiteaves.
See Trans. Royal Soc. Canada, vol. 3, 1885, p. 76.

——— Contributions to Canadian Palæontology. Vol. 1. Montreal, 1885-1892.
Part 2. On some fossils from the Hamilton Formation of Ontario, etc.
Dalmanites helena Hall, *D. boothii* Green. *Phacops rana* Green. *Cythere? (Beyrichia) punctulifera* Hall.
Part 3. The fossils of the Devonian rocks of the Mackenzie River Basin.
Primitia scitula Jones. *Aparchites mitis* Jones. *Isochilina bellula* Jones. *Proetus haldemani* Hall.
Part 4. The fossils of the Devonian rocks of the islands, shores or immediate vicinity of lakes Manitoba and Winnepegosis.
Isochilina dawsoni Jones. *Elpe tyrrellii* Jones. *Leperditia? exigua* Jones. *Bronteus manitobensis* n. sp. *Lichas (Terataspis)* n. sp. *Cyphaspis bellula* n. sp. *Proetus mundulus* n. sp.

——— Description of a new genus and species of Phyllocarid Crustacea from the Middle Cambrian of Mt. Stephen, British Columbia.
In Canadian Record of Science, vol. 5, 1892, pp. 205-208.
Anomalocaris n. g., *A. canadensis* n. sp.

Whitfield (R. P.) Description of new species of fossils.
In Rep. Reconnoissance from Carrol, Montana Territory, on the Upper Missouri, to the Yellowstone National Park and return, Ludlow's report, Washington, 1876.
Crepicephalus (Loganellus) montanensis n. sp. *Arionellus tripunctatus* n. sp.
See, also, Rep. Chief Eng. U. S. A., 1876, Appendix NN.

—— Preliminary descriptions of new species of fossils from the lower geological formation of Wisconsin.
In Ann. Rep. Geol. Survey Wisconsin, 1877, pp. 50–89.
Conocephalites calymenoides n. sp. *Crepicephalus onustus* n. sp. *Ptychaspis granulosa* Owen, *P. striata* n. sp., *P. minuta* n. sp. *Agraulos (Bathyurus?) woosteri* n. sp. *Arionellus (Agraulos) convexus* n. sp. *Ellipsocephalus curtus* n. sp. *Dicellocephalus barbuensis* n. sp., *D. eatoni* n. sp. *Illænurus convexus* n. sp. *Illænus pterocephalus* n. sp. *Bronteus laphami* n. sp.

—— Preliminary report on the palæontology of the Black Hills, containing descriptions of new species of fossils from the Potsdam, Jurassic and Cretaceous formations of the Black Hills of Dakota.
In U. S. Geol. Survey Rocky Mountain Region, Washington, 1877.
Crepicephalus (Loganellus) centralis n. sp., *C. (Loganellus) planus* n. sp.

—— Description of new species of fossils from the Palæozoic formations of Wisconsin.
In Ann. Rep. Geol. Survey Wisconsin, 1879, pp. 44–71.
Conocephalites? quadratus n. sp., *C. (Ptychaspis?) explanatus* n. sp. *Crepicephalus? gibbsi* n. sp. *Ptychaspis striata* n. sp. *Dicellocephalus lodensis* n. sp. *Aglaspis eatoni* n.sp. *Asaphus triangulatus* n.sp. *Illænus niagriensis* n. sp.

—— New forms of fossil Crustaceans from the Upper Devonian rocks of Ohio, with descriptions of new genera and species.
In Am. Jour. Sci., 3d series, vol. 19, 1880, p. 33.
Echinocaris n. g. *Palæopalæmon* n. g.
A plate and explanation accompany the author's edition.
Echinocaris sublævis n. sp., *E. pustulosa* n. sp., *E. multinodosa* n. sp. *Palæopalæmon newberryi* n. sp. *Aristozoe canadensis* Whitf.

―――― Palæontology of the Black Hills of Dakota.
In U. S. Geog. and Geol. Survey Rocky Mountain Region. Rep. Geol.
and Resources of the Black Hills of Dakota, with atlas; by Henry Newton and Walter Jenney, Washington, 1880, p. 325, pl. 2.
Crepicephalus planus n. sp., *C. centralis* n. sp.

―――― Palæontology.
In Geol. Wisconsin, vol. 4, 1882, pt. 3, pp. 163-363, pls. 1-27.
Potsdam Sandstone: (*Dicellocephalus* zone.) *Conocephalites calymenoides* Whitf. *Dicellocephalus lodensis* Whitf., *D. crassimarginatus* n. sp. *Conocephalites quadratus* n. sp., *C. (Ptychaspis?) explanatus* n. sp. *Crepicephalus onustus* Whitf., *C.? gibbsi* n. sp. *Ptychaspis granulosa* Owen, *P. striata* n. sp., *P. minuta* Whitf. *Dicellocephalus minnesotensis* Owen, *Agraulos (Bathyurus?) woosteri* Whitf. *Arionellus convexus* Whitf. *Ellipsocephalus curtus* Whitf. *Aglaspis eatoni* n. sp.
Species from the Lower Magnesian Limestone: *Dicellocephalus barabuensis* Whitf., *D. eatoni* Whitf. *Illanurus convexus* Whitf.
Species from the Trenton Limestone: *Asaphus susae* Calvin MS., *A. homalonotoides* Wal. *Illænus ovatus* Conrad.
Species from the Niagara Group, Ill.: *Illænus ioxus* Hall, *I. insignis* Hall, *I. imperator* Hall, *I. madisonianus* n. sp., *I. pterocephalus* Whitf. *Bronteus laphami* Whitf. *Sphærexochus romingeri* Hall.
Devonic: *Leperditia alta* Conrad. *Phacops rana* Green.

―――― Description of new species of fossils from Ohio, with remarks on some of the geological formations in which they occur.
In Annals New York Acad. Sci., vol. 2, Nos. 7, 8, 1882, p. 193.
Eurypterus criensis n. sp. *Leperditia angulifera* n. sp. *Plumulites newberryi* n. sp.

―――― On the fauna of the Lower Carboniferous limestones of Spergen Hill, Indiana, with a revision of the descriptions of its fossils heretofore published, and illustrations of the species from the original type species.
In Bull. Am. Mus. Nat. Hist. New York, vol. 1, No. 3, 1882, pl. 9.
Leperditia carbonaria Hall. *Cytherina glandella* n. sp.
Plates republished in 12th Rep. Geol. Nat. Hist. Indiana, 1882.

―――― List of Wisconsin fossils.
In Geol. Wisconsin, vol. 1, 1883, p. 362; also illustrations of typical fossils of the Wisconsin formations.
Fig. 15 is an entire specimen of *Dicellocephalus minnesotensis* Owen.

────── Notice of some new species of Primordial fossil in the collection of the Museum, and corrections of previously described species.
In Bull. Am. Mus. Nat. Hist. New York, vol. 1, No. 5, 1884, p. 139, pls. 14, 15.
Nothozoe vermontana n. sp. *Conocephalites verrucosus* n. sp. *Arionellus quadrangularis* n. sp. *Angelina hitchcocki* n. sp. *Dikelocephalus? marcoui* n. sp. *Olenellus thompsoni* Hall.

────── Notice of a very large species of *Homalonotus* from the Oriskany sandstone formation.
In Bull. Am. Mus. Nat. Hist., New York, vol. 1, No. 6, 1885, p. 193, pl. 22.
Homalonotus major n. sp.

────── Notice of geological investigations along the eastern shore of Lake Champlain, conducted by Prof. H. M. Seely and Prest. Ezra Brainerd, of Middlebury College, with descriptions of the new fossils discovered.
In Bull. Am. Mus. Nat. Hist. New York, vol. 1, No. 8, 1886, p. 293.
Sao? lamottensis n. sp. *Lichas champlainensis* n. sp. *Asaphus canalis* Conrad.

────── Observations on some imperfectly known fossils from the calciferous sandrock of Lake Champlain, and descriptions of several new forms.
In Bull. Am. Mus. Nat. Hist. New York, vol. 2, No. 2, 1889, p. 41, 13 pls.
Primitia gregaria n. sp., *P.? cristata* n. sp., *P. seelyi* n. sp. *Bathyurus conicus* n. sp., *B. seelyi* n. sp.

────── Additional notes on *Asaphus canalis* Conrad.
In Bull. Am. Mus. Nat. Hist. New York, vol. 2, No. 2, 1889, p. 64, pls. 11, 12.

────── Description of a new form of fossil, Balanoid Cirripede, from the Marcellus shale of New York.
In Bull. Am. Mus. Nat. Hist. New York, vol. 2, No. 2, 1889, p. 66.
Protobalanus n. g., *P. hamiltonensis* n. sp., pl. 13, fig. 22.

────── Contributions to Invertebrate Palæontology.
In Annals N. Y. Acad. Sci., vol. 5, 1890, p. 505, 15 pls.
Leperditia alta Conrad, *L. angulifera* Whitf. *Eurypterus eriensis* Whitf. *Plumulites newberryi* Whitf. *Echinocaris sublævis* Whitf., *E. pustulosa* Whitf., *E. multinodosa* Whitf. *Aristozoe canadensis* Whitf. *Palæopalæmon newberryi* Whitf.

—— Discovery of the second example of the Macrouran Decapod Crustacean *Palæopalæomon newberryi*.
In the American Geologist, vol. 9, 1822, p. 237.
See Hall (James) and Whitfield (R. P.)

Wigand (Georg von). Ueber die Trilobiten der silurischen Geschiebe in Mecklenburg.
In Zeitschr. Deutsch. Geol. Gessell., vol. 40, 1888, p. 39, pls. 6–10.
Phacops stokesi Milne-Edw., *P. downingiæ* Murch, *P. dubius* Steinh., *P. exilis* Eichw., *P. panderi* Schm., *P. recurvus* Linrs., *P. bucculenta* Sjögr., *P. wrangeli* Schm., *P. maxima* Schm., *P. macroura* Sjögr., *P. conicophthalma* S. & B., *P. wesenbergensis* Schm., *P.* cf. *eichwaldi* Schm., *P. tumida* Ang., *P. marginata* Schm.
Lichas illænoides Nieszk., *L.* aff. *illænoides* Nieszk., *L. holmi* Schm., *L. (Hoplolichas) tricuspidata* Beyr., *L. (H.)* aff. *proboscidea* Dames, *L.* cf. *pachyrhina* Dalm., *L. deflexa* Sjögr., *L.* cf. *cicatricosa* Lovén, *L. nasuta* n. sp., *L.* cf. *gibba* Ang., *L. illæniformis* n. sp., *L. triconica* Dames.
Illænus chiron Holm., *I. crassicauda* Wahl., *I. parvulus* Holm., *I. sinuatа* Holm., *I. fallax* Holm., *I. linnarssonni* Holm., *I. centrotus* Dalm., *I.* cf. *schmidti* Nieszk., *I.* sp.? (cf. *I. fallax* Holm.)
Cheirurus exsul Beyr., *C. spinulosus* Nieszk., *C. (Cyrtometopus) pseudohemicranium* Nieszk., *C. (C.)* cf. *affinis* Ang., *C. (Pseudosphærexochus) hemicranium* Kut., *C. (P.)* cfr. *granulatus* Ang., *C. (Nieszkowskia)* cf. *tumidus* Ang., *C. (N.) cephaloceros* Nieszk., *C. (N.) variolaris* Linrs.
Sphærexochus mirus Beyr. *Amphion fischeri* Eichw. *Cybele bellatula* Dalm., *C.* cf. *coronata* Schm., *C. grewingki* Schm., *C.* cf. *wörthi* Eichw. *Encrinurus punctatus* Brünn, *E.* cf. *obtusus* Ang., *E. lævis* Ang. *Acidaspis mutica* Emm., *A.* cf. *ovata* Emm.

Wiik (F. G.) Om fossilierna i Olans silur-kalksten jemförda med de i Sverige och Estland förekommande.
In Bidrag till Kännedom af Finlands Natur och Folk, vol. 35, Helsingfors, 1881, p. 21.
Calymene blumenbachii Brong. *Phacops conicophtalmus* Boeck. *Lichas* sp.? Not descriptive.

Wilckens (C. F.) Sendschreiben, worin wahrscheinlich dargethan wird, dass die Conchilioligisten eben keine Ursache mehr haben, das Petrefact, welches bisher unter der Benennung eines *Conchitæ trilobi rugosi* bekannt geworden ist, als einen Theil ihrer Wissenschaft anzusehen. (*)
In Stralsundisches Mag., vol. 1, 1768, p. 267.

―――― Nachricht von seltenen Versteinerungen, vornemlich des Thier-Reiches, welche bisher noch nicht genau genug beschieben und erklärt worden. Mit Kupfern, in drei Sendschreiben an seine Gönner und Freunde. Berlin und Stralsund, 1769, pl. 8.

The species described in 'this work have been referred to the modern genera. *Calymene blumenbachii* Brong., varieties *pulchella* Dalm. and *tuberculosa* Dalm. *Proetus concinna* Dalm. (Tab. 1). *Encrinurus. Phacops sclerops* Dalm. and *Asaphus* (Tab. 2). *Encrinurus punctatus* Brünn. *Asaphus angustifrons* Dalm. (Tab. 3). *Illænus* (Tab. 4). *Dalmanites hausmanni* Brong. *Ceraurus* sp.? (Tab. 5). *Trinucleus* sp.? *Lichas* sp.? (Tab. 6). *Trinucleus ornatus* Sternb. *Agnostus pisiformis* Linné. *Beyrichia wilckensiana* Jones (Tab. 7).

Williams (H. S.) On the occurrence of *Proetus longicauda* Hall in Madison county, Kansas.

In Am. Jour. Sci., 3d series, vol. 21, 1881, p. 156.

―――― Notice of a new Limuloid Crustacean from the Devonian.

In Am. Jour. Sci., 3d series, vol. 30, 1885, p. 45, 3 figs.

Prestwichia eriensis n. sp.

This species was used by A. S. Packard (Mem. Nat. Acad. Sci., vol. 3, 1888, p. 150) for the type of his new genus *Protolimulus*.

―――― Notice of a new Limuloid Crustacean from the Devonian formation of Erie county, Pennsylvania.

In Nature, vol. 32, London, 1885, p. 350.

Prestwichia eriensis n. sp.

Winchell (Alexander). Descriptions of fossils from the Marshall and Huron groups of Michigan.

In Proc. Acad. Nat. Sci. Phila., vol. 14, 1862, p. 405.

Cythere crassimarginata n. sp.

―――― Description of fossils from the yellow sandstone lying beneath the Burlington limestone at Burlington, Iowa.

In Proc. Acad. Nat. Sci. Phila., vol. 15, 1863, p. 2.

Phillipsia insignis n. sp., *P. meramecensis* Shumard.

―――― Notice of a small collection of fossils from the Potsdam sandstone of Wisconsin and Lake Superior sandstone of Michigan.

In Am. Jour. Sci., 2d ser., vol. 37, 1864, p. 226.
Dicellocephalus minnesotensis Owen, *D. pepinensis* Owen. *Ptychaspis barabuensis* n. sp.

―――― Description of new species of fossils from the Marshall group of Michigan and its supposed equivalent in other States; with notes on some fossils of the same age previously described.
In Proc. Acad. Nat. Sci. Phila., vol. 17, 1865, p. 109.
Phillipsia rockfordensis n. sp., *P. doris* (Hall) Winchell.

―――― and **Marcy** (O.) Enumeration of the fossils collected in the Niagara limestone at Chicago, Ill.; with descriptions of some new species.
In Mem. Boston Soc. Nat. Hist., vol. 1, 1866, p. 81, pls. 2, 3.
Lichas pugnax n. sp., *L. decipiens* n. sp. *Bronteus occasus* n. sp. *Illænus (Bumastus) worthenanus* n. sp. *Acidaspis ida* n. sp.

Winchell (N. H.) Fossils from the red quartzite at Pipestone.
In 13th Rep. Geol. Nat. Hist. Survey Minnesota, 1884, p. 65, pl.
Paradoxides barberi n. sp.

Woltersdorf (J. L.) Systema minerale.
Berlin, 1748, p. 42.
The author classes the Trilobites with the Mollusca, calling them "*Conchites trilobus*."

Woodward (Henry). On a nearly perfect specimen of *Eurypterus lanceolatus* (Salter) from the Upper Ludlow rocks at Lesmahagow, etc.
In Geol. Mag., vol. 1, London, 1864, p. 107, pl. 5.

―――― On the *Eurypteridæ* and *Xiphosura*.
In Rep. 34th Meeting Brit. Assoc. Adv. Sci., 1864, Sec. C; Geol. Mag., vol. 1, London, p. 239.
Eurypterus, Pterygotus, Slimonia, Stylonurus scoticus n. sp. *Hemiaspis limuloides* n. sp.

―――― Descriptions of some new Palæozoic Crustacea.
In Geol. Mag., vol. 1, London, 1864, p. 196, pl. 10, wood cuts.
Stylonurus logani n. sp., *S. ensiformis* n. sp. *Pterygotus minor* n. sp. *Eurypterus brewsteri* (Powrie MS.)

―――― On some Crustacean teeth from the Carboniferous and Upper Ludlow rocks of Scotland.
In Geol. Mag., vol, 2, London, 1865, p. 401, pl. 11.
The author illustrates *Ceratiocaris papilio* Salt. *Dithyrocaris scouleri* McCoy.

―――― On some new species of Crustacea belonging to the order *Eurypterus*.
In Geol. Mag., vol. 2, London, 1865, p. 219; Quart. Jour. Geol. Soc. London, vol. 21, 1865, p. 482, pl. 13.
Stylonurus powriei Page, *S. scoticus* H. W., *S. (Eurypterus) symondsii* Salt., *S. (E.) megalops* Salt., *S. logani* H. W., *S. ensiformis* H. W.

―――― On the discovery of a new genus of *Cirripediæ* in the Wenlock limestone, Dudley.
In Quart. Jour. Geol. Soc. London, vol. 21, 1865, p. 486, pl. 14; Geol. Mag., vol. 2, London, 1865, p. 319.
Turrilepas n. g., *T. wrightii* n. sp.

―――― On a new genus of *Eurypteridæ* from the Lower Ludlow rocks of Leintwardine, Shropshire.
In Quart. Jour. Geol. Soc. London, vol. 21, 1865, p. 490, pl. 14.
Hemiaspis n. g.
The author illustrates *Hemiaspis limuloides* H. W., and notes *Hemiaspis tuberculata* Salt. MS., *H. optata* Salt. MS., *H. sperata* Salt. MS., *H. salweyi* Salt. MS.

―――― and Salter (J. W.) Chart of the genera of fossil Crustacea.
See Salter (J. W.) and Woodward (Henry).

―――― Shield bearing Crustacea, recent and fossil.
In Intellectual Observer, vol. 8, 1865, p. 321. pl.
The author illustrates *Limulus moluccanus* Fabr., *L. polyphemus* Latr., *L. walchii* Desmar. *Belinurus rotundatus* Prestw., *B. reginæ* Baily, *B. trilobitoides* König. *Hemiaspis limuloides* H. W. *Harpes ungula* Sternb. *Dithyrocaris scouleri* McCoy. *Apus cancriformis* Schæffer.

―――― First report on the structure and classification of fossil Crustacea.
In Rep. 35th Meeting Brit. Assoc. Adv. Sci., 1865, p. 320.
Turrilepas wrightii, figs 1-3.

―――― On a new Phyllopodous Crustacean.
In Rep. 35th Meeting Brit. Assoc. Adv. Sci., 1865, Trans. of Sec., p. 78.
Discinocaris n. g., *D. browniana* n. sp.
For description and illustration of *Discinocaris*, see Quart. Jour. Geol. Soc., vol. 22, 1866.

―――― On the occurrence of *Ceratiocaris* in the Wenlock limestone of England.
In Geol. Mag., vol. 3, London, 1866, p. 203, pl. 10.
Ceratiocaris papilio Salt., *C. stygius* Salt., *C. inornatus* McCoy, *C. murchisoni* McCoy. *C. leptodactylus* McCoy, *C. robustus* Salt., *C. decorus* Phill., *C. ensis* Salt., *C. cassia* Salt.

―――― Note on some fossil Crustacea and a Chilognathous Myriapod from the Coal Measures of the west of Scotland.
In Trans. Geol. Soc. Glasgow, vol. 2, 1866, p. 234, pl. 3, wood cuts.

―――― On a new genus of Phyllopodous Crustacean from the Moffat shales, Dumfrieshire.
In Quart. Jour. Geol. Soc. London, vol. 22, 1866, p. 503. pl. 25.
Discinocaris n. g. H. W., *D. browniana*. *Peltocaris aptychoides* Salt.
See, also, Rep. 35th Meeting Brit. Assoc. Adv. Sci., 1865, Trans. of Sec., p. 78.

―――― Second report on the structure and classification of the fossil Crustacea.
In Rep. 36th Meeting Brit. Assoc. Adv. Sci., 1866, p. 179.
Eurypterida, Xiphosura.

―――― On some points in the structure of the *Xiphosura*, having reference to their relationship with the *Eurypteridæ*.
In Quart. Jour. Geol. Soc. London, vol. 23, 1867, p. 28, pls. 1, 2.
Belinurus reginæ Baily. *Hemiaspis limuloides* H.W. *Prestwichia* n. g., *P. rotundata* Prest., *P. anthrax* Prest. *Bunodes linula* Eichw. *Pseudoniscus aculeatus* Nieszk. *Exapinurus schrenkii* Nieszk. *Eurypterus remipes* Dekay.

―――― On a new Limuloid Crustacean *(Neolimulus falcatus)*, from the Upper Silurian of Lesmahagow, Lanarkshire.
In Geol. Mag., vol. 5, London, 1868, p. 1, pl. 1.

―――― On a newly discovered long-eyed *Calymene* from the Wenlock limestone, Dudley.
In Geol. Mag., vol. 5, London, 1868, p. 489, pl. 21.
Calymene ceratophthalma n. sp., *C. blumenbachii* Brong. *Asaphus expansus* Linné, *A. kowalewskii* Lawr. *Encrinurus variolaris* Brong. *Ceratocephalus grayianus* White MS.

―――― On some new species of Crustacea from the Upper Silurian rocks of Lanarkshire, etc.; and further observations on the structure of *Pterygotus*.
In Quart. Jour. Geol. Soc. London, vol. 24, 1868, p. 289, pls. 9, 10.
Eurypterus (Pterygotus) punctatus Salt., *E. scorpioides* n. sp., *E. obesus* n. sp. *Pterygotus raniceps* n. sp.

―――― Fourth report on the structure and classification of the fossil Crustacea.
In Rep. 38th Meeting Brit. Assoc. Adv. Sci., 1868, p. 72, pl. 2.
Neolimulus n. g., *N. falcatus* H. W. *Cyclus radialis* Phill., *C. rankini* n. sp.

―――― On the occurrence of *Stylonurus* in the "cornstone" of Herefordshire.
In Rep. 39th Meeting Brit. Assoc. Adv. Sci., 1869, Trans. of Sec., p. 103.

―――― Note on the palpus and other appendages of an *Asaphus* from the Trenton limestone in the British Museum.
In Quart. Jour. Geol. Soc. London, vol. 26, 1870, p. 486, wood cut.

―――― On the remains of a giant Isopod, *Præarcturus gigas*, from the Old Red Sandstone of Rowlestone quarry, Herefordshire.
In Trans. Woolhope Nat. Field Club, 1870, p. 266, 3 pls.

―――― On *Necrogrammarus salweyi*, Lower Ludlow, Leintwardine.
In Trans. Woolhope Nat. Field Club, 1870, p. 271.

―――― On a new species of *Eurypterus (E. brodicei)* from Perton, near Stoke Edith, Herefordshire.
In Quart. Jour. Geol. Soc. London, vol. 27, 1871, p. 261, figs.

―――― On *Eurypterus brodicei*, from Perton, Herefordshire.
In Trans. Woolhope Field Club, 1871, p. 276, 2 pls.

―――― Notes on fossil Crustacea.
In Rep. 40th Meeting Brit. Assoc. Adv. Sci., 1870, Trans. of Sec., p. 91.
Eurypterus, *Ceratiocaris*, *Cyclus*, *Dithyrocaris*.

―――― Contributions to British fossil Crustacea.
In Geol. Mag., vol. 7, London, 1870, p. 554, pl. 22.
Cyclus rankini H. W., *C. radialis* Phill., *C. bilobatus* n. sp., *C. torosus* n. sp., *C. wrightii* n. sp., *C. harknessi* n. sp., *C. (Halicyne) laxus* H. v. Meyer, *C. (Halicyne) agnotus* H. v. Meyer, *C. brongniartianus* DeKoninck, (Hypostome of *Phillipsia ?*). *Cyclus jonesianus* n. sp.

―――― On the structure of Trilobites.
In Geol. Mag., vol. 8, London, 1871, p. 289, pl. 8.
Asaphus platycephalus Stokes.

―――― On a new fossil Crustacean, from the Devonian of Canada.
In Canadian Naturalist, new series, vol. 6, 1871, p. 18.
Dithyrocaris ? belli n. sp.

―――― On some new Phyllopodous Crustaceans from the Palæozoic rocks.
In Geol. Mag., vol. 8, 1871, p. 104, pl. 3.
Extract published in the Canadian Naturalist, new series, vol. 6, 1871, p. 18.
Ceratiocaris ludensis n. sp., *C. oretonensis* n. sp. *Dithyrocaris tenuistriatus* McCoy, *D. belli* H.W. *Ceratiocaris truncatus* n. sp.
Dithyrocaris tenuistriatus is identical with *Avicula paradoxides* DeKoninck.

―――― Fifth report on the structure and classification of the fossil Crustacea, etc.
In Rep. 41st Meeting Brit. Assoc. Adv. Sci., 1871, p. 53; also, Geol. Mag., vol. 8, London, 1871, p. 521.

―――― Note on some British Palæozoic Crustacea belonging to the order Mesostomata.
In Geol. Mag., vol. 9, London, 1872, p. 433, pl. 10.
Hemiaspis limuloides H.W. *Pseudoniscus aculeatus* Nieszk. *Exapinurus schrenkii* Nieszk. *Bunodes lunula* Eichw. *Hemiaspis speratus* Salt. MS., *H. horridus* H.W., *H. salweyi* Salt. *Belinurus königianus* n. sp. *Prestwichia birtwelli* n. sp

—— Further remarks on the relationship of the Xiphosura to the Eurypteridæ, and to the Trilobita, etc.
In Quart. Jour. Geol. Soc. London, vol. 28, 1872, p. 46, figs.; also, Annals Mag. Nat. Hist. London, 4th series, vol. 9, London, 1872, p. 406.
The author illustrates *Trinucleus ornatus* Sternb., *Sao hirsuta* Barr., *Limulus polyphemus* Latr.

—— Sixth report on the structure and classification of the fossil Crustacea, etc.
In Rep. 42d Meeting Brit. Assoc. Adv. Sci., 1872, p. 321; also, Geol. Mag., vol. 9, London, 1872, p. 563.

—— Life forms of the past and present.
In Popular Sci. Review, vol. 11, London, 1873, p. 391, pls. 90, 91.
The author figures: *Pterygotus anglicus* Agas. *Slimonia acuminata* H.W. *Stylonurus logani* H.W. *Eurypterus scouleri* Hibbert. *Hemiaspis limuloides* H.W. *Prestwichia rotundata* H.W. *Belinurus königianus* H.W. *Neolimulus falcatus* H.W. *Prestwichia birtwelli* H.W. *Dithyrocaris scouleri* McCoy. *Trinucleus ornatus* Sternb. *Sao hirsuta* Barr. *Limulus polyphemus* Latr.

—— On a new Trilobite from the Cape of Good Hope.
In Quart. Jour. Geol. Soc. London, vol. 29, 1873, p. 32, pl. 2.
Encrinurus crista-galli n. sp.

—— Seventh report of the committee appointed for the purpose of continuing researches in fossil Crustacea, consisting of Prof. P. Martin Duncan, Henry Woodward and Robert Etheridge.
In 43d Meeting Brit. Assoc. Adv. Sci., 1873, p. 304; also, Geol. Mag., vol. 10, London, 1873, p. 520.

—— and **Etheridge, Jr.** (Robert). On some specimens of *Dithyrocaris* from the Carboniferous limestone series, East Kilbride, and the Old Red Sandstone of Lanarkshire.
In Geol. Mag., vol. 10, London, 1873, p. 482, pl. 16.
Dithyrocaris testudineus Scouler, *D. tricornis* Scouler.

—— and **Etheridge, Jr.** (Robert). On some specimens of *Dithyrocaris* from the Carboniferous limestone series, East Kilbride, and from the Old Red Sandstone of Lanarkshire.

In Rep. Brit. Assoc., 1873, Trans. of Soc., p. 92.
Geol. Mag., decade 2, vol. i, No. 3, 1874, pl. 5.
Dithyrocaris ovalis W. & E., *D. granulata* W. & E., *D. glabra* W. & E., *D.? striata* W. & E., *D. tricornis* Scouler.
See, also, Mem. Geol. Survey, Expl. Sheet 23 (Scotland), 1873, pp. 99-100.

—————— Eighth report on fossil Crustacea.
In Rep. 45th Meeting Brit. Assoc. Adv. Sci., 1875; also, Geol. Mag., new series, decade 2, vol. 2, 1875, p. 620.

—————— Two lectures on recent and fossil Crustacea delivered at the Royal Institution on April 29 and May 6, 1876.
In Proc. Royal Inst. Gr. Brit., 1876; Times, Morning Post, Illustrated News, 1876.

—————— A monograph of the British fossil Crustacea of the order Mesostomata.
In Palæont. Soc. London, pt. 1, 1866; pt. 2, 1869; pt. 3, 1871; pt. 4, 1872; pt. 5, 1877, 36 pls.
Part I, pp. 1-44, pls. 1-9. *Pterygotus anglicus* Agassiz.
Part II, 1869, pp. 45-70, pls. 10-15. *Pterygotus bilobus* Salt., also var. 1 *inornatus*, 2 var. *crassus*, 3 var. *perornatus*, 4 var. *accidens*.
Part III, pp. 71-120, pls. 16-20. *Pterygotus raniceps* H. W., *P. taurinus* Salt., *P. ludensis* Salt., *P. banksii* Salt., *P.? stylops* Salt., *P. arcuatus* Salt., *P. gigas* Salt., *P. problematicus* Salt. *Slimonia acuminata* Salt.
Part IV, pp. 121-180, pls. 21-30. *Stylonurus powriei* Page, *S. megalops* Salt., *S. symondsii* Salt., *S. ensiformis* H. W., *S. scoticus* H. W., *S. logani* H. W. *Eurypterus scouleri* Hibbert, *E. lanceolatus* Salt, *E. pygmæus* Salt., *E. linearis* Salt., *E. abbreviatus* Salt., *E. hibernicus* Baily, *E. brewsteri* H. W., *E. scorpioides* H. W., *E. punctatus* Salt., *E. obesus* H. W., *E. brodiei* H. W., *E.? mammatus* Salt. *Hemiaspis limuloides* H. W., *H. speratus* Salt., *H. horridus* H. W., *H. salweyi* Salt.
Part V, pp. 81-263, pls. 31-36. *Belinurus königianus* H. W., *B. kiltorkensis* Baily, *B. bellurus* König, *B. reginæ* Baily, *B. arcuatus* Baily. *Prestwichia birtwelli* H. W., *P. anthrax* Prest., *P. rotundata* Prest.. *Neolimulus falcatus* H. W. *Cyclus bilobatus* H. W., *C. torosus* H. W., *C. wrightii* H. W., *C. radialis* Phill., *C. harknessi* H. W., *C. jonesianus* H. W., *C. rankini* H. W., *C. (Halicyne) laxus* H. v. Meyer, *C. (H.) agnotus* H. v. Meyer.
Plate XXXII illustrates the Silurian and Carboniferous Crustacea.
The author figures *Agnostus bibullatus* Barr., *A. nudus* Beyr. *Sao hirsuta* Barr. *Trinucleus ornatus* Sternb., *T. fimbriatus* Murch. *Cyclus rankini* H. W., *C. radialis* Phill., *C. harknessi* H. W., *C. bilobatus* H. W., *C. jonesiana* H. W., *C. wrightii* H. W., *C. (Halicyne) laxus* H. v. Meyer, *C. (Halicyne) agnotus* H. v. Meyer.

―――― Third report on the structure and classification of the fossil Crustacea.
In Rep. 37th Meeting Brit. Assoc. Adv. Sci., 1867, p. 44.

―――― Recent and fossil Crustacea; being a lecture delivered before the Geological Association, December 12, 1867.
In Geol. Mag., vol. 5, London, 1868, p. 33; Rep. Geol. Assoc., 1867; notice of lecture.

―――― A catalogue of British fossil Crustacea, with their synonyms and the range in time of each genus and order. Published by order of the Trustees of the British Museum, 1877.
A review of the book was published in the Geol. Mag., decade 2, vol. 4, 1877.

―――― Article "Crustacea."
In Ency. Britannica, vol. 6, 1877, 86 wood. cuts.
The author follows the classification of Audouin of 1821 and places the Trilobites with the Isopoda.

―――― Discovery of the remains of a fossil crab *(Decapoda-Brachyura)* in the Coal Measures of the environs of Mons, Belgium.
In Geol. Mag., decade 2, vol. 5, London, 1878, p. 433, pl. 11.
Brachypyge carbonis n. sp.

―――― On *Necroscilla wilsoni*, a supposed Stomapod Crustacean, from the Middle Coal Measures of Cossall, near Ikeston.
In Quart. Jour. Geol. Soc., London, vol. 35, 1879, p. 551.

―――― Note on the genus *Anthrapalæmon (Palæocarabus)* of Salter from the Coal Measures.
In Geol. Mag., decade 2, vol. 6, 1879, p. 193.

―――― Note on the Palæozoic Crustacea.
In Geol. Mag., decade 2, vol. 6, 1879, p. 196, pl. 5.
Eurypterus scouleri Hibbert.

―――― A new genus of Trilobites *Onycopyge liversidgei*, from South Wales.
In Geol. Mag., decade 2, vol. 7, 1880, p. 97, fig.

────── Note on a new English *Homalonotus* from the Devonian, Torquay, South Devon.
In Geol. Mag., decade 2, vol. 8, London, 1881, p. 489, pl. 13.
Homalonotus champernownei n. sp.

────── Contributions to the study of fossil Crustacea.
In Geol. Mag., decade 2, vol. 8, London, 1881, p. 529, pl. 14.
Eryon perroni n. sp., *Palæocaris burnetti* n. sp., fig., p. 533, pl. 14, fig. 3. *Palæocaris typus* M. & W. *Ananthotelson stimpsoni* M. & W.

────── Additional note on *Homalonotus* from the Devonian.
In Geol. Mag., decade 2, vol. 9, London, 1882, p. 154, pl. 4.
Homalonotus goniopygæus n. sp., *H. champernownei* H. W.

────── Note on *Ellipsocaris dewalquei*, a new Phyllopod from the Upper Devonian of Belgium.
In Geol. Mag., decade 2, vol. 9, London, 1882, p. 444, figs.
The author illustrates the shields of *Peltocaris*, *Discinocaris*, *Apthycopsis* and *Ellipsocaris*.
Ellipsocaris dewalquei n. sp.

────── On a series of Phyllopod Crustacean shields from the Upper Devonian of the Eifel; and on one from the Wenlock shales of Wales.
In Geol. Mag., new series, decade 2, vol. 9, London, 1882, p. 385, pl. 9.
Cardiocaris n. g.   *Pholadocaris* n. g.
1, *Cardiocaris romeri* n. sp.   2, *C. veneris* n. sp.   3, *C. lata* n. sp.   4, *C. bipartita* n. sp.   5, *Pholadocaris leei* n. sp.   6, *Aptychospis salteri* n. sp.

────── Synopsis of the genera and species of Carboniferous limestone Trilobites.
In Geol. Mag., new series, decade 2, vol. 10, London, 1883, pp. 445, 481, 534, pls. 11-13.
Proetus? levis n. sp. *Phillipsia derbyensis* Martin, *P. colei* McCoy, *P. gemmulifera* Phill., *P. truncatula* Phill., *P. eichwaldi* Fisch., *P. eichwaldi* var. *mucronata* McCoy, *P. quadrilimba* Phill., *P. laticaudata* n. sp. *Griffithides seminiferus* Phill., *G. globiceps* Phill., *G. acanthiceps* H. W., *G. longiceps* Portl., *G. platyceps* Portl., *G. obsoletus* Phill., *G. longispinus* Portl., *G. calcaratus* McCoy, *G. moriceps* n. sp.   *Brachymetopus maccoyi* Portl., *B. discors* McCoy, *B. hibernicus* n. sp., *B. ouralicus* DeVern.
Note on the nature of certain pores observed in the cephalon or head shield of some Trilobites.

——— Synonyms of *Phillipsia gemmulifera*.
In Geol. Mag., decade 3, vol. 1, London, 1884, p. 22.
The author states that the first figure of the species was given in Brongniart's Crust. Foss., 1822, p. 145, pl. 4, fig. 12, where it is called *Asaphus*. It was named by Phillips in Geol. Yorkshire, 1836, p. 240, pl. 22, fig. 11, *Asaphus gemmuliferus*. It is not *Tril. pustulatus* Schloth, which is a true *Phacops latifrons* Broun, from the Devonian at Eifel, Germany.

——— Monograph of the British Carboniferous Trilobites.
In Palæont. Soc. London, pt. 1, 1883, pp. 1-38, pls. 1-6. Pt. 2 (conclusion), 1884, pp. 39-83, pls. 7-10.
Phillipsia derbyensis Martin, *P. colei* McCoy, *P. gemmulifera* Phill., *P. truncatula* Phill., *P. eichwaldi* Fisch., *P. eichwaldi* var. *mucronata* McCoy, *P. quadrilimba* Phill.
Griffithides seminiferus Phill., *G. globiceps* Phill., *G. ananthiceps* n. sp., *G. longiceps* Portl., *G. platyceps* Portl., *G. obsoletus* Phill., *G. longispinus* Portl., *G. calcaratus* McCoy, *G. brevispinus* n. sp., *G. moriceps* H.W., *G. glaber* n. sp., *G.? carringtonensis* Ether. MS.
Phillipsia laticaudata n. sp., *P. scabra* n. sp., *P. carinata* Salt. MS. *Brachymetopus ouralicus* DeVern., *B. maccoyi* Portl., *B. discors* McCoy, *B. hibernicus* H.W. *Proetus ? lævis* H.W. *Phillipsia leei* n. sp., *P. æqualis?* H. v. Meyer, *P. minor* n. sp., *P. cliffordi* n. sp., *P. articulosa* n. sp. *Dalmanites ? cuyahoga* Claypole.

——— On the structure of Trilobites.
In Geol. Mag., decade 3, vol. 1, London, 1884, p. 78.
Asaphus.

——— Appendages of Trilobites.
In Geol. Mag., decade 3, vol. 1, London, 1884, p. 162.
Asaphus megistos Locke, 2 figs. *Calymene senaria* Conrad.

——— Note on the remains of Trilobites from South Australia.
In Geol. Mag., decade 3, vol. 1, London, 1884, p. 342.
Dolichometopus tatei n. sp. *Conocephalites australis* n. sp.

——— Synopsis of the genera and species of Carboniferous limestone Trilobites.
In Geol. Mag., decade 3, vol. 1, London, 1884, p. 484, pl. 16.
Phillipsia laticaudata H. W., *P. scabra* H. W., *P. carinata* H. W. *Griffithides brevispinus* H.W., *G. glaber* H.W., *G.? carringtonensis* H.W.

—— On the discovery of Trilobites in the Culm shales of southeast Devonshire.
In Geol. Mag., decade 3, vol. 1, 1884, p. 534.
Phillipsia leei H. W., *P. minor* H. W., *P. cliffordi* H. W., *P. articulosa* H. W.

—— and **Jones** (T. Rupert). Notes on the Phyllodiform Crustacea, referable to the genus *Echinocaris*, from the Palæozoic rocks.
In Geol. Mag., decade 3, vol. 1, London, 1884, p. 393, pl. 13.
Echinocaris wrightiana Dawson, *E. armata* Hall, *E. sublœvis* Whitf., *E. pustulosa* Whitf.

—— On some Palæozoic Phyllopod shields and on Nebalia and its allies.
In Geol. Mag., decade 3, vol. 2, London, 1885, p. 345, pl. 9.

—— Notes on the British species of Ceratiocaris.
In Geol. Mag., decade 3, vol. 2, London, 1885, p. 385, pl. 10.
1, *Ceratiocaris murchisoni* Agas. 2, *C. ludensis* H. W. 3, *C. papilo* Salt. 4, *C. stygia* Salt.

—— Notice of a new limuloid Crustacean from the Devonian, by H. S. Williams.
In Geol. Mag., decade 3, vol. 2, London, 1885, p. 427.

—— On the discovery of Trilobites in the Upper Green (Cambrian) slates of the Penrhyn quarries, Bethesda, near Bangor, North Wales.
In Quart. Jour. Geol. Soc. London, vol. 44, 1888, p. 74, pl. 6.
Conocoryphe viola n. sp.

—— On the discovery of Trilobites in the Upper Green (Cambrian) slates, etc., North Wales.
In Rep. 57th Meeting Brit. Assoc. Adv. Sci., 1887, p. 696.
Conocoryphe viola H.W.

—— On a new species of *Eurypterus* from the Lower Carboniferous shale, Eskdale, Scotland.
In Rep. 57th Meeting Brit. Assoc. Adv. Sci., 1887, p. 696.
See Jones (T. Rupert) and Woodward (Henry).
See Etheridge (Robert) and Woodward (Henry).

Worthen (A. H.) Description of two new species of Crustacea, fifty-one species of mollusca, and three species of Crinoids from the Carboniferous formation of Illinois and adjacent States.
In Bull. No. 2, State Mus. Nat. Hist. Illinois, 1884.
Colpocaris chesterensis n. sp. *Solenocaris st. ludovici* n. sp.
See also Meek (F. B.) and Worthen (A. H.)

—— Geology and Palæontology of Illinois, vol. 7, Springfield, 1890, pp. 728. Appendix of 152 pp., 78 pls.
Solenocaris st. ludovici Worthen. *Colpocaris chesterensis* Worthen.

Wyatt-Edgell (H. G.) On the genera of Trilobites *Asaphus* and *Ogygia*, and the subgenus *Ptychopyge*.
In Geol. Mag., vol. 4, 1867, p. 14.
The author illustrates the following species: *Asaphus (Ptychopyge) corndensis* Murch. *Ogygia scutatrix* Salt. *Asaphus tyrannus* Murch. *Ogygia peltata* Salt.
The English species of these genera are arranged as follows:
Asaphus (Basilicus) tyrannus Murch., *A. (B.) peltastes* Salt., *A. (B.) powisii* Murch., *A. (B.) marstoni* Salt., *A. (B.) laticostatus* McCoy, *A. (B.) radiatus* Salt.
Asaphus (Ptychopyge) corndensis Murch.
Asaphus (Isotelus) gigas Dekay, *A. (I.) affinis* McCoy, *A. (I.) homfrayi* Salt.
Asaphus (Brachaspis) rectifrons Salt.
Asaphus (Cryptonymus) scutalis Salt.
Ogygia buchii Brong., *O.* var. *convexa*, *O.* var. *angustissima*, *O. scutatrix* Salt., *O. peltata* Salt., *O. selwynii* Salt., *O. ? hydrida* Salt.

Young (J.) Descriptive notes of several new and rare forms of Entomostraca.
In Trans. Geol. Soc. Glasgow, vol. 5, pt. 2, 1877.

—— On new forms of Crustacea from the Silurian rocks at Girvan.
In Proc. Nat. Hist. Soc. Glasgow, vol. 1, pt. 1, 1868, p. 169; also in Proc. Royal Phys. Soc. Edinburgh, vol. 4, 1878, p. 167, pl. 1.
Solenocaris n. g. (not *Solenocaris* Meek), *S. solenoides* n. sp.

Zenker (J. C.) Beiträge zur Naturgeschichte der Urwelt. Organische Reste (Petrefacten) aus der Altenburger Braunkohlen-Formation, dem Blankenburger

Quadersandstein, jenaischen bunten Sandstein und böhmischen Uebergangsgebirge, mit 6 illuminirten Kupfertafeln. Jena, 1833.

Olenus longicaudatus n. sp., *O. pyramidalis* n. sp., *O. latus* n. sp. *Otarion* n. g., *O. diffractum* n. sp., *O.? squarrosum* n. sp. *Conocephalus* n. g., *C. costatus* n. sp. *Ellipsocephalus* n. g., *E. ambiguus* n. sp.

The genus *Otarion* was made up from the head of *Cyphaspis burmeisteri* Barr., connected with the pygidium of *Cromus beaumonti* Barr. The second species, described as *Otarion? squarrosum*, is referred by Barrande to *Cheirurus quenstedti*.

Zenker uses two generic names for his new genus *Conocephalus*, the first on p. 48 in describing the genus and the second in his description of plate 5, where he calls it *Trigonocephalus*.

Both these generic terms had been used in natural history for other genera, the first by Thurnburg in 1812 for a genus of the Orthoptera, and the second by Oppel in 1811 for the Reptiles

The typical species described as *Conocephalus costatus* on p. 49, and also in the explanation of pl. 5, figs. G–K, p. 51, as *Trigonocephalus costatus* has been referred by Barrande to an amended genus, *Conocephalites* (Zenker) Barr., and the species to the older name of *sulzeri*. The other new genus, *Ellipsocephalus*, has been used by all subsequent authors, the species taking the older name of *Ellipsocephalus hoffi* Schloth.

Zeno (Franz). Von Seeversteinerungen und Fossilien, welche bei Prag zu finden sind.

In Neue phys. Belustigungen, etc., vol. 1, 1770, p. 65. Continued same vol., p. 362, pl. 1, fig. 1.

Under the names of *Cacada* or *Käfermuschel* Zeno describes the pygidium of *Dalmanites hausmanni* and the head of *Phacops latifrons* Bronn.

Zittel (Karl A.) Handbuch der Palæontologie. München und Leipzig, 1881–85.

The Crustacean are described in I, Band. II, Abtheilung. III, Lieferung, 1885.

The class is arranged in the following order:

A. Entomostraca. 1, Cirripedia. 2, Copepoda. 3, Ostracoda. 4, Phyllopoda. 5, Trilobitae.

B. Merostomata. 6, Xiphosura. 7, Gigantostraca.

C. Malacostraca. 8, Phyllocarida. 9, Isopoda. 10, Amphipoda. 11, Stomatopoda. 12, Cumacea. 13, Schizopoda. 14, Decapoda.

The author gives a diagonis of each genus, with numerous illustrations; also, tabulated stratigraphic distribution of the genera.

Zippe (F. X.) See **Sternberg** (Kasper Graf von).

PART II.

CATALOGUE OF TRILOBITES.

For the geological horizons the author has used the term "Cambrian" for the original Cambrian of Sedgwick, 1836 (Lower Silurian, Murchison). The term "Taconic" is used for the zone with the Olenellus, Paradoxides, Olenus and Dicellocephalus faunæ. "Silurian" is used for Murchison's Upper Silurian.

A SYSTEMATIC CLASSIFICATION OF THE GENERA.

AGNOSTIDÆ.

Agnostus Brongniart, 1822. Hist. Crust. Foss., p. 38. Type, *Agnostus pisiformis* Linné. Taconic. Cambrian.
    Sec. 1. Parvifrontes. Type, *Agnostus brevifrons* Ang.
    Sec. 2. Longifrontes. Type, *A. punctuosus* Ang.
    Sec. 3. Limbati. Type, *A. rex* Barr.
    (B) Fallaces. Type, *A. fallax* Linrs.
    Sec. 4. Lævigati. Type, *A. lævigatus* Dalm.
    Sec. 5. Arthrorhachis. Type, *A. tardus* Barr.
Shumardia Billings, 1862. Pal. Foss., vol. 1, p. 92. Type, *Shumardia granulosa* Bill. Taconic.
Microdiscus Emmons, 1856. Am. Geology, vol. 1, pt. 2, p. 116. Type, *M. speciosus* Ford. Taconic.

TRINUCLEIDÆ.

Trinucleus Lhlwyd, 1698 (Cryptolithus Green, 1832). Murchison's Silurian Syst., 1839, p. 659. Type, *Trinucleus fimbriatus* Murch.
    Cambrian.
Ampyx Dalman, 1826. Palæad., p. 53. Type, *Ampyx nasutus* Dalm.
    Cambrian. Silurian.
    Sec. 1. Brevifrontes. Type, *Ampyx nudus* Murch.
    Sec. 2. Longifrontes. Type, *A. nasutus* Dalm.
    Sec. 3. Lonchodomas. Type, *L. domatus* Ang.
Dionidæ Barrande, 1847. Neues Jahrb. für Mineral., 1847, pt. 4, p. 391;
  • Syst. Sil. Bohême, vol. 1, p. 540. Type, *Dionidæ formosa* Barr.
    Cambrian.
    1. Salteria Wyv. Thompson, 1864. Mem. Geol. Survey United Kingdom, decade xi, pl. 7, p. 1. Type, *Salteria primæva* Wyv. Thom.
    2. Endyminia Billings, 1862. Pal. Foss., vol. 1, p. 93. Type, *Endyminia meeki* Bill. Cambrian.

OLENELLIDÆ.
Olenellus Hall, 1861. Fifteenth Rep. New York State Cab. Nat. Hist., p. 114. Type, *Olenellus thompsoni* Hall. Twelfth Rep. New York State Cab. Nat. Hist., p. 59, figure. Taconic.

PARADOXIDÆ.
Paradoxides Brongniart, 1822. Hist. Crust. Foss., p. 30. Type, *Paradoxides tessini* Brong. Taconic.
Anopolenus Salter, 1864. Geol. Soc. London Jour., vol. 20, p. 236. Type, *Anopolenus henrici* Salt. Taconic.
Bathynotus Hall, 1860. Thirteenth Rep. New York State Cab. Nat. Hist., p. 117. Type, *Bathynotus holopyga* Hall. Taconic.
Dicellocephalus Owen, 1852. Rep. Geol. Survey Wisconsin, Iowa and Minnesota, p. 573. Type, *Dikelocephalus minnesotensis* Owen. Taconic.
Dolichometopus Angelin, 1854. Palæont. Scand., p. 72. Type, *Dolichometopus svecicus* Aug. Palæont. Scand., p. 73, pl. 37, fig. 9. Taconic.
Hydrocephalus Barrande, 1846. Notice Prélim. Syst. Sil. Bohême, p. 18. Type, *Hydrocephalus carens* Barr. Taconic.

REMOPLEURIDÆ.
Remopleurides Portlock, 1843. Geol. Rep. Londonderry, etc., p. 254. Type, *Remopleurides colbii* Portl.

OLENIDÆ.
Protolenus Matthew, 1892. Bull. Nat. Hist. Soc. N. B., No. 10, p. 2. Type, *P. elegans* Matt. Taconic.
Olenus Dalman, 1826. Palæad., p. 56. Type, *Olenus gibbosus* Wahl. Taconic.
Zecanthoides Walcott, 1887. Am. Jour. Sci., 3d series, vol. 36, p. 165. Type, *Olenoides spinosus* Wal. Taconic.
Parabolina Salter, 1849. Mem. Geol. Survey United Kingdom, decade 2. Type, *Olenus spinulosa* Wahl. Taconic.
Peltura Milne-Edwards, 1840. Hist. Nat. Crust., vol. 3, p. 344. Type, *Olenus scarabæoides* Wahl. Taconic.
Cyclognathus Linnarsson, 1875. Geol. Föreningens Stockholm Förhandl., vol. 2, No. 12, p. 500. Type, *Cyclognathus micropygus* Linrs. Taconic.
Triarthrus Green, 1832. Mon. Trilobites North America, p. 87. Type, *Triarthrus beckii* Green. Cambrian.
Leptoplastus Angelin, 1854. Palæont. Scand., p. 46. Type, *Leptoplastus stenotus* Ang. Taconic.
 1. Sphærophthalmus Angelin, 1854. Palæont. Scand., p. 49. Type, *Sphærophthalmus flagellifer* Aug. Taconic.
 2. Ctenopyge Linnarsson, 1880. Geol. Föreningens Stockholm Förhandl., vol. 5, p. 145. Type, *Olenus (Sphærophthalmus) pecten* Salt. Taconic.

Bœckia Brögger, 1882. Die silurischen Etagen 2 und 3, p. 122. Type,
    Bœckia hirsuta Brögg. Taconic.
Ceratopyge Corda, 1847. Prodr., p. 161. Type, Olenus forficula Sars.
    Taconic.

CONOCORYPHIDÆ.

Atops Emmons, 1844. Taconic System, p. 20. Type, Trilobite sulzeri
    Schloth. Taconic.
Ptychoparia Corda, 1847. Prodr., p. 26. Type, Conocephalus striatus
    Emm. Taconic. Cambrian.
Ctenocephalus Corda, 1847. Prodr., p. 26. Type, Ctenocephalus barrandi
    Cord. Taconic.
Solenopleura Angelin, 1854. Palæont. Scand., p. 26. Type, Solenopleura
    holometopa Ang. Taconic.
Liostracus Angelin, 1854. Palæont. Scand., p. 27. Type, Liostracus
    aculeatus Ang. Taconic.
Aneuacanthus Angelin, 1852. Palæont. Scand., p. 5. Type, Aneuacanthus acutangulus Ang.
Anomocare Angelin, 1854. Palæont. Scand., p. 24. Type, Anomocare
    aculeatus Ang. Taconic.
Angelina Salter, 1847. Mem. Geol. Survey United Kingdom, decade xi.
    Type, Angelina sedgwicki Salt. Taconic.
Agraulos Corda, 1847. Prodr., p. 26. Agraulus certicephalus Barr.
    Taconic.
Olenoides Meek, 1877. U. S. Geol. Expl. 40th Par., vol. 4, p. 25. Type,
    Olenoides nevadensis Meek. Taconic.
Bathyurus Billings, 1859. Canadian Naturalist, vol. 4, p. 364. Type,
    Asaphus extans Hall. Cambrian.
Bathyurella Billings, 1865. Pal. Foss., vol. 1, p. 262. B. marginatus,
    B. nitidus, B. expansus Bill. Taconic. Cambrian.
Chariocephalus Hall, 1863. Sixteenth Rep. New York State Cab. Nat.
    Hist., p. 175. Type, Chariocephalus whitfieldi Hall. Taconic.
Corynexochus Angelin, 1854. Palæont. Scand., p. 59. Type, Corynexochus
    spinulosa Ang. Taconic.
Ellipsocephalus Zenker, 1833. Beiträge Nat. Urwelt, p. 31. Type, Ellipsocephalus ambiguus Zenker=Ellipsocephalus hoffi Schloth. Taconic.
Ptychaspis Hall, 1863. Sixteenth Rep. New York State Cab. Nat. Hist.,
    p. 170. Type, Dikelocephalus miniscœnsis Owen. Taconic.
Sao Barrande, 1846. Notice Prélim. Syst. Sil. Bohême, p. 13. Type,
    Sao hirsuta Barr. Taconic.

BOHEMILLIDÆ.

Bohemilla Barrande, 1872. Syst. Sil. Bohême, vol. 1, Suppl., p. 137.
    Type, Bohemilla stupenda Barr. Cambrian.

CALYMENIDÆ.

Calymene Brongniart, 1822. Crust. Foss., p. 11. Type, Calymene tuberculata Brünn. Cambrian. Silurian.

Homalonotus Koenig, 1825.   Icones Foss. Sectiles, p. 7.   Type, *Homa-*
lonotus knighti Koenig.   Cambrian.   Silurian.   Devonian.
Bavarilla Barrande, 1868.   Faune Sil. des environs de Hof en Bavière,
p. 75.   Type, *Bavarilla hofensis* Barr.   Taconic.

ASAPHIDÆ.

Asaphus Brongniart, 1822.   Crust. Foss., p. 17.   Type, *Asaphus expansus*
Linné.   Cambrian.
Ptychopyge Angelin, 1854.   Palæont. Scand., p. 51.   Type, *Ptychopyge*
appanata Ang.   Cambrian.
Brachyaspis Salter, 1866.   Mon. Brit. Trilobites, p. 167.   Type, *Isotelus*
rectifrons Portl. and *Asaphus lævigatus* Ang.   Cambrian.
Megalaspis Angelin, 1852.   Palæont. Scand., p. 15.   Type, *Megalaspis*
gigas Ang.   Cambrian.
Ogygia Brongniart, 1822.   Crust. Foss., p. 26.   Type, *Asaphus buchi*
Broug.   Cambrian.
Barrandia McCoy, 1849.   Annals Nat. Hist., 2d series, vol. 4, p. 409.
Type, *Barrandia cordai* McCoy.   Cambrian.
Niobe Angelin, 1852.   Palæont. Scand., p. 13.   Type, *Asaphus frontalis*
Dalm.   Cambrian.
Asaphellus Callaway, 1877.   Quart. Jour. Geol. Soc. London, vol. 33,
p. 663.   Type, *Asaphellus homfrayi* Salt.   Taconic.
Nileus Dalman, 1826.   Palæad., p. 49.   Type, *Nileus armadillo* Dalm.
                                                      Cambrian.

STYGINIDÆ.

Styginia Salter, 1853.   Rep. 22d Meeting Brit. Assoc. Adv. Sci., p. 59.
Type, *Asaphus latifrons* Portl.   Cambrian.

ILLÆNIDÆ.

Illænus Dalman, 1826.   Palæad., p. 51.   Type, *Illænus crassicauda* Wahl.;
Nova Acta Reg. Soc. Sci. Upsal., vol. 8, p. 27, pl. 2, figs. 3, 5, 6.
(Not pl. 7, figs. 5, 6=*Illænus esmarki*).   Cambrian.   Silurian.
Dysplanus Burmeister, 1843.   Organization Trilobites, p. 105 (Ray Soc.
Ed., 1846).   Type, *Illænus centrotus* Dalm.   Cambrian.
Bumastus Murchison, 1839.   Sil. Syst., p. 656.   Type, *Bumastus barrien-*
sis Murch.   Cambrian.   Silurian.
Panderia Volborth, 1863.   Mem. Acad. St. Petersburg, 7th series, vol. 6,
No. 2, p. 31.   Type, *Panderia triquetra* Volb.   Cambrian.

ÆGLINIDÆ.

Æglina Barrande, 1852.   Syst. Sil. Bohême, vol. 1, p. 663.   Type, *Æglina*
rediviva Barr.   Cambrian.

ILLÆNURIDÆ.

Illænurus Hall, 1863.   Sixteenth Rep. New York State Cab. Nat. Hist.,
p. 176.   Type, *Illænurus quadratus* Hall.   Taconic.

CATALOGUE OF TRILOBITES. 257

BRONTIDÆ.
Bronteus Goldfuss, 1843. Beiträge Petrefact., Nova Acta Physico-Med., vol. 19, p. 360. Type, *Brontes flabellifer* Goldf.
Cambrian. Silurian. Devonian.

DALMANITIDÆ.
Phacops Emmrich, 1839. De Trilob., etc., p. 18. Type, *Phacops latifrons* Bronn. Cambrian. Silurian. Devonian.
Dalmanites Emmrich, 1844. Zur Nat. Trilobiten, etc., p. 15. Barrande's Syst. Sil. Bohême, vol. suppl., p. 27. Type, *Dalmanites caudata* Brünn. Cambrian. Silurian.
Coronura Hall, 1888. Pal. New. York, vol. 7, p. xxxii. Type, *Coronura aspectens* Con. Devonian.
Cryphæus Green, 1837. Am. Jour. Sci., 1st series, vol. 32, p. 343. Type, *Cryphæus boothi* Green. Devonian.
Chasmops McCoy, 1849. Annals Nat. Hist., 2d series, vol. 4; also Contrib. to Palæont., p. 143. Type, *Calymene odini* Eichw. Cambrian.

CERAURIDÆ.
Ceraurus Green, 1832 (*Cheirurus* Beyrich, 1845). Mon. Trilobites N. A., p. 83. Type, *Ceraurus pleurexanthemus* Green.
Cambrian. Silurian. Devonian.
Sphærexochus Beyrich, 1845. Ueber einige böhm. Trilobiten, p. 19. Type, *Sphærexochus mirus* Beyr. Cambrian.
Areia Barrande, 1872. Syst. Sil. Bohême, vol. 1, suppl., p. 96. Type, *Areia bohemica* Barr. Cambrian.
Deiphon Barrande, 1852. Syst. Sil. Bohême, vol. 1, p. 814. Type, *Deiphon forbesi* Barr. Silurian.
Placoparia Corda, 1847. Prodr., p. 128. Type, *Placoparia zippi* Boeck. Cambrian.
Staurocephalus Barrande, 1852. Syst. Sil. Bohême, vol. 1, p. 810. Type, *Staurocephalus murchisoni* Barr. Cambrian. Silurian.
Onycopyge Woodward, 1880. Geol. Mag., decade 2, vol. 7, p. 97. Type, *Onycopyge liversidgei* H. W. Cambrian.

AMPHIONIDÆ.
Amphion Pander, 1850. Beiträge Geog. Russischen Reiches, p. 439. Type, *Amphion fischeri* Eichw. Cambrian.
Cromus Barrande, 1852. Syst. Sil. Bohême, vol. 1, p. 821. Type, *Cromus intercostatus* Barr. Silurian.

ENCRINURIDÆ.
Encrinurus Emmrich, 1844 (Cryptonymus Eichwald). Zur Nat. Trilobiten, p. 16. Type, *Trilobus punctatus* Brünn. Cambrian. Silurian.
Cybele Loven, 1845. Svenska Akad. Förhandl., 1845, p. 110. Type, *Calymene bellatula* Dalm. Cambrian.
Dindymene Corda, 1847. Prodr., p. 119. Type, *Dindymene frederico-augusti* Cord. Cambrian.

ACIDASPIDÆ.

Acidaspis Murchison, 1839 (Odontopleura Emmrich, 1839). Silurian System, p. 558.   Type, *Acidaspis brightii* Murch.
            Cambrian. Silurian. Devonian.

LICHASIDÆ.

Lichas Dalman, 1826. Palæad., p. 53. Type, *Lichas laciniatus* Wahl.
             Cambrian. Silurian.
Terataspis Hall, 1863. 16th Rep. New York State Cab. Nat. Hist., p. 223. Type, *Lichas (Terataspis) grandis* Hall.   Devonian.

PROETIDÆ.

Proetus Steininger, 1831. Mém. Soc. Géol. France, vol. 1, p. 355. Type, *Proetus curieri* Stein.
       Cambrian. Silurian. Devonian. Carboniferous.
Dechenella Kayser, 1880. Zeitschr. Deutsch. Geol. Gesell., Berlin, vol. 32, p. 703. Type, *Eonia verticalis* Burm.   Devonian.
Prionopeltis Corda, 1847 (Phæton Barr., Phæthonides Angelin). Prodr., p. 121. Type, *Prionopeltes polydorus* Cord. Devonian.
Xiphogominm Corda, 1847. Prodr., p. 70. Type *(X. sieberianum* Cord.) *Proetus loveni* Barr.   Silurian.
Celmus Angelin, 1852. Palæont. Scand., p. 23. Type, *C. granulatus* Ang.   Cambrian.
Brachymetopus McCoy, 1847. Annals Mag. Nat. Hist., 1st series, London, vol. 20, p. 229. Type, *Phillipsia maccoyi* Portl. Carboniferous.
Cyphaspis Burmeister, 1843. Org. Trilobiten (Ray Soc. Ed., p. 98). *Phacops ceratophthalma* Goldf. Cambrian. Silurian. Devonian.
Arethusina Barrande, 1852. Syst. Sil. Bohême, vol. 1, p. 494. *Arethusina konincki* Barr.   Cambrian. Silurian. Devonian.
Harpides Beyrich, 1846. Untersuch. Trilobiten, p. 34. Type, *Harpides hospes* Beyr.   Taconic. Cambrian.
Carmon Barrande, 1872. Syst. Sil. Bohême, vol. 1, suppl., p. 19. Type, *Carmon multilus* Barr.   Cambrian.
Cyphoniscus Salter, 1852. Rep. 22d Meeting Brit. Assoc. Adv. Sci., Trans. Sec., p. 95. Type, *Cyphoniscus socialis* Salt. Cambrian.

PHILLIPSIDÆ.

Phillipsinella Novák, 1885. Sitzungsber. k. böhm. Gesell. Wiss., 1885. Author's edition, p. 4. Type, *Phillipsia parabola* Barr.
                   Cambrian.
Phillipsia Portlock, 1843. Geol. Rep. Londonderry, etc., p. 305. Type, *Phillipsia kelli* Portl. = *Asaphus gemmulifera* Phill.
              Carboniferous. Permian.
Griffithides Portlock, 1843. Geol. Rep. Londonderry, etc., p. 310. Type, *Griffithides longiceps* Portl.   Carboniferous.

HARPESIDÆ.

Harpes Goldfuss, 1839. Nova Acta Physico-Med., vol. 19, p. 358. Type, *Harpes ungula* Sternb.   Cambrian. Silurian. Devonian.

A CATALOGUE OF NORTH AMERICAN SPECIES OF TRILOBITES.

Acantholoma Con., 1840; 3d Ann. Rep. Pal. Dept. New York Geol. Survey, p. 205. (See *Acidaspis tuberculatus*.)
—— *spinosa* Con. (See *Acidaspis tuberculatus*.)
ACIDASPIS Murch., Silurian System, 1839, p. 658. Cambrian and Silurian.
  Used as a subgroup by Clarke. Notes on the genus *Acidaspis*, 1892, p. 9; for species with a single large straight median spine on the occipital ring, *Acidaspis brightii* Murch.
Ceratocephala Ward., 1838; Am. Jour. Sci., 1st series, vol. 34, p. 377.
  Used in the generic sense by Clarke. Notes on the genus *Acidaspis*, 1892, p. 9; for species with two straight divergent spines on the occipital ring, *Ceratocephala goniata* Ward.
Odontopleura Emm., 1839; De Tril., p. 53.
  Used as a subgenus by Clarke. Notes on the genus *Acidaspis*, 1892, p. 9; for species having the occipital ring smooth or with central tubercle, *Odontopleura ovata* Emm.
Arges Goldf. (in part), Neues Jahrb. für Mineral, 1843, pp. 544, 556.
Ceraurus (Green) Lovén, 1845, Kong. Svenska Vet. Akad., p. 46.
Polyeris Rouault, 1846; Bull. Soc. Géol. France, vol. 4, p. 320.
Selenopeltis Cord, 1847; Prodr., p. 34.
  Used as a subgenus by Clarke. Notes on the genus *Acidaspis*, 1892, p. 9; for species with confluent glabella lobes, oblique thoracic, pleural and spineless pygidium, *Selenopeltis buchi* Barr.
Trapelocera Cord., 1847; Prodr., p. 158.
Dicranurus Con., 1841; 5th Ann. Rep. Pal. Dept. New York Geol. Survey, p. 48.
  Used as a subgenus by Clarke. Notes on the genus *Acidaspis*, 1892, p. 9; for species with two recurved spines of great size on the occipital ring, *Dicranurus hamatus* Conrad.
Ancyropyge n. subgen. Clarke, 1892. Note on the genus *Acidaspis*, p. 9.
  Used as a subgenus to *Ceratocephala*, type *Acidaspis roemingeri* Hall.
—— ANCHORALIS Miller, 1875; Cincinnati Quart. Jour. Sci., vol. 2, p. 349, figs 2-4. cf. *Acidaspis ceralepta*. Hudson.
Ceratocephala (Acidaspis) anchoralis Clarke, 1892. Note on *Acidaspis*, pp. 11 and 12, syn. to *Ceratocephala ceralepta*.
—— *armatus*. (See *Acidaspis eriops*.)
—— CALLICERA Hall, 1888; Pal. New York, vol. 7, pp. 69, 224, pl. 16 b, figs. 1-3. Clarksville, N. Y. Upper Helderberg.
Ceratocephala (Odontopleura) CALLICERA Clarke, 1892. Note on *Acidaspis*, p. 13.
Ceratocephala CERALEPTA Anth., 1838; Am. Jour. Sci., 1st series, vol. 34, p. 379; figs. 1-2. Cincinnati, Ohio. Hudson.
—— CERALEPTA Meek, 1873; Pal. Ohio, vol. 1, p. 169, pl. 14, figs. 8-9.

ACIDASPIS.
—— CERALEPTA Miller, 1874; Cincinnati Quart. Jour. Sci., vol. 1, p. 130.
Syn., *Acidaspis anchoralis* Miller.
Ceratocephala (Acidaspis) CERALEPTA Clarke, 1892. Note on *Acidaspis*, p. 12.
—— CINCINNATIENSIS Meek, 1873; Pal. Ohio, vol. 1, p. 167, pl. 14, fig. 3. Cincinnati, Ohio. Hudson.
—— CINCINNATIENSIS Miller, 1874; Cincinnati Quart. Jour. Sci., vol. 1, p. 130.
Ceraurus CROSOTUS Locke, 1842; Am. Jour. Sci., 1st series, vol. 44, p. 347, fig. Cincinnati, Ohio. Hudson.
—— CROSOTUS James, 1871; Catalogue Fossils Cincinnati Group, p. 14.
—— CROSOTUS Meek, 1873; Pal. Ohio, vol. 1, p. 165, pl. 14, figs. 10 a–b.
—— CROSOTUS Miller, 1874; Cincinnati Quart. Jour. Sci., vol. 1, p. 129.
Ceratocephala (Odontopleura) CROSSOTA Clarke, 1892. Note on *Acidaspis*, p. 13.
—— *danai* Hall, 1862; Geol. Wisconsin, vol. 1, p. 432. Niagara.
Acidaspis ida (Winchell and Marcy) Hall; 20th Rep. New York State Cab. Nat. Hist., p. 389.
—— *Danai* Hall, 1867; 20th Rep. New York State Cab. Nat. His., p. 333, pl. 21, figs. 8, 9. Bridgeport, Ill.
—— *Danai* Hall, 1870; 20th Rep. New York State Cab. Nat. Hist., p. 423, pl. 21, figs. 8, 9 (rev. ed.) This species with *A. ida* are synonymous to *Acidaspis goniata*.
—— (*Terataspis*) ERIOPS Hall, 1876; Illus. Devonian Foss. Crust., pl. 19, figs. 4–7, 10, 11, 12. (See *Lichas (Conolichas) eriops* Hall.) Schoharie, N. Y. Upper Helderberg.
—— FIMBRIATA Hall, 1879; Description of New Species from the Niagara Group, p. 20. Waldron, Ind. Niagara.
—— FIMBRIATA Hall, 1882; Trans. Albany Inst., vol. 10, p. 76.
—— FIMBRIATA Hall, 1883; 11th Rep. Geol. Nat. Hist. Indiana, p. 334, pl. 33, fig. 11.
—— (*Terataspis*) GRANDIS Hall, 1876; Illus. Devonian Foss. Crust., pl. 17, figs. 1–8; pl. 18, figs. 1–4. (See *Lichas (Terataspis) grandis* Hall.) Schoharie, N. Y.
Ceratocephala goniata Ward., 1838; Am. Jour. Sci., 1st series, vol. 34, p. 377.
Ceratocephala goniata Clarke, 1892; Note on *Acidaspis*, p. 12. Niagara. The original species was from Springfield, Ohio; Clarke places as syn. *A. danai* Hall and *A. ida* from Bridgeport, Ill.
—— HALLI Shum., 1855; 1st and 2d Rep. Geol. Survey Missouri, pt. 2, p. 200, pl. B, figs. 7 a, b, c. Cape Girardeau County, Missouri.
Silurian.
Ceratocephala (Odontopleura) HALLI Clarke, 1892. Note on *Acidaspis*, p. 13.

CATALOGUE OF TRILOBITES. 261

ACIDASPIS.
Dicranurus HAMATA Con., 1841; 5th Ann. Rep. Pal. Dept. New York Geol. Survey, p. 39, pl. 1, fig. 1. Albany and Schoharie counties.
Lower Helderberg.
—— (*Dicranurus*) HAMATA Hall, 1862; 15th Rep. New York State Cab. Nat. Hist., pl. 2, fig. 1. Republication of pl. 2, 5th Ann. Rep. Pal. Dept. New York Geol. Sur.
—— HAMATA Hall, 1859; Pal. New York, vol. 3, p. 371, pl. 79, figs. 15-19.
—— HAMATA M. & W., 1868; Geol. Survey Illinois, vol. 3, p. 390, pl. 7, fig. 17.
Ceratocephala (Dicranurus) HAMATUS Clarke, 1892. Note on *Acidaspis*, p. 13.
—— HORANI Bill., 1863; Geol. Sur. Canada, p. 190, fig. 190. Bay of St.' Paul, St. Lawrence River and Canada. Trenton.
—— *ida* W. & M., 1865; Mem. Boston Soc. Nat. Hist., vol. 1, p. 106, pl. 3, fig. 13. See *Acidaspis goniata*.
—— O'NEALLI Miller, 1875; Cincinnati Quart. Jour., vol. 2, p. 86, fig. 9. Near Lebannon, Ohio. Hudson.
Ceratocephala (Odontopleura) O'NEALLI Clarke, 1892. Note on *Acidaspis*, p. 13.
—— ORTONI Foerste 1887; Bull. Denison Univ., vol. 2, p. 90, pl. 8, fig. 1. Near New Carlisle, Ohio. Clinton.
Ceratocephala (Odontopleura) ORTONI Clarke, 1892. Note on *Acidaspis*, p. 13.
—— PARVULA Wal., 1877; Advanced sheets 31st Rep. New York State Mus., p. 16. Trenton Falls, N. Y. Trenton.
—— PARVULA Wal., 1879; 31st Rep. New York State Mus., p. 69.
Ceratocephala (Odontopleura) PARVULA Clarke, 1892. Note on *Acidaspis*, p. 13.
—— ROMINGERI Hall, 1888; Pal. New York, vol. 7, pp. 71, 224, pl. 16 b, figs. 15-18. Little Traverse Bay, Mich. Used for the type of the subgenus *Ancyropyge* Clarke, 1892. Note on *Acidaspis*, p. 10.
Hamilton.
—— *spiniger*. (See *Bathyurus spiniger*.)
—— TRENTONENSIS Hall, 1847; Pal. New York, vol. 1, p. 240, pl. 64, figs 4 a-f. Bay of Quinta on Lake Ontario. Trenton.
—— TRENTONENSIS Emmons, 1855; American Geology, vol. 1, p. 216, fig. 73.
Ceratocephala (Odontopleura) TRENTONENSIS Clarke, 1892. Note on *Acidaspis*, p. 13.
—— TUBERCULATUS Con., 1840; 3d Ann. Rep. Pal. Dept. New York Geol. Survey, p. 205. Lower Helderberg.
Syn., *Acantholoma* Con., 1840; 3d Ann. Rep. Pal. Dept. New York Geol. Survey, p. 205, fig. 3.
Acantholoma spinosa Con., 4th Ann. Rep. Pal. Dept. New York Geol. Survey, p. 39, pl. 1, fig. 3.

ACIDASPIS. AGNOSTUS.

—— TUBERCULATUS Hall, 1859; Pal. New York, vol. 3, p. 368, pl. 79, figs. 1-14. Albany and Schoharie counties, N. Y.
—— TUBERCULATUS Hall, 1861; 15th Rep. New York State Cab. Nat. His., pl. 2, fig. 3.

Ceratocephala (Acidaspis) TUBERCULATUS Clarke, 1892. Note on *Acidaspis*, p. 13.

AGLASPIS Hall, 1862; Canadian Naturalist, vol. 7, p. 445, figure. Hall, 1863; 16th Rep. New York State Cab. Nat. Hist., p. 181.

—— BARRANDI Hall, 1863; 16th Rep. New York State Cab. Nat Hist., p. 181, pl. 11, figs. 7-16. Miniska, Minn., and Mazomania, Wis.
Taconic.

—— EATONI Whit., 1880; Ann. Rep. Geol. Survey, Wisconsin, p. 52.
——- EATONI Whit., 1882; Geol. Wisconsin, vol. 4, p. 192, pl. 10, fig. 11. Lodi, Wis. Taconic.

Arctinurus De Castelnau, 1843; Sil. Syst. de l'Amérique, p. 21. (See *Lichas boltoni*.)

AGNOSTUS Brong., 1822; Hist. Nat. Crust. Foss., p. 38. Taconic. Cambrian.
Syn., *Battus* Dalm., 1826; Palæad., p. 33.
  Trinodus McCoy, 1846; Sil. Foss. Ireland, p. 56.
  Phalacroma Cord. (in part), 1847; Prodr., p. 158.
  Mesospheniscus Cord., 1847; Prodr., p. 46 (*Agnostus integer*).
  Diplorrhina Cord., 1847; Prodr., p. 46.
  Condylopyge Cord., 1847; Prodr., p. 50 (*Agnostus rex*).
  Lejopyge Cord., 1847; Prodr., p. 51 (*Agnostus lævigatus*).
  Arthrorhachis Cord., 1847; Prodr., p. 114 (*Agnostus tardus*).
  Peronopsis Cord., 1847; Prodr., p. 115 (*Agnostus integer*).
  Pleuroctenium Cord., 1847; Prodr., p. 116 (*Agnostus granulatus*).

Sec. 1. PARVIFRONTES. The glabella is only partly developed in this section. Type *Agnostus brevifrons* Ang.

Sec. II. LONGIFRONTES. This section is distinguished by its strongly projecting glabella and also by the axis of the pygidium, which is generally moderately long. Crust smooth. The cheeks grooved. The crust on the cheeks and pygidium is provided with raised points. Limb generally narrow. The cheeks in front of the glabella and side lobes of the pygidium, behind the axis, divided by a groove. Type *Agnostus punctuosus* Aug., *A. pisiformis* Linné.

Sec. III. LIMBATI. General form subquadrate, head has a broad limb, basal lobes large. The cheeks in front of the glabella are not divided by a central groove or grooved at the sides. The pygidium is usually produced into lateral spines. Type *Agnostus rex* Barr.

Series A (REGII). Distinguished by its broad limb, diminishing cheeks and side lobes of the pygidium. Both the anterior lobe of the glabella and the posterior lobe of the pygidium expand. Type *Agnostus rex* Barr.

AGNOSTUS.

Series B (FALLACES). This series has a smaller head, and moderately broad limb. Cheeks large; basal lobes rather large with a broad posterior lobe to the axis of the pygidium. Type *Agnostus fallax* Linrs.

Sec. IV. LÆVIGATI. The dorsal grooves marking the glabella and axis of the thorax and pygidium are wanting, or faintly indicated. Crust smooth, sometimes with slight indications of striæ. Limb on the head disappearing; on the pygidium it becomes broad. Type, *Agnostus lævigatus* Dalm.

Sec. V. ARTHRORHACHIS. The glabella is prominent, long and marked with small basal lobes; the axis of the pygidium is short, as in *Agnostus tardus* Barr.

—— ACADICUS Hartt, 1868; Acadian Geology, p. 655, fig. 229 (3d ed.) St. John, N. B.   Taconic.

Syn., *similis* Hartt, 1868; Acadian Geology, p. 656, (3d. ed.)
similis Hartt (Walcott), 1884; Bull. U. S. Geol. Survey, No. 10, p. 22.

—— ACADICUS Wal., 1884; Bull. U. S. Geol. Survey, No. 10, p. 22, pl. 2, figs. 2, 2 a, c.

—— ACADICUS Vogd., 1892; American Geol., vol. 4, p. 386, pl. 9, fig. 7. Sec. Limbati.

—— ACADICUS var. DECLIVIS Matt., 1885; Tran. Royal. Soc. Canada, vol. 3, p. 70, pl. 7, figs. 6 a-b.   St. John.

—— ACADITUS var. DECLIVIS, Vogd., 1892; American Geol., vol. 4, p. 387, pl. 9, fig. 8. Sec. Limbati.

—— ACUTILOBUS Matt., 1885; Trans. Royal Soc. Canada, vol. 3, p. 73, pl. 7, fig. 10.   St. John.

—— ACUTILOBUS Vogd., 1892; American Geol., vol. 4, p. 384, pl. 9, fig. 1. Sec. Longifrontes.

—— AMERICANUS Bill., 1860; Canadian Naturalist, vol. 5, p. 302, fig. 1. Point Levis, Quebec.   Taconic.

—— AMERICANUS Bill., 1865; Pal. Foss., vol. 1, p. 395, fig. 372.

—— AMERICANUS Brögg., 1886; Geol. Fören. Stock. Förhandl., vol. 8, p. 207. cf. *Agnostus punctuosus* Ang. var.

—— AMERICANUS Vogd., 1892; American Geol., vol. 4, p. 389, pl. 9, fig. 5. Sec. Longifrontes.

—— BISECTUS Matt., 1891; Trans. Royal Soc. Canada, vol. 9, p. 50, pl. 13, figs. 2 a b. Navy Island, St. John, N. B.   Taconic.

—— BIDENS Meek, 1873; 6th Ann. Rep. U. S. Geol. Survey Territories, p, 463. Gallatin City, Montana.   Taconic.

—— BIDENS Wal., 1884; Pal. Eureka Dist., Mon. U. S. Geol. Survey, vol. 8, p. 26, pl. 9, figs. 13, 13 a.

—— BIDENS Brögg., 1866; Geol. Fören. Stock. Förhandl., vol. 8, p. 207. cf. *Agnostus parvifrons* Linrs.

AGNOSTUS.
—— bidens Vogd., 1892; American Geol., vol. 4, p. 392, pl. 10, fig. 5.
Sec. Limbati.
—— canadensis Bill., 1860; Canadian Naturalist, vol. 5, p. 304, fig. 3.
Point Levis, Quebec. Taconic.
—— canadensis Bill.; 1865; Pal. Foss., vol. 1, p. 397, fig. 374.
—— canadensis Brögg., 1886; Geol. Fören. Stock. Förhandl., vol. 8, p. 207. cf. *Agnostus hybridus* Brögg.
—— canadensis Vogd., 1892; American Geol., vol. 4, p. 390, pl. 9, fig. 9. Sec. Longifrontes.
—— coloradoensis Shum., 1861; Am. Jour. Sci., 2d series, vol. 32, p. 218. Burnet County, Texas. Taconic.
—— coloradoensis Vogd., 1892; American Geol., vol. 4, p. 390, pl. 9, fig. 16. cf. *Agnostus neon*.
—— communis H. & W., 1877; U. S. Geol. Expl. 40th Par., vol. 4, p. 228, pl. 1, figs. 28, 29. White Pine, Nevada. Taconic.
—— communis Wal., 1884; Pal. Eureka Dist., Mon. U. S. Geol. Survey, vol. 8, p. 27.
—— communis Brögg., 1866; Geol. Fören. Stock. Förhandl., vol. 8, p. 207. cf. *Agnostus cyclopyge* Tullb. Olenus zone.
—— communis Vogd., 1892; American Geol., vol. 4, p. 390, pl. 9, fig. 15. Sec. Longifrontes.
—— desideratus Wal., 1889; Proc. Natl. Mus., vol. 12, p. 39. Salem, N. Y. Taconic.
—— desideratus Wal., 1891; 10th Rep. U. S. Geol. Survey, p. 629, pl. 80, fig. 5. Olenellus zone.
—— desideratus Vogd., 1892; American Geol., vol. 4, p. 383, pl. 10, fig. 7. Sec. Parvifrontes.
—— disparlis Hall, 1863; 16th Rep. New York State Cab. Nat. Hist., p. 179, pl. 10, figs. 25-27. Oseola Mills, Wis. Taconic.
—— disparlis Vogd., 1892; American Geol., vol. 4, p. 394, pl. 10, fig. 15. Sec. Lævigati.
—— fabius Bill., 1865; Pal. Foss., vol. 1, p. 298, fig. 289. Table Head and Pistolet Bay, N. F. Cambrian?
—— fabius Vogd., 1892; American Geol., vol. 4, p. 396, pl. 9, fig. 10. Sec. Arthrorhachis.
—— glabba Bill., 1865; Pal. Foss., vol. 1, p. 297, fig. 288. Table Head and Pistolet Bay, N. F. Cambrian?
—— glabba Vogd., 1892; American Geol., vol. 4, p. 395, pl. 9, fig. 6. Sec. Arthrorhachis.
—— interstrictus White, 1874; Prelim. Rept. Invert. Foss., p. 7. Antelope Springs, Utah. Taconic.
—— interstrictus White, 1877; U. S. Geog. and Geol. Survey W. 100th Meridian, p. 38, pl. 2, figs. 5 a, 5 b.
—— interstrictus Wal., 1886, Bull. U. S. Geol. Survey, No. 30, p. 149 pl. 16, figs. 6, 6 a.

AGNOSTUS.
— INTERSTRICTUS Wal., 1888; Am. Jour. Sci., 3d series, vol. 36, p. 166.
— INTERSTRICTUS Brögg., 1866; Geol. Fören. Stock. Förhandl., vol. 8, p. 207 cf. *Agnostus fallax* Linrs. Paradoxides zone.
— INTERTRICTUS Vogd., 1892; American Geol., p. 393, pl. 10, fig. 6. Sec. Limbati.
— sp.? Rominger, 1887; Proc. Acad. Nat. Sci. Phila., p. 18.
— JOSEPHA Hall, 1863; 16th Rep. New York State Cab. Nat. Hist., p. 178, pl. 6, figs. 54, 55. Lake Pepin, Wis. Taconic.
— JOSEPHA Vogd., 1892; American Geol., vol. 4, p. 391, pl. 9, fig. 17. Sec. Longifrontes.
— *lobatus* Hall. (See *Microdiscus lobatus*.)
— MALADENSIS Meek, 1873; 6th Annual Rep. U. S. Geol. Survey Territories, p. 464. Malade City, Utah. Taconic.
— *maladensis* Vogd., 1892; American Geol., vol. 4, p. 392. This species is a syn. of *Agnostus josepha* Hall.
— *neon* H. & W., 1877; U. S. Geol. Expl. 40th Par., vol. 4, p. 229, pl. 1, figs. 26, 27. Eureka, Nevada. Taconic.
— *neon* Wal., 1884; Pal. Eureka Dist., Mon. U. S. Geol. Survey, vol. 8, p. 27. See *Agnostus coloradoensis*.
— *nobilis* Ford, 1872; Am. Jour. Sci., 3d series, vol. 3, p. 421, figs. 1, 2. Troy, N. Y. Taconic.
— *nobilis* Wal., 1886; Bull. U S. Geol. Survey, No. 30, p. 150, pl. 16, fig. 7.
— ? *nobilis* Wal., 1891; 10th Rep. U. S. Geol. Survey, p. 629, pl. 80, fig. 7, 7 a. Olenellus zone.
— *nobilis* Vogd., 1892; American Geol., vol. 4, p. 383. Probably a *Microdiscus*.
— OBTUSILOBUS Matt., 1888; Trans. Roy. Soc. Canada, vol. 3, p. 72, pl. 7 fig. 9. Porter's and Hanford brooks, St. Martins.
— OBTUSILOBUS Vogd., 1892; American Geologist, vol. 4, p. 385, pl. 9, fig. 3. Sec. Longifrontes. St. John. Paradoxides zone.
— ORION Bill., 1860; Canadian Naturalist, vol. 5, p. 304, fig. 2. Point Levis, Quebec. Taconic.
— ORION Bill., 1865; Pal. Foss., vol. 1, p. 397, fig. 393.
— ORION Shum., 1863; Trans. Acad. Sci. St. Louis, vol. 2, p. 105.
— ORION Vogd., 1892; American Geol., vol. 4, p. 391, pl. 9, fig. 12. Sec. Longifrontes. This species differs from *A. pisiformis* by the absence of the median node, which may be due to its state of preservation. Syn. of *A. pisiformis* Linné.
— PABILIS Hall, 1863; 16th Rep. New York State Cab. Nat. Hist., p. 179, pl. 11, figs. 23, 24. Lake Pepin, Wis. Taconic.
— PABILIS Brögg., 1886; Geol. Fören. Stock. Förhandl, vol. 8, p. 207. cf. *Agnostus laevigatus* Dalm. and *A. nudus* Beyr. Paradoxides zone.

AGNOSTUS.
— PARILIS Vogd., 1892; American Geol., vol. 4, p. 394, pl. 10, fig. 4. Sec. Lævigati.
— PARTITUS Matt., 1888; Trans. Roy. Soc. Canada, vol. 3, p. 68, pl. 7, fig. 2 a–b.
— PISIFORMIS (Linné) Matt., 1891; Trans. Roy. Soc. Canada, vol. 9, p. 59, pl. 13, fig. 1 a–b. (Variety.) St. John.
— PROLONGUS H. & W., 1877; U. S. Geol. Expl. 40th Par., vol. 4, p. 230, pl. 1, figs. 30, 31. Eureka, Nevada. Taconic.
— PROLONGUS Wal., 1884; Pal. Eureka Dist., Mon. U. S. Geol. Survey, vol. 8, p. 28.
— PROLONGUS Brögg., 1886; Geol. Fören Stock. Förhandl, vol. 8, p. 207. cf. *Agnostus nudus* Beyr. var.
— PROLONGUS Vogd., 1892; American Geol., vol. 4, p. 394. pl. 10, fig. 10. Sec. Lævigati.
— REGULARIS Matt., 1888; Trans. Roy. Soc. Canada, vol. 3, p. 67, pl. 7, fig. 1 a–c. Portland. Hanford and Porter's brooks, etc.
— REGULARIS Vogd., 1892; American Geol., vol. 4, p. 387, pl. 10, fig. 11. St. John. Sec. Limbati.
— *richmondensis* Wal., 1884; Pal. Eureka Dist., Mon. U. S. Geol. Survey, vol. 8, p. 24, pl. 9, fig. 10. Prospect Mountain, Nevada.
Syn. AGNOSTUS AMERICANUS Vogd., 1892; American Geol., vol. 4, p. 392, pl. 9, fig. 11.
— RICHMONDENSIS Brögg., 1886; Geol. Fören Stock. Förhandl, vol. 8, p. 207. cf. *Agn. nathorsti* Brögg.
— SECLURUS Wal., 1884; Pal. Eureka Dist., Mon. U. S. Geol. Survey, vol. 8, p. 25, pl. 9, fig. 14. Secret Cañon, Nevada. Taconic.
— SECLURUS Brögg., 1886; Geol. Fören Stock. Förhandl, vol. 8, p. 207. cf. *Agnostus parvifrons* Linrs. Paradoxides zone.
— SECLURUS Vogd., 1892; American Geol., vol. 4, p. 395, pl. 10, fig. 16.
— TESSELLA Matt., 1888; Trans. Roy. Soc. Canada, vol. 3, p. 71, pl. 7, figs. 7, 7 a–b. Porter's brook, St. Martins.
— TESSELLA Vogd., 1892; American Geol., vol. 4, p. 385, pl. 9, fig. 4. *Agnostus tessella* is a Brevifrons with an indistinct anterior lobe to the glabella.
— UMBO Matt., 1888; Trans. Roy. Soc. Canada, vol. 3, p. 71, pl. 7, fig. 8, 8 a–b. Porter's brook.
— UMBO Vogd., 1892; American Geol., vol. 4, p. 387, pl. 10, fig. 9. Sec. Parvifrontes.
— VIR Matt., 1888; Trans. Roy. Soc. Canada, vol. 3, p. 69, pl. 7, fig. 3. Portland and Hanford brook, etc.
— also, var. CONCINNUS, p. 70, pl. 7, figs. 4, 4 a–c. Taconic.
— VIR Vogd., 1892; American Geol., vol. 4, p. 388, pl. 10, fig. 14. Sec. Limbati.
— also, var. CONCINNUS, p. 389, pl. 9, fig. 13.
— *similis* Hartt. (See *Agnostus acadicus*.)

AGNOSTUS. AGRAULOS.
—— TUMIDOSUS H. & W., 1877; U. S. Geol. Expl. 40th Par., vol. 4, p. 201, pl. 1, fig. 32. Eureka, Nevada. Taconic.
—— TUMIDOSUS Brögg., 1886; Geol. Fören Stock. Förhandl, vol. 8, p. 207. cf. *Agnostus quadratus* Tullb. Paradoxides zone.
—— TUMIDOSUS Vogd., 1892; American Geol., vol. 4, p. 393, pl. 10, fig. 8. Sec. Limbati.

AGRAULOS Cord., 1847; Prodr., p. 23. Taconic.
Syn., *Arion* Barr., 1846; Notice Prélim., p. 13.
Ellipsocephalus Barr., 1846; Notice Prélim., p. 12 (not *Ellipsocephalus* Zenk. or *Elliptocephalus* Emm.)
Arionellus Barr., 1852; Syst. Sci. Bohême, p. 404.
—— AFFINIS Bill., 1874; Pal. Foss., vol. 2, p. 72. Branch St. Mary's Bay, N. F. Taconic.
—— AFFINIS Matt., 1886; Trans. Royal Soc. Canada, vol. 2, p. 153, figs. 2, 2 a-b.
—— ? *articephalus* Matt., 1885; Trans. Roy. Soc. Canada, vol. 3, p. 75, pl. 7, figs. 14, a-b. Hanford Brook, St. Martin's. St. John Group. (See *Ellipsocephalus anticephalus*.)
Arionellus BIPUNCTATUS Shum., 1863; Trans. Acad. Sci. St. Louis, vol. 2, p. 101. Lawrence Creek, Minnesota. Taconic.
Arionellus BIPUNCTATUS Hall., 1863; 16th Rep. New York State Cab. Nat. Hist., p. 169, pl. 7, figs. 50-57.
Arionellus (*Agraulos*) CONVEXUS Whitf., 1878; Ann. Rep. Geol. Survey Wisconsin, 1877, p. 57. Ironton, Sauk county, Wis. Taconic.
Arionellus CONVEXUS Whitf., 1882; Geol. Wisconsin, vol. 4, p. 190, pl. 1, fig. 17.
Arionellus cylindricus Bill., 1860; Canadian Naturalist, vol. 5, p. 314, fig. 14. Point Levis, Quebec.
Arionellus cylindricus Bill., 1865; Pal. Foss., vol. 1, p. 406, fig. 385. (See *Ptychaspis cylindricus*.)
—— ? GLOBOSUS Wal., 1884; Pal. Eureka Dist., Mon. U. S. Geol. Survey, vol. 8, p. 61, pl. 9, fig. 23. Eureka, Nevada. Taconic.
—— (*Strenuella* ?) HALLIANUS Matt., 1887; Trans. Roy. Soc. Canada, vol. 5, p. 132, pl. 1, fig. 2. Portland, Simonds and St. Martin's. St. John Group.
—— ? HOLOCEPHALUS Matt., 1890; Trans. Roy. Soc. Canada, vol. 8, p, 138, pl. 11, fig. 5, 5 a-d. Torryburn Cove, Kennebecasis. St. John Group.
—— OWENI Meek and Hayden, 1861; Proc. Acad. Nat. Sci. Phila., 1861, p. 436. Head of Powder River, Big Horn Mts. The species was also referred to the genera *Arionellus* and *Crepicephalus*.
Taconic.
Arionellus OWENI Meek and Hayden, 1862; Am. Jour. Sci., 2d series, vol. 33, p. 74.

AGRAULOS.
Arionellus oweni Meek and Hayden; Pal. Upper Missouri, p. 9, figs. a, b, c, pl. 1, fig. 4. Referred by Walcott, 1884, Pal. Eureka Dist., Mon. U. S. Geol. Survey, vol. 8, p. 55, pl. 10, to *Ptychoparia oweni*.

Arionellus (*Bathyurus*) planus Shum., 1861;. Am. Jour. Sci., 2d series, vol. 32, p. 219. Morgan's Creek, Burnet county, Texas.
Taconic.

Arionellus pustulatus Wal., 1877; Advanced sheets 31st Rep. New York State Mus. Nat. Hist., p. 15. Chazy, N. Y.

Arionellus pustulatus Wal., 1877; 31st Rep. New York State Mus. Nat. Hist., p. 68.

Arionellus quadrangularis Whit., 1884; Bull. Mus. Nat. Hist, New York, vol. 1, p. 147, pl. 14, fig. 8. Braintree, Mass. Taconic.

—— quadrangularis Wal., 1884; Bull. U. S. Geol. Survey, No. 10, p. 48, pl. 7, fig. 1.

—— redpathi Wal., 1891; 10th Rep. U. S. Geol. Survey, p. 654, fig. 69. St. Simonds, Quebec. Taconic.

—— socialis Bill., 1874; Pal. Foss., vol. 2, p. 71 fig. 40. Chapel Arm., Trinity Bay, N. F. Taconic.

—— socialis Bill., 1882; Rep. of Progress Geol. Survey Newfoundland, p. 10, fig. 4.

—— socialis Matt., 1886; Trans. Royal Soc. Canada. vol. 2, p. 151, fig. 1 a-b.

—— strenuus Bill., 1874; Pal. Foss., vol. 2, p. 71, fig. 41. Topsail Head and Brigus, Conception Bay. Taconic.

—— strenuus Bill., 1882; Rep. of Progress Geol. Survey of Newfoundland, p. 10, fig. 5.

—— strenuus Matt., 1886; Trans. Royal Soc. Canada, vol. 2, p. 153, figs. 3, 3 a-b.

—— strenuus Wal., 1891; 10th Rep. U. S. Survey, p. 655, pl. 96, figs. 1 a-c; also, var. *nasutus*, p. 654, pl. 97, figs. 2, 2 a-c. Olenellus zone. Walcott refers *Ptychoparia mucronatus* S. & F. to this species. Matthew uses the species for the type of the subgenus *Strenuella*, Trans. Roy. Soc. Canada, 1886, p. 154.

—— *subclavatus* Bill. (See *Ptychaspis subclavatus*.)

Arionellus (*Bathyurus*) texanus Shum., 1861; Am. Jour. Sci., 2d series, vol. 32, p. 218. Clear Creek, Burnet County, Texas.
Taconic.

Arionellus tripunctatus Whitf., 1876; Rep. Recon. Upper Missouri, p. 141, pl. 1, figs. 3-5. Moss Agate Springs near Camp Baker, Montana. Taconic.

—-? whitfieldianus Matt., 1887; Trans. Roy. Soc. Canada, vol. 5, p. 130, pl. 1, figs. 1, 1 a-f.

CATALOGUE OF TRILOBITES. 269

AGRAULOS. AMPYX.
—— ? WHITFIELDIANUS Matt., 1890; Trans. Roy. Soc. Canada, vol. 8, p. 138, pl. 11, figs. 6, 6 a-e? Div. 1, c. 1. St. John group.
—— (*Bathyurus?*) WOOSTERI Whitf., 1878; Ann. Rep. Geol. Survey Wisconsin, 1877, p. 56. Ettrick and Eau Claire, Wis. Taconic.
—— (*Bathyurus?*) WOOSTERI Whitf., 1882; Geol. Wisconsin, vol. 4, p. 189, pl. 1, figs. 19-21.

AMPHION Pand., 1830; Beiträge zur Geol. russischen Reiches, p. 139. Cambrian.
—— BARRANDI Bill., 1865; Pal. Foss., vol. 1, p. 288, fig. 277. Point Rich, Table Head, etc., N. F.
—— CANADENSIS Bill., 1859; Can. Nat., vol. 4, p. 381, fig. 12, a, b. Chazy Limestone, Mingan Islands.
—— CANADENSIS Bill., 1863; Geol. Survey Canada, p. 133, fig. 69.
—— CANADENSIS Bill., 1865; Pal. Foss., vol. 1, p. 288, fig. 278.
—— CONVEXUS Bill., 1865; Pal. Foss., vol. 1, p. 322. Stanbridge, range 6, lot 20.
—— INSULARIS Bill., 1865; Pal. Foss., vol. 1, p. 290. Port aux Choix, N. F.
—— JULIUS Bill., 1865; Pal. Foss., vol. 1, p. 290, fig. 279. Cow Head, N. F.
—— MATUTINA Hall, 1863; 16th Rep. New York State Cab. Nat. Hist., p. 222, pl. 5 a, fig. 6. Trempaleau, Wis. Taconic.
—— *multisegmentatus*. (See *Encrinurus multisegmentatus*.)
—— NEVADENSIS Wal., 1884; Pal. Eureka Dist., Mon. U. S. Geol. Survey, vol. 8, p. 94, pl. 12, fig. 13. Bellevue Peak, etc., Eureka District, Nevada. Cambrian.
—— SALTERI Bill., 1861; Canadian Naturalist, vol. 6, p. 322, fig. 6. Township of Oxford, Canada. Calciferous.
—— SALTERI Bill., 1865; Pal. Foss., vol. 1, p. 352, fig. 309.
—— WESTONI Bill., 1865; Pal. Foss., vol. 1, p. 321, fig. 307. Stanbridge, range 6, lot 20.

AMPYX Dalm., 1826; Palæad., p. 53 (German edition.) Cambrian. Silurian.
—— AMERICANUS Saff. and Vogd., 1889; Proc. Phila. Acad. Nat. Sci., p. 166, fig. Near Bull's Gap on the road to Russelville, Tenn. Trenton.
—— AMERICANUS Vogd., 1893; American Geol., vol. 11, p. 106, fig. 4.
—— HALLI Bill., 1861; Pal. Foss., vol. 1, p. 24, figs. 25, a, b, c.
—— HALLI Bill., 1862; Geol. Survey Vermont, vol. 2, p. 959, fig. 365. Chazy Limestone. St. Dominique, Canada, and Highgate Springs, Vt.
—— HALLI Vogd., 1893; American Geol., vol. 11, p. 106, fig. 5.
—— LÆVIUSCULUS Bill., 1865; Pal. Foss., vol. 1, p. 295, fig. 285. Table Head, N. F.
—— NORMALLIS Bill., 1865; Pal. Foss., vol. 1, p. 295, fig. 286. Table Head and Pistolet Bay, N. F.
—— NORMALLIS Vogd., 1893; American Geol., vol. 11, p. 107, fig. 6.

AMPYX. ANOMOCARE.

—— RUSTILIUS Bill., 1865; Pal. Foss., vol. 1, p. 296.   4 miles NE. from Portland Creek, N. F.
—— RUSTILIUS Vogd., 1893; American Geol., vol. 11, p. 108.
—— SEMICOSTATUS Bill., 1865; Pal. Foss., vol. 1, p. 297, fig. 287. Table Head and Pistolet Bay, 4 miles NE. from Portland Creek, N. F. This species differs so slightly from *A. rustilius*, which occurs in the same bed, that the author hints that they should be united under *A. rustilius*.
—— SEMICOSTATUS Vogd., 1893; American Geol., vol. 11, p. 108, fig. 6.
ANGELINA Salt., 1864; Mem. Geol. Survey United Kingdom, decade xi, pl. 7. Type, *Angelina sedgwicki* Salt.   Taconic.
—— hitchcocki Whitf., 1884; Bull. Am. Mus. Nat. Hist. New York, vol. 1, p. 148, pl. 14, fig. 13. (See *Protypus hitchcocki*.) Braintree, Mass.
ANOMOCARE Ang., 1852; Palæont. Scand., p. 24.   Taconic.
ANOMOCARE CHIPPEWAENSIS.

Lonchocephalus CHIPPEWAENSIS Owen, 1852; Rep. Geol. Survey Wisconsin, Iowa and Minnesota, p. 576, pl. 8, figs. 6, 14; pl. 1 a, fig. 9. Red Cedar or Menomonie branch of the Chippewa.

Conocephalites CHIPPEWAENSIS (Owen) Shumard, 1863; Trans. St. Louis Acad. Sci., vol. 2, p. 104.

Lonchocephus HAMULUS Owen, 1852; Geol. Survey Wisconsin, Iowa and Minnesota, p. 576, pl. 1 a, figs. 8–12. Near the mouth of the Miniskah River at Mountain Island.

—— undet. Trilobite, Owen, 1848; Geol. Reconnoissance Chippewa Land Dist., p. 15, pl. 7, fig. 5.

Conocephalites HAMULUS Shum., 1863; Acad. Sci. St. Louis, vol. 2, p. 104.

Conocephalites HAMULUS Hall, 1863; 16th Rep. New York State Cab. Nat. Hist., p. 166, pl. 7, figs. 43, 44; pl. 8, figs. 25, 26.

—— HAMULUS Dames, 1884; China (Richthofen), vol. 4, p. 24.

Conocephalites PATERSONI Hall, 1863; 16th Rep. New York State Cab. Nat. Hist., p. 159, pl. 7, figs. 45, 46. Trempleau, Wis. Dicellocephalus zone.

—— PATERSONI Dames, 1884; China (Richthofen), vol. 4, p. 24.

—— ? PARVUM Wal., 1884; Pal. Eureka Dist., Mon. U. S. Geol. Survey, vol. 8, p. 59, pl. 9, fig. 17. West slope of Prospect Mountain, Eureka District, Nev. Olenellus zone.

—— ? PARVUM Wal., 1886; Bull. U. S. Geol. Survey, No. 30, p. 209, pl. 25, fig. 1.

—— ? PARVUM Wal., 1890; 10th Rep. U. S. Geol. Survey, p. 653, pl. 96, fig. 2. Prof. Matthew in a letter to the author says: "The long prominent glabella and volutted frontal limb, with furrows between it and the glabella, are characters of *Strenuella;* also, the prominent eye lobes and strong occipital ring."

ANOMOCARE. ASAPHISCUS.
—— SPINIGER Matt., 1891; Trans. Royal Soc. Canada, vol. 9, p. 61, pl. 13, fig. 4 a-e. Long Island, Kennebecasis River. Referred to *Leptoplastus spiniger*, Can. Rec. Sci., October, 1889.
—— STENOTOIDES Matt., 1891; Trans. Royal Soc. Canada, vol. 9, p. 61, pl. 13, figs. 3 a-d. Referred to *Leptoplastus* in Can. Rec. Sci., Oct., 1889. Long Island, Kennebecasis River. St. John group.
Crepicephalus? WISCONSENSIS Owen, 1852; Rep. Geol. Survey Wisconsin, Iowa and Minnesota, pl. 1, fig. 13. (The upper figure only; the other figured specimen belongs to *Dicellocephalus granulosus*.)
Crepicephalus WISCONSENSIS Shum., 1863; Trans. Acad. Sci. St. Louis, vol. 2, p. 103.
Conocephalites WISCONSENSIS Hall, 1863; 16th Rep. New York State Cab. Nat. Hist., p. 164, pl. 7, figs. 39-41; pl. 8, figs. 22, 23, 24, 27, 28. Trempleau, Wis., mouth of Chippewa River in Minnesota. Syn., *Dikelocephalus latifrons* (Shum.) Hall, 1863; 16th Rep. New York State Cab. Nat. Hist., p. 165, pl. 7, fig. 40.
—— WISCONSENSIS Dames, 1884; China (Richtofen), vol. 4, p. 24. Dicellocephalus zone.
ANOPOLENUS Salt., 1864; Geol. Mag., vol. 1, p. 236.
—— VENUSTUS Bill., 1874; Pal. Foss., vol. 2, p. 73, fig. 42. Chapel Arm, Trinity Bay, N. F. Taconic.
—— VENUSTUS Bill., 1882; Report of Progress Geol. Newfoundand, p. 11, fig. 6. Paradoxides zone.
ARETHUSINA Barr., 1852; Syst. Sil. Bohême, vol. 1, p. 493.
Cambrian. Silurian. Devonian.
—— AMERICANA Wal., 1884; Pal. Eureka Dist., Mon. U. S. Geol. Survey, vol. 8, p. 62, pl. 9, fig. 27. Eureka District, Nevada.
Upper Cambrian of Walcott's classification.
Arges. (See *Lichas*.)
ARGES, subgenus *Lichas* Hall, 1888; Pal. New York, vol. 7, p. lx.
—— *phlyctanodes* Green. (See *Lichas phlyctanodes*.)
Arionellus. (See *Agraulos*.)
ASAPHISCUS Meek, 1873; 6th Ann. Rep. U. S. Geol. Survey Territories, p. 485 (foot-note).
Bathyurellus (*Asaphiscus*) BRADLEYI Meek, 1873; 6th Ann. Rep. U. S. Geol. Survey Territories, p. 484. Malade City, Utah. Taconic.
—— BRADLEYI Meek, 1873; 6th Ann. Rep. U. S. Geol. Survey, Territories, p. 485.
—— WHEELERI Meek, 1873; 6th Rep. U. S. Geol. Survey Territories p. 485 (foot-note). Antelope Springs, Utah. Taconic.
—— WHEELERI White, 1875; Geog. and Geol. Survey West 100th Meridian, vol. 4, p. 43, pl. 2, figs. 1 a-f.
—— WHEELERI Wal., 1886; Bull. U. S. Geol. Survey, No. 30, p. 220, pl. 31, figs. 3, 3 a.

Asaphus Broug., 1822; Hist. Nat. Crust. Foss., p. 18, pl. 2, fig. 1 a, b.
Cambrian.
Syn., *Cryptonymus* Eich., 1825; (in part) not *Cryptonymus* Eichw., 1840.
Isotelus Dekay, 1822; Annals Lyceum Nat. Hist. New York, vol. 1, p. 174.
Hemicrypturus Green, 1832; Mon. Tril. North America, p. 20.
—— *acantholeurus*. (See *Dalmanites acantholeurus*.)
—— ALACER Bill., 1866; Catalogue Sil. Foss. Anticosti, p. 26, fig. 9.
Charleton Point. Hudson.
—— *aspectans*. (See *Dalmanites aspectans*.)
—— ASTRAGALOTES Green, 1834; Am. Jour. Sci., 1st series, vol. 25, p. 335.
—— ASTRAGALOTES Green, 1835; Suppl. Mon. Tril. N. A., p. 11, cast 37.
Probably *Phacops rana* Green.
—— *barrandi*. (See *Ogygia barrandi*.)
—— CANADENSIS Chap., 1856; Canadian Jour., vol. 1, p. 482. Georgean Bay, Ontario. Utica Slate.
—— CANADENSIS Chap., 1858; Canadian Jour., vol. 2, p. 47.
—— CANADENSIS Chap., 1858; Canadian Jour., vol. 3, p. 230, fig.
—— CANADENSIS Chap., 1859; Canadian Jour., vol. 4, p. 1, fig.
—— CANADENSIS Chap., 1859; Annals. Nat. Hist., 3d series, vol. 2, p. 9, fig. 1.
—— CANADENSIS Bill., 1863; Geol. Survey Canada, p. 204, fig. 201.
Isotelus CANALIS Conrad MSS. Chazy.
Isotelus CANALIS Hall, 1847; Pal. New York, vol. 1, p. 25, pl. 4 *bis*, figs. 17–19.
—— CANALIS Bill., 1865; Pal. Foss., vol. 1, p. 255; also p. 352, fig. 340 (not Conrad sp.)
—— CANALIS Bill., 1862; Geol. Vermont, vol. 1, p. 299, pl. 12, fig. 5.
This species is probably *Asaphus platycephalus*.
—— CANALIS Whitf., 1886; Bull. Am. Mus. Nat. Hist. New York, vol. 1, p. 336, pl. 34, figs. 1–8. Fort Cassin, Vermont. Calciferous.
—— CANALIS Whitf., 1889; Bull. Am. Mus. Nat. Hist. New York, vol. 2, p. 64, pls. 11, 12.
—— CARIBOUENSIS Wal., 1884; Pal. Eureka Dist., Mon. U. S. Geol. Survey, vol. 8, pl. 98; pl. 12, figs. 7-7 a, b. Caribou Hill, Eureka Dist., Nevada. Pogonip.
—— *caudatus*. (See *Dalmanites limulurus*.)
—— *cordieri*. (See *Dalmanites limulurus*.)
—— *corycæus*. (See *Proetus corycæus*.)
—— *crypturus*. (See *Homalonotus crypturus*.)
—— CURIOSUS Bill., 1865; Pal. Foss., vol. 1, p. 318, fig. 305. Stanbridge, range 6, lot 20.
—— ? CURIOSUS Wal., 1884; Pal. Eureka Dist., Mon. U. S. Geol. Survey, vol. 8, p. 98, pl. 12, fig. 15. Caribou Hill, Eureka District, Nev.

CATALOGUE OF TRILOBITES. 273

ASAPHUS.
—— *denticulatus.* (See *Dalmanites denticulatus.*)
—— *dilmarsiæ* Honeyman, 1879; Proc. Nova Scotia Inst. Nat. Sci., vol. 5, p. 18. (See *Homalonotus crypturus* Green.)
—— *diurus* Green, 1839; Am. Jour. Sci., 1st series, vol. 37, p. 40. (See *Dalmanites diurus.*)
—— *edwardi.* (See *Dalmanites limulurus.*)
—— EMORYI Hall, 1857; Rep. U. S. Mexican Boundary Survey, vol. 1, pl. 20, fig. 5. Not defined.
—— *extans.* (See *Bathyurus extans.*)
—— *gigas.* (See *Asaphus platycephalus*).
—— (*Megalaspis?*) GONIOCERCUS Meek, 1873; 6th Ann. Rep. U. S. Geol. Survey Territories, p. 480. Near Malade City, Utah. Cambrian.
—— GONIURUS Bill., 1860; Canadian Naturalist, vol. 5, p. 324. Point Levis, Quebec.
—— GONIURUS Bill., 1865; Pal. Foss., vol. 1, p. 415.
—— *halli.* (See *Cryphæus boothi.*)
—— HALLI Chap., 1858; Annals Nat. Hist., 3d series, vol. 2, p. 14, fig. 2. Trenton and Utica.
—— HALLI Chap., 1858; Canadian Jour., 2d series, vol. 3, p. 236, fig. This species is the same as *Asaphus canadensis* Chap.
—— *hincksii* Chap., 1859; Canadian Jour. Industry, Sci. and Arts, vol. 4, p. 2, fig. (See *Asaphus canadensis* Chap.) Utica.
—— HOMALONOTOIDES Wal., 1877; Advanced sheets 31st Rep. New York State Mus. Nat. Hist., p. 20. Two miles above Dunlith, Illinois. Trenton.
—— HOMALONOTOIDES Wal., 1879; 31st Rep. New York State Mus. Nat. Hist., p. 71.
—— HOMALONOTOIDES Whitf., 1882; Geol. Wisconsin, vol. 4, p. 237. (See *Asaphus triangulatus.*)
—— HUTTONI Bill., 1865; Pal. Foss., vol. 1, p. 271, fig. 256. Table Head, N. F.
—— ILLÆNOIDES Bill., 1860; Canadian Naturalist, vol. 5, p. 323. Point Levis, Quebec.
—— ILLÆNOIDES Bill., 1865; Pal. Foss., vol. 1, p. 414.
Isotelus IOWENSIS Owen, 1852; Rep. Geol. Survey Wisconsin, Iowa and Minnesota, p. 577, pl. 2 a, figs. 1-7. On Turkey River, Iowa. Trenton.
—— *laticostatus.* (See *Dalmanites anchiops.*)
——? LATIMARGINATA Hall, 1847; Pal. New York, vol. 1, p. 253, pl. 66, figs. 4 a-b. Watertown, Jefferson County, N. Y. Utica.
—— *limulurus.* (See *Dalmanites limulurus.*)
—— MARGINALIS Hall, 1847; Pal. New York, vol. 1, p. 24, pl. 4 *bis*, fig. 15. Chazy Village, N. Y. Chazy.

18

ASAPHUS.
—— MARGINALIS Emmons, 1855; Am. Geol., vol. 1, pt. 2, p. 235, pl. 3, fig. 16.
Isotelus MAXIMUS Locke, 1838; 2d Ann. Rep. Geol. Survey Ohio, p. 246, figs. 8, 9. Trenton and Hudson.
Syn., *Isotelus megistos* Locke, 1841; Trans. Am. Assoc. Geol. Nat., p. 221, pl. 6.
Isotelus megistos Locke, 1842; Am. Jour. Sci., 1st series, vol. 42, p. 366, pl. 3.
Isotelus megistos Meek, 1873; Pal. Ohio, vol. 1, p. 157, pl. 14, fig. 13.
Isotelus megistos Miller, 1874; Cincinnati Quart. Jour. Sci., vol. 1, p. 137.
Asaphus megistos Wal., 1884; Science, March 7, vol. 3, p. 200, fig. 1.
Asaphus megalopthalmus Troost, 1840; 5th Geol. Rep. Tennessee, p. 57. This is a species of the genus *Phacops*.
—— MICROPLEURUS Green, 1835; Suppl. Mon. Tril. North America, p. 21, cast No. 41.
—— *micrurus*. (See *Dalmanites micrurus*.)
—— *morrisi* Bill., 1865; Pal. Foss., vol. 1, p. 272, fig. 257, syn. to *Asaphus huttoni* Bill.
—— *murchisoni* Castelnau, Essai Sys. Sil., 1843, p. 19, pl. 4, fig. 3. *Bathyurus? murchisoni*. Trenton.
—— *myrmecophorus*. (See *Dalmanites myrmecophorus*.)
—— *nasutus*. (See *Dalmanites nasutus*.)
—— *? nodostriatus* Hall, 1847; Pal. New York, vol. 1, p. 248, pl. 61, figs. 1 a, b. (See *Bathyurus extans* Hall.)
—— NOTANS Bill., 1866; Catalogue Silurian Foss. Anticosti, p. 25, fig. 8. cf. *Asaphus maximus* Locke. English Head. Hudson.
—— ? OBTUSUS Hall, 1847; Pal. New York, vol. 1, p. 24, pl. 4 bis, fig. 14. Chazy village, Clinton county, N. Y.
—— OBTUSUS Emm., 1855; Am. Geol., vol. 2, pt. 2, p. 236, pl. 3, fig. 14.
—— PELOPS Bill., 1865; Pal. Foss., vol. 1, p. 317, fig. 304. Bedford, in the bed of Pike River, and upper end of the island of Orleans.
Quebec.
PLATYCEPHALUS Stokes, 1823; Trans. Geol. Soc. London, 2d series, vol. 1, p. 208, pl. 27. Canada, New York, Kentucky, etc.
Trenton.
Syn., *Isotelus gigas* Dekay, 1824; Annals Lyceum Nat. Hist. New York, vol. 1, p. 176, pl. 12, fig. 1.
Isotelus planus Dekay, 1824; Annals Lyceum Nat. Hist. New York, vol. 1, p. 178, pl. 13, fig. 7.
Asaphus gigas Dalm., 1826; Palæad., p. 70.
Asaphus gigas and *planus* Dalm., 1826; Palæad., p. 70.

CATALOGUE OF TRILOBITES. 275

ASAPHUS.
Isotelus gigas Green, 1832; Monthly Am. Jour. Geol., p. 560.
Isotelus gigas Green, 1832; Mon. Tril. N. A., p. 71, casts 21, 22.
Isotelus planus Green, 1832; Monthly Am. Jour. Geol., p. 560;
   Mon. Tril. N. A., p. 68, cast 23.
Isotelus stegops Green, 1832; Mon. Tril. N. A., p. 71, casts 26, 27.
Brongniartia isotelea Eaton, 1832; Geol. Text Book, p. 33, pl. 2,
   fig. 22.
—— PLATYCEPHALUS Bronn, 1835; Lethæa Geogn., vol. 1, p. 115, pl. 9, fig. 8.
   Syn., *Isotelus gigas* Milne-Edw., 1840; Crust., vol. 3, p. 298.
—— PLATYCEPHALUS Buckl., 1840; Bridgw. Treatise, vol. 2, p. 76, pl. 63,
   fig. 12.
—— PLATYCEPHALUS Burm., 1843; Org. Trilobites, p. 110, pl. 2, fig. 12
   (Ray Soc. ed., 1846).
   Syn., *Isotelus gigas* Portl., 1843; Geol. Rep. Londonderry, p. 295, pl.
      7, fig. 1; pl. 8, fig. 1.
   Isotelus planus Portl., 1843; Geol. Rep. Londonderry, p. 295,
      pl. 7, figs. 2, 3.
   Isotelus gigas Hall, 1847; Pal. New York, vol. 1, p. 231, pls. 60-63.
   Isotelus gigas Emmons, 1855; Am. Geol., vol. 1, pt. 2, p. 215,
      pl. 16, fig. 12.
   Isotelus PLATYCEPHALUS Roemer, 1851-56; Lethæa Geogn., vol. 1, p.
      632, pl. 9, fig. 8; pl. 9¹, fig. 5.
—— PLATYCEPHALUS Nieszk., 1857; Mon. Sil. Ostseeprovinz, Tril., p. 37.
   Syn., *Isotelus gigas* Rogers, 1858; Geol. Survey Pennsylvania, vol. 2,
      p. 819, fig. 610.
—— PLATYCEPHALUS Bill., 1863; Geol. Survey Canada, p. 184, fig. 183.
   Syn., *Asaphus (Isotelus) gigas* Salt., 1864; Mon. Brit. Tril., p. 161, pl.
      24, figs. 1-5; pl. 25, fig. 1 (var.)
   Asaphus (Isotelus) gigas Salt., 1864; Mem. Geol. Survey United
      Kingdom, decade xi, pl. 3.
—— PLATYCEPHALUS Bill., 1870; Quart. Jour. Geol. Soc. London, vol.
   26, p. 486, pls. 31, 32.
   Syn., *Isotelus gigas* Miller, 1874; Cincinnati Quart. Jour. Sci., vol. 1,
      p. 138.
—— *platyleurus* Green, 1837; Am. Jour. Sci., 1st series, vol. 32, p. 169.
   (See *Dalmanitis platyleurus*.)
—— POLYPLEURUS Green, 1838; Am. Jour. Sci., 1st series, vol. 34, p. 380.
   On the banks of the Mississippi, near the mouth of Des Moines
   River.
—— QUADRATICAUDATUS Bill., 1865; Pal. Foss., vol. 1, p. 272, fig. 258.
   Table Head, N. F.
—— ROEMINGERI Wal., 1876; 28th Rep. New York State Mus. Nat. Hist.,
   p. 96. Russia, Herkimer county, N. Y.; Quimby's Mill, Lafayette
   county, Wis. Trenton.

ASAPHUS. ATOPS.

— *selenurus.* (See *Dalmanites selenurus.*)
— *stokesi.* (See *Proetus stokesi.*)
— susæ Calvin; MSS.
— susæ Whitf., 1882; Geol. Wisconsin, vol. 4, p. 236, pl. 5, fig. 3; pl. 10, fig. 8. Apple River, just across the Illinois line. Trenton.
— tetragonocephalus Green, 1834; Am. Jour. Sci., 1st series, vol. 25, p. 135.
— tetragonocephalus Green, 1835; Suppl. Mon. Tril. N. A., p. 12, cast 38. Found in boulder of bituminous shaly limestone in New York State.
Paradoxides tetragonocephalus Emm., 1839; De Tril., etc., p. 34.
— *trentonensis.* (See *Lichas trentonensis.*)
— *triangulatus* Whitf., 1880; Ann. Rep. Geol. Survey Wisconsin, 1879, p. 59. This species is referred to *Asaphus homalonotoides* Wal., Final Report, vol. 4, p. 237, pl. 5, fig. 4. Sec. 5 T., 5 R., 5 W., Grant county, Wisconsin. Trenton.
— trimblii Green, 1837; Acad. Nat. Sci. Phila., Jour., vol. 7, p. 224. Huntingdon county, Penn. Devonian.
— trimblii Green, 1837; Am. Jour. Sci., 1st series, vol. 32, p. 348.
Ogygia ? vetustus Hall, 1847; Pal. N. Y., vol. 1, p. 227, pl. 60, fig. 1. Mohawk Valley, New York. Trenton. Hudson.
Ogygia ? vetustus Emmons, 1855; Am. Geol., vol. 1, pt. 2, p. 216, fig. 72.
— (*Isotelus*) vigilans M. & W., 1870; Proc. Acad. Nat. Sci. Phila., p. 53. Hudson.
— (*Isotelus*) vigilans M. & W., 1875; Geol. Illinois, vol. 6, p. 497, pl. 23, fig. 6. Carroll county, near Mount Carroll; Oswego, Kendall county, Illinois.
— wisconsensis Wal., 1876; 28th Rep. N. Y. State Mus. Nat. Hist., p. 97. Russia, Herkimer county, N. Y. Mineral Point and Plattsville, Wis. Trenton.

Aspidolites Con., 1841; 4th Ann. Rep. Pal. Dept. New York Geol. Survey. p. 48.

Atops Emmons, 1844; Taconic System, p. 20.
Syn., *Conocoryphe* Cord., 1847; Prodr., p. 139, pl. 2, fig. 10.
Conocephalites baileyi Hartt., 1868; Acadian Geol., p. 645 (3d ed.) St. John's group at Ratcliff's Millstream, N. B.
Conocoryphe baileyi Matt., 1884; Trans. Royal Soc. Canada, vol. 1, p. 111, pl. 1, figs. 22–27.
Conocoryphe (Bailiella) baileyi Matt., 1884; Trans. Royal Soc. Canada, vol. 1, p. 1, fig. 22.
Conocoryphe (Salteria) baileyi Wal., 1884; Bull. U. S. Geol. Survey, No. 10, p. 32, pl. 4, figs. 3, 3 a; pl. 5, figs. 7, 7 a. The author herein proposed the subgenus *Salteria* (corrected to *Bailiella*) for

ATOPS.

Erinnys Salt., a preoccupied generic name; the same is true of *Salteria*, used for a genus of fossil Crustacea.

Conocoryphe BAILEYI Matt., 1890; Trans. Royal Soc. Canada, vol. 8, p. 135, pl, 11, fig. 10. Paradoxides zone.

Conocephalites CALCIFERA Wal., 1879; 32d Rep. New York State Mus. Nat. Hist., p. 129. Saratoga County, N. Y. Calciferous.

Conocephalites CORDILLDRÆ Rominger, 1887; Proc. Acad. Nat. Sci., Phila., p. 17, pl. 1, fig. 7. Mount Stephens, B. C. Taconic.

Conocephalites ELEGANS Hartt, 1868; Acadian Geol. (3d ed.),. p. 650. Radcliff's Millstream, N. B.

Conocoryphe ELEGANS Wal., 1884; Bull. U. S. Geol. Survey, No. 10, p. 33, pl. 4, figs. 2, 2 a–b.

—— *gemini-spinosus* Hartt. (Syn., *Atops matthewi*.)

Conocephalites HARTTI Wal., 1879; 32d Rep. New York State Mus. Nat. His., p. 130. Saratoga County, N. Y. Calciferous.

Conocephalites MATTHEWI Hartt, 1868; Acadian Geol., p. 646, fig. 224. Radcliffe's Millstream and St. John.

Conocephalites MATTHEWI Matt., 1884; Trans. Royal Soc. Canada, vol. 1, p. 103, pl., figs. 6-21.

Syn., *Harttia matthewi* Wal., 1884; Bull. U. S. Geol. Survey, No. 10, p. 19, pl. 1, fig. 3.

Conocoryphe MATTHEWI Wal., 1884; Bull. U. S. Geol. Survey, No. 10, p. 28, pl. 4, figs. 1, 1 a–b.

Conocephalites OPTATUS Hall, 1863; 16th Rep. New York State Cab. Nat. Hist., p. 222, pl. 5 a, fig. 7. Trempaleau, Wisconsin. Dikelocephalus zone.

—— TRILINEATUS Emmons, 1844; Taconic System, p. 20, fig. 1, pl. 2, fig. 3. Reynolds Inn, 7 miles north of Union Village, Washington County, N. Y. Taconic.

Conformity to the rules governing nomenclature, which are intended to render justice to every investigator, the genus *Atops* Emmons, 1844, should replace that of *Conocoryphe* Corda described in 1847. The original figure of this species was the first primodial fossil described from American rocks; this species with *Olenellus asaphoides* Emm. clearly defined the Taconic System in 1844 and its position below Murchison's Lower Silurian of 1837. The fact that the original Cambrian was the same as the Lower Silurian cannot be recalled and in justice to Sedgwick it must be given to that system.

—— TRILINEATUS Emmons, 1847; Nat. Hist. New York, Agriculture, vol. 1, p. 64, fig. 8, pl. 14, fig. 3.

Syn., *Calymene beckii* Hall, 1847; Pal. New York, vol. 1, p. 252, pl. 67, figs. 4 a, 4 e.

—— TRILINEATUS Haldeman, 1848; Am. Jour. Sci., 2d series, vol. 5, p. 107.

ATOPS.

—— TRILINEATUS Emmons, 1849; Proc. Am. Assoc Adv. Sci., 1st Meeting, Phila., 1848, pp. 16, 17.
Calymene beckii Green, *Atops trilineatus* (Emm.) Fitch, 1849; Trans. Agric. Soc. New York, vol. 9, p. 865 (not *Calymene beckii* Green).
—— TRILINEATUS Emmons, 1855; Am. Geol., vol. 1, pt. 2, p. 115, pl. 1, fig. 16.
Syn., *Atops punctatus* Emm., 1860; Manual Geol., p. 88, fig. 71; also p. 280.
—— TRILINEATUS Barr., 1861; Bull. Soc. Géol. France, 2d series, vol. 18, p. 269, pl. 5, fig. 1.
—— *punctatus* Barr., 1861; Bull. Soc. Géol. France, 2d series, vol. 18, p. 271, pl. 5, fig. 3.
Conocephalites (*Atops*) TRILINEATUS Ford, 1871; Am. Jour. Sci., 3d series, vol. 2, p. 33.
Conocephalites TRILINEATUS Ford, 1873; Am. Jour. Sci., 3d series, vol. 6, p. 135.
Conocephalites (*Atops*) TRILINEATUS Ford, 1879; Am. Jour. Sci., 3d series, vol. 9, p. 205.
Triarthrus TRILINEATUS Miller, 1877; Catalogue Am. Pal. Foss., p. 223.
Conocoryphe TRILINEATUS Ford, 1880; Am. Jour. Sci., 3d series, vol. 19, p. 132.
Triarthrus becki Wal., 1879; Trans. Albany Inst., vol. 10, p. 23.
Ptycopharia TRILINEATUS Wal., 1886; Bull. U. S. Geol. Survey, No. 30, p. 203, pl. 27, fig. 1 a-c.
The figures of this species given by Dr. Emmons are all imperfect. Ford's figure given in Bull. U. S. Geol. Survey, No. 30, pl. 27, fig. 1 a-b, shows a glabella and a pygidium, which is nearly semi-circular in form; the axis strong, composed of 7 segments, with a row of obtuse spines along the middle; lateral lobes composed of 5 segments, each with distinct intervening grooves; marginal rim nearly flat, widest towards the front and distinctly and regularly notched or denticulated all around. This may prove to be a species of the genus *Olenoides*.
Conocoryphe TRILINEATUS Wal., 1887; Am. Jour. Sci., 3d series, vol. 34, p. 197, pl. 1, fig. 7 a-b.
Conocoryphe TRILINEATA Wal., 1890; 10th Rep. U. S. Geol. Survey, p. 647, pl. 94, figs. 5, 5 a-e. Olenellus zone.
Conocoryphe (*Bailiella*) WALCOTTI Matt., 1884; Trans. Royal Soc. Canada, vol. 1, p. 119, pl. 1, figs. 36, 36 b. At St. John, in Division 1 c. Tacouic.
Conocoryphe WALCOTTI Wal., 1884; Bull. U. S. Geol. Survey, No. 10, p. 30.
Conocoryphe WALCOTTI, 1890; Trans. Roy. Soc. Canada, vol. 8, p. 134, pl. 11, fig. 7, 7 a-c. Paradoxides zone.

AVALONIA. BATHYURELLUS.

AVALONIA Wal., 1889; Proc. Nat. Mus., vol. 12, p. 44.
— MANUELENSIS Wal.; Proc. Nat. Mus., vol. 12, p. 44. Manuel's Brook, Conception Bay, N. F. See, also, 10th Rep. U. S. Geol. Survey, p. 646, pl. 95, figs. 3, 3 a. Taconic.

Bailiella Matt., 1884; Trans. Roy. Soc. Canada, vol. 1, p. 124 (used as a subgenus). (See *Conocoryphe (Bailiella) Baileyi.*)

Barrandia Hall, 1860; 13th Rep. New York State Cab. Nat. Hist., p. 115 (not *Barrandia* McCoy). (See *Olenellus.*)
— *thompsoni.* (See *Olenellus thompsoni.*)
— *vermontana.* (See *Olenellus vermontana* and *Mesonasis vermontana.*)

BARRANDIA McCoy, 1849; Annals Nat. Hist., 2d series, vol. 4, p. 409.
— ? MACCOYI Wal., 1884; Pal. Eureka Dist., Mon. U. S. Geol. Survey, vol. 8, p. 96, pl. 12, fig. 5. Hamburg Ridge, Eureka Dist., Nev. Pogonip group.

BATHYNOTUS Hall, 1860; 13th Rep. New York State Cab. Nat. Hist., p. 117.
Peltura (Olenus) HOLOPYGA Hall, 1859; 12th Rep. New York State Cab. Nat. Hist., p. 61, fig. Georgia, Vermont. Taconic.
Peltura (Olenus) HOLOPYGA Hall, 1859; Pal. New York, vol. 3, p. 528, fig.
Syn., *Paradoxides ? quadrispinosus* Emmons, 1860; Manual Geol., pp. 80 and 280, fig. 57.
— HOLOPYGA Hall, 1860; 13th Rep. New York State Cab. Nat. Hist., p. 118, fig. 3.
— HOLOPYGA Hall, 1861; Geol. Vermont, vol. 1, p. 371, pl. 13, fig. 3.
Peltura (Olenus) HOLOPYGA Barr., 1861; Bull. Soc. Géol. France, 2d series, vol. 18, p. 278, pl. 5, figs. 9–10.
— HOLOPYGA Wal., 1886; Bull. U. S. Geol. Survey, No. 30, p. 191, pl. 31, figs. 1, 1 a.
— HOLOPYGA Wal., 1890; 10th Rep. U. S. Geol. Survey, p. 646, pl. 95, fig. 1 a. Olenellus zone.

BATHYURELLUS Bill., 1865; Pal. Foss., vol. 1, p. 262.
— ABRUPTUS Bill., 1865; Pal. Foss., vol. 1, p. 263, fig. 247. Port aux Choix and Keppel Island, N. F. Calciferous.
— *(Asaphiscus) bradleyi* Meek. (See *Asaphiscus bradleyi.*)
— EXPANSUS Bill., 1865; Pal. Foss., vol. 1, p. 318, figs. 306 a-b. cf. *B. nitidus* and *B. marginatus* Bill. Stanbridg, Range 6, Lot 20.
— FORMOSUS Bill., 1865; Pal. Foss., vol. 1, p. 266, fig. 250. Cow Head, N. F.
— FRATERNUS Bill., 1865; Pal. Foss., vol. 1, p. 267, fig. 251. Cow Head, N. F. The head of this species may belong to Salter's genus *Stygina.*
— LITORECS Bill., 1865; Pal. Foss., vol. 1, p. 320. Point Levis, Quebec.

BATHYURELLUS. BATHYURUS.
— MARGINATUS Bill., 1865; Pal. Foss., vol. 1, p. 261, fig. 248. cf. *B. nitidus* and *B. expansus* Bill. Keppel Island, Port aux Choix and Table Head, N. F. Calciferous.
— NITIDUS Bill., 1865; Pal. Foss., vol. 1, p. 265, fig. 249. Cow Head, N. F. On page 320 Mr. Billings says that it may be found necessary to unite *B. nitidus*, *B. marginatus* and *B. expansus* into one species.
— RARUS Bill., 1865; Pal. Foss., vol. 1, p. 320, no fig. Point Levis, Quebec.
— (*Dicellocephalus*) TRUNCATUS Meek, 1873; 6th Ann. Rep. Geol. Survey Territories, p. 465 (no description given).
— VALIDUS Bill., 1865; Pal. Foss., vol. 1, p. 268, fig. 252. Point Rich, N. F.
— (*Asaphiscus*) *wheeleri*. (See *Asaphiscus wheeleri*.)
BATHYURISCUS Meek, 1873; 6th Ann. Rep. U. S. Geol. Survey Territories, p. 484.
Bathyurus? HAYDENI Meek, 1873; 6th Ann. Rep. U. S. Geol. Survey Territories, p. 482. Gallatin River, Gallatin City, Mont.
— HAYDENI Wal., 1886; Bull. U. S. Geol. Survey, No. 30, p. 215.
— HOWELLI Wal., 1886; Bull. U. S. Geol. Survey, No. 30, p. 216, pl. 30, figs. 2, 2 a. Chisholm Mine, near Pioche, Nev. Taconic.
Syn., *Embolimus rotundatus* (Rominger) Walcott, 1888; Am. Jour. Sci., 3d series, vol. 36, p. 165, referred to *Bathyuriscus howelli*.
— *Ogygia* PARABOLA H. & W., 1877; U. S. Geol. Expl. 40th Par., vol. 4, p. 245, pl. 2, fig. 35. Oquirrh Mts., Utah. *Niobe parabola*.
Taconic.
Ogygia PRODUCTA H. & W., 1877; U. S. Geol. Expl. 40th Par., vol. 4, p. 244, pl. 2, figs. 31-34. Oquirrh Mts., Utah. Taconic.
— PRODUCTA Wal., 1886; Bull. U. S. Geol. Survey, No. 30, p. 217, pl. 30, figs. 1, 1 a-c. Big Cottonwood Cañon, City Creek, Wahsatch Mts., etc., Utah.
BATHYURUS Bill., 1859; Canadian Naturalist, vol. 4, p. 364.
— AMPLIMARGINATUS Bill., 1857; Canadian Naturalist, vol. 4, p. 365, fig. 12 a-b. Mingan Islands in the Calciferous.
— AMPLIMARGINATUS Bill., 1863; Geol. Canada, p. 122, fig. 41.
— AMPLIMARGINATUS Bill., 1865; Pal. Foss., vol. 1, p. 353, fig. 341 a.
— ANGELINI Bill., 1859; Canadian Naturalist, vol. 4, p. 467, fig. 37. cf. *B. extans* Hall. Grenville and Fitzroy Harbor. Chazy group.
— ANGELINI Bill., 1863; Geol. Canada, p. 133, fig. 68.
— ARCUATUS Bill., 1865; Pal. Foss., vol. 1, p. 205, fig. 190. St. Antoine, Quebec.
— ARMATUS Bill., 1860; Canadian Naturalist, vol. 5, p. 319, fig. 23. Point Levis, Quebec.
— ARMATUS Bill., 1865; Pal. Foss., vol. 1, p. 411, fig. 392.

BATHYURUS.

—— ARMATUS Wal., 1879; 32d Rep. New York State Mus. Nat. Hist., p. 131. *Dorypyge ?* ARMATUS Vogd., Bull. U. S. Geol. Survey, No. 63, p. 97.

—— BITUBERCULATUS Bill., 1860; Canadian Naturalist, vol. 5, p. 319, fig. 22; also, Geol. Canada, 1863, p. 238, fig. 270. Point Levis, Quebec. Taconic.

—— BITUBERCULATUS Bill., 1865; Pal. Foss., vol. 1, p. 410, fig. 391. This species cannot be included within the limits of *Bathyurus*. As we know of no other to which it bears a resemblance, the new generic term of *Lloydia* is proposed, in honor of the first author on Trilobites, Edward Lhwyd. Diagnosis: Glabella tumid and extending to the frontal limb; basal lobes elongate-oval, pointed at both ends, separated from the glabella by shallow, obscure grooves; eyes opposite the mid length of the glabella; thorax unknown.

—— BREVICEPS Bill., 1865; Pal. Foss., vol. 1, p. 262, fig. 246. Table Head, N. F.

—— CAPAX Bill., 1860; Canadian Naturalist, vol. 5, p. 318, fig. 20. Point Levis, Quebec.

—— CAPAX Bill., 1863; Geol. Canada, p. 238, fig. 271.

—— CAPAX Bill., 1865; Pal. Foss., vol. 1, p. 409, fig. 389, A similar species was referred by Mr. R. P. Whitfield (Geol. Survey Wisconsin, vol. 4, pl. 4, fig. 6) to *Dicellocephalus*. cf. *Niobe insignis* Brögg.

—— CAUDATUS Bill., 1865; Pal. Foss., vol. 1, p. 261, fig. 245. Port aux Choix, N. F. Calciferous.

—— CONICUS Bill., 1859; Canadian Naturalist, vol. 4, p. 366, fig. 12 d. St. Timothy Beaucharnois Canal, in the Upper Calciferous.

—— CONICUS Bill., 1863; Geol. Canada, p. 122, fig. 42.

—— CONICUS Bill., 1865; Pal. Foss., vol. 1, p. 353, fig. 341 b.

—— CONICUS Whitf., 1889; Bull. Am. Mus. Nat. Hist., vol. 2, No. 2, p. 61, pl. 13, figs. 15-21.

——? CONGENERIS Wal., 1884; Pal. Eureka Dist., Mon. U. S. Geol. Survey, vol. 8, p. 92, pl. 12, fig. 8. Wood cone and below Bellevue Peak, Fish Creek Mts., Eureka Dist.

—— CORDAI Bill., 1860; Canadian Naturalist, vol. 5, p. 321, fig. 26. Cow Head, Point Levis, etc. Calciferous.

—— CORDAI Bill., 1863; Geol. Canada, p. 238, fig. 269.

—— CORDAI Bill., 1865; Pal. Foss., vol. 1, p. 259, fig. 242; p. 412, fig. 395.

——? CROTALIFRONS Dwight, 1884; Am. Jour. Sci., 3d series, vol. 27, p. 253, figs. 4, 4 a, 5, 6. Rockdale, N. Y. Calciferous.

—— CYBELE Bill., 1859; Canadian Naturalist, vol. 4, p. 366, fig. 12 c. Mingan Islands, Beauharnois and in the counties of Leeds and Greenville, Canada. Calciferous.

—— CYBELE Bill., 1863; Geol. Canada, p. 122, fig. 43.

BATHYURUS.
—— CYBELE Bill., 1865; Pal. Foss., vol. 1, p. 352, fig. 341 c.
—— DUBIUS Bill., 1860; Canadian Naturalist, vol. 5, p. 319, fig. 21. Point Levis, Quebec.
—— DUBIUS Bill., 1865; Pal. Foss., vol. 1, p. 410, fig. 390.
Asaphus? EXTANS Hall, 1847; Pal. New York, vol. 1, p. 228, pl. 60, figs. 2 a–c. Great Bend, Jefferson County, N. Y. Trenton.
Asaphus? EXTANS Hall, 1850; 3d Rep. New York State Cab. Nat. Hist., p. 174, pl. 3, figs. 1 a–c.
Syn., *Asaphus nodostriatus* Hall, 1850; 3d Rep. New York State Cab. Nat. Hist., p. 17.
—— EXTANS Bill., 1863; Geol. Canada, p. 153, fig. 114.
—— *gregarius* Bill., 1865; Pal. Foss., vol. 1, p. 363, fig. 349. St. Mary's Bay, N. F. Mr. Walcott (Bull. U. S. Geol. Survey, No. 10, p. 36) refers this species to the genus *Solenopleura* Ang., and remarks that many of the Cambrian species of the genus *Bathyurus* described in the Canadian reports might be referred to that genus. This species occurs with *Paradoxides bennetti* Salt. Taconic.
—— ? *haydeni*. (See *Bathyuriscus haydeni*.)
—— LONGISPINUS Wal., 1876; 28th Rep. N. Y. State Cab. Nat. Hist., p. 94. Russia, Herkimer Co., N. Y.; Plattsville, Wis. Trenton.
—— MINGANENSIS Bill., 1865; Pal. Foss., vol. 1, p. 353. Mingan Islands.
Calciferous.
—— *nero*. (See *Solenopleura nero*.)
—— OBLONGUS Bill., 1865; Canadian Naturalist, vol. 5, p. 321, fig. 25.
—— OBLONGUS Bill., 1865; Pal. Foss., vol. 1, p. 411, fig. 394. Point Levis, Quebec.
—— OBLONGUS Bill., 1861; Geol. Vermont, vol. 1, p. 952, fig. 361.
—— PARVULUS Bill., 1861; Pal. Foss., Canada, p. 16, fig. 21. Taconic.
—— PARVULUS Bill., 1861; Geol. Vermont, vol. 2, p. 952, fig. 361.
—— PARVULUS Bill., 1865; Pal. Foss., vol. 1, p. 16, fig. 21.
—— PERPLEXUS Bill., 1865; Pal. Foss., vol. 1, p. 364, fig. 350. Bonne Bay, N. F. Taconic.
—— PERSPICATOR Bill., 1865; Pal. Foss., vol. 1, p. 205, fig. 191. Near St. Antoine, Quebec.
—— POGONIPENSIS Hall & Whit., 1877; U. S. Geol. Expl. 40th Par., vol. 4, p. 243. pl. 1, figs. 33, 34. Cambrian.
—— QUADRATUS Bill., 1860; Canadian Naturalist, vol. 5, p. 321, fig. 27. Point Levis, Quebec. Taconic.
—— QUADRATUS Bill., 1865; Pal. Foss., vol. 1, p. 412, fig. 396.
—— SAFFORDI Bill., 1860; Canadian Naturalist, vol. 5, p. 321, fig. 24. Point Levis, Quebec. Taconic.
—— SAFFORDI Bill., 1865; Pal. Foss., vol. 1, p. 259, fig. 341; p. 411, fig. 393.
—— SEELYI Whitf., 1886; Bull. Am. Mus. Nat. Hist. New York, vol. 1, No. 8, p. 339, pl. 33, figs. 12–18. Fort Casin, Vermont.
Calciferous.

BATHYURUS.

—— SEELYI Whitf., 1889; Bull. Am. Mus. Nat. Hist. New York, vol. 2, No. 2, p. 62, pl. 13, figs. 8-14.
—— SENECTUS Bill., 1861; Pal. Foss., Canada, p. 16.
—— SENECTUS Bill., 1861; Geol. Vermont, vol. 2, p. 953, figs. 359, 360. Highgate, Vermont. Taconic.
—— SENECTUS Bill., 1863; Geol. Canada, p. 286, figs. 298 a, b; Pal. Foss., vol. 1, p. 15 (1865), figs. 19, 20. Anse au Loup, north shore of Straits of Belle Isle. Referred to *Protypus senectus* Wal., 1886; Bull. U. S. Geol. Survey, No. 30, p. 213, pl. 31, figs. 2, 2 a-c.
—— SERRATUS Meek, 1873; 6th Ann. Rep. U. S. Geol. Survey Territories, p. 480. Also referred to *Corynexochus serratus* p. 482. Probably a species of *Olenoides*. Gallatin River above Gallatin City, Montana. Taconic.
—— ? SIMILLIMUS Wal., 1884; Pal. Eureka Dist., Mon. U. S. Geol. Survey, vol. 8, p. 93, pl. 12, fig. 11. North slope of White Mountain, Eureka Dist. Pogonip group.
—— SMITHI Bill., 1863; Geol. Canada, p. 153, fig. 115. Petersborough. Trenton.
—— SMITHI Bill., 1865, Pal. Foss., vol. 1, p. 56.
—— SOLITARICUS Bill., 1865; Pal. Foss., vol. 1, p. 362. Hare Bay, N. F. *Acidaspis* SPINIGER Hall, 1847; Pal. New York, vol. 1, p. 241, pl. 64, fig. 5. Mohawk Valley, New York. Trenton.
—— STONEMANI Vogd., 1884; 12th Ann. Rep. Geol. Nat. Hist. Minnesota, p. 8. Minneapolis, Minn. Trenton.
—— STRENUUS Bill., 1865; Pal. Foss., vol. 1, p. 204, fig. 188. St. Antoine, Quebec. Trenton?
—— TAURIFRONS Dwight, 1884; Am. Jour. Sci., 3d series, vol. 27, p. 252, pl. 7, figs. 1-3. Rochdale, N. Y. Calciferous.
—— TIMON Bill., 1865; Pal. Foss., vol. 1, p. 261, fig. 244. Port au Choix, N. F. Calciferous.
—— ? TUBERCULATUS Wal., 1884; Pal. Eureka Dist., Mon. U. S. Geol. Survey, vol. 8, p. 12, fig. 9. Wood cone Eureka Dist., Nevada, etc. Pogonip group.
—— VETULA Bill., 1865; Pal. Foss., vol. 1, p. 365. Bonne Bay, N. F. Taconic.

Agraulos (Bathyurus?) WOOSTERI Whitf., 1878; Ann. Rep. Geol Survey, Wisconsin, p. 56. Ettrick and Eau Claire, Wis. Taconic.

Agraulos (Bathyurus?) WOOSTERI Whitf., 1882; Geol. Wisconsin, vol. 4, p. 189, pl. 1, figs. 19-21.

Brongniartia Eaton, Am. Jour. Sci., 1st series, vol. 22, p. 166, and Geol. Text-Book, p. 32 (Syn., *Asaphus* in part), not *Brongniartia* Salt. (subgenus *Homalonotus*.)
—— CARCINOIDES Eaton. (See *Triarthrus carcinoides*.)
—— *isoletea*. (See *Asaphus platycephalus*.)
—— *platycephus* Eaton. (See *Homalonotus delphinocephalus*.)

BRACHYMETOPUS.   BRONTEUS.

BRACHYMETOPUS McCoy, 1847; Ann. Nat. Hist., vol. 20, p. 220.
—— ARMATUS Vogd., 1891; Trans. St. Louis Acad. Sci., vol. 5, pl. 15, figs. 4 and 5.   Sedalia, Missouri.   Waverly.
   Phæthonides ? IMMATURUS Herr., 1889; Bull. Denison Univ., vol. 4, p. 59, pl. 1, figs. 9-15. Newark, Ohio, and also in the upper layers at Cuyahoga Falls. cf. *Brachymetopus lodiensis*. Kinderhook.
   Phillipsia (Griffithides) LODIENSIS Meek., 1875; Pal. Ohio, vol. 2, p. 323, pl. 18, fig. 3. Lodi, Medina county, and Loudonville, Ohio.   Waverly.
—— LODIENSIS Vogd., 1888; Annals New York Acad. Sci., vol. 4, p. 102.
   Phæthonides (?) LODIENSIS (Meek.) Herr., 1889; Bull. Denison Univ., vol. 4, p. 59. For other references, see *Brachymetopus lodiensis* Meek.
   Phæthonides OCCIDENTALIS Herr., 1889; Bull. Denison Univ., vol. 4, p. 57, pl. 1, figs. 10 a-b. Granville and Newark, Ohio. Waverly.
   Phæthonides SPINOSUS Herr., 1889; Bull. Denison Univ., vol. 4, p. 58, pl. 1, figs. 4-5. Licking and Ashland counties, Ohio. Waverly.
   Phillipsia (Brachymetopus) ornata Hall. (See *Cyphaspis ornata*.)
BRONTEUS Goldf., 1839; Beiträge Petrefactenkunde, Nova Acta Physico-Med., vol. 19, p. 360, pl. 33, fig. 3.
   Syn., *Goldius* DeKon., 1841; Mém. Crust. Foss. Belgique, p. 5.
—— ACAMAS Hall, 1865; 20th Rep. New York State Cab. Nat. Hist., p. 332, pl. 21, figs. 19-20. Racine, Wis.   Niagara.
—— ACAMAS Hall, 1870; 20th Rep. New York State Cab. Nat. Hist., p. 422, pl. 21, figs. 19-20 (rev. ed.)
—— BARRANDI Hall, 1859; Pal. New York, vol. 3, p. 350, pl. 63, figs. 1-4. Schoharie county, N. Y.   Lower Helderberg.
—— CANADENSIS Logan, 1844; Geol. Canada, p. 54, 2 figs. Bay Chaleur Gaspé.   Lower Helderberg.
—— INSULARIS Bill., 1866; Catalogue Sil. Foss. Anticosti, p. 66. Southwest Point, Div. 4. Anticosti group).
—— LAPHAMI Whitf., 1878; Ann. Rep. Geol. Survey Wisconsin (1877), p. 88. Kewaunee, Wis.   Niagara.
—— LAPHAMI Whitf., 1882; Geol. Wisconsin, vol. 4, p. 310, pl. 22, figs. 1-4.   Trenton.
—— LUNATUS Bill., 1863; Geol. Canada, p. 188, fig. 187. Murray Bay at Les Ecorchés.
—— MANITOBENSIS Whiteaves, 1892; Contrib. Canadian Palæont., vol. 1, p. 347, pl. 46, figs. 5, 6 and 7. Whiteaves Point, Dawson Bay, Lake Winnipegosis, Manitoba.   Devonian.
—— ? NIAGARENSIS Hall, 1852; Pal. New York, vol. 2, p. 314, pl. 70, fig. 3. Limestone in the Niagara River, below the Canada Falls. Niagara.
—— OCCASUS W. & M., 1866; Mem. Boston Soc. Nat. Hist., vol. 1, p. 104, pl. 3, fig. 12. Bridgeport, Chicago, Ill.   Niagara.

BRONTEUS. CALYMENE.
—— OCCASUS (W. & M.) Hall, 1867; 20th Rep. New York State Cab. Nat.
Hist., p. 400.
—— POMPILIUS Bill., 1863; Proc. Portland Soc. Nat. Hist., p. 123, pl. 1,
fig. 25. Square Lake, Maine, also at Port Daniel, Bay Chaleur
and Lower Helderberg rock at Cape Gaspé.
—— SENESCENS Clarke, 1889; 42d Ann. Rep. N. Y. State Mus., p. 403, fig.
Italy Hollow, Yates county, N. Y. Chemung.
—— TULLIUS Hall, 1888; Pal. New York, vol. 7, p. 12, pl. 8, figs. 34–36.
In the Tully limestone, Kingsley's Hill, northeast of Obisco and
Borodino, Onondaga county, N. Y.

BUMASTUS Murch., 1839; Sil. Syst., p. 656.
—— GRAFTONENSIS M. & W., 1870; Proc. Acad. Nat. Sci. Phila., p. 54.
For other references see *Illænus graftonensis*.
—— INSIGNIS Hall, 1867; 20th Rep. N. Y. State Cab. Nat. Hist., p. 331,
figs. 5, 6. For other references in *Illænus insignis*.
—— IOXUS Hall, 1867; 20th Rep. N. Y. State Cab. Nat. His., p. 378, pl. 22,
figs. 4–10. For other references see *Illænus ioxus*.
—— TRENTONENSIS Emm., 1842; Geol. N. Y. 3d Geol. Dist., p. 390, fig.
1. For other references see *Illænus trentonensis*.
—— WORTHENANUS W. & M., 1866; Mem. Boston Soc. Nat. Hist., vol. 1,
p. 105, pl. 3, fig. 13. For other references see *Illænus worthenanus*.

CALYMENE Brong., 1822; Crust. Fos., p. 9.
Syn., *Prionocheilus* Ron., 1846; Bull. Soc. Géol. France, vol. 4, p. 309.
—— anchiops. (See *Dalmanites anchiops*.)
—— beckii. (See *Triarthrus becki*.)
—— BLUMENBACHI Brong., 1822; Crust. Foss., p. 11, pl. 1, figs. 1 a–d.
—— BLUMENBACHI Barr., 1852; Syst. Sil. Bohême, pls. 19 and 45.
—— BLUMENBACHI Salt., 1863; Mon. Brit. Trilobites, pl. 8, figs. 7–11,
13–16; pl. 9, fig. 1. This species should take the older name of
Calymene tuberculata Bünn.
—— BLUMENBACHI (Green) not Brong.; Mon. Tril. N. A., p. 28, cast 1.
—— BLUMENBACHI? Foerste, 1885; Bull. Denison Univ., vol. 1, p. 100, pl.
13, fig. 25.
—— BLUMENBACHII var. VOGDESII Foerste, 1889; Proc. Boston Soc. Nat.
His., vol. 24, p. 265. For other references, see *Calymene vogdesii*.
—— bucklandi. (See *Ceraurus pleurexanthus* Green.)
—— bufo. (See *Phacops bufo* Green.)
—— bufo var. rana. (See *Phacops rana* Green.)
—— CALLICEPHALA Green, 1832; Mon. Tril. N. A., p. 30, cast 2.
—— CALLICEPHALA Burm., 1843; Org. Trilobiten, p. 83, pl. 2, figs. 9–10
(Ray Soc. ed.)

CALYMENE.
— CALLICEPHALA Miller, 1882; Jour. Cincinnati Soc. Nat. Hist., vol. 5, p. 117, pl. 5, fig. 8. The cast of this species (from a dark yellowish limestone of Hampshire county, Virginia), issued with Dr. Green's monograph, exhibits a *Calymene* with a broad frontal margin. The other specimens referred to by Dr. Green from Ohio and Indiana are species of the well-known *Calymene senaria* Conrad.
— CAMERATA Con., 1842; Jour. Acad. Nat. Sci. Phila., vol. 8, p. 278. Schoharie, N. Y. Niagara.
— CAMERATA Hall., 1852; Pal. New York, vol. 2, p. 337, pl. 78, figs. 10 a-f.
— — CHRISTYI Hall, 1860; 13th Rep. New York State Cab. Nat. Hist., p. 119. Oxford, Ohio. Hudson.
— — — CHRISTYI Hall, 1862; 15th Rep. New York State Cab. Nat. Hist., pl. 10, figs. 1, 2 and 5.
— — — CHRISTYI H. & W., 1875; Pal. Ohio, vol. 2, p. 107, pl. 4, figs. 13-15.
— — CHRISTYI Miller, 1874; Cincinnati Quart. Jour. Sci., vol. 1, p. 141.
Hemicrypturis CLINTONI Vanuxem, 1842; Geol. New York, 3d Geol. Dist., p. 79, fig. 3; also, p. 80. Clinton.
— CLINTONI Hall, 1852; Pal. New York, vol. 2, p. 298, pl. 66 a, figs. 5 a-d. Herkimer, Wayne and Cayuga counties, N. Y.
— — CLINTONI Rogers, 1858; Geol. Survey Pennsylvania, vol. 2, p. 823, fig. 637.
— CLINTONI Vogd., 1880; Proc. Acad. Nat. Sci. Phila., p. 178, fig. 4 (not fig. 3).
— *clintoni* Vogd., 1880; Des. New Species Foss. Crust. Clinton Group, p. 5, fig. 3. (See *Calymene vogdesii* Foerste.)
— CONRADI Emmons, 1855; Am. Geology, vol. 1, pt. 2, p. 236. Locality not given. Hudson.
— *crassimarginata*. (See *Proetus crassimarginata*.)
— — *diops* Green, 1832; Mon. Tril. N. A., p. 37, pl. 1, fig. 2, cast 8. (See *Proetus diops*.)
— *macrophthalma* Green, 1832; Mon. Tril. N. A., p. 39, cast 9. (See *Phacops bufo*.)
— MAMMILLATA Hall, 1861; Geol. Wisconsin, p. 50. Makoqueta Creek, 12 miles west of Dubuque, Iowa. Trenton.
— MAMMILLATA Hall, 1862; Geol. Wisconsin, vol. 1, p. 432, figs. 1-2.
— *marginatus*. (See *Proetus marginatus*.)
— — *microps* Green, 1832; Mon. Tril. N. A., p. 34, cast 6. Ripley, Ohio.
— MULTICOSTA Hall, 1847; Pal. New York, vol. 1, p. 228, pl. 60, fig. 3. Trenton.
— NASUTA Ulrich, 1879; Jour. Cincinnati Soc. Nat. Hist., vol. 2, p. 131, fig. 3. Osgood, Ripley county, Indiana. Niagara.
— NIAGARENSIS Hall, 1843; Geol. New York, 4th Dist., p. 102, fig. 3 (on p. 101). Lockport, Rochester, etc., N. Y. Niagara.

CALYMENE.

—— *blumenbachii* var. NIAGARENSIS Hall, 1852; Pal. New York, vol. 2, p. 307, pl. 67, figs. 11-12.
—— *blumenbachii* Roemer, 1860; Sil. Fauna W. Tenn., p. 79, pl. 5, fig. 22.
—— NIAGARENSIS H. & W., 1875; Pal. Ohio, vol. 2, p. 153, pl. 7, figs. 14–15.
—— NIAGARENSIS Hall; 20th Rep. New York State Cab. Nat. Hist., p. 334; rev. ed., p. 425.
—— NIAGARENSIS Hall, 1876; Illus. Devon. Foss. Crust., pl. 1, fig. 10.
—— NIAGARENSIS Hall, 1879; 28th Rep. N. Y. State Mus. Nat. Hist., p. 187, pl. 32, figs. 8-15.
—— NIAGARENSIS Hall, 1883; 11th Ann. Rep. Geol. Nat. Hist. Indiana, 1881, p. 331, pl. 34, figs. 8-15. cf. *C. brevicapitata* (Portl.) McCoy; Brit. Pal. Foss., pl. 1 F, fig. 4.
—— *nupera*. (See *Phacops rana*.)
—— *odontocephala*. (Syn., *Dalmanites selenurus*.)
—— *phlyctainodes* Green, 1837; Am. Jour. Sci., 1st series, vol. 32, p. 167. Dr. Green compares this species with *Encrinurus variolaris*.
—— PLATYS Green, 1832; Mon. Tril. N. A., p. 32, casts 4-5.
                  Upper Heldeberg.
—— PLATYS Hall, 1861; Des. New Species Foss., p. 54.
—— PLATYS Hall, 1862; 15th Rep. N. Y. State Cab. Nat. Hist., p. 82.
—— PLATYS Hall, 1876; Illus. Devon. Foss. Crust., pl. 1, figs. 1-9.
—— PLATYS Hall, 1888; Pal. New York, vol. 7, p. 1, pl. 1, figs. 1-9; pl. 25, figs. 1-2. Schoharie, Knox and Albany counties, etc., N. Y. Fall of the Ohio.
—— *rana*. (See *Phacops rana* Green.)
—— ROSTRATA Vogd., 1879; Am. Jour. Sci., 3d series, vol. 18, p. 477. Taylor's Ridge, near Catoosa Station, Georgia.   Clinton.
—— ROSTRATA Vogd., 1880; Proc. Acad. Nat. Sci. Phila., p. 176, figs. 1-2.
—— ROSTRATA Vogd., 1886; Des. New Species Foss. Crust. Clinton Group, p. 2, figs. 1-2.
—— ROSTRATA Foerste, 1889; Proc. Boston Soc. Nat. Hist., vol. 24, p. 267.
—— *rowi*. (See *Phacops rowi* Green.)
—— RUGOSA Shum., 1855; 1st and 2d Reps. Geol. Missouri, p. 200, pl. B, fig. 14. One mile below Birmingham, Missouri.
                 Lower Helderberg.
—— *selenocephala*. (See *Calymene senaria* Conrad.)
—— SENARIA Con., 1841; 4th Ann. Rep. Pal. Dept. N. Y. Geol. Survey, p. 49. Hudson group at Turin, Lowville and elsewhere, also in Trenton limestone of New York, Ohio, Indiana, etc.
—— SENARIA Emmons, 1842; Geol. New York, 2d Geol. Dist., p. 390, fig.2. Syn , *selenocephala* Green, 1832; Mon. Tril. N. A., p. 31, cast 3.
  blumenbachii Green, 1832; Mon. Tril. N. A., p. 28, cast 1. (Not Brougniart's species.)
  callicephala Green, 1832; Mon. Tril. N. A., p. 31. (Specimens from Ohio and Indiana only.)

CALYMENE. CERATOLICHAS.

brevicapitata Salt., 1848; Mem. Geol. Surv. United Kingdom, vol. 2, p. 341, pl. 11, figs. 1-2. (Not of Portlock, 1843; Geol. Rep. Londonderry, etc., p. 286, pl. 3, fig. 3.

brevicapitata (Portl.) Salt.; Mem. Geol. Survey United Kingdom, vol. 3, pl. 17, figs. 11-12. (Not of Portlock.)

forcipata (McCoy) Salt.; Palæon. Soc. London, vol. 17, p. 97. (Not of McCoy, Sil. Foss. Ireland, pl. 4, fig. 14.)

—— SENARIA Emmons, 1855; Am. Geology, vol. 1, pt. 2, p. 213, pl. 16, fig. 9.

——— SENARIA Hall, 1847; Pal. New York, vol. 1, p. 238, pl. 64, figs. 3 a-u.

—— SENARIA Owen, 1852; Rep. Geol. Wis., Iowa and Minn., p. 624, pl. 2 a, fig. 12.

—— (*blumenbachii*) var. SENARIA Hall, 1852; Pal. New York, vol. 2, p. 299, pl. 66 a, figs. 6 a-e. Clinton.

—— SENARIA var. BLUMENBACHII Salt., 1865; Mon. Brit. Tril., p. 97, pl. 9, figs. 6-10 (not figs. 7, 8, *Calymene brevicapitata* Portl.)

—— SENARIA Meek., Pal. Ohio, vol. 1, p. 173, pl. 14, figs. 14 a-f.

—— SENARIA Miller, 1874; Cincinnati Quart. Jour. Sci., vol. 1, p. 140.

Syn., *callicephala* (Green) Miller, 1877; Catalogue Am. Pal. Foss., p. 213.

—— SENARIA White, 1880; 2d Ann. Rep. Dept. Statistics Geol. Indiana, p. 493, pl. 2, figs. 1-2.

Syn., *callicephala* (Green) Miller, 1882; Jour. Cincinnati Soc. Nat. Hist., vol. 5, p. 117, pl. 5, fig. 8.

—— SENARIA Wal., 1884; Science, March 7, 1884, vol. 3, p. 201, figs. 2-3.

—— SENARIA Wal., 1887; Notes on Some Sections of Trilobites, pl. 1.

—— *spinifera* Con., 1842; Jour. Acad. Nat. Sci. Phila., vol. 8, p. 277. cf. *Cyphaspis girardeauensis*.

—— *trisulcata*. (See *Phacops trisulcata*.)

——— VOGDESII Foerste, 1887; Bull. Denison Univ., vol. 2, p. 95, pl. 8, figs. 12-16. Allen's Quarry, Soldiers' Home Quarry, Ohio; Catoosa Station, Georgia. Clinton.

—— ? *blumenbachi* Foerste, 1886; Bull. Denison Univ., vol. 1, p. 190, pl. 13, fig. 24.

Syn., *clintoni* Vog., 1882; Proc. Acad. Nat. Sci. Phila., p. 178, fig. 3 (not fig. 4).

clintoni Vog., 1886; Des. New Species Foss. Crust. Clinton Group, p. 5, fig. 3 (not fig. 4).

Ceratocephala Warder, 1838. (See *Acidaspis*.)

——— *ceralepta* Anthony. (See *Acidaspis ceralepta*.)

—— *goniata* Warder, 1838; Am. Jour. Sci., 1st series, vol. 34, p. 378, fig. (See *Acidaspis goniata*.)

CERATOLICHAS Hall, 1888; Pal. New York, vol. 7, p. xl. Subgenus *Lichas*. (See *Lichas (Ceratolichas) gryps* Hall. *Lichas (Ceratolichas) dracon* Hall.)

CERAURUS Green, 1832; Monthly Am. Jour. Geol., vol. 1, p. 560, pl., fig. 6.
Mon. Tril. N. A., p. 83, cast 33, pl., fig. 10.
—— Hall, 1847; Pal. New York, vol. 1, p. 242.
- Syn., *Otarion* (in part) Zenk., 1833; Beiträge Nat. Urwelt, p. 44.
Arges (in part) Goldf., 1843; Neues Jahrbuch für Mineral, p. 5.
Cyphaspis (in part) Burm., 1843; Org. Trilobiten, p. 104.
Amphion (in part) Portl., 1843; Geol. Sur. Londonderry, p. 272.
Cyphaspis (in part) McCoy, 1846; Sil. Foss. Ireland, p. 44.
Cheirurus Beyr., 1846; Unters. Trilobiten, etc., II, p. 3.
Eccoptochile Cord., 1847; Prodr., p. 245.
Actinopeltes Cord., 1847; Prodr., p. 247.
Crotalocephalus Salt., 1853; Mem. Geol. Survey United Kingdom, decade 7, pl. 2, p. 10.
Cyrtometopus Ang., 1854; Pal. Scand., p. 32.
Sphærocoryphe Ang., 1854; Pal. Scand., p. 65.
Pseudophærozochus Schm., 1881; Rev. Baltic Sil. Tril., p. 130.
Nieszkowskia Schm., 1881; Rev. Baltic Sil. Tril., p. 130.
Cheirurus APPOLLO Bill., 1860; Canadian Naturalist, vol. 5, p. 322, fig. 28. Point Levis, Quebec. Limestone No. 2.
Cheirurus APPOLLO Bill., 1865; Pal. Foss., vol. 1, p. 413, fig. 397.
Cheirurus bimucronatus (Murch.) Roemer, Die Sil. Fauna W. Tenn., p. 80, pl. 5, fig. 19. Not Murchison's species. (See *Ceraurus niagarensis* Hall.)
crosotus Lock. (See *Acidaspis crosotus*.)
Cheirurus ERYX Bill., 1860; Canadian Naturalist, vol. 5, p. 322, fig. 30. Point Levis, Quebec. Limestone No. 2, also Phillipsburgh.
Cheirurus ERYX Bill., 1865; Pal. Foss., vol. 1, p. 413, fig. 399.
Cheirurus GLAUCUS Bill., 1865; Pal. Foss., vol. 1, p. 323, fig. 308. Staubridge Range 6, lot 20, Canada.
Cheirurus ICARUS Bill., 1860; Canadian Naturalist, vol. 5, p. 67, fig. 14. Anticosti. Hudson.
Cheirurus ICARUS Bill., 1863; Geol. Canada, p. 219, fig. 231.
—— ICARUS Meek, 1873; Pal. Ohio, vol. 1, p. 162, pl. 14, fig. 11 a, b, c.
—— ICARUS Miller, 1874; Cincinnati Quart. Jour. Sci., vol. 1, p. 133.
Cheirurus INSIGNIS (Beyr.) Hall, 1852; Pal. New York, vol. 2, p. 300; also p. 306, pl. 67, figs. 9, 10. Not Beyrich's species.
—— *insignis*. (See *Ceraurus niagarensis* Hall.)
—— MERCURIUS Bill., 1865; Pal. Foss., vol. 1, p. 285, fig. 272. Cow Head, N. F.
—— NIAGARENSIS Hall, 1867; 20th Rep. New York State Cab. Nat. Hist., p. 376, pl. 21, fig. 10. Wauwatosa, Wis. Niagara.
—— NIAGARENSIS Hall, 1870; 20th Rep. New York State Cab. Nat. Hist. (2d ed.), p. 427, pl. 21, figs. 10, 11.
Syn., *Sphærexochus romingeri*? Hall; Doc. Ed. 28th Rep. New York State Mus. Nat. Hist. Explanation of pl. 32, fig. 16 (erroneous reference).

CERAURUS.
— (*Cheirurus*) NIAGARENSIS Hall, 1879; 28th Rep. New York State Mus. Nat. His., p. 189, pl. 32, fig. 16.
  Syn., *Cheirurus bimucronatus* (Murch.) Roemer; Die Sil. Fauna W. Tenn., p. 80, pl. 5, fig. 19.
— NIAGARENSIS Hall, 1882; 12th Ann. Rep. Geol. Nat. Hist. Indiana, 1881, p. 335; pl. 34, fig. 16; pl. 33, fig. 10.
Cheirurus NUMITOR Bill, 1866; Catalogue Sil. Foss. Anticosti, p. 27, fig. 11. English Head. Hudson.
Cheirurus NUPERUS Bill., 1866; Catalogue Sil. Foss. Anticosti, p. 61, fig. 20. East Point Div. 3. Anticosti group.
Cheirurus PERFORATOR Bill., 1865; Pal. Foss., vol. 1, p. 287, fig. 275. All the species described from the Quebec group of Newfoundland by Billings are probably of other genera. The genus *Ceraurus* belongs to the Trenton, Hudson and Niagara periods. Table Head, N. F.
— PLEUREXANTHEMUS Green, 1832; Monthly Am. Jour. Geol., vol. 1, p. 560, pl., fig. 10. Trenton and Hudson.
— PLEUREXANTHEMUS Green, 1832; Mon. Tril. N. A., p. 84, fig. 10, pl. 1, cast 33.
  Syn., *Calymene bucklandi* Anthony, 1838; Am. Jour. Sci., 1st series, vol. 36, p, 106, figs. 1, 2.
— PLEUREXANTHEMUS Hall, 1847; Pal. New York, vol. 1, p. 242, pl. 65, figs. 1 a-n; pl. 66, figs. 1, 1 h. Middleville, Trenton Falls, etc., N. Y.
— PLEUREXANTHEMUS Emmons, 1855; Am. Geol., vol. 1, p. 217, pl. 15, figs. 1 a-k.
Cheirurus PLEUREXANTHEMUS Bill., 1863; Geol. Canada, p. 188, fig. 188.
— PLEUREXANTHEMUS Miller, 1874; Cincinnati Quart. Jour. Sci., vol. 1, p. 132.
— PLEUREXANTHEMUS Wal., 1875; Annals Nat. Hist. New York, vol, 11, p. 155, pl. 11.
— PLEUREXANTHEMUS Wal., 1881; Bull. Mus. Comp. Zool. Harvard Coll., vol. 8, p. 211, pl. 5, figs. 1-6.
Cheirurus POLYDORUS Bill., 1865; Pal. Foss., vol. 1, p. 286, fig. 274. Table Head, Portland Creek, N. F.
Cheirurus POMPILIUS Bill., 1865; Pal. Foss., vol. 1, p. 181, fig. 162. Large Island, Mingan Island. Chazy and Hudson.
Cheirurus PROLIFICUS Bill., 1865; Pal. Foss., vol. 1, p. 285, fig. 273; p. 325, fig. 311. Cow Head, N. F.
— ? PUSTULOSA Hall, 1847; Pal. New York, vol. 1, p. 246, pl. 61, fig. 2 a-h. Watertown, N. Y. Trenton.
— RARUS Wal., 1877; Advanced sheets 31st Rept. New York State Mus. Nat. Hist., p. 15. Beloit, Wis. Trenton.

CERAURUS. CONOCORYPHE.
—— rarus Wal., 1879; 31st Rep. New York State Mus. Nat. Hist., p. 68.
Cheirurus satyrus Bill., 1865; Pal. Foss., vol. 1, p. 324, fig. 309.
Montreal, Canada. Chazy.
Cheirurus sol Bill., 1865; Pal. Foss., vol. 1, p. 288, fig. 276. Table Head, Portland Creek, N. F.
Cheirurus solitarius Bill., 1865; Pal. Foss, vol.* 1, p. 206. St. Antoine, Quebec.
Cheirurus tarquinius Bill., 1863; Proc. Portland Soc. Nat. Hist., vol. 1, p. 121, pl. 1, fig. 22. Port Daniel on the Bay Chaleur.
Upper Silurian.
—— vigilans. (See Encrinurus vigilans.) Middleville, N. Y.
Cheirurus vulcanus Billings, 1865; Pal. Foss., vol. 1, p. 284, fig. 27; p. 324, fig. 310. Cow Head, N. F.

CHARIOCEPHALUS Hall, 1863; 16th Rept. New York State Cab. Nat. Hist., p. 175.
—— tumifrons H. & W., 1877; U. S. Geol. Expl. 40th Par., vol. 4, p. 224, pl. 2, figs. 38, 39. Pogonip Mt., White Pine District, Nevada.
Taconic.
——? tumifrons Wal., 1884; Pal. Eureka Dist., Mon. U. S. Geol. Survey, vol. 8, p. 61, pl. 10, fig. 16.
—— whitfieldi Hall, 1863; 16th Rep. New York State Cab. Nat. Hist., p. 175, pl. 6, figs. 49-53; pl. 10, fig. 20. Trempaleau, Wis. Taconic.

CHASMOPS McCoy, 1849; Annals Mag. Nat. Hist., 2d series, vol. 4, p. 403.
  (See Dalmanites (Chasmops) anchiops, also var. armatus, and var. sobrinus, D. (Chasmops) calypso, D. (Chasmops) erina, D. (Chasmops) macrops.
Dalmanites troosti Saff., 1869; Geol. Tennessee, p. 290. Trenton.
—— troosti Saff. & Vogd., 1889; Proc. Acad. Nat. Sci. Phila., p. 167, fig.

Conocephalus Zenk., 1833; Beiträge Nat. Urwelt, p. 48.

Conocephalites Barr., 1852; Syst. Sil. Bohème, vol. 1, p. 415. For species previously classed under these genera, see Conocoryphe, Ptychoparia and Crepicephalus.
——? contiguus Matt., 1891; Trans. Royal. Soc. Canada, vol. 9, p. 58, pl. 13, figs. 14 a-b. St. John.

CONOCORYPHE Cord., 1847; Prodr., p. 139, pl. 2, fig. 10.
  Conocephalites baileyi Hartt, 1868; Acadian Geol., p. 645 (3d ed.)
  Ratcliff's Millstream. St. John.
—— baileyi Matt., 1884; Trans. Royal Soc. Canada, vol. 1, p. 111, pl. 1, figs. 22-27.
—— (Bailiella) baileyi Matt., 1884; Trans. Royal Soc. Canada, vol. 1, pl. 1, fig. 22.

CONOCORYPHE.   CORYCEPHALUS.
— *(Salteria)* BAILEYI Wal., 1884; Bull. U. S. Geol. Survey, No. 10, p. 32, pl. 4, figs. 3, 3 a; pl. 5, figs. 7, 7 a. The author herein proposed the subgenus *Salteria* (corrected to *Bailiella*) for *Erinnys* Salter, a preoccupied generic name; the same is true of *Salteria*, which was also used for a genus of fossil Crustacea.
Conocephalites CALCIFERA Wal., 1879; 32d Rept. New York State Mus. Nat. Hist., p. 129. Saratoga County, N. Y.   Calciferous.
Conocephalites CORDILLERÆ Rominger, 1887; Proc. Acad. Nat. Sci. Phila., p. 17, pl. 1, fig. 7. Mount Stephen, British Columbia.
                                                                  Taconic.
Conocephalites ELEGANS Hartt, 1868; Acadian Geol. (3d ed.), p. 650. Ratcliff's Millstream.   St. John.
— ELEGANS Wal., 1884; Bull. U. S. Geol. Survey, No. 10, p. 33, pl. 4, figs. 2, 2 a, b.
— *gemini-spinosus* Hartt. (See *Conocoryphe matthewi*.)
Conocephalites HARTTI Wal., 1879; 32d Rept. New York State Mus. Nat. Hist., p. 130. Saratoga County, N. Y.   Calciferous.
Conocephalites MATTHEWI Hartt, 1868; Acadian Geology, p. 646, fig. 224. Ratcliff's Millstream and St. John.
Conocephalites MATTHEWI Matt., 1884; Trans. Royal Soc. Canada, vol. 1, p. 103, pl., figs. 6-21.
Harttia matthewi Wal., 1884; Bull. U. S. Geol. Survey, No. 10, p. 19, pl. 1, fig. 3.
— MATTHEWI Wal., 1884; Bull. U. S. Geol. Survey, No. 10, p. 28, pl. 4, figs 1, 1 a, b.
— *Conocephalites* OPTATUS Hall, 1863; 16th Rep. New York State Cab. Nat. Hist., p. 22, pl. 5 a, fig. 7. Trempleau, Wis.   Taconic.
Conocephalites trilineatus. (See *Atops trilineatus* Emmons.)
— *trilineatus.* (See *Atops trilineatus* Emmons.)
— *(Bailella)* WALCOTTI Matt., 1884; Trans. Royal Soc. Canada, vol. 1, p. 119, pl. 1, figs. 36-36 b. St. John, N. B.
— WALCOTTI Wal., 1884; Bull. U. S. Geol. Survey, No. 10, p. 30.

CONOLICHAS Dames, 1877; Zeitschr. Deutsch. Geol. Gesell., Berlin, 1877, p. 806. *Conolichas,* subgenus *Lichas* Hall, 1888; Pal. New York, vol. 7, p. xxxix. (See *Lichas (Conolichas) hispidus* Hall and *Lichas (Conolichas) eriops* Hall.

CORONURA Hall, 1888; Pal. New York, vol. 7, p. xxxii. (See *Dalmanites (Coronura) aspectans, D. (Coronura) myrmecophorus* and *D. (Coronura) emarginatus*).

CORYCEPHALUS Hall, 1888; Pal. New York, vol. 7, p. xxxiv. See *Dalmanites (Corycephalus) regalis, D. (Corycephalus) pygmæus,* and *D. (Corocephalus) dentatus*.)

CRYPHÆUS.
Crepicephalus Owen, 1852; Rept. Geol. Survey Wisconsin, Iowa and Minnesota, p. 576. (See *Lonchocephalus*.) There is hardly a question that Dr. D. D. Owen had before him when he wrote the description of his genus *Crepicephalus* a true *Ptychoparia*. He refers the following figures to the genus: Plate 1, fig. 8, *Crepicephalus?*, pygidium from the Miniskah Trilobite grit. Plate 1, fig. 13, *Crepicephalus? wisconensis* (n. s.), part of cephalothorax and a portion of the cephalic shield of *D. granulosus?* Plate 1 a, fig. 10, glabella of *Crepicephalus* n. g., near Miniskah. Plate 1 a, fig. 14, pygidium of *Crepicephalus? miniscaenis* n. s., from Miniskah grits. Plate 1 a, fig. 16, pygidium and portion of *Crepicephalus?* and portions of cephalothorax of that genus, from the Miniskah Trilobite grit. Plate 1 a, fig. 18, glabella of *Crepicephalus*, near Miniskah, Mississippi River.

—— (*Bathyurus?*) *angulatus* H. & W. (See *Ptychoparia angulatus*.)
—— (*Loganellus*) *anytus* H. & W. (See *Ptychoparia anytus*.)
—— (*Loganellus*) *centralis* Whitf. (See *Ptychoparia oweni*.)
——? GIBBSI Whitf., 1880; Ann. Rept. Geol. Survey Wisconsin, 1879, p. 50. Berlin, Green Lake Co., Wis.　　Taconic.
—— GIBBSI Whitf., 1880; Geol. Wisconsin, vol. 2, p. 67.
—— GIBBSI Whitf., 1882; Geol. Wisconsin, vol. 4, p. 184, pl. 10, figs. 12, 13.
—— (*Loganellus*) *granulosa* H. & W. (See *Ptychoparia granulosa*.)
—— (*Loganellus*) *haguei*. (See *Lisostracus haguei* and *Ptychoparia haguei*.)
—— (*Loganellus*) *maculosus* H. & W. (See *Ptychoparia maculosus*.)
—— (*Loganellus*) *montanensis* Whitf. (See *Ptychoparia montanensis*.)
—— (*Loganellus*) *nitidens* H. & W. (See *Ptychoparia nitidens*.)
—— ONUSTUS Whitf., 1878; Ann. Rept. Geol. Survey Wisconsin, 1877, p. 53. Ettrick and Eau Claire, Wis.　　Taconic.
—— ONUSTUS Whitf., 1882; Geol. Wisconsin, vol. 4, p. 182, pl. 1, figs. 22, 23.
—— *oweni*. (See *Agraulos oweni* and *Ptychoparia oweni*.)
—— (*Loganellus*) *planus* Whitf. (See *Ptychoparia planus*.)
—— (*Loganellus*) *quadrans* H. & W. (See *Ptychoparia quadrans*.)
—— (*Loganellus*) *simulator* H. & W. (See *Ptychoparia nitidens*.)
—— (*Loganellus*) *unisulcatus* H. & W. (See *Ptychoparia unisulcatus*.)
——? *wisconsensis* Owen. (See *Anomocare wisconsensis*.)

Crepicephalus Wal., 1884; Bull. U. S. Geol. Survey, No. 10, p. 35. Type, *Dikelocephalus? iowensis* Owen. (See *Lonchocephalus*.)

CRYPHÆUS Green, 1837; Am. Jour. Sci., 1st series, vol. 32, p. 343.
　　Dalmanites (*Cryphœus*) BARRISI Hall, 1888; Pal. New York, vol. 7, p. 48, pl. 16 a, fig. 18. Near New Buffalo, Searstown and Iowa City, Iowa.　　Hamilton.
—— BOOTHI Green, 1837; Am. Jour. Sci., 1st series, vol. 32, p. 343, fig. Huntington, Penn.　　Hamilton.

CRYPHÆUS. CRYPTOLITHUS.

Syn., *Cryphæus greeni* Conrad, 1839; 2d Ann. Rep. Pal. Dept. New York Geol. Survey, p. 66.
Cryphæus calliteles Conrad, 1839; 2d Ann. Rep. Pal. Dept. New York Geol. Survey, p. 62.
Asaphus halli Conrad, 1840; 3d Ann. Rep. Pal. Dept. New York Geol. Survey, p. 204.
Cryphæus calliteles Hall, 1843; Geol. New York, pt. 4, p. 201.
Cryphæus calliteles De Verneuil, 1850; Bull. Soc. Géol. France, 2d series, vol. 7, p. 164, pl. 1, fig. 3.
Phacops pectinatus F. A. Roemer, 1850; Beiträge zur Kenntn. Harzgeb., p. 62, pl. 9, fig. 27.
Dalmania calliteles Emmons, 1860; Manual Geol., p. 138, fig. 124 (7).
Dalmania BOOTHI Hall, 1861; Des. New Species of Fossils, p. 63.
Dalmania BOOTHI Hall, 1862; 15th Rep. New York State Cab. Nat. Hist., p. 91.
Dalmanites BOOTHI Hall, 1876; Illus. Devonian Foss., pl. 16, figs 1-6, 9-11, 13, 15, 16.
Syn., *Cryphæus stellifer* Kayser, 1878; Abhandl. Geol. Specialkarte Pr., vol 2, pt. 4, p. 33 (in part).
Cryphæus laciniatus Kayser 1878; idem., p. 34.
Dalmanites (Cryphæus) BOOTHI Hall, 1888; Pal. New York, vol. 7, p. 42, pl. 16, figs. 1-4; pl. 16 a, figs. 3-8.
—— CALLITELUS Green, 1837; Am. Jour. Sci., 1st series, vol. 32, p. 346. Huntingdon, Penn. Hamilton.
Dalmanites BOOTHI Hall, 1876; Illus. Devonian Foss., pl. 16, figs. 7, 8, 12, 14.
—— (?) CALLITELES Kayser, 1878; Abhandl. Geol. Specialkarte Pr., vol. 2, pt. 4, p. 32, pl. 3, fig. 10.
Dalmanites (Cryphæus) BOOTHI var. CALLITELES Hall, 1888; Pal. New York, vol. 7, p. 45, pl. 16, figs. 5-22; pl. 16 a, figs. 9-17.
Dalmanites (Cryphæus) COMIS Hall, 1888; Pal. New York, vol. 7, p. 41, pl. 16 a, fig. 1. Walpole, Ontario. Corniferous.
Dalmania PLEIONE Hall, 1861; Des. New Species Fossils, p. 62. Fall of the Ohio. Hamilton.
Dalmania PLEIONE Hall, 1862; 15th Rep. New York State Cab. Nat. Hist., p. 90.
Dalmanites PLEIONE Hall, 1876; Illus. Devonian Foss, pl. 16, fig. 17.
—— PLEIONE Kayser, 1878; Abhandl. Geol. Specialkarte Pr., vol. 2, pt. 4, p. 33.
Dalmanites (Cryphæus) PLEIONE Hall, 1888; Pal. New York, vol. 7, p. 41, pl. 16 a, fig. 2.

CRYPTOLITHUS Green, 1832; Mon. Tril. North America, p. 72. This genus should replace that of *Trinucleus* Murch., 1839.

CTENOPYGE. CYPHASPIS.

Cryptonymus Eichw., 1825; Syn. to *Asaphus*, *Illænus* and *Niobe*.
Cryptonymus Eichw., 1840. (See *Encrinurus*.)
CTENOPYGE Linrs., 1880; Geol. Föreningens Stockholm Förhandl., vol. 5, p. 145.
—— FLAGELLIFERA (Ang.) Matt., 1891; Trans. Roy. Soc. Canada, vol. 9, p. 56, pl. 13, fig. 12 a-b. St. John.
—— SPECTABILIS (Brögg.) Matt., 1891; *idem*., p. 57, pl. 13, fig. 13 a-b. St. John.
—— PECTEN (Salt.) Matt., 1891; *idem*., p. 58.
CYBELE Löven, 1845; Svenska Vetensk. Akad. Förhandl., p. 110.
—— *punctata* Hall. (See *Encrinurus ornatus*.)
CYPHASPIS Burm., 1843; Org. Trilobiten, p. 98.
—— BELLULA Whiteaves, 1892; Contrib. Canadian Pal., vol. 1, p. 349, pl. 46, figs. 9, 9 a. Lake Winnipegosis, Manitoba. Devonian.
—— CHRISTYI Hall, 1863; Trans. Albany Inst., vol. 4, p. 220. Waldron, Indiana. Niagara.
—— CHRISTYI Hall, 1875; 28th Rep. New York State Mus. Nat. Hist., pl. 32, figs. 5-7.
—— CHRISTYI Hall, 1879; 28th Rep. New York State Mus. Nat. Hist., p. 188, pl. 32, figs. 5-7.
—— CHRISTYI White, 1880; 2d Ann. Rep. Dept. Statistics and Geol. Indiana, p. 493, pl. 3. fig. 9.
—— CHRISTYI Hall, 1882; 11th Ann. Rep. Dep. Geol. Nat. Hist. Indiana, 1881, p. 333, pl. 34, figs. 5-7.
—— CLINTONI Foerste, 1889; Proc. Boston Soc. Nat. Hist., vol. 24, p. 272, pl. 4, fig. 22. Anticosti, also Cumberland Gap, etc.
—— CŒLEBS Hall, 1888; Pal. New York, vol. 7, p. 151, pl. 24, fig. 1. Clarksville and Schoharie, New York. Lower Helderberg.
—— CRASPEDOTA Hall, 1888; Pal. New York, vol. 7, p. 148, pl. 24, figs. 15-20. Near Centerfield, Township of Canandaigua, N. Y. Hamilton.
—— DIADEMA Hall, 1888; Pal. New York, vol. 7, p. 144, pl. 24, fig. 13. Upper Helderberg.
—— GIRARDEAUENSIS Shum., 1855; 1st and 2d Rep. Geol. Missouri, p. 197, pl. B, fig. 11 a-b. Silurian.
—— GIRARDEAUENSIS Emmons, 1860; Manual Geol., p. 159, fig. 145 (2).
—— HYBRIDA Hall, 1888; Pal. New York, vol. 7, p. 144, pl. 24, fig. 14. In boulders Canandaigua, Ontario Co., N. Y. Upper Helderberg.
Phillipsia LÆVIS Hall, 1876; Illus. Devonian Foss., pl. 21, fig. 29.
—— LÆVIS Hall, 1888; Pal. New York, vol. 7, p. 150, pl. 21, fig. 29. Chemung County, N. Y.
Phillipsia MINUSCULA Hall, 1876; Illus. Devonian Foss., pl. 20, fig. 17. Clarksville, Schoharie, Phelps, Le Roy, etc., N. Y. Upper Helderberg.

CYPHASPIS. DALMANITES.
—— MINUSCULA Hall, 1888; Pal. New York, vol. 7, p. 140, pl. 20, fig. 17; pl. 24, figs. 7-12.
Phillipsia? (*Brachymetopus?*) ORNATA Hall, 1876; Illus. Devonian Foss., pl. 21, fig. 1. 18 Mile Creek, Erie Co., N. Y. Hamilton.
Syn., *Phillipsia coronata* Miller, 1877; Catalogue Am. Pal. Foss., p. 221.
Phillipsia coronata? Wal., 1884; Pal. Eureka Dist., Mon. U. S. Geol. Survey, vol. 8, p. 211.
—— ORNATA Hall, 1888; Pal. New York, vol. 7, p. 145, pl. 21, fig. 1; pl. 24, fig. 21.
—— ORNATA var. BACCATA Hall, 1888; Pal. New York, vol. 7, p. 146, pl. 24, figs. 22, 23.
—— STEPHANOPHORA Hall, 1888; Pal. New York, vol. 7, p. 142, pl. 24, figs. 2-6. Canandaigua, East Bloomfield, N. Y. Upper Helderberg.
Dalmania Emm., 1844; Zur. Naturgeschichte Trilobiten, p. 15. This name having been used in 1830 for a genus of the Diptera, J. Barrande suggested that of *Dalmanites*.
DALMANITES (Emm.) Barr., 1872; Syst. Sil. Bohême, Suppl., vol. 1, p. 27.
Syn., *Pleuracanthus* Milne-Edw., 1840; Crust., vol. 3, p. 329.
Odontochile Cord., 1847; Prodr., p, 208, pl. 5, fig. 56.
Osteropyge Cord., 1847; Prodr., p. 241, pl. 6, fig. 66.
Metacanthus Cord., 1847; Prodr., p. 242, pl. 6, fig. 67.
Chasmops McCoy, 1849; Annals Nat. Hist., 2d series, vol. 4.
Cryphæus Green, 1837; Am. Jour. Sci., 1st series, vol. 32, p. 343.
Odontocephalus Con., 1840; 3d Rep. Pal. Dept. New York Geol. Survey, p. 204.
—— (*Hausmannia*) Hall, 1888; Pal. New York, vol. 7, p. xxxi.
—— (*Coronura*) Hall, 1888; Pal. New York, vol. 7, p. xxxii.
—— (*Corycephalus*) Hall, 1888; Pal. New York, vol. 7, p. xxxiv.
Asaphus? *acantholeurus* Con., 1841; 5th Rep. Pal. Dept. New York Geol. Survey, p. 48. Onondaga.
Syn., *Dalmanites myrmecophorus*.
Asaphus? *acantholeurus* (Con.) Hall, 1862; 15th Rep. New York State Cab. Nat. Hist., p. 113. Near Schoharie.
—— *acantholeurus* Hall, 1876; Illus. Devonian Foss. Crust., pl. 19, fig. 14.
—— *acantholeurus* Hall, 1876; Illus. Devonian Foss. Crust., pl. 19, fig. 14.
—— ACHATES Bill., 1860; Canadian Naturalist, vol. 5, p. 63, fig. 9. City of Ottawa. Trenton.
—— ACHATES Bill., 1863; Geol. Canada, p. 187, fig. 186.
—— ÆGERIA Hall., 1861; Des. New Species Fossils, p. 57. Williamsville, Clarence Hollow, etc., N. Y. Upper Helderberg.
—— ÆGIRIA Hall, 1862; 15th Rep. New York State Cab. Nat. Hist., p. 85.
—— ÆGIRIA Hall, 1876; Illus. Devonian Foss. Crust., pl. 12, figs. 1, 2, 7 (not figs. 1, 2, 7).

DALMANITES.

—— ÆGIRIA Hall, 1876; Illus. Devonian Foss. Crust., p. 12, figs. 3, 6, 8.
  Odontocephalus ÆGERIA M. & W., 1868; Geol. Illinois, vol. 3, p. 417, pl. 10, figs. 4 a, b, c. cf. *D. bifida.*
—— *(Odontocephalus)* ÆGERIA Hall, 1888; Pal. New York, vol. 7, p. 53. pl. 11 b, figs. 1-11.
  Calymene ANCHIOPS Green, 1832; Monthly Am. Jour. Geol., vol. 1, p. 558. Upper Helderberg.
  Calymene ANCHIOPS Green, 1832; Mon. Tril. N. A., p. 35, cast 7. Vicinity of Albany, N. Y.
    Syn., *Asaphus laticostatus* Green, 1832; Monthly Am. Jour. Geol., vol. 1, p. 559.
      Asaphus laticostatus Green, 1832; Mon. Tril. N. A., p. 45, cast 13.
  Phacops ANCHIOPS Burm., 1843; Org. Trilobiten, p. 90 (Ray. Soc. edition).
  Phacops ANCHIOPS Hall, 1851; Rep. Geol. Lake Superior Land Dist. (Foster & Whitney), pt. 2, p. 224, pl. 35, figs. 3 a, 3 b. (See *Dalmanites anchiops* var. *armatus.*)
  Dalmania ANCHIOPS Hall, 1861; Des. New Species Fossils, p. 55. Schoharie and Helderberg Mountain, N. Y.
  Dalmania ANCHIOPS Hall, 1862; 15th Rep. New York State Cab. Nat. Hist., p. 83.
—— ANCHIOPS Hall, 1876; Illus. Devonian Foss. Crust., p. 9, figs. 1, 3-6, 9, 10, 12, 13; pl. 10, figs. 6-14.
—— *(Chasmops)* ANCHIOPS Hall, 1888; Pal. New York, vol. 7, p. 59, pl. 9, figs. 1-6, 10, 13; pl. 10, figs. 1-14.
  Dalmania ANCHIOPS var. ARMATUS Hall, 1861; Des. New Species Foss., p. 56.
  Dalmania ANCHIOPS var. ARMATUS Hall, 1862; 15th Rep. New York State Cab. Nat. Hist., p. 84.
—— ANCHIOPS var. ARMATUS Hall, 1876; Illus. Devonian Foss. Crust., pl. 9, figs. 2, 7, 8.
—— *(Chasmops)* ANCHIOPS var. ARMATUS Hall, 1888; Pal. New York, vol. 7, p. 62, pl. 9, figs. 7-9; pl. 10, fig. 14. Schohariegnt, Albany and Schoharie counties, N. Y.
—— *(Chasmops)* ANCHIOPS var. SOBRINUS Hall, 1888; Pal. New York, vol. 17, p. 62, pl. 9, fig. 11.
  Odontocephalus arenarius M. & W., 1868; Geol. Survey Illinois, vol. 3, p. 417, pl. 9, fig. 10. (See *Dalmanites selenurus.*) Four miles west of Jonesboro, Ill.
  Asaphus aspectans Con., 1841; 5th Ann. Rep. Pal. Dept. New York Geol. Survey, p. 49, fig. 9. Upper Helderberg.
    Syn., *Asaphus! denticulatus* Con.; *idem.*, p. 48.
  Dalmania aspectans Hall, 1861; Des. New Species Foss., p. 60 (in error *D. adspectans*). Schoharie and Helderberg Mountains, N.Y.

DALMANITES.

Dalmania ASPECTANS Hall, 1862; 15th Rep. New York State Cab. Nat.
Hist., p. 88 (in error *D. adspectans*).
Syn., *Dalmania helena* Hall, 1861; Des. New Species Foss., p. 61.
Dalmania helena Hall, 1862; 15th Rep. New York State Cab.
Nat. Hist., p. 89.
Dalmanites ohioensis Meek, 1871; Proc. Acad. Nat. Sci. Phila.,
p. 91.
Dalmanites ohioensis Meek, 1873; Pal. Ohio, vol. 1, p. 234, pl.
23, fig. 1.
Dalmanites denticulatus Hall, 1876; Illus. Devonian Foss. Crust.,
pl. 10, fig. 1.
—— *aspectans* Hall, 1876; Illus. Devonian Foss. Crust., pl. 13, figs. 6 8.
Syn., *helena* Hall, 1876; Illus. Devonian Foss. Crust., pl. 13, figs.
11–14.
—— (*Coronura*) *aspectans* Hall, 1888; Pal. New York, vol. 7, p. 33, pl. 13,
figs. 1–11, 13. (See *Dalmanites diurus*.)
—— *Cryphæus harrisi* Hall. (See *Cryphæus harrisi* Hall.)
—— BEBRYX Bill., 1860; Canadian Naturalist, vol. 5, p. 62, fig. 8. City of
Ottawa. Trenton.
—— BEBRYX Bill., 1863; Geol. Canada, p. 187, fig. 185.
Dalmania BIFIDA Hall, 1861; Des. New Species Foss., p. 63.
Upper Helderberg.
Dalmania BIFIDA Hall, 1862; 15th Rep. New York State Cab. Nat.
Hist., p. 91. Stafford.
—— (*Odontocephalus*) BIFIDA Hall, 1888; Pal. New York, vol. 7, p. 53,
pl. 11 b, figs. 22–25. Upper Helderberg. Phelps, Ontario
county, N. Y., etc. cf. *Dalmanites ægeria* Hall. Corniferous.
Dalmania boothi. (See *Cryphæus boothi* Green.)
—— BICORNIS Hall, 1875; 28th Rep. New York State Mus. Nat. Hist.
Expl., pl. 33, fig. 18. Waldron, Indiana. Niagara.
—— BICORNIS Hall, 1879; 28th Rep. New York State Mus. Nat. Hist., p.
196, pl. 33, fig. 18.
—— BICORNIS Hall, 1882; 11th Ann. Rep. Geol. Nat. Hist. Indiana, 1881,
p. 342, pl. 35, fig. 18.
—— BREVICEPS Hall, 1866; 24th Rep. New York State Mus. Nat. Hist., p.
223, pl. 8, figs. 15–16. Lebanon, Ohio. Hudson.
—— BREVICEPS Miller, 1874; Cincinnati Quart. Jour. Sci., vol. 1, p. 143.
—— BREVICEPS H. & W., 1875; Pal. Ohio, vol. 2, p. 108, pl. 4, figs. 15–16.
Phacops CALLICEPHALUS Hall, 1847; Pal. New York, vol. 1, p. 247, pl.
65, figs. 3 a–3 i. Middleville and Watertown, N. Y. Trenton.
Phacops CALLICEPHALUS Hall, 1851; Rep. Geol. Lake Superior Land
Dist. (Foster & Whitney), vol. 2, p. 212, pl. 3, figs. 3 a, 3 b.
Phacops CALLICEPHALUS Emmons, 1855; Am. Geol., vol. 1, pt. 2, p.
214, pl. 15, figs. 7, 7 a.

DALMANITES.

Dalmania callitelea. (See *Cryphæus callitelus.*)
Dalmania CALYPSO Hall, 1861; Des. New Species Foss., p. 61. Falls of the Ohio. Upper Helderberg.
—— *Dalmania* CALYPSO Hall, 1862; 15th Rep. New York State Cab. Nat. Hist., p. 89.
—— CALYPSO Hall, 1876; Illus. Devonian Foss. Crust., pl. 13, figs. 1-2.
—— (*Chasmops*) CALYPSO Hall, 1888; Pal. New York, vol. 7, p. 4, pl. 16 a, fig. 1.
—— ? CAYAHOGOE Claypole, 1884; Geol. Mag., vol. 1, p. 305, fig. Akron, Ohio. Carboniferous.
—— ? CAYAHOGOE H. W., 1884; Mon. Brit. Carboniferous Tril., p. 79, fig.
—— CARLEYI Meek, 1872; Am. Jour. Sci., 3d series, vol. 3, p. 424. Cincinnati, Ohio. Hudson.
—— CARLEYI Meek, 1873; Pal. Ohio, vol. 1, p. 170, pl. 14, figs. 2 a, 2 d.
—— CARLEYI Miller, 1874; Cincinnati Quart. Jour. Sci., vol. 1, p. 142.
Dalmania caudata (Broug.) Roemer. (See *Dalmanites limulurus* Green.)
—— (*Cryphæus*) *comis.* (See *Cryphæus comis* Hall.)
—— CONCINNUS Hall, 1876; Illus. Devonian Foss. Crust., pl. 10, figs. 3-5. Schoharie.
—— (*Hausmannia*) CONCINNUS Hall, 1888; Pal. New York, vol. 7, p. 30, pl. 11 a, figs. 9-11. Schoharie, near Clarksville, Albany county, etc., N. Y.
—— (*Hausmannia*) CONCINNUS var. SERRULA Hall, 1888; Pal. New York, vol. 7, p. 30, pl. 11 a, fig. 12.
—— *Dalmania* CORONATUS Hall, 1861; Des. New Species Foss., p. 58. Upper Helderberg.
—— *Dalmania* CORONATUS Hall, 1862; 15th Rep. New York State Cab. Nat. Hist., p. 86. In Hamilton Group, near Skeneateles Lake, N. Y.
—— CORONATUS Hall., 1876; Illus. Devonian Foss. Crust., pl. 12, figs. 15-17; pl. 13, fig. 5.
—— (*Odontocephalus?*) CORONATUS Hall, 1888; Pal. New York, vol. 7, p. 54, pl. 11 b, figs. 12-14.
—— DANÆ M. & W., 1865; Proc. Acad. Nat. Sci. Phila., p. 264. Two miles above Thebes, Illinois. Niagara.
—— DANÆ M. & W., 1868; Pal. Illinois, vol. 3, p. 363, pl. 6, figs. 1 a, 1 f.
—— DENTATA Barrett, 1876; Am. Jour. Sci., 3d series, vol. 11, p. 200, pl. Lower Helderberg.
—— DENTATA Barrett, 1876; Am. Jour. Sci., 3d series, vol. 12, p. 70.
—— (*Corycephalus*) DENTATUS Hall, 1888; Pal. New York, vol. 7, p. 58, pl. 11 a, figs. 4-6. Port Jervis, Becraft's Mt. and Hudson, N. Y.
Asaphus DENTICULATUS Conrad, 1841; 5th Rep. Pal. Dept. New York Geol. Survey, p. 48. Lower Helderberg.
—— DENTICULATUS Hall, 1876; Illus. Devonian Foss. Crust., pl. 10, fig. 1.

DALMANITES.
— (*Phæton*) DENTICULATUS Meek, 1877; U. S. Geol. Expl. 40th Par., vol. 4, p. 49, pl. 1, figs. 10 a, b.
Asaphus diurus Green is referred to *Coronura diurus* by Clarke. (See note on *Coronura aspectans*, p. 5.)
Syn., *Asaphus aspectans* Conrad; also, *Dalmanites ohioensis* Meek, and *D. helena* Hall, all from the Corniferous limestone.
—— EMARGINATUS Hall, 1876; Illus. Devonian Foss. Crust., pl. 10, fig. 2.
Upper Helderberg.
— (*Coronura?*) EMARGINATUS Hall, 1888; Pal. New York, vol. 7, p. 40, pl. 11 a, figs. 7, 8. Schoharie, N. Y.
— EPICRATES Bill., 1863; Proc. Portland Soc. Nat. Hist., vol. 1, p. 119, pl. 1, fig. 21. Stair Falls, Maine. Silurian.
Dalmania ERINA Hall, 1861; Des. New Species Foss., p. 62. Williamsville, N. Y. Upper Helderberg.
Dalmania ERINA Hall, 1862; 15th Rep. New York State Cab. Nat. Hist., p. 90.
— ERINA Hall, 1876; Illus. Devonian Foss. Crust., pl. 13, figs. 3–4.
— (*Chasmops*) ERINA Hall, 1888; Pal. New York, vol. 7, p. 67, pl. 11, figs. 16–18.
Dalmania helena. (See *Dalmanites aspectans*.)
—— INTERMEDIUS Wal., 1877; advance sheets 31st Rep. New York State Cab. Nat. Hist., p. 17. Two miles north of Dunlith, Ill., etc.
Trenton.
—— INTERMEDIUS Wal., 1879; 31st Rep. New York State Cab. Nat. Hist., p. 69.
Phacops? LATICAUDA Hall, 1847; Pal. New York, vol. 1, p. 248, pl. 64, fig. 3.
Asaphus LIMULURUS Green, 1832; Monthly Am. Jour. Geol., vol. 1, p. 559. Niagara.
Asaphus LIMULURUS Green, 1832; Mon. Tril. N. A., p. 48, cast 16.
Phacops LIMULURUS Hall, 1852; Pal. New York, vol. 2, p. 303, pl. 67, figs. 1–8. Lockport, N. Y.
Syn., *Dalmania caudata* (Broug.) Roemer, 1860; Die Sil. Fauna W. Tenn., p. 82, pl. 5, fig. 21.
Dalmania LOGANI Hall, 1860; Canadian Naturalist, vol. 5, p. 156, fig. 18. Nova Scotia. Silurian.
Dalmania MACROPS Hall, 1861; Des. New Species Foss., p. 59.
Schoharie, N. Y. Upper Helderberg.
Dalmania MACROPS Hall, 1862; 15th Rep. New York State Cab. Nat. Hist., p. 87.
—— MACROPS, 1876; Illus. Devonian Foss. Crust., pl. 13, figs. 9, 10.
— (*Chasmops*) MACROPS Hall, 1888; Pal. New York, vol. 7, p. 68, pl. 11 a, figs. 14–15. Schoharie, N. Y.

DALMANITES.
—— MACROURA? (Sjogr.) Bill., 1866; Catalogue Sil. Foss. Anticosti, p. 61. Billings proposes the following classification of this variable species: *Dal. macroura* (Swedish), *D. affinis* (English), *D. anticostientis* n. sp. (Canadian).
Dalmania META Hall, 1862; Geol. Wisconsin, vol. 1, p. 433. Trenton.
—— MEEKI Wal., 1884; Pal. Eureka Dist. Mon. U. S. Geol. Survey, vol. 8, p. 207, pl. 17, figs. 5-5 a, b (not fig. 5 c). Comb's Peak, Reese and Berry Cañon, Nev.
—— (undetermined sp.) Meek, 1877; U. S. Geol. Expl. 40th Par., vol. 4, p. 48, pl. 1, figs. 11, 11 a.
—— (*Hausmannia*) MEEKI Hall, 1888; Pal. New York, vol. 7, p. 32, pl. 11 a, figs. 28-30.
Asaphus MICRURUS Green, 1832; Monthly Am. Jour. Geol., p. 559, pl. 14, fig. 3. Lower Helderberg.
Asaphus MICRURUS Green, 1832; Mon. Tril. N. A., p. 56, cast 19, pl., fig. 3.
Dalmania MICRURUS Hall, 1859; Pal. New York, vol. 3, p. 359, pl. 75, figs. 13-18.
Asaphus MYRMECOPHORUS Green, 1835; Am. Jour. Sci., 1st series, vol. 23, p. 307. Upper Helderberg.
Asaphus MYRMECOPHORUS Green, 1835; Suppl. Mon. Tril. N. A., p. 16, cast 40*.
Dalmania MYRMECOPHORUS Hall, 1861; Des. New Species Foss., p. 60.
Dalmania MYRMECOPHORUS Hall, 1862; 15th Rep. New York State Cab. Nat. Hist., p. 88.
—— MYRMECOPHORUS Hall, 1876; Illus. Devonian Foss. Crust., pl. 13, figs. 15-16.
—— (*Coronura*) MYRMECOPHORUS Hall, 1888; Pal. New York, vol. 7, p. 37, pl. 11 a, fig. 13; pl. 13, fig. 12; pl. 14, figs. 1-6; pl. 15, figs. 1-4. Clarksville, Schoharie, etc., N. Y.
Asaphus NASUTUS Con., 1841; 5th Rep. Pal. Dept. New York Geol. Survey, p. 48. Lower Helderberg.
Dalmania NASUTUS Hall, 1859; Pal. New York, vol. 3, p. 362, pl. 75, fig. 2; pl. 76, figs. 1-9. Schoharie, N. Y.
Dalmanites ohioensis. (See *Dalmanites aspectans*.)
(*Hausmannia*) PHACOPTYX Hall, 1888; Pal. New York, vol. 7, p. 31, pl. 11 a, figs. 23-29. Cayuga, Province of Ontario.
Upper Helderberg.
Dalmania pleione. (See *Cryphæus pleione*.) Fall of the Ohio.
Asaphus PLEUROPTERYX Green, 1832; Monthly Jour. Geol., vol. 1, p. 559. Lower Helderberg.
Asaphus PLEUROPTERYX Green, 1832; Mon. Tril. N. A., p. 55, cast 18.
Dalmania PLEUROPTERYX Hall, 1859; Pal. New York, vol. 3, p. 356, pl. 74, figs. 5 (?), 9 (?) (not figs. 1-4, 6-8, 10-12), pl. 75, fig. 1?

DALMANITES.
—— (*Hausmannia*) PLEUROPTERYX Hall, 1888; Pal. New York, vol. 7, p. 28,
  pl. 11 a, figs. 1-3. Clarksville, Schoharie; also in Oriskany sand-
  stone, Township of Walpole; Corniferous limestone of Clarence
  Hollow, etc., N. Y.
—— (*Corycephalus*) PYGMÆUS Hall, 1888; Pal. New York, vol. 7, p. 56, pl.
  11, figs. 5-8. Boulders, Canandaigua, N. Y.
                                              Corniferous. Upper Helderberg.
- - - REGALIS Hall, 1876; Illus. Devonian Foss. Crust., pl. 11, figs. 1-4.
  Knox, Albany county, N. Y.   Schoharie.   Upper Helderberg.
—— (*Corycephalus*) REGALIS Hall, 1888; Pal. New York, vol. 7, p. 55, pl.
  11, figs. 1-4.
Asaphus SELENURUS Eaton, 1832; Geological Text Book, p. 31, pl. 1,
  fig. 1.   Upper Helderberg.
Asaphus SELENURUS Green, 1832; Mon. Tril. N. A., p. 46, casts 14
  and 15.
  Syn., *Calymene?* *odontocephala* Green, 1835; Suppl. Mon. Tril. N. A.,
    p. 9, cast 36.
      Calymene? *odontocephala* Green, 1835; Am. Jour. Sci., 1st series,
        vol. 25, p. 334.
Odontocephalus SELENURUS Con., 1840; 3d Rep. Pal. Dept. New York
  Geol. Survey, p. 204.
Odontocephalus SELENURUS Vanuxem, 1842; Geol. New York, 3d Geol.
  Dist., pp. 139, 140, fig. 1.
Odontocephalus SELENURUS Hall, 1843; Geol. New York, 4th Geol.
  Dist., p. 175, fig. 1.
Dalmania SELENURUS Hall, 1859; 12th Rep. New York State Cab.
  Nat. Hist., p. 88.
Dalmania SELENURUS Hall, 1861; Des. New Species Foss., p. 56.
  Helderberg Mountains.
Dalmania SELENURUS Hall, 1862; 15th Rep. New York State Cab. Nat.
  Hist., p. 84. Helderberg Mts., Schoharie and Auburn, N. Y.
  Syn., *Odontocephalus?* *Dalmanites* (*Odontocephalus*) *arenarius* M. &
    W., 1868; Geol. Survey Illinois, vol. 3, p. 416, pl. 9, fig. 10.
—— SELENURUS Hall, 1876; Illus. Devonian Foss. Crust., pl. 12, figs. 12-
  14.
—— (*Odontocephalus*) SELENURUS Hall, 1888; Pal. New York, vol. 7, p. 49,
  pl. 11 b, figs. 15-21; pl. 12, figs. 1-13. Marbletown, Clarksville,
  Schoharie, etc., N. Y.
Dalmania TRIDENS Hall, 1859; Pal. New York, vol. 3, p. 361, pl. 75,
  figs. 3-6. Schoharie, N. Y.   Lower Helderberg.
Dalmania TRIDENTIFERA Shumard, 1855; 1st and 2d Geol. Reps.
  Missouri, p. 199, pl. B, figs. 8 a-c. Bailey's Landing, Birming-
  ton, Missouri.   Lower Helderberg.
- TRIDENTIFERA M. & W., 1868. Pal. Illinois, vol. 3, p. 391, pl. 7, fig.
  16.

DALMANITES. DICELLOCEPHALUS.
— *troosti* Safford, 1869; Geol. Tennessee, p. 290. (See *Chasmops troosti* Safford.) Nashville.
Dalmania VERRUCOSUS Hall, 1863; Trans. Albany Inst., vol. 4, p. 218. (Abstract published May, 1863, p. 24.) Waldron, Indiana.
Niagara.
— VERRUCOSUS Hall, 1875; 28th Rep. New York State Mus. Nat. Hist., pl. 23, figs. 5-17, 13-15. Museum ed., 1879, p. 195, pl. 23, figs. 5-17; pl. 34, figs. 13-15.
— VERRUCOSUS Hall, 1882; 11th Ann. Rep. Dept. Geol. Nat. Hist. Indiana, p. 341, pl. 35, figs. 5-17; pl. 36, figs. 13-15.
— VIGILANS Hall, 1861; Rep. Progress Wisconsin Geol. Survey, p. 51.
Niagara.
Dalmania VIGILANS Hall, 1867; 20th Rep. New York State Cab. Nat. Hist., p. 335, pl. 21, figs. 16-18. Waukesha, and other localities in Wisconsin.
Dalmania VIGILANS Hall, 1870; 20th Rep. New York State Cab. Nat. Hist. (2d ed.), p. 426, pl. 21, figs. 16-18.
— VIGILANS Hall, 1875; 28th Rep. New York State Mus. Nat. Hist. Expl., pl. 33, figs. 1-4.
— VIGILANS Hall, 1879; 28th Rep. New York State Mus. Nat. Hist. (Mus. ed.), p. 193, pl. 33, figs. 1-4.
— VIGILANS Hall, 1882; 11th Ann. Rep. Geol. Nat. Hist. Indiana, 1881, p. 339, pl. 35, figs. 1-4; pl. 33, fig. 9.
— WERTHNERI Foerste, 1885; Bull. Denison Univ., vol. 1, p. 116. Soldiers' Home Quarry, Ohio. Clinton.
— WERTHNERI Foerste, 1887; Bull. Denison Univ., vol. 2, p. 109, pl. 8, figs. 22-25.

DICELLOCEPHALUS Owen, 1852; Rep. Geol. Survey Wisconsin, Iowa and Minnesota, p. 573.
Syn., *Centropleura* Aug., 1854; Pal. Scand., p. 87, pl. 41, figs. 9-11.
— AFFINIS Bill., 1865; Pal. Foss., vol. 1, p. 197, fig. 183 a, b. Point Levis, Quebec. Taconic.
— ? ANGUSTIFRONS Wal., 1884; Pal. Eureka Dist. Mon. U. S. Geol. Survey, vol. 8, p. 42, pl. 10, figs. 1, 1 a, b. Secret Cañon, etc., Nevada.
Taconic.
— BELLI Bill., 1860; Canadian Naturalist, vol. 5, p. 311, fig. 7. Point Levis, Quebec. Taconic.
— BELLI Bill., 1865; Pal. Foss., vol. 1, p. 403, fig. 378.
— BARABUENSIS Whitf., 1878; Ann. Rep. Geol. Survey Wisconsin, 1877, p. 63. Devil's Lake region, 7 miles east of Baraboo.
Lower Magnesian.
— BARABUENSIS Whitf., 1882; Geol Wisconsin, vol. 4, p. 201, pl. 4, figs. 6-9. cf. *Bathyurus capax* Bill.
— BILOBATUS H. & W., 1877; U. S. Geol. Expl. 40th Par., p. 226, pl. 2, fig. 36. Eureka Dist., Nevada. Taconic.

DICELLOCEPHALUS.

— BILOBATUS Wal., 1884; Pal. Eureka Dist. Mon. U. S. Geol. Survey, vol. 8, p. 40. East side Sierra Valley, east of Pinnacle Peak, Eureka Dist., Nevada.

— CRASSIMARGINATUS Whitf., 1882; Geol. Wisconsin, vol. 4, p. 344, pl. 27, fig. 14. Lagrange Mt., Minnesota. Potsdam. Syn., *pepinensis* (Hall); 16th Rep. New York State Cab. Nat. Hist. Expl., pl. 11, fig. 2.

— ? *corax* Bill., 1865; Pal. Foss., vol. 1, p. 334, figs. 322 a, b. Point Levis, Quebec. Referred to *Olenoides? corax* Wal., 1884; Bull. U. S. Geol. Survey, No. 30, p. 184. Cf. *parabolina spinulosa* Wahl.

— CRISTATUS Bill., 1860; Canadian Naturalist, vol. 5, p. 312, fig. 10.
Taconic.

— CRISTATUS Bill., 1865; Pal. Foss., vol. 1, p. 404, fig. 384. Point Levis, Quebec. This species might be referred to the genus *Olenoides*.

— DEVINEI Bill., 1865; Pal. Foss., vol. 1, p. 195, figs. 180, 181. Point Levis, Quebec. Probably of the genus *Ptychoparia*.

— EATONI Whitf., 1878; Ann. Rep. Geol. Survey Wisconsin, 1877, p. 65. Devil's Lake region, Baraboo, Wis. Lower Magnesian.

— EATONI Whitf., 1882; Geol. Wisconsin, vol. 4, p. 202, pl. 4, figs. 11–17; pl. 10, figs. 4, 5. Devil's Lake Dist., Baraboo, Wis.

— ? EXPANSUS Wal., 1884; Pal. Eureka Dist. Mon. U. S. Geol. Survey, vol. 8, p. 89, pl. 9, fig. 19. Prospect Mt., Secret Cañon, etc., Nev.
Taconic.

— FINALIS Wal., 1884; Pal. Eureka Dist. Mon. U. S. Geol. Survey, vol. 8, p. 89, pl. 12, figs. 12 a, b. Hamburg Ridge, Eureka Dist., Nev., Pogonip group.

— FLABELLIFER H. & W., 1877; U. S. Geol. Expl. 40th Par., vol. 4, p. 227, pl. 2, figs. 29, 30. Pogonip Mt. and Eureka Dist., Nev.
Potsdam.

— *flagicaudus* White. (See *Olenoides? flagricauda*.)
— *granulosus* Owen. (See *Ptychaspis granulosa*.)
— ? *gothicus* H. & W. (See *Dorypyge gothicus*.)
— HISINGERI Bill., 1865; Pal. Foss., vol. 1, p. 196, fig. 182. Point Levis, Quebec. Probably of the genus *Ptychoparia*.

— INEXPECTANS Wal., 1884; Pal. Eureka Dist. Mon. U. S. Geol. Survey, vol. 8, p. 90, pl. 1, fig. 10. Hamburg Ridge, Nev. Pogonip.

— ? *iowensis* Owen. (See *Lonchocephalus iowensis* (Owen).)
— IOLE Wal., 1884; Pal. Eureka Dist. Mon. U. S. Geol. Survey, vol. 8, p. 43, pl. 10, fig. 19. Prospect Mt., Bullwhacker Mine, Nev.
Taconic.

— LATIFRONS Shum., 1863; Trans. Acad. Sci. St. Louis, vol. 2, p. 101.
Potsdam.

CATALOGUE OF TRILOBITES.   305

DICELLOCEPHALUS.
—— LODENSIS Whitf., 1880; Ann. Rep. Geol. Survey Wisconsin, 1879, p. 51.
    Potsdam.
—— LODENSIS Whitf., 1882; Geol. Wisconsin, vol. 4, p. 188, pl. 10, fig. 14;
    p. 341, pl. 27, figs. 12-13.
—— MAGNIFICUS Bill., 1860; Canadian Naturalist, vol. 5, p. 307, fig. 5.
    This species has not the glabella or pygidium of a *Dicellocephalus*.
—— MAGNIFICUS Bill,, 1865; Pal. Foss., vol. 1, p. 399, figs. 376. Point
    Levis, Quebec.
—— ? *marcoui* Whitf., 1884; Bull. Am. Mus. Nat. Hist. New York, vol. 1,
    p. 150, pl. 14, fig. 7. Referred to *Olenoides marcoui* Wal., 1886;
    Bull. U. S. Geol. Survey, No. 30, p. 186, pl. 26, figs. 5, 5 a, b.
    Potsdam.
—— MARICA Wal., 1884; Pal. Eureka Dist. Mon. U. S. Geol. Survey, vol.
    8, p. 44, pl. 10, fig. 13. Prospect Mt., Secret Cañon, etc., Nev.
    Taconic.
—— MEGALOP Bill., 1860; Canadian Naturalist, vol. 5, p. 311, fig. 9.
    Point Levis, Quebec.
—— MEGALOP Bill., 1865; Pal. Foss., vol. 1, p. 403, fig. 380.
—— *miniscaensis*. (See *Ptychaspis miniscaensis*.)
—— MINNESOTENSIS Owen. A very large sp. *Asaphus* Owen, 1848; Rep.
    Geol. Reconn. Chippewa Land Dist., p. 15, pl. 7, figs. 2, 3.
    Taconic.
—— MINNESOTENSIS Owen, 1852; Rep. Geol. Survey Wis., Iowa and Minn.,
    p. 574, pl. 1, figs. 2, 10; pl. 10, figs. 3, 6.
—— MINNESOTENSIS Hall, 1863; 16th Rep. New York State Cab. Nat. Hist.,
    p. 138, pl. 9, figs. 5-10; pl. 10, figs. 10-12; pl. 11, figs. 1-4, 6.
    Mazomania, Lagrange Mountain, Minnesota; Madison, Wis.
—— MINNESOTENSIS var. LIMBATUS Hall, 1863; 16th Rep. New York State
    Cab. Nat. Hist., p. 141, pl. 9, fig. 12. Lagrange Mt., Minnesota.
—— MINNESOTENSIS var. Hall, 1863; 16th Rep. New York State Cab. Nat.
    Hist., p. 141, pl. 9, fig. 11; pl. 10, fig. 9. Lagrange Mt., Still-
    water, Minn.
—— MINNESOTENSIS Winchell, 1864; Am. Jour. Sci., 2d series, vol. 37, p.
    229.
—— MINNESOTENSIS Whitf., 1882; Geol. Wisconsin, vol. 4, p. 167, pl. 3,
    fig. 1.
—— MISA Hall, 1863; 16th Rep. State Cab. Nat. Hist., p. 144, pl. 8, fig. 15;
    pl. 10, figs. 4-8. Trempleau and Miniska, Wis.   Potsdam.
—— MISSISQUOI Bill., 1865; Pal. Foss., vol. 1, p. 199. Phillipsburgh. cf.
    D. magnificus.
—— MULTICINCTUS H. & W., 1877; U. S. Geol. Expl. 40th Par., vol. 4, p.
    226, pl. 2, fig. 37. Pogonip Mt., White Pine Dist., Nev.
—— NASUTUS Wal., 1884; Pal. Eureka Dist. Mon. U. S. Geol. Survey, vol.
    8, p. 40, pl. 10, fig. 15. Prospect Mt., Secret Cañon, etc., Nev.
    Taconic.
20

DICELLOCEPHALUS. DOLICHOMETOPUS.
—— OSCEOLA Hall, 1863; 16th Rep. New York State Cab. Nat. Hist., p. 146,
pl. 7, fig. 49; pl. 10, fig. 18. Osceola Mills, Wis. Potsdam.
—— OSCEOLA Wal., 1884; Pal. Eureka Dist. Mon. U. S. Geol. Survey, vol.
8, p. 40, pl. 9, fig. 25. Prospect Mt., etc., Nevada.
—— OWENI Bill., 1860; Canadian Naturalist, vol. 5, p. 310, fig. 8. Point
Levis, Quebec.
—— OWENI Bill., 1865; Pal. Foss., vol. 1, p. 402, fig. 379.
—— PAUPER Bill., 1865; Pal. Foss., vol. 1, p. 200. cf. *D. sesostris*. Point
Levis, Quebec.
—— PEPINENSIS Owen, 1852; Rep. Geol. Survey Wis., Iowa and Minn., p.
574, pl. 1, figs. 9, 9 a; pl. 1 a, fig. 7. Potsdam.
—— PEPINENSIS Hall, 1863; 16th Rep. New York State Cab. Nat. Hist., p.
142, pl. 9, figs. 1-4; pl. 10, figs. 13-17. (Expl., pl. 11, fig. 2. See
Dicellocephalus crassimarginatus). Lagrange Mt., Minnesota.
—— PEPINENSIS Winchell, 1864; Am. Jour. Sci., 2d series, vol. 37, p. 229.
—— PLANIFRONS Bill., 1860; Canadian Naturalist, vol. 5, p. 309, fig. 6.
Point Levis, Quebec.
—— PLANIFRONS Bill., 1865; Pal. Foss., vol. 1, p. 401, fig. 377. Point
Levis.
—— *quadriceps* H. & W. *Barrandia quadriceps*?
—— RICHMONDENSIS Wal., 1884; Pal. Eureka Dist. Mon U. S. Geol. Survey, vol. 8, p. 41, pl. 10, fig. 7. Prospect Mt. and Secret Cañon,
Nev. Taconic.
—— ROEMERI Shum., 1861; Am. Jour. Sci., 2d series, vol. 32, p. 220.
Clear Creek, Burnet county, Texas. Potsdam.
—— SELECTUS Bill., 1865; Pal. Foss., vol. 1, p. 199. Point Levis, Quebec.
—— SELECTUS Bill., 1865; Pal. Foss., vol. 1, p. 198. (See *Ptychaspis selectus*.)
—— *sesostris* Bill. (See *Oryctocephalus sesostris*.)
—— SPINIGER Hall, 1863; 16th Rep. New York State Cab. Nat. Hist., p.
143, pl. 10, figs. 1, 2, 3 (?). Trempleau, Wis. Potsdam.
—— *wahsatchensis*. (See *Olenoides wahsatchensis*.)
DIONIDE Barr., 1847; Neues Jahrbuch für Mineral., vol. 4, p. 391.
Syn., *Dione* Barr.; Notice Prélim. Sil. Syst. Bohême, p. 33.
Polytomurus Cord., 1847; Prodr., p. 153.
—— ? PERPLEXA Bill., 1866; Catalogue Sil. Foss. Anticosti, p. 67. Jumpers Div. 4.
DIPLEURA Green, 1832; Mon. Tril. North America, p. 79. (See *Homalonotus dekayi*.)
DISCRANURUS Con., 1841; 5th Rep. Pal. Dept. New York Geol. Survey, p. 48. (See *Acidaspis hamata*.)
DOLICHOMETOPUS Ang., 1854; Palæont. Scand., p. 72.
—— ? CONVEXUS Bill., 1865; Pal. Foss., vol. 1, p. 269, fig. 253. Port aux Choix, N. F.

DOLICHOMETOPUS.
——? GIBBERULUS Bill., 1865; Pal. Foss., vol. 1, p. 269, fig. 254. Port aux Choix, N. F.
——? RARUS Bill., 1865; Pal. Foss., vol. 1, p. 352, fig. 308. Oxford, Canada.
Dorypyge Dames, 1884; China (Richthofen), vol. 4, p. 23. (See *Olenoides*.)
DYSPLANUS Burm., 1843; Organization of Trilobites (Ray. Soc. ed., p. 205).
ELLIPSOCEPHALUS Zenk., 1833; Beiträge Naturgeschichte der Urwelt, p. 51.
—— ARTICEPHALUS Matt., 1892; Canadian Record of Sci., Oct., 1892, p. 250, fig. 5 a-b. Hanford Brook, St. John county. Taconic.
—— CURTUS Whitf., 1878; Ann. Rep. Geol. Wisconsin, 1877, p. 58. Hudson, Wis. Potsdam.
—— CURTUS Whitf., 1882; Geol. Wisconsin, vol. 4, p. 191, pl. 1, fig. 18.
—— GALEATUS Matt., 1892; Canadian Record of Sci., Oct., 1892, p. 249, fig. 4 a-e. Hanford Brook, St. John county, N. B. Taconic.
ELIPTOCEPHALUS Emmons, 1844; Taconic Syst., p. 21, figs. 1-3.
—— *asaphoides*. (See *Olenellus asaphoides* Emmons.)
Embolimus Rominger, 1887; Proc. Acad. Nat. Sci. Phila., p. 16. (See *Zacanthoides* Wal.)
—— *rotundata*. (See *Bathyuriscus howelli*.)
—— *spinosa*. (See *Olenoides spinosus*.)

ENCRINURUS.
ENCRINURUS Emm., 1844; Zur Naturg. Trilobiten, p. 16.
—— AMERICANUS Vogd., 1886; Des. New Crust. Clinton Group, Georgia, p. 1. Taylor's Ridge, Catoosa Station, Georgia. Clinton.
—— AMERICANUS Foerste, 1887; Bull. Denison Univ., vol. 2, p. 102.
—— DELTOIDEUS Shum., 1855; Geol. Missouri, p. 198, pl. B, fig. 10. Cape Gararideau, Missouri. Silurian.
Cryptonymus DELTOIDEUS Vogd., 1878; Mon.Genera Zethus, etc., p. 21.
—— DELTOIDEUS Foerste, 1887; Bull. Denison Univ., vol. 2, p. 102.
—— EGANI Miller, 1879; Jour. Cincinnati Soc. Nat. Hist., vol. 2, p. 254, pl. 15, figs. 1, 1 a, b. Joliet, Illinois. Niagara.
—— ELEGANTULUS Bill., 1866; Catalogue Sil. Foss. Anticosti, p. 62. The Jumpers, Div. 4.
—— ELEGANTULUS Foerste, 1887; Bull. Denison Univ., vol. 2, p. 102.
—— *excedrensis* Saff., 1869; Geol. Tennessee, p. 290. (See *Encrinurus varicostatus* Wal.)
—— MIRUS Bill., 1865; Pal. Foss., vol. 1, p. 292, fig. 282. Table Head and Pistolet Bay, N. F.
Cryptonymus (Atractopyge) MIRUS Vogd., 1878; Mon. Genera Zethus, etc., p. 34.
Amphion MULTISEGMENTATUS Portl., 1843; Geol. Londonderry, etc., p. 291, pl. 3, fig. 6.
—— MULTISEGMENTATUS Salt.; Mem. Geol. Survey United Kingdom, decade 7, pl. 4.

ENCRINURUS.
— MULTISEGMENTATUS Bill., 1866; Catalogue Sil. Foss. Anticosti, p. 61. Junction Cliff, Div. 1.
  Cryptonymus MULTISEGMENTATUS Vogd., 1878; Mon. Genera Zethus, etc., p. 29.
— NEREUS Hall, 1867; 20th Rep. New York State Cab. Nat. Hist., p. 375, pl. 21, fig. 15. Racine, Wis. Niagara.
— NEREUS Hall, 1870; 20th Rep. New York State Cab. Nat. Hist., p. 425, pl. 21, fig. 15.
  Cryptonymus NEREUS Vogd., 1878; Mon. Genera Zethus, etc., p. 24, pl. 3, fig. 17.
— ORNATUS H. & W., 1875; Pal. Ohio, vol. 2, p. 154, pl. 7, fig. 16. Eaton and Yellow Springs, Ohio. Niagara.
  Syn., *Cybele punctata* Hall, 1852; Pal. New York, vol. 2, p. 297, pl. A 66, figs. 1 a–k.
  Cryptonymus ORNATUS Vogd., 1878; Mon. Genera Zethus, etc., p. 23.
— ORNATUS Foerste, 1887; Bull. Denison Univ., vol. 2, p. 102.
— PUNCTATUS (Brünnich) Bill., 1866; Catalogue Sil. Foss. Anticosti, p. 61. East Point and Jumpers, Div. 4.
— THRESHERI Foerste, 1887; Bull. Denison Univ., vol. 2, p. 101, pl. 8, fig. 26. Clinton.
— TRENTONENSIS Wal., 1877; advance sheets 31st Rep. New York State Mus. Nat. Hist., p. 15. Clifton, Grant county, Wis., and Dunleth, Ill. Trenton.
— TRENTONENSIS Wal., 1879; 31st Rep. New York State Mus. Nat. Hist., p. 68.
  Cryptonymus TRENTONENSIS Vogd., 1878; Mon. Genera Zethus, etc., p. 28.
— VARICOSTATUS Wal., 1877; advance sheets 31st Rep. New York State Mus. Nat. Hist., p. 16. Mineral Point, Beloit and Janesville, Wis. Trenton.
— VARICOSTATUS Wal., 1879; 31st Rep. New York State Cab. Nat. Hist., p. 69.
  Cryptonymus VARICOSTATUS Vogd., 1878; Mon. Genera Zethus, etc., p. 27.
— VARICOSTATUS Saff. & Vogd., 1889; Proc. Acad. Nat. Sci. Phila., p. 167, fig.
  Ceraurus VIGILANS Hall, 1847; Pal. New York, vol. 1, p. 245, pl. 65, figs. 2 a, 2 h. Middleville, N. Y. Trenton.
  Ceraurus VIGILANS Emmons, 1855; Am. Geology, vol. 1, pt. 2, p. 217, pl. 15, figs. 2 a, 2 c.
  Cryptonymus VIGILANS Vogd., 1878; Mon. Genera Zethus, etc., p. 29, pl. 2, figs. 2 a, 2 h.
Endymion Bill., 1862; Pal. Foss., vol. 1, p. 94. Changed by author to *Endymionia*.

CATALOGUE OF TRILOBITES. 309

ENDYMIONIA. GRIFFITHIDES.
ENDYMIONIA Bill., 1865; Pal. Foss., vol. 1, p. 281.
—— MEEKI Bill., 1862; Pal. Foss., vol. 1, pp. 94, 281, fig. 84. Point Levis, Quebec.
GRIFFITHIDES Portl., 1843; Geol. Rep. Londonderry, etc., p. 310.
Phillipsia (Griffithides) BUFO M. & W., 1870; Proc. Acad. Nat. Sci. Phila., p. 52. Warsaw, Illinois. Keokuk.
Phillipsia (Griffithides) BUFO M. & W., 1873; Geol. Survey Illinois, vol. 5. p. 528, pl. 19, fig. 5.
—— BUFO Vogd., 1887; Annals New York Acad. Sci., vol. 4, p. 95, pl. 3, figs. 4, 5, 10.
Proetus GRANULATUS Wetherby, 1881; Jour. Cincinnati Soc. Nat. Hist., vol. 4, p. 31, pl. 2, figs. 8, 8 a, 9, 9 a. Pulaski county, Ky. Chester.
—— GRANULATUS Vogd., 1887; Annals New York Acad. Sci., vol. 4, p. 101.
Phillipsia (Griffithides?) lodiensis Meek. (See *Brachymetopsis lodiensis*.)
Phillipsia (Griffithides) PORTLOCKI M. & W., 1865; Proc. Acad. Nat. Sci. Phila., p. 268. Warsaw, Illinois. Keokuk.
Phillipsia (Griffithides) PORTLOCKI M. & W., 1873; Geol. Survey Illinois, vol. 5, p. 525, pl. 19, fig. 6.
—— PORTLOCKI Wal., 1884; Pal. Eureka Dist. Mon. U. S. Geol. Survey, vol. 8, p. 266, pl. 24, figs. 4, 4 a, b.
—— PORTLOCKI Vogd., 1887; Annals New York Acad. Sci., vol. 4, p. 93, pl. 3, fig. 9.
Phillipsia (Griffithides?) SANGAMONENSIS M. & W., 1865; Proc. Acad. Nat. Sci. Phila., p. 270. Springfield, Ill. Coal Measures.
Phillipsia (Griffithides?) SANGAMONENSIS M. & W., 1873; Geol. Survey Illinois, vol. 5, p. 615, pl. 32, fig. 4.
Phillipsia (Griffithides?) SANGAMONENSIS White, 1883; 13th Rep. Dept. Geol. Nat. Hist. Indiana, p. 174, pl. 39, figs. 4, 5.
Phillipsia SANGAMONENSIS Herr., 1887; Bull. Denison Univ., vol. 2, p. 61, pl. 5, fig. 13.
—— SANGAMONENSIS Vogd., 1887; Annals New York Acad. Sci., vol. 4, p. 99, pl. 3, figs. 7, 8.
Phillipsia (Griffithides) SCITULA M. & W., 1865; Proc. Acad. Nat. Sci. Phila., p. 270. Springfield, Ill.; Platte's mouth, Neb.
Coal Measures.
Phillipsia (Griffithides) SCITULA M. & W., 1873; Geol. Survey Illinois, vol. 5, p. 612, pl. 32, fig. 3.
Phillipsia (Griffithides) SCITULA White, 1883; 13th Ann. Rep. Dept. Geol. Nat. Hist. Indiana, 1882, p. 173, pl. 39, figs. 6-9.
Phillipsia SCITULA Meek, 1872, U. S. Geol. Survey Territories, Final Rep. Nebraska, p. 238, pl. 6, fig. 9.
Phillipsia SCITULA, Herr., 1887; Bull. Denison Univ., vol. 2, p. 62.
—— SCITULA Vogd., 1887; Annals New York Acad. Sci., vol. 4, p. 97, pl. 2, figs. 11-13.

GRIFFITHIDES.   HOMALONOTUS.
—— ? SEDALIENSIS Vogd., 1888; Trans. New York Acad. Sci., vol. 7, p. 249. Sedalia, Missouri. Waverly.
Phæthonides SEDALIENSIS (Vogd.) Herr., 1889; Bull. Denison Univ., vol. 4, p. 57.
HARPES Goldf., 1839; Nova Acta Acad. Leop. Carol., vol. 19, p. 358.
—— ANTIQUATUS Bill., 1859; Canadian Naturalist, vol. 4, p. 469, fig. 38. Mingan Islands.   Chazy.
—— ANTIQUATUS Bill., 1863; Geol. Canada, p. 133, fig. 67.
—— CONSUETUS Bill., 1866; Catalogue Sil. Foss. Anticosti, p. 64. South West Point, Div. 4.
—— DENTONI Bill., 1863; Canadian Naturalist, vol. 8, p. 36, fig. Trenton.
—— DENTONI Bill., 1865; Pal. Foss., vol. 1, p. 183, fig. 166. City of Ottawa, Canada.
—— ECANABIÆ Hall, 1851; Rep. Geol. Lake Superior Land Dist. (Foster & Whitney), pt. 2, p. 211, pl. 27, fig. 2 a. Escanaba River, below Indian Creek.   Trenton.
—— GRANTI Bill., 1865; Pal. Foss., vol. 1, p. 326, fig. 314. Stanbridge Range 6, lot 20.
—— OTTAWENSIS Bill., 1865; Pal. Foss., vol. 1, p. 182, fig. 165. Ottawa City, Canada.   Trenton.
HARPIDES Beyr., 1846; Untersuch. Trilobiten, p. 34.
—— ATLANTICUS Bill., 1865; Pal. Foss., vol. 1, p. 281, fig. 267. Portland Creek, N. F.
—— CONCENTRICUS Bill., 1865; Pal. Foss., vol. 1, p. 282, fig. 268. Portland Creek, N. F.
—— ? DESERTUS Bill., 1865; Pal. Foss., vol. 1, p. 333, fig. 321. Bedford, Canada.
HAUSMANNIA Hall, 1888; Pal. New York, vol. 7, p. xxxi. (Used as a subgenus to *Dalmanites*.)
Hemicrupturus Greene, 1832; Mon. Tril. N. A., p. 20.
—— *clintoni*. (See *Calymene clintoni* Vanuxem.)
HOLOMETOPUS Ang., 1854; Palæont. Scand., p. 58.
—— ANGELINI Bill., 1862; Pal. Foss., vol. 1, p. 95, fig. 85; also p. 281. Point Levis, Quebec.
HOMALONOTUS Koenig, 1825; Icones Foss. Sectiles, p. 4.
Syn., *Brongniartia* Eaton, 1832; Geol. Text Book, p. 32, pl. 2, fig. 20 (in part).
Dipleura Green, 1832; Monthly Am. Jour. Geol., p. 560; Mon. Tril. N. A., p. 78.
Trimerus Green, 1832; Monthly Am. Jour. Geol., p. 560; Mon. Tril. N. A., p. 81.
Plæsiacomia Cord., 1847; Prodr., p. 171.
Brongniartia Salt., 1865; Mon. Brit. Tril., p. 104 (subgenus).
Burmeisteria Salt., 1865; Mon. Brit. Tril., p. 105 (subgenus).
Koenigia Salt., 1865; Mon. Bri. Tril., p. 104 (subgenus).

HOMALONOTUS.

Asaphus CRYPTURUS Green, 1834; Trans. Geol. Soc. Pennsylvania, vol. 1, p. 37, pl. 4. Moose River at Anapolis Basin, Nova Scotia.
Devonian.

Asaphus? CRYPTURUS Green, 1835; Suppl. Mon. Tril. N. A., p. 18, cast 41.

Syn., *Asaphus ditmarsiæ* Honeyman, 1879; Proc. Nova Scotia Inst., vol. 5, p. 18.

Asaphus? CRYPTURUS Honeyman, 1888; Proc. Nova Scotia Inst., vol. 7, pt. 1, p. 63.

—— DAWSONI Hall, 1860; Canadian Naturalist, vol. 5, p. 155. Nova Scotia.
Silurian.

—— DAWSONI Daw., 1878; Acadian Geol. (3d ed.), p. 606, fig. 214.

Dipleura DEKAYI Green, 1822; Mon. Tril. N. A., p. 79, casts 30, 31, pl. 1, figs. 8, 9. Hamilton.

Syn., *Nuttainia sparsa* Eaton, 1832; Geol. Text Book, p. 34.

Nuttainia sparsa Green, 1832; Mon. Tril. N. A., p. 89, cast 35.

—— *Dipleura* DEKAYI Vanuxem, 1842; Geol. New York, 3d Geol. Dist., p. 150, fig. 1.

—— *Dipleura* DEKAYI Hall, 1843; Geol. New York, 4th Geol. Dist., p. 205, fig. 1.

—— DEKAYI Emmons, 1860; Manual Geol., pp. 146, 147, figs. 134, 135.

—— DEKAYI Hall, 1861; Des. New Species Foss, p. 85.

—— DEKAYI Hall, 1862; 15th Rep. New York State Cab. Nat. Hist., p. 113.

—— DEKAYI Hall, 1877; Illus. Devonion Foss. Crust., pls. 2-5.

—— DEKAYI Hall, 1888; Pal. New York, vol. 7, p. 7, pl. 2, figs. 1-12; pl. 3, figs. 1-5; pl. 4, figs. 1-7; pl. 5, figs. 1-4. Marcellus shales at Flint Creek, Ontario county; Hamilton shales at Bear's Gulf, Schoharie county, and elsewhere in New York.

Trimurus DELPHINOCEPHALUS Green, 1832; Monthly Am. Jour. Geol., vol. 1, p. 559, pl. 0, fig. 1. Niagara.

Trimerus DELPHINOCEPHALUS Green, 1832; Mon. Tril. N. A., p. 82, pl. 1, fig. 1, cast 32.

Trimerus DELPHINOCEPHALUS Harlan, 1835; Trans. Geol. Soc. Pennsylvania, vol. 1, p. 105.

Syn., *Brongniartia platycephala* Eaton, 1832; Geol. Text Book, p. 32, pl. 2, fig. 20.

Brongniartia platycephala Green, 1832; Mon. Tril. N. A., p. 91.

Ogygies latissima Eaton, 1832; Am. Jour. Sci., 1st series, vol. 21, p. 136 (foot-note).

—— DELPHINOCEPHALUS Murch., 1839; Sil. Syst., p. 651, pl. 7 bis, figs. 1 a, 1 b.

Syn., *ahrendii* Roemer, 1843; Verst. Harzgebirges, p. 39, pl. 11, fig. 5.

—— DELPHINOCEPHALUS Hall, 1843; Geol. New York, 4th Geol. Dist., p. 103, fig. 34.

HOMALONOTUS. ILLÆNUS.

— DELPHINOCEPHALUS Verneuil, Note sur la parallélisme, etc., p. 47.
  Syn., *atlas* Castelneau, 1843; Syst. Sil. de l'Amér., p. 20, pl. 4, fig. 4.
    giganteus Castelneau, 1843; Syst. Sil de l'Amér., p. 20, pl. 3, fig. 1.
    herculaneus Castelneau, 1843; *idem*, p. 20, pl. 4, fig. 5.
— DELPHINOCEPHALUS Hall, 1852; Pal. New York, vol. 2, p. 309, pl. 68, figs. 1-14. Lockport and Rochester, N. Y.
— DELPHINOCEPHALUS Salt., 1865; Mon. Brit. Tril., p. 113, pl. 11, figs. 1-10.
— DELPHINOCEPHALUS Hall, 1883; 12th Ann. Rep. Dept. Geol. Nat. Hist. Indiana, 1882, p. 332, pl. 34, figs. 17, 18.
— DELPHINOCEPHALUS? Hall, 1875; 28th Rep. New York State Mus. Nat. Hist., pl. 34, figs. 17, 18.
— DELPHINOCEPHALUS Hall, 1879; 28th Rep. New York State Mus. Nat. Hist., p. 187, pl. 32, figs. 17, 18.
  Trimerus JACKSONI Green, 1837; Am. Jour. Sci., 1st series, vol. 32, p. 347. Huntiugdon county, Penn. Devonian.
— MAJOR Whitf., 1885; Bull. Am. Mus. Nat. Hist., vol. 1, No. 6, p. 193, pl. 22. 4th Binnewater and 5th Binnewater, Ulster county, N. Y. Oriskany.
— MAJOR Hall, 1888; Pal. New York, vol. 7, p. 4, pl. 5 a, fig. 1.
— TRENTONENSIS Simpson, 1890; Trans. Amer. Philos. Soc., p. 460, fig. 31, Reedsville, Miffin county, Penn. Trenton.
— VANUXEMI Hall, 1859; Pal. New York, vol. 3, p. 352, p. 73, figs. 9-14. Kingston, Ulster county, N. Y. Lower Helderberg.
— VANUXEMI Hall, 1888; Pal. New York, vol. 7, p. 11, pl. 5 b, figs. 1, 2.
HOPOLICHAS Dames, 1877; Zeitschr. Deutsch. Geol. Gesell., p. 794. Type, *Lichas (Hopolichas) tricuspidata* Beyr. For a list of the species see *Lichas*.
ILLÆNURUS Hall, 1863; 16th Rep. New York State Cab. Nat. Hist., p. 176.
— CONVEXUS Whitf., 1878; Ann. Rep. Geol. Survey Wisconsin, 1877, p. 66. Devil's Lake region, Baraboo, Wis. Lower Magnesian.
— CONVEXUS Whitf., 1882; Geol. Wisconsin, vol. 4, p. 203, pl. 4, figs. 3-5.
— EUREKENSIS Wal., 1884; Pal. Eureka Dist., Mon. U. S. Geol. Survey, vol. 8, p. 97, pl. 12, figs. 4, 4 a. East slope of the ridge east of Hamburg Ridge, etc., Nev. Pogonip.
— QUADRATUS Hall, 1863; 16th Rep. New York State Cab. Nat. Hist., p. 176, pl. 7, figs. 52-57. Osceola Mills, St. Croix River, Wis.
Potsdam.

ILLÆNUS Dalm., 1826; Palæd., p. 50.
  Syn., *Actinolobus* Eichw., 1860.
    Alceste Cord., 1847.
    Cryptonymus Eichw., 1825 (in part), not *Cryptonymus* Eichw., 1840.
    Deucalion Chtchegloff, 1827.

ILLÆNUS.
    Ectillænus Salt., 1866 (subgenus).
    Hydrolænus Salt., 1866 (subgenus).
    Illænopsis Salt., 1865 (subgenus).
    Octillænus Salt., 1866 (subgenus).
    Thaleops Con., 1843.
—— AMBIGUUS Foerste, 1885; Bull. Denison Univ., vol. 1, p. 106, pl. 14, figs. 9 a, b; figs. 10 a, b, c; fig. 11. Dayton, Huffman's Quarry, Ohio.   Clinton.
—— AMBIGUUS Foerste, 1877; Bull. Denison Univ., vol. 2, p. 94.
—— AMBIGUUS Foerste, 1886; 15th Ann. Rep. Geol. Nat. Hist. Survey Minnesota, p. 480, fig. 3.
—— AMBIGUUS Foerste, 1889; Proc. Boston Soc. Nat. Hist., vol. 24, p. 267.
—— AMERICANUS Bill., 1859; Canadian Naturalist, vol. 4, p. 371. Ottawa City, Canada.   Trenton.
—— AMERICANUS Bill., 1865; Pal. Foss., vol. 1, p. 329, fig. 316 a–d, fig. 318.
—— ANGUSTICOLLIS Bill., 1859; Canadian Naturalist, vol. 4, p. 377, fig. 10 a–d. Island of St. Joseph, Grant's Island, Lake Huron, also at the Falls of La Petite Chaudiere.   Trenton.
—— ARCTURUS Hall, 1847; Pal. New York, vol. 1, p. 23, pl. 4 *bis*, fig. 12. Chazy village, N. Y.   Chazy and Trenton.
—— ARCTURUS Emmons, 1855; Am. Geology, vol. 1, pt. 2, p. 235, pl. 3, fig. 12.
—— ARCTURUS (Hall) Bill., 1859; Canadian Naturalist, vol. 4, p. 379. Mingan Islands.
—— ARCTURUS Bill., 1865; Pal. Foss., vol. 1, p. 279, fig. 265.
—— ARMATUS Hall, 1867; 20th Rep. New York State Cab. Nat. Hist., p. 330, pl. 22, figs. 1, 2. Bridgeport, Grafton and Racine, Wis.
    Niagara.
—— ARMATUS Hall, 1870; 20th Rep. New York State Cab. Nat. Hist., p. 418, pl. 22, figs. 1, 2; pl. 23, fig. 22.
—— ARMATUS? Hall, 1875; 28th Rep. New York State Mus. Nat. Hist., pl. 32, figs. 19, 20.
—— ARMATUS? Hall, 1879; 28th Rep. New York State Mus. Nat. Hist., p. 189, pl. 32, figs. 19, 20.
—— ARMATUS Hall, 1883; 12th Ann. Rep. Dept. Geol. Nat. Hist. Indiana, 1882, p. 335, pl. 34, figs. 19, 20; pl. 33, fig. 12.
—— BARRIENSIS (Murch.) Hall; Trans. Albany Inst., vol. 4, p. 227; abstract, p. 33, May, 1863.
—— BARRIENSIS (Murch.) Roemer, 1860; Die Sil. Fauna W. Tennessee, p. 83, pl. 5, fig. 23.   Niagara.
    Bumastus BARRIENSIS (Murch.), 1839; Sil. Syst., p. 656, pl. 7 *bis*, figs. 3 a–d, pl. 14, fig. 7 (in part).
—— BARRIENSIS Salt., 1859; Murchison's Siluria (2d ed.), pl. 17, figs. 9–11.

ILLÆNUS.

—— *Bumastus* BARRIENSIS Hall, 1867; 20th Rep. New York State Cab. Nat. His., p. 332, pl. 22, figs. 4-11. Racine, Waukesha, Wauwatosa and other places.

—— BAYFIELDI Bill., 1859; Canadian Naturalist, vol. 4, p. 369, figs. 4-6. Mingan Islands. Chazy.

—— CLAVIFRONS Bill., 1859; Canadian Naturalist, vol. 4, p. 379. Mingan Islands. Chazy and Trenton.

—— *conradi* Bill., 1859; Canadian Naturalist, vol. 4, p. 372, figs. 7-9. (See *Panderia conradi*.) La Petite Chaudiere, Township of Hull. Trenton.

—— CONSIMILIS Bill., 1865; Pal. Foss., vol. 1, p. 277, figs. 263 a-c; p. 331, fig. 317. (cf. *Illænus esmarki* Schloth.) Point Rich and Table Head, N. F.

—— CONSOBRINUS Bill., 1865; Pal. Foss., vol. 1, p. 280, fig. 268; p. 332, fig. 320. Cow Head, N. F.

—— CORNIGERUS Hall, 1872; 24th Rep. New York State Cab. Nat. Hist., p. 186, pl. 13, figs. 20, 21. Bear-grass Creek near Louisville, Ky. Clinton.

Entomostracetes CRASSICAUDA Wahl., 1821; Nova Acta. Soc. Scient. Upsal., vol. 8, p. 27, pl. 2, figs. 5, 6 (not pl. 7, figs. 5, 6).

—— CRASSICAUDA Holm, 1882; Svenska Vet.-Akad. Handl., vol. 7, No. 3, pl. 2, figs. 21-27.

—— CRASSICAUDA? Hall, 1847; Pal. New York, vol. 1, p. 24, pl. 4 *bis*, fig. 13; pl. 60, figs. 4 a-d. Carlisle, Penn.

—— CRASSICAUDA (Wahl.) M. & W., Geol. Survey Illinois, vol. 3, p. 322, pl. 3, figs. 1 a, 1 b. cf. *Illænus comsimilis* and *I. esmarkii*.

—— CUNICULUS Hall, 1867; 20th Rep. New York State Cab. Nat. Hist., p. 377, pl. 22, fig. 12. Wauwatosa, Wis. Niagara.

—— CUNICULUS Hall, 1870; 20th Rep. New York State Cab. Nat. Hist., p. 421, pl. 22, fig. 12 (rev. ed.).

—— DAYTONENSIS H. & W., 1875; Pal. Ohio, vol. 2, p. 119, pl. 5, figs. 14-16. Dayton, Ohio. Clinton.

—— DAYTONENSIS Foertse, 1885; Bull. Denison Univ., vol. 1, p. 104, pl. 14, figs. 4 a, b; figs. 6, 7, 7 a, b, c.

—— DAYTONENSIS Foerste, 1887; Bull. Denison Univ., vol. 2, p. 93, pl. 8, figs. 1-7.

—— FRATERNUS Bill., 1865; Pal. Foss., vol. 1, p. 276, figs. 262 a, b. Point Rich, Table Head, etc., N. F.

—— GLOBOSUS Bill., 1859; Canadian Naturalist, vol. 4, p. 367, figs. 1-3. Mingan Island, also Island of Montreal. Chazy.

—— (*Bumastus*) GRAFTONENSIS M. & W., 1870; Proc. Acad. Nat. Sci. Phila., p. 54. Grafton, Illinois. Niagara.

—— (*Bumastus*) GRAFTONENSIS M. & W., 1875; Geol. Survey Illinois, vol. 6, p. 508, pl. 25, fig. 4.

CATALOGUE OF TRILOBITES. 315

ILLÆNUS.
—— GRANDIS Bill., 1859; Canadian Naturalist, vol. 4, p. 380. Anticosti.
Hudson River.
—— HERRICKI Foerste, 1886; 15th Ann. Rep. Geol. Nat. Hist. Minnesota, p. 479, fig. 2. Minneapolis, Minnesota. Trenton.
—— IMPERATOR Hall, 1861; Rep. Progress Geol. Survey Wisconsin, p. 49.
Niagara.
—— IMPERATOR Hall, 1867; 20th Rep. New York State Cab. Nat. His., p. 332, pl. 22, figs. 15-17; pl. 23, figs. 2, 3. Racine and Waukesha.
—— IMPERATOR Hall, 1870; 20th Rep. New York State Cab. Nat. Hist., p. 420, pl. 22, figs. 15-17; pl. 23, figs. 2, 3.
—— IMPERATOR Whitf., 1882; Geol. Wisconsin, vol. 4, p. 306, pl. 21, figs. 4, 5.
—— INCERTUS Bill., 1865; Pal. Foss., vol. 1, p. 332, fig. 319. Stanbridge Range 6, lot 20.
—— INDETERMINATUS Wal., 1877; Advanced sheets 31st Rep. New York State Mus. Nat. His., p. 19. Russia, N. Y., Plattsville, Wis.
Trenton.
—— INDETERMINATUS Wal., 1879; 31st Rep. New York State Mus. Nat. His., p. 70.
—— INSIGNIS Hall, 1867; 20th Rep. New York State Cab. Nat. Hist., p. 331, figs. 5, 6; pl. 22, figs. 13, 14. Waukesha and Milwaukee, Wis. Niagara.
—— (Bumastus) INSIGNIS Salt., 1867; Mon. Brit. Tril., p. 207.
—— INSIGNIS Hall, 1870; 20th Rep. New York State Cab. Nat. Hist., p. 419, figs. 10, 11; pl. 22, figs. 13, 14.
—— (Bumastus) INSIGNIS Meek, 1873; Pal. Ohio, vol. 1, p. 189, pl. 15, figs. 5 a-c.
—— INSIGNIS Holm., 1882; De Svenska Art. Tril. Illænus, p. 127.
—— INSIGNIS Whitf., 1882; Geol. Wisconsin, vol. 4, p. 305, pl. 21, figs. 6-10.
—— INSIGNIS Foerste, 1886; 15th Ann. Rep. Geol. Nat. Hist. Survey Minnesota, p. 481.
—— IOXUS Hall, 1867; 20th Rep. New York State Cab. Nat. Hist., p. 378, pl. 22, figs. 4-10. cf. Illænus barriensis. Racine, etc., Wis.
Niagara.
—— (Bumastus) IOXUS Hall, 1870; 20th Rep. New York State Cab. Nat. Hist., p. 420, pl. 22, figs. 4-10.
—— (Bumastus) IOXUS Hall, 1883; 12th Ann. Rep. Dep. Geol. Nat. Hist. Indiana, 1882, p. 335, pl. 38, figs. 13, 14.
—— IOXUS Whitf., 1882; Geol. Wisconsin, vol. 4, p. 304, pl. 21, figs. 11, 12.
—— (Bumastus) IOXUS Hall, 1883; Trans. Albany Inst., vol. 10, p. 76.
—— IOXUS Foerste, 1889; Proc. Boston Soc. Nat. Hist., vol. 24, p. 268.
—— LATIDORSATA Hall, 1847; Pal. New York, vol. 1, p. 230, pl. 60, figs. 6 a, b. Near Watertown, N. Y. Trenton.

ILLÆNUS.
— MADISONIANUS Whitf., 1882; Geol. Wisconsin, vol. 4, p. 307, pl. 20, figs. 8, 9. Locality not given. (See *P. niagarensis*.) Niagara.
— MADISONIANUS Foerste, 1885; Bull. Denison Univ., vol. 1, p. 106, pl. 14, figs. 1 a, b and 2 a, b. Soldiers' Home Quarry, Ohio.
— MADISONIANUS Foerste, 1887; Bull. Denison Univ., vol. 2, p. 93, pl. 8, figs. 8-10. Clinton.
— MILLERI Bill., 1859; Canadian Naturalist, vol. 4, p. 375, fig. 10. City of Ottawa, etc., Canada. Trenton.
— MILLERI Wal., 1877; Advanced sheets 31st Rep. N. Y. State Mus. Nat. Hist., p. 20. Russia, N. Y., and Plattsville, Wis.
— MILLERI Wal., 1879; 31st Rep. N. Y. State Mus. Nat. Hist., p. 71.
— (*Nileus*) MINNESOTENSIS Foerste, 1886; 15th Ann. Rep. Geol. Minnesota, p. 478, fig. 1. Minneapolis, Minnesota. Trenton.
— *niagarensis* Whitf., 1880; Ann. Rep. Geol. Wisconsin, 1879, p. 68. Name changed by request of the Wisconsin Geological Survey, from *Illænus niagarensis* to *Illænus madisonianus*. Explanation omitted in the Final Report Pal. Wisconsin, p. 307. Niagara.
— ORBICAUDATUS Bill., 1859; Canadian Naturalist, vol. 4, p. 379. Gemache Bay, Anticosti. Hudson.
Thaleops OVATUS Con., 1843; Proc. Acad. Nat. Sci. Phila., vol. 1, p. 332. Trenton.
Thaleops (*Illænus*) OVATUS Hall, 1847; Pal. New York, vol. 1, p. 259, pl. 67, figs. 6 a-b. Mineral Point, Wis.
— OVATUS Whitf., 1882; Geol. Wisconsin, vol. 4, p. 238, pl. 5, figs. 1, 2. Beloit, Wis.
— PTEROCEPHALUS Whitf., 1878; Ann. Rep. Geol. Wisconsin, 1877, p. 87. Pewaukee, Wis. Niagara.
— PTEROCEPHALUS Whitf., 1882; Geol. Wisconsin, vol. 4, p. 309, pl. 20, figs. 10-12.
— STIMULATOR Bill., 1865; Pal. Foss., vol. 1, p. 337, figs. 315 a, b. Stanbridge Range 6, lot 20.
— TAURUS Hall, 1861; Rep. Prog. Geol. Survey Wisconsin, p. 49. Beloit, Mineral Point, Wis.; Rockford, Illinois. Niagara.
— TAURUS M. & W., 1865; Geol. Survey Illinois, vol. 3, p. 320, pl. 3, fig. 2.
Bumastus TRENTONENSIS Emmons, 1842; Geol. New York, 3d Geol. Dist., p. 390, fig. 1. Hogansburg, Franklin, N. Y. Trenton.
Bumastus TRENTONENSIS, 1854; 7th Rep. New York State Cab. Nat. Hist., p. 64, fig. 1.
(*Bumastus*) TRENTONENSIS Emmons, 1855; Am. Geology, vol. 1, pt. 2, p. 315, pl. 15, fig. 13.
— TRENTONENSIS Hall, 1847; Pal. New York, vol. 1, p. 230, pl. 60, fig. 5.
— TUMIDIFRONS Bill., 1865; Pal. Foss., vol. 1, p. 278, fig. 264. Cow Head, N. F.

CATALOGUE OF TRILOBITES. 317

ILLÆNUS. LICHAS.
—— VINDEX Bill, 1865; Pal. Foss., vol. 1, p. 179, fig. 160. Mingan Islands.
Chazy.
—— (*Bumastus*) WORTHENANUS W. & M., 1866; Mem. Boston Soc. Nat.
His., vol. 1, pt. 1, p. 105, pl. 3, fig. 13 (referred by Hall, 20th Rep.
New York State Cab. Nat. Hist., to *Illænus armatus* Hall).
Isoletus De Kay, 1825; Annals Lyceum Nat. Hist. New York, vol. 1, p. 174.
—— *canalis*. (See *Asaphus canalis*.)
—— *cyclops*. (See *Asaphus platycephalus*.)
—— *gigas*. (See *Asaphus platycephalus*.)
—— *maximus*. (See *Asaphus maximus*.)
—— *megalops* Green, 1832; Mon. Tril. N. A., p. 70, cast 75.
—— *megistos*. (See *Asaphus maximus*.)
—— *platycephalus*. (See *Asaphus platycephalus*).
—— *planus*. (See *Asaphus platycephalus*.)
—— *stegops*. (See *Asaphus platycephalus*.)
—— *vigilans*. (See *Asaphus vigilans*.)
LEPTOPLASTUS Ang., 1854; Palæont. Scand., p. 46.
—— LATUS Matt., 1891; Canadian Record of Sci., 1891, p. 462, figs, 1-3.
—— LATUS Matt., 1891; Trans. Royal Soc. Canada, vol. 9, p. 54, pl. 13,
figs. 10 a, b. St. John, N. B. For other species see *Anomocare*.
LICHAS Dalm., 1826; Palæad., pp. 53, 71, pl. 6, fig. 1.
Syn., *Platynotus* Con., 1838; Geol. Survey New York, Rep. Pal Dept.,
p. 118.
Arges Gold., 1839 (in part).
Metopias Eichw., 1842; Die Urwelt Russ., p. 62, pl. 3, figs. 21-23.
Arctinurus Cast., 1843; Syst. Sil. de l'Amér., p. 21.
Nuttainia Portl. & Emm.
Corydocephalus Cord., 1847; Prodr., p. 255.
Dicranopeltis Cord., 1847; Prodr., p. 257.
Acanthopyge Cord., 1847; Prodr., p. 260.
Dicranogmus Cord., 1847; Prodr., p. 261.
Subgenera, *Hoplolichas* Dames, 1877; Zeitschr. Deutsch. Geol. Gesell.,
p. 794. Type, *Hoplolichas tricuspidata* Beyr.
Conolichas Dames, 1877; Zeitschr. Deutsch. Geol. Gesell., p.
806. Type, *Conolichas æquiloba* Steinh.
—— *armatus* Hall. (See *Acidaspis eriopis* Hall.)
—— (*Hoplolichas*) BICOMES Ulreth, 1892; The Amer. Geol., vol. 10, p. 272,
fig. Spring Valley, Minn. Hudson.
—— BIGSBYI Hall, 1859; Pal. New York, vol. 3, p. 364, pl. 77, figs. 1-8.
Coeymans, Schoharie and Carlisle. Lower Helderberg.
—— (*Conolichas*) BIGSBYI ? Hall, 1888; Pal. New York, vol. 7, p. 80, pl. 19
a, fig. 1.
Paradoxides BOLTONI Bigsby, 1825; Jour. Acad. Nat. Sci. Phila.,
vol. 4, p. 365, plate. Lockport, N. Y. Niagara.

LICHAS.
Paradoxides BOLTONI Green, 1832; Monthly Am. Jour. Geol., etc., vol. 1, p. 560, pl. 1, fig. 5.
Paradoxides BOLTONI Green, 1832; Mon. Tril. North America, p. 60, pl. 1, fig. 5.
Paradoxides BOLTONI Harlan, 1834; Trans. Geol. Soc. Penn., vol. 1, p. 103.
Paradoxides BOLTONI Harlan, 1835; Med. and Phys. Researches, p. 303.
Platynotus BOLTONI Con., 1838, Rep. Pal. Dept. Geol. Survey New York, p. 118.
Arctinurus BOLTONI Cast., 1843; Sys. Sil. de l'Amér., p. 21, pl. 3, fig. 3.
—— BOLTONI Hall, 1852; Pal. New York, vol. 2, p. 311, pl. 69; pl. 70, figs. 1 a–i.
—— BOLTONI M. & W., 1875, Geol. Survey Illinois, vol. 6, p. 508, pl. 25, fig. 5.
—— BOLTONI var. OCCIDENTALIS Hall, 1863; Trans. Albany Inst., vol. 4, p. 223; abstract, p. 29.
—— BOLTONI var. OCCIDENTALIS Hall, 1875; 28th Rep. New York State Mus. Nat. Hist. Expl., p. 34, figs. 8–11.
—— BOLTONI var. OCCIDENTALIS Hall, 1879; 28th Rep. New York State Mus. Nat. Hist., p. 198, pl. 34, figs. 8–11.
—— BOLTONI var. OCCIDENTALIS Hall, 1883; 12th Ann. Rep. Dept. Geol. Nat. Hist. Indiana, 1882, p. 334, pl. 36, figs. 8, 11, 12 (?).
—— BOLTONI var. OCCIDENTALIS Foerste, 1889; Proc. Boston Soc. Nat. Hist., vol. 24, p. 272.
—— BREVICEPS Hall, 1863; Trans. Albany Inst., vol. 4, p. 222; abstract p. 28. New York and Indiana. Niagara.
—— BREVICEPS H. & W., 1875; Pal. Ohio, vol. 2, p. 156, pl. 6, fig. 17.
—— BREVICEPS Hall, 1875; 28th Rep. New York State Mus. Nat. Hist. Expl., pl. 34, figs. 1–7.
—— BREVICEPS Hall, 1879; 28th Rep. New York State Mus. Nat. Hist., p. 197, pl. 34, figs. 1–7.
—— BREVICEPS Hall, 1883; 12th Ann. Rep. Dept. Geol. Nat. Hist. Indiana, 1882, p. 343, pl. 36, figs. 1–7.
—— BREVICEPS Foerste, 1885; Bull. Denison Univ., vol. 1, p. 112, pl. 13, figs. 2, 6 a, b. Clinton.
—— BREVICEPS Foerste, 1887; Bull. Denison Univ., vol. 2, p. 98, pl. 8, figs. 18, 19. Not *Lichas breviceps* Hall, 1867; 20th Rep. New York State Cab. Nat. Hist., p. 377, pl. 21, figs. 12–14; also rev. ed., p. 424, pl. 21, figs. 12–14. (See *Lichas emarginatus*.)
—— CANADENSIS Bill., 1866; Catalogue Sil. Foss. Anticosti, p. 65, fig. 22. East Point, Div. 3.
—— CHAMPLAINENSIS Whitf., 1886; Bull. Am. Mus. Nat. Hist. New York, vol. 1, No. 8, p. 342, pl. 23, figs. 6–8. Isle La Motte, Lake Champlain, N. Y. Birdseye.

LICHAS.

— (*Arges*) CONTUSUS Hall, 1888; Pal. New York, vol. 7, p. 83, pl. 19 b, figs. 3-6. Clarksville, Cherry Valley, etc., N. Y.
Upper Helderberg.
— CUCULLUS M. & W., 1865; Proc. Acad. Nat. Sci. Phila., p. 266. Alexander county, Illinois.
Trenton.
— CUCULLUS M. & W., 1866; Geol. Survey Illinois, vol. 3, p. 299, pl. 1, figs. 6 a–c.
— DECIPIENS W. & M., 1866; Mem. Boston Soc. Nat. Hist., vol. 1, pt. 1, p. 104, pl. 3, fig. 11. Bridgeport, Illinois.
Niagara.
— (*Ceratolichas*) DRACON Hall, 1888; Pal. New York, vol. 7, p. 85, pl. 19, figs. 14-17. Schoharie and Le Roy, N. Y. Hall, 1876; Illus. Devonian Foss. Crust., pl. 19, figs. 2, 3.
Upper Helderberg.
— EMARGINATUS Hall, 1879; 28th Rep. New York State Mus. Nat. Hist., p. 199. Waldron, Indiana.
Niagara.
Syn., *Lichas breviceps* Hall; 20th Rep. New York State Cab. Nat. Hist., p. 377, pl. 21, figs. 12-14, rev. ed., p. 424, pl. 21, figs. 12-14.
— ERIOPIS, Hall.
Corniferous and Upper Helderberg.
Syn., *Lichas armatus* Hall, 1861; Des. New Species Foss., p. 81.
Lichas armatus Hall, 1862; 15th Rep. New York State Cab. Nat. Hist., p. 109.
— ERIOPS Hall, 1863; 16th Rep. New York State Cab. Nat. Hist., p. 226.
Acidaspis (*Terataspis*) ERIOPS Hall, 1876; Illus. Devonian Foss. Crust., pl. 19, figs. 4-7, 10, 11.
Acidaspis (*Terataspis?*) sp.? Hall, 1876; Illus. Devonian Foss. Crust., pl. 18, fig. 12.
— (*Conolichas*) ERIOPS Hall, 1888; Pal. New York, vol. 7, p. 78, pl. 19 a, figs. 2-13, 15, 16. Schoharie, also in boulders Canandaigua, N. Y.
— GRANDIS Hall, 1861; Des. New Species Foss., p. 82. Schoharie, N. Y.
Upper Helderberg.
— GRANDIS Hall, 1862; 15th Rep. New York State Cab. Nat. Hist., p. 110.
— GRANDIS Hall, 1863; 16th Rep. New York State Cab. Nat. Hist., p. 223. (Subgenus *Terataspis*.)
Syn., *Lichas superbus* Bill., 1875; Canadian Naturalist, vol. 7, p. 239.
Acidaspis (*Terataspis*) GRANDIS Hall, 1876; Illus. Devonian Foss. Crust., pls. 17, 18.
— (*Terataspis*) GRANDIS Hall, 1888; Pal. New York, vol. 7, p. 73, pl. 17, figs. 1-6; pl. 18, figs. 1, 2; pl. 19, figs. 1-7. Schoharie, Thompson's Lake, Knox, and near Clarksville, N. Y., Cayuga, Ontario.
— (*Ceratolichas*) GRYPS Hall, 1888; Pal. New York, vol. 7, p. 84, pl. 19 b, figs. 7-13. Schoharie county, N. Y.
Upper Helderberg. Corniferous.
(*Acidaspis* n. sp.?) Hall, 1876; Illus. Dev. Crust., pl. 19, fig. 1 (not figs. 2, 3).

LICHAS. LLOYDIA.

—— HARRISI Miller, 1878; Jour. Cincinnati Soc. Nat, Hist., vol. 1, p. 106, pl. 3, fig. 9. Waynesville, Ohio. Hudson.

—— (*Conolichas*) HISPIDUS Hall, 1888; Pal. New York, vol. 7, p. 77, pl. 19 a, figs. 14, 17, 18. Upper Helderberg group, Schoharie grit near Clarksville, Corniferous limestone, Le Roy, N. Y. Syn., *Acidaspis* (*Terataspis*) *eriops* Hall, 1876; Illus. Dev. Foss., pl. 19, figs. 8, 9. Upper Helderberg.

—— (*Hoplolichas*) HYLÆUS Hall, 1898; Pal. New York, vol. 7, p. 81, pl. 19 b, figs. 1, 2. Province of Ontario. Upper Helderberg.

—— JUKESI Bill., 1865; Pal. Foss., vol. 1, p. 282, fig. 260; also p. 335, fig. 323. Cow Head, N. F.

—— MINGANENSIS Bill., 1865; Pal. Foss., vol. 1, p. 181, figs. 163 a, b. Large Island, Mingan Islands. Chazy and Trenton.

—— NEREUS Hall, 1863; 16th Rep. New York State Cab. Nat. Hist., p. 226. No description or locality mentioned.

—— OBVIUS Hall, 1870; 20th Rep. New York State Cab. Nat. Hist., p. 424, pl. 25, fig. 19. Lyons, Iowa. Niagara.

Calymene PHLYCTANODES Green, 1837; Am. Jour. Sci., 1st series, vol. 32, p. 167. Niagara.

Arges PHLYCTANODES Hall, 1852; Pal. New York, vol. 2, p. 314, pl. 70, figs. 2 a, c. Near Albion, Orleans Co., N. Y.

—— (*Dicranogmus*) PTYONURUS Hall, 1888; Pal. New York, vol. 7, p. 86, pl. 19 b, figs. 19-21. Schoharie, N. Y. Niagara.

—— PUGNAX W. & M., 1866; Mem. Boston Soc. Nat. Hist., vol. 1, pt. 1, p. 103, pl. 3, fig. 10. Bridgeport, Ill. Niagara.

—— PUGNAX Hall, 1867; 20th Rep. New York State Cab. Nat. Hist., p. 393 (rev. ed., p. 424, pl. 25, fig. 20).

—— PUSTULOSUS Hall, 1859; Pal. New York, vol. 3, p. 366, pl. 77, figs. 9-12; pl. 78, figs. 1-6, fig. 7 (?). Albany and Schoharie Co., N. Y. Lower Helderberg.

—— PUSTULOSUS Hall, 1876; Illus. Devonian Foss. Crust., pl. 19, fig. 13.

—— (*Conolichas*) PUSTULOSUS Hall, 1888; Pal. New York, vol. 7, p. 80, pl. 19, figs. 8, 10, 11.

—— *Hoplolichas* ROBBINSI Ulrich, 1892; The Amer. Geol., vol. 10, p. 271, fig. Near Wykoff, Minn. Trenton.

—— *superbus* Bill., 1875; Canadian Naturalist, new series, vol. 7, p. 239. (See *Lichas grandis*.) Corniferous.

Asaphus? TRENTONENSIS Con., 1842; Jour. Acad. Nat. Sci. Phila., vol. 8, p. 277, pl. 16, fig. 16. Trenton.

Platynotus TRENTONENSIS Hall, 1847; Pal. New York, vol. 1, p. 235, pl. 64, figs. 1 a-e. Middleville, etc., N. Y.

—— TRENTONENSIS Miller, 1874; Cincinnati Quart. Jour. Sci., vol. 1, p. 127.

LLOYDIA Vogd., 1889 (generic name, with *Bathyurus bituberculatus* Bill. for its type).

LONCHOCEPHALUS.

Loganellus Devine, 1863; Canadian Naturalist, vol. 8, p. 93. (See *Ptychoparia, Crepicephalus* and *Olenus*.)

—— *logani* Devine, 1863; Canadian Naturalist, vol. 8, p. 95. See *Ptychoparia logani*.) Point Levis, Quebec.

Olenus? logani Bill., 1865; Pal. Foss., vol. 1, p. 201, figs. 185, 186.

—— *quebecensis* Bill., 1865; Pal. Foss., vol. 1, p. 203. (See *Ptychoparia logani*.)

LONCHOCEPHALUS Owen, 1852; Rep. Geol. Survey Wisconsin, Iowa, and Minnesota, p. 576. Dr. Owen in his report, Geol. Survey, Wis., Iowa and Minnesota, described four new genera of trilobites under the names of *Dikelocephalus, Lonchocephalus, Crepicephalus* and *Menocephalus*. The second genus is now classed as a synonym to *Anomocare* with *Anomocare chippewœnsis*, and *A. hamulus*, described in the report. There is a third species mentioned by Owen on p. 624, and also plate 1 a, fig. 15, Pygidium of *Lonchocephalus* (?) with long, slender, divergent caudal spines; also a similar pygidium, figured plate 1 a, fig. 13, as *Dikelocephalus (?) iowensis*. Walcott, Bull. U. S. Geol, Survey, N. 10, p. 36, places the latter species under a subgenus to *Ptychoparia*, using Dr. Owen's *Crepicephalus* for it on account of the peculiar pygidium, overlooking, however, the fact that Owen has referred a similar pygidium directly to his genus Lonchocephalus on p. 624 of his report. Under the law of priority the genus *Ptychoparia* replaces *Crepicephalus*, and the later name should be cancelled *in toto*, and not be retained in a modified sense. *Lonchocephalus* forms an exception to this rule, if we divide the genus into two sections, retaining *Anomocare* for *hamulus* and *chippewœnsis*, and *Lonchocephalus* for the species described by Dr. Owen with long, slender, divergent caudal spines which includes *Dikelocephalus (?) iowensis*.

—— sp.? Owen, 1852; Rep. Geol. Survey, Wisconsin, Iowa and Minnesota, p. 624.

CREPICEPHALUS ANGUSTA Wal., 1886; Bull. U. S. Geol. Survey, No. 30, p. 208, pl. 28, figs. 2, 2 a, b. Pioche, also 11 miles north of Bennet's Springs, Highland Range, Nevada.   Taconic.

CREPICEPHALUS ANGUSTA Wal., 1890; 10th Rep. U. S. Geol. Survey, p. 655, pl. 96, figs. 9, 9 a, b.

Dicellocephalus? IOWENSIS Owen, 1852; Rep. Geol. Wisconsin, Iowa and Minnesota, p. 575, pl. 1, fig. 4; pl. 1 a, fig. 13. Mountain Island and near the mouth of Black River.

Undetermined Trilobite, Owen, 1848; Rep. Geol. Reconnoissance Chippewa Land Dist., p. 14, pl. 7, fig. 1.   Taconic.

Conocephalites IOWENSIS Shum., 1863; Trans. Acad. Sci. St. Louis, vol. 2, p. 102.

LONCHOCEPHALUS. MICRODISCUS.

Conocephalites IOWENSIS Hall, 1863; 16th Rep. New York State Cab. Nat. Hist., p. 162, pl. 7, fig. 33; pl. 8, figs. 10-12 and 30.

Crepicephalus IOWENSIS Wal., 1884; Bull. U. S. Geol. Survey, No. 10, p. 36.

Crepicephalus IOWENSIS Wal., 1886, Bull. U. S. Geol. Survey, No. 30, p. 207.

—— sp.? *undet.* Owen, 1852; Rep. Geol. Wisconsin, Iowa and Minnesota, p. 624, No. 8, table 1 a, fig. 15. Near Mountain Island.

Crepicephalus LILIANA Wal., 1886; Bull. U. S. Geol. Survey, No. 30, p. 209, pl. 28, figs. 3, 3 a, c. Pioche, also 11 miles north of Bennet Springs, etc., Nevada. Taconic.

Crepicephalus LILIANA Wal., 1890; 10th Rep. U. S. Geol. Survey, p. 653; pl. 96, fig. 4, 4 a. *Olenellus* zone.

MEGALASPIS Ang., 1852; Palæont. Scand., p. 15.

—— BELEMNURUS White, 1874; Rep. Invert. Foss., etc., p. 11.

—— BELEMNURUS White, 1877; Geol. Survey. W. 100th Meridian, Pal., vol, 4, p. 59, pl. 3, fig. 9. Queen Spring Hill, Schell Creek, Nev. *Asaphus* (*Meglaspis*) GONIOCERUS Meek, 1873; 6th Ann. Rep. Geol. Survey Territories, 1872, p. 480. Quebec.

MENOCEPHALUS Owen, 1852; Rep. Geol. Survey Wisconsin, Iowa and Minnesota, p. 577. Dr. D. D. Owen refers one species to this genus, *Menocephalus minnesotensis*, pl. 1, fig. 11, which has a highly arched and hemispherical glabella.

—— GLOBOSUS Bill., 1860; Canadian Naturalist, vol. 5, p. 317, figs. 17-19. Point Levis. Taconic.

—— GLOBOSUS Bill., 1865; Pal. Foss., vol. 1, p. 408, figs. 388 a-c.

—— MINNESOTENSIS Owen, 1852; Rep. Geol. Survey Wisconsin, Iowa and Minnesota, p. 577, pl. 1, fig. 11. From Miniskah. Taconic.

—— SALTERI Bill., 1863; Canadian Naturalist, vol. 8, p. 210, fig.

—— SALTERI Bill., 1865; Pal. Foss., vol. 1, p. 203, fig. 187. Point Levis, Quebec. Taconic.

—— SALTERI? Rominger, 1887; Proc. Acad. Nat. Sci. Phila., p. 16, pl. 1, fig. 6. Mount Stephens, N. W. Terr., Canada. Taconic.

—— SEDGWICKI Bill., 1860; Canadian Naturalist, vol. 5, p. 316, fig. 16. Point Levis, Quebec. Taconic.

—— SEDGWICKI Bill., 1865; Pal. Foss., vol. 1, p. 407, fig. 387.

MESONACIS Wal., 1885; Am. Jour. Sci., 3d series, vol. 29, p. 329. Mesonacis is used as a subgenus to *Olenellus* by Walcott in the 10th Report U. S. Geol. Survey, p. 637. (See *Olenellus* (*Mesonacis*) *vermantana*.

MICRODISCUS Emmons, 1855; Am. Geology, vol. 1, pt. 2, p. 116.

—— BELLI-MARGINATUS S. & F., 1888; Bull. Mus. Comp. Zool. Harvard Coll., vol. 16, p. 35, pl. 2, fig. 19. North Attleborough, Mass.
Taconic.

MICRODISCUS.
— BELLI-MARGINATUS Wal., 1890; 10th Rep. U. S. Geol. Survey, p. 630, pl. 81, fig. 2, 2 a-b. *Olenellus* zone.
— CONNEXUS Wal., 1887; Am. Jour. Sci., 3d series, vol. 34, p. 194, pl. 1, figs. 4, 4 b. Washington Co., N. Y. Taconic.
— CONNEXUS Wal., 1890; 10th Rep. U. S. Geol. Survey, p. 631, pl. 80, fig. 9, 9 a-b. *Olenellus* zone.
— DAWSONI Hartt, 1868; Acadian Geol., p. 654, fig. 228. St. John, N. B. Mr. Hartt originally described this species under the new generic name of *Dawsonia*, but on Mr. E. Billings' authority the species was referred to *Microdiscus* in the Acadian Geol. Taconic.
— DAWSONI Whiteaves, 1878; Am. Jour. Sci., 3d. series, vol. 16, p. 225.
— DAWSONI Wal., 1884; Bull. U. S. Geol. Survey, No. 10, p. 23, pl. 2, figs. 3, 3 a.
— DAWSONI Matt., 1855; Trans. Roy. Soc. Canada, vol. 3, p. 75, pl. 7, figs. 11 a-c.
— HELENA Wal., 1889; Proc. Nat. Mus., vol. 12, p. 41.
— HELENA Wal., 1890; 10th Rep. U. S. Geol. Survey, p. 632, pl. 81, figs. 1, 1 a. *Olenellus* zone.
Agnostus LOBATUS Hall, 1847; Pal. New York, vol. 1, p. 258, pl. 67, figs. 5 a-f. One mile east of Troy, N. Y. Taconic.
Agnostus LOBATUS Rogers, 1858; Geol. Survey Pennsylvania, vol. 2, p. 820, fig. 614 (1-4).
— LOBATUS Ford, 1873; Am. Jour. Sci., 3d series, vol. 6, p. 135 (footnote).
— LOBATUS Wal., 1886; Bull. U. S. Geol. Survey, No. 30, p. 156, pl. 16, figs. 1, 1 a, b. Troy, N. Y.
— LOBATUS S. & F., 1888; Bull. Mus. Comp. Zool. Harvard Coll., vol. 16, p. 36, pl. 2, fig. 13.
— LOBATUS Wal., 1890; 10th Rep. U. S. Geol. Survey, p. 632, pl. 81, figs. 4 a-b.
— MEEKI Ford, 1876; Am. Jour. Sci., 3d series, vol. 11, p. 371. Troy, N. Y. Taconic.
— MEEKI Wal., 1886; Bull. U. S. Geol. Survey, No. 30, p. 155, pl. 16, fig. 4.
— MEEKI Wal., 1890; 10th Rep. U. S. Geol. Survey, p. 632, pl. 81, fig. 3. *Olenellus* zone.
— PARKERI Wal., 1886; Bull. U. S. Geol. Survey, No. 30, p. 157, pl. 16, figs. 2, 2 a. Township of Georgia, Vt. Taconic.
— PARKERI Wal., 1890; 10th Rep. U. S. Geol. Survey, p. 632, pl. 80, figs. 7, 7 a. *Olenellus* zone.
— PULCHELLUS (Hartt MS.) Wal., 1884; Bull. U. S. Geol. Survey, No. 10, p. 24. Ratcliff's Millstream, N. B.; Manuel's Brook, near Conception Bay, N. F. Taconic.
— PUNCTATUS Salt., 1864; Quart. Jour. Geol. Soc. London, vol. 20, p. 237, pl. 13, fig. 11. St. John, N. B.

MICRODISCUS. OGYGIA.
— PUNCTATUS Whiteaves, 1878; Am. Jour. Sci., 3d series, vol. 16, p. 225.
— PUNCTATUS var. PRECURSOR Matt., 1885; Trans. Roy. Soc. Canada, vol. 3, p. 75, pl. 7, fig. 13. *Paradoxides* zone.
— PUNCTATUS Wal., 1884; Bull. U. S. Geol. Survey, No. 10, p. 24, pl. 2, figs. 1, 1 a-c.
— QUADRICOSTATUS Emmons, 1855; Am. Geol., vol. 1, pt. 2, p. 116, pl. 1, fig. 8. Augusta county, Va. Taconic.
— QUADRICOSTATUS Emmons, 1860; Manual Geol., p. 88, fig. 73.
— QUADRICOSTATUS Barr., 1861; Bull. Soc. Géol. France, 2d series, vol. 18, p. 280, pl. 5, figs. 13 a, b.
— SPECIOSUS Ford, 1873; Am. Jour. Sci., 3d series, vol. 6, p. 137, figs. 2 a, b. Troy, N. Y.
— SPECIOSUS Ford, 1877; Am. Jour. Sci., 3d series, vol. 13, p. 147.
— SPECIOSUS Ford, 1879; New York Tribune Extra, Sept. 2, 1879, fig. 2.
— SPECIOSUS Wal., 1886; Bull. U. S. Geol. Survey, No. 30, p. 154, pl. 16, figs. 3, 3 a-c.
— SPECIOSUS Wal., 1890; 10th Rep. U. S. Geol. Survey, p. 632, pl. 81, figs. 5, 5 a-c. *Olenellus* zone.
NIOBE Ang., 1852; Palæont. Scand., p. 13.
— PRODUCTA (H. & W.) Brögg.; Om alderen af Olenellus zonæ i N. A., p. 211. Oquirrh Mts., Utah. Taconic.
NILEUS Dalm., 1826; Palæad., p. 246 (German trans., p. 49).
— AFFINIS Bill., 1865; Pal. Foss., vol. 1, p. 275, figs. 261 a, b. Cow Head and Island of Orleans. Cambrian.
— MACROPS Bill., 1865; Pal. Foss., vol. 1, p. 273, fig. 259. Table Head, N. F. Cambrian.
— SCRUTATOR Bill., 1865; Pal. Foss., vol. 1, p. 274, fig. 260. Table Head, Portland Creek, N. F. Cambrian.
Nuttania Eaton, 1832 (in part); Geol. Text Book, p. 33. (See *Trinucleus* and *Homalonotus*.)
— *concentrica* Eaton, 1832; Geol. Text Book, p. 34, pl. 1, fig. 2.
— *sparsa* Eaton, 1832; Geol. Text Book, p. 34.
Odontocephalus Con., 1840; 3d Ann. Rep. Pal. Dept. New York Geol. Survey, p. 204.
Dalmanites (*Odontocephalus*) Hall, 1888; Pal. New York, vol. 7, p. xxxiii.
Odontocephalus. (See *Dalmanites* (*Odontocephalus*) *selenurus*, D. (*Odontocephalus*) *bifidens*, D. (*Odontocephalus*) *ægeria*, and D. (*Odontocephalus*) *coronatus*.)
OGYGIA Broug., 1822; Crust. Foss., p. 26.
Asaphus BARRANDI Hall, 1851; Rep. Geol. Lake Superior Land Dist. (Foster & Whitney), pt. 2, p. 210, pl. 27, figs. 1 a-d; pl. 28. St. Mary's River, also near Plattsville, Wisconsin. cf. *Ogygia cornensis* Murch. Trenton.
Asaphus BARRANDI Hall., 1862; Geol. Wisconsin, vol. 1, p. 41, fig. 4.

CATALOGUE OF TRILOBITES. 325

OGYGIA. OLENELLUS.
— KLOTZI Rominger, 1887; Proc. Acad. Nat. Sci. Phila., p. 12, pl. 1,
fig. 1. Mount Stephens, British Columbia. Taconic.
— KLOTZI Wal., 1888; Am. Jour. Sci., 3d series, vol. 36, p. 166. In the
Proc. U. S. Natl. Mus., vol. 11, 1888, p. 446, Mr. Walcott proposes
the new generic name of *Ogygopsis* for *Ogygia klotzi*.
Oyygies latissimus Eaton, 1832; Am. Jour. Sci., vol. 21, p. 136, footnote.
Ogygies latissimus Eaton, 1832; Geol. Text Book, p. 33 (See *Homalonotus delphinocephus*).
— PARABOLA H. & W., 1877; U. S. Geol. Expl. 40th Par., vol. 4, p. 245,
pl. 2, fig. 35. East Cañon, Oquirrh Mts., Utah. Taconic.
— ? PROBLEMATICA Wal., 1884; Pal. Eureka Dist. Mon. U. S. Geol. Survey, vol. 8, p. 63, pl. 10, figs. 2 a, b, 4.
— PRODUCTA H. & W., 1877; U. S. Geol. Expl. 40th Par., vol. 4, p. 234,
pl. 2, figs. 31-34; also p. 245, pl. 2, fig. 35. East Cañon, Oquirrh
Mts., Utah. Taconic.
Niobe PRODUCTA Brögg., 1886; Om alderen af Olenellus zonæ i Nordamerika, p. 211.
— (*Bathyuriscus*) PRODUCTUS Wal., 1886; Bull. U. S. Geol. Survey, No.
30, p. 217, pl. 30, figs. 1, 1 a-i.
Syn., *Ogygia parabola* (H. & W.) Wal., 1886; *idem.*, p. 217.
— SERRATA Rominger, 1887; Proc. Acad. Nat. Sci. Phila., p. 13, pl. 1,
figs. 2, 2 a. Mount Stevens, British Columbia. Taconic.
— ? *spinosa* Wal., 1884; Pal. Eureka Dist. Mon. U. S. Geol. Survey, vol.
8, p. 63, pl. 9, fig. 22. (See *Olenoides spinosa* Wal.)
— ? VETUSTUS Hall, 1847; Pal. New York, vol. 1, p. 227, pl. 60, fig. 1.
Mohawk Valley, N. Y. Trenton.
— VETUSTUS Emmons, 1855; American Geology, vol. 1, pt. 2, p. 216,
fig. 72.
OLENELLUS Hall, 1861; Des. N. Species of Foss., published in advance,
15th Rep. State Cab., p. 86. Hall, 1862; 15th Rep. State Cab.
Nat. Hist. Albany, p. 114.
Syn., *Holmia* Matt., 1890; Trans. Roy. Soc. Canada, vol. 7, p. 160.
Schmidtia Wal., 1890; Am. Geol., vol. 5, p. 363.
Cephalacanthus Lapworth, 1890; Wal., 10th Rep. U. S. Geol.
Survey, p. 641.
Elliptocephala ASAPHOIDES Emmons, 1844; Taconic System, p. 21,
figs. 1-3. Reynolds Inn, NE. of Bald Mountain, Washington
county, N. Y. Taconic.
Elliptocephala ASAPHOIDES Emmons, 1846; Agric. New York, vol. 1,
p. 65, figs. 1-3.
Olenus ASAPHOIDES Hall, 1847; Pal. New York, vol. 1, p. 256, pl.
67, figs. 2 a, c.
Olenus ASPAHOIDES Fitch, 1849; Trans. N. Y. State Agric. Soc., vol.
9, p. 865.

OLENELLUS.

Elliptocephalus ASAPHOIDES Emmons, 1849; Proc. Am. Assoc. Adv. Sci., 1st meeting, p. 18.

Elliptocephalus ASAPHOIDES Emmons, 1855; Am. Geology, vol. 1, pt. 2, p. 114, figs. 1-3, pl. 1, fig. 18.

Paradoxides ASAPHOIDES Emmons, 1860; Manual Geol., pp. 87, 280.

Paradoxides ASAPHOIDES Barr., 1861; Bull. Soc. Géol. France, 2d series, vol. 18, p. 273, pl. 5, figs. 4, 5.

—— ASAPHOIDES Ford, 1871; Am. Jour. Sci., 3d series, vol. 2, p. 33.

—— (*Elliptocephalus*) ASAPHOIDES Ford, 1877; Am. Jour. Sci., 3d series, vol. 13, p. 266, plate, figs. 1-10. Troy, N. Y., also one mile below Schodack Landing, N. Y.

—— ASAPHOIDES Ford; Am. Jour. Sci., 3d series, vol. 15, p. 129.

—— ASAPHOIDES Ford, 1881; Am. Jour. Sci., 3d series, vol. 22, p. 250.

—— ASAPHOIDES Wal., 1866; Bull. U. S. Geol. Survey, No. 30, p. 168, pl. 17, figs. 4, 8, 10; pl. 20, figs. 3, 3 a, b; pl. 25, fig. 8.

—— (*Mesonacis*) ASAPHOIDES Wal., 1890; 10th Rep. U. S. Geol. Survey, p. 637, pls. 86, 88, 89 and 90. *Olenellus* zone.

—— BRÖGGERI Wal., 1889; Proc. Nat. Mus., vol. 12, p. 41. Manuel Brook, Conception Bay, N. F.   Taconic.

—— (*Holmia*) BRÖGGERI Wal., 1890; Rep. U. S. Geol. Survey, p. 638, pls. 91 and 92. *Olenellus* zone.

—— GILBERTI (Meek MS.) White, 1874; Prelim. Rep. Invt. Foss. U. S. Geog. and Geol. Survey West 100th Mer., p. 7.

—— (*Olenus*) GILBERTI Gilbert, 1875; U. S. Geog. Survey West 100th Mer., vol. 3, p. 182.

—— GILBERTI White, 1877; U. S. Geog. Survey West 100th Mer., vol. 4, p. 44, pl. 2, figs. 1, 3 a-e.

—— GILBERTI Wal., 1884; Pal. Eureka Dist., Mon. U. S. Geol. Survey, vol. 8, p. 29, pl. 9, figs. 16, 16 a; pl. 21, fig. 13.

—— GILBERTI Wal., 1886; Bull. U. S. Geol. Survey, No. 30, p. 170, pl. 18, figs. 1, 1 a-c; pl. 19, fig. 2, 2 a-k; pl. 20, figs. 1, 1 a, 1 b, 4. In Nevada at Pioche, Highland Ridge, eight miles north of Bennet's Spring, Groome District, etc. In Utah in the Oquirrh Range, Big Cottonwood Cañon, Wahsatch Mountains, one mile below Argenta; also Kicking Horse Lake, B. C.

—— GILBERTI Wal., 1890; Rep. U. S. Geol. Survey, p. 636, pls. 84, 85 and 86.

—— HOWELLI (Meek MS.) White, 1874; Prelim. Rep. Invt. Foss. U. S. Geog. and Geol. Survey West 100th Mer., p. 8.

—— (*Olenus*) HOWELLI Gilbert, 1875; U. S. Geog. Survey West 100th Mer., vol. 3, p. 183.

—— *howelli* White, 1877; U. S. Geog. Survey West 100th Mer., vol. 4, p. 47, pl. 2, figs. 4 a, b.

—— *howelli* Wal., 1884; Pal. Eureka Dist. Mon. U. S. Geol. Survey, vol. 8, pp. 29, 30, pl. 9, figs. 15 a-b, 16-16 a; pl. 21, figs. 1-9.

OLENELLUS.

—— *howelli* Wal., 1886; Bull. U. S. Geol. Survey, p. 171. Referred to the older name of *O. gilberti* Meek.
—— IDDINGSI Wal., 1884; Pal. Eureka Dist. Mon. U. S. Geol. Survey, vol. 8, p. 28, pl. 9, fig. 12.
—— IDDINGSI Wal., 1886; Bull. U. S. Geol. Survey No. 30, p. 170, pl. 19, fig. 1. Prospect Peak, Eureka Dist., Nev., south end of the Timpahute Range, Eastern Nev.
—— IDDINGSI Wal., 1890; 10th Rep. U. S. Geol. Survey, p. 636, pl. 84, fig. 2.

Paradoxides MACROCEPHALUS Emmons, 1860; Manual Geology, p. 88, fig. 70.
Olenus THOMPSONI Hall, 1859; 12th Rep. New York State Cab. Nat. Hist., p. 57, fig. Township of Georgia, Vermont. Taconic.
Olenus THOMPSONI Hall, 1859; Pal. New York, vol. 3, p. 525, fig.
Barrandia THOMPSONI Hall, 1860; 13th Rep. New York State Cab. Nat. Hist., p. 116, fig.
Paradoxides THOMPSONI Emmons, 1860; Manual Geology, p. 280, note A.
Paradoxides THOMPSONI Barr., 1861; Bull. Soc. Géol. France, 2d series, vol. 18, p. 276, pl. 5, fig. 6.
Paradoxides THOMPSONI Bill., 1861; Geol. Vermont, vol. 2, p. 950.
Barrandia THOMPSONI Hall, 1861; Geol. Vermont, vol. 2, p. 369, pl. 13, fig. 1.
Paradoxides THOMPSONI Bill., 1861; Pal. Foss., vol. 1, p. 11 (pamphlet published in advance). Bonne Bay, N. F., l'Anse au Loup, on the north shore of the Straits of Belle Isle, Big Harbor, on the St. Lawrence River, below Quebec.
—— THOMPSONI Hall, 1862; 15th Rep. New York State Cab. Nat. Hist., p. 114.
—— THOMPSONI Bill., 1863; Geol. Canada, p. 953.
—— THOMPSONI Bill., 1865; Pal. Foss., vol. 1, p. 11.
—— - THOMPSONI Whitf., 1884; Bull. Am. Mus. Nat. Hist. New York, vol. 1, p. 151, pl. 15, figs. 1–4.
—— THOMPSONI Wal., 1886; Bull. U. S. Geol. Survey No. 30, p. 167, pl. 17, figs. 1, 2, 4, 9; pls. 22, 23, fig.
—— THOMPSONI Wal., 1890; 10th Rep. U. S. Geol. Survey, p. 635, pls. 82, 83.
—— WALCOTTI Wal., 1890; 10th Rep. U. S. Geol. Survey, p. 636, pl. 88, fig. 2. The author therein refers *Paradoxides walcotti* Shaler & Foerste to the genus *Olenellus*. North Attleborough, Mass.

Olenus VERMONTANA Hall, 1859; 12th Rep. New York State Cab. Nat. Hist., p. 60, fig. Township of Georgia, Vermont. Taconic.
Olenus VERMONTANA Hall, 1859; Pal. New York, vol. 3, p. 527, fig.

OLENELLUS. OLENOIDES.

Paradoxides VERMONTI Emmons, 1860; Manual Geology, p. 280 (note A).
Paradoxides VERMONTANA Barr., 1861; Bull. Soc. Géol. France, 2d series, vol. 18, p. 277, pl. 5, fig. 8.
Paradoxides VERMONTANA Bill., 1861; Geology Vermont, vol. 2, p. 950.
Barrandia VERMONTANA Hall, 1861; 13th Rep. New York State Cab. Nat. Hist., p. 117, fig.
Barrandia VERMONTANA Hall, 1861; Geol. Vermont, vol. 1, p. 370, pl. 13, figs. 2, 4, 5.
Paradoxides VERMONTANA Bill., 1863; Geol. Canada, p. 953.
—— VERMONTANA Hall, 1862; 15th Rep. New York State Cab. Nat. Hist., p. 114.
—— VERMONTANA Bill., 1865; Pal. Foss., vol. 1, p. 10.
—— VERMONTANA Whitf., 1884; Bull. Am. Mus. Nat. Hist. New York, vol. 1, p. 152, pl. 15, figs. 2-4.
—— (*Mesonacis*) VERMONTANA Wal., 1885; Am. Jour. Sci., 3d series, vol. 29, p. 329, figs. 1, 2.
—— (*Mesonacis*) VERMONTANA Wal., 1885; Nature, vol. 32, p. 68.
—— (*Mesonacis*) VERMONTANA Wal., 1886; Bull. U. S. Geol. Survey No. 30, p. 158, pl. 24, figs. 1 a, b.
—— (*Mesonacis*) VERMONTANA Wal., 1890; 10th Rep. U. S. Geol. Survey, p. 637, pl. 87, fig. 1, 1 a-b.

OLENOIDES Meek, 1877; U. S. Geol. Expl. 40th Par. Pal., vol. 4, p. 25.
Dorypyge Dames, 1884; China (Richthofen), vol. 4, p. 23.
Corynexochus Aug., 1854; Palæont. Scand., p. 59, pl. 33, fig. 11 (not pl. 33, figs. 9, 10).
Bathyurus ARMATUS Bill., 1860; Canadian Naturalist, vol. 5, p. 319, fig. 23. Point Levis, Quebec. Taconic.
Bathyurus ARMATUS Bill., 1865; Pal. Foss., vol. 1, p. 411, fig. 392.
Bathyurus ARMATUS Wal., 1879; 32d Rep. New York State Mus. Nat. Hist., p. 131.
Dorypyge ? ARMATUS Vogd.; Bull. U. S. Geol. Survey No. 63, p. 97.
Dicellocephalus cristatus Bill., 1860; Canadian Naturalist, vol. 5, p. 312, fig. 10. Point Levis, Quebec.
Dicellocephalus cristatus Bill., 1865; Pal. Foss., vol. 1, p. 405, fig. 384. This might be referred to the pygidium of *D. armatus*.
—— CURTICEI Wal., 1888; Proc. U. S. Nat. Mus., vol. 11, p. 443, fig.
—— CURTICEI Wal., 1889; Dictionary of Fossils (Lesley), vol. 1, p. 492, fig.
—— CURTICEI Wal., 1890; 10th Rep. U. S. Geol. Survey, p. 752, pl. 94, fig. 1. Coosa Valley, Georgia. Olenellus zone.
—— (*Dorypyge*) DESIDERATA Wal., 1890; 10th Rep. U. S. Geol. Survey, p. 644, fig. 67. Highgate Springs, Vermont. Taconic.
—— ELLSI Wal., 1890; 10th Rep. U. S. Geol. Survey, p. 642, fig. 66. Four miles below Quebec. Taconic.

OLENOIDES.
— FORDI Wal., 1887; Am. Jour. Sci., 3d series, vol. 34, p. 195, pl. 1,
figs. 5, 5 b. Taconic.
— FORDI Wal., 1890; 10th Rep. U. S. Geol. Survey, p. 641, pl. 94, figs.
3, 3 a, b. Washington county, N. Y. Olenellus zone.
— *flagricaudus*. (See *Zacanthoides flagricaudus*.)
Dicellocephalus? GOTHICUS H. & W., 1877; U. S. Geol. Expl. 40th
Par., vol. 4, p. 242, pl. 1, fig. 36. Taconic.
Dorypyge GOTHICUS Dames, 1884; China (Richthofen), vol. 4, p. 23.
Dicellocephalus? GOTHICUS Wal., 1884; Pal. Eureka Dist. Mon. U. S.
Geol. Survey, vol. 8, p. 45, pl. 9, fig. 24.
— GOTHICUS Wal., 1886; Bull. U. S. Geol. Survey, vol. 30, p. 187, pl.
29, figs. 1, 1 a–c.
— *levis* Wal. (See *Zacanthoides levis*.)
— ? MARCOUI Wal., 1886; Bull. U. S. Geol. Survey No. 30, p. 186, pl. 26,
figs. 5, 5 a, b. Georgia, Vermont. Taconic.
Dicellocephalus ? MARCOUI Whitf., 1884; Bull. Am. Mus. Nat. Hist.
New York, vol. 1, p. 150. pl. 14, fig. 7.
— MARCOUI Wal., 1890; 10th Rep. U. S. Geol. Survey, p. 642, pl. 94,
figs. 2, 2 a, b.
Paradoxides? NEVADENSIS Meek, 1870; Proc. Acad. Nat. Sci. Phila.,
vol. 22, p. 62. House Range, Antelope Springs, Utah. Taconic.
Paradoxides? NEVADENSIS Meek, 1877; U. S. Geol. Expl. 40th Par.
Pal., vol. 4, p. 23, pl. 1, fig. 5.
— NEVADENSIS Meek, 1877; U. S. Geol. Expl. 40th Par. Pal., vol. 4, p. 25.
— NEVADENSIS Wal., 1886; Bull. U. S. Geol. Survey No. 30, p. 181, pl.
25, fig. 7; also in Proc. U. S. Nat. Mus., vol. 11, 1888, p. 443.
— NEVADENSIS Wal., 1888; Am. Jour. Sci., 3d series, vol. 36, p. 165.
Syn., *Ogygia serrata* (Rominger) Wal., 1888; Am. Jour. Sci., 3d
series, vol. 36, p. 165.
Dicellocephalus QUADRICEPS H. & W., 1877; U. S. Geol. Expl. 40th
Par., vol. 4, p. 240, pl. 1, figs. 37–40. Ute Peak, Wahsatch
Range, Utah. Taconic.
Dicellocephalus QUADRICEPS Wal., 1884; Pal. Eureka Dist. Mon.
U. S. Geol. Survey, vol. 8, p. 45, pl. 9, fig. 24.
Dorypyge QUADRICEPS Dames, 1884; China (Richthofen), vol. 4, p. 23.
Referred to
— QUADRICEPS Wal., 1886; Bull. U. S. Geol. Survey No. 30, p. 187, pl.
29, figs. 1, 1 a–c.
— QUADRICEPS Wal., 1890; 10th Rep. U. S. Geol. Survey, p. 646, pl. 94,
figs. 4, 4 a–d.
— *spinosus*. (See *Zacanthoides spinosus*.)
— STISSINGENSIS Dwight, 1889; Am. Jour. Sci., vol. 38, p. 147, pl. 6,
figs. 9–15. Mount Stissing, New York. Taconic.
— *typicalis*. (See *Zacanthoides typicalis*.)

OLENOIDES. PARABOLINA.

Dicellocephalus WAHSATCHENSIS H. & W., 1877; U. S. Geol. Expl. 40th Par. Pal., vol. 4, p. 241, pl. 1, fig. 35.
—— WAHSATCHENSIS Wal., 1886; Bull. U. S. Geol. Survey No. 30, p. 189, pl. 29, figs. 2, 2 a. Box Elder Cañon, Wahsatch Mts., etc., Utah. Taconic.
Syn., *Dicellocephallus? gothicus* (H. & W.) Wal., 1886; Bull. U. S. Geol. Survey No. 30, p. 189.
—— WAHSATCHENSIS Wal., 1888; Am. Jour. Sci., 3d series, vol. 36, p. 165.

OLENUS Dalm., 1826; Palæad., p. 220.
—— *asaphoides*. (See *Olenellus asaphoides*.)
—— *holopyga*. (See *Bathynotus holopyga*.)
—— *logani*. (See *Ptychoparia logani*.)
—— *thompsoni*. (See *Olenellus thompsoni*.)
—— UNDULOSTRIATUS Hall, 1847; Pal. New York, vol. 1, p. 258, pl. 67, figs. 3 a, b. Snake Hill, Saratoga, N. Y. Hudson.
—— UTAHENSIS H. W., 1873; Rep. 43d Meeting Brit. Assoc. Adv. Sci.
—— UTAHENSIS H. W., 1873; Geol. Mag., vol. 10, p. 523. Utah. The author gives the following note: "It shows evidence of a median axis apparently corresponding with the so-called straight alimentary canal noticed by Barrande. The matrix is composed of a hydrated silicate of magnesia."
—— *vermontana*. (See *Olenellus vermontana* Hall.)

Crepicephalus (*Loganella*) QUADRANS H. & W., 1877; U. S. Geol. Expl. 40th Par., Palæont., vol. 4, p. 238, pl. 2, figs. 11-13.
Ptychoparia QUADRANS Wal., 1886; Bull. U. S. Geol. Survey, No. 30, p. 199, pl. 29, figs. 4, 4 a, b. Above Call's Fort, north of Box Elder Cañon and one mile below Argenta, etc., Utah. Taconic.
—— QUADRANS Brögg., 1886; Om Alderen af Olenelluszonen i Nordamerika, p. 213.

ORYCTOCEPHALUS Wal., 1886; Bull. U. S. Geol. Survey, No. 30, p. 210.
—— PRIMUS Wal., 1886; Bull. U. S. Geol. Survey, No. 30, p. 210, pl. 29, figs. 3, 3 a. East of Pioche, Nev. Taconic.
—— PRIMUS Wal., 10th Rep. U. S. Geol. Survey, p. 653, pl. 95, figs. 4, 4 a. Olenellus zone.

Dicellocephalus SESOSTRIS Bill. This species appears to be near the genus *Oryctocephalus*, according to Matthews, Diffusion of the Cambrian fauna, p. 11, note.

PANDERIA Volb.; Mém. Acad. Sci. St. Petersburg, vol. 6, p. 31.
Illænus CONRADI Bill., 1859; Canadian Naturalist, vol. 4, p. 372, figs. 7-9. Falls of La Petite Chaudiere, Township of Hull, Canada. For other references, see *Illænus conradi* Bill. Trenton.

PARABOLINA Salt., 1849; Mem. Geol. Survey, decade 2.
—— SPINULOSA (Wahl.) Matt., 1891; Trans. Roy. Soc. Canada, vol. 9, p. 51, pl. 13, figs. 5 a-d. St. John, N. B.

PARABOLINA. PARADOXIDES.

—— HERES (Brögg.) Matt., 1891; Trans. Roy. Soc. Canada, vol. 9, p. 51, At Germaine street, St. John, N.B. Also var. *lata*, pl. 13, figs. 6 a-f, var. *grandis*, p. 52, pl. 13, fig. 7. At Navy Island, St. John Harbor, N. B. Taconic.

Dicellocephalus? corax Bill., shows some resemblance to *Parabolina spinulosa* Wahl. See Matthew's note, p. 11. Diffusion of the Cambrian fauna.

PARADOXIDES Brong., 1822; Crust. Foss., p. 30, pl. 4, fig. 1.
  Syn., *Plutonia* Salt., 1869; Quart. Jour. Geol. Soc. London, vol. 25, p. 52.

—— ABENACUS Matt., 1885; Trans. Roy. Soc. Canada, vol. 3, p. 78, pl. 7, figs. 17, 17 a-d. Paradoxides zone. St. John, N. B.

—— ACADICUS Matt., 1882; Trans. Roy. Soc. Canada, vol. 1, p. 103, pl. 9, figs. 16-18. Portland, N. B., Hanford Brook, St. Martin's.

—— ACADICUS Matt., 1884; Trans. Roy. Soc. Canada, vol. 2, p. 99, pl. 1, fig. 1.

—— ACADICUS Wal., 1884; Bull. U. S. Geol. Survey, No. 10, p. 25, pl. 6, figs. 3, 3 a.

—— ACADICUS var. SURICUS Matt., 1885; Trans. Roy. Soc. Canada, vol. 3, p. 77, pl. 7, fig. 16. Paradoxides zone.

—— *arcuatus* Harlan. (See *Triarthrus beckii*.)

—— BARBERI Winch., 1884; 13th Rep. Geol. Nat. His. Survey Minn., p. 67, pl. 1, fig. 7. Pipestone county, Minnesota.

—— *beckii*. (See *Triarthrus beckii*.)

—— BENNETII Salt., 1859; Quart. Jour. Geol. Soc. London, vol. 15, p. 353, fig. St. Mary's and Placentia Bay, N. F. Paradoxides zone.

—— BENNETII S. & W., 1865; Chart of Foss. Crust., fig. 25.

—— *boltoni*. (See *Lichas boltoni*.)

—— DECORUS Bill., 1865; Pal. Foss., vol. 2, p. 75. Chapel Arm, Trinity Bay, N. F. cf. *P. tenellus*. Paradoxides zone.

—— *eatoni*. (See *Triarthrus beckii*.)

—— ETEMINICUS Matt., 1882; Trans. Roy. Soc. Canada, vol. 1, p. 92, pl. 9, figs. 7-12. Portland, also at Radcliff's Stream. St. John. Var. *suricoides* (figs. 4-6), var. *breviatus*, var. *malicitus* (fig. 13), var. *pontificalis* (figs. 15, 15 a), var. *quacoensis* (figs. 14, 14 a).

—— ETEMINICUS Wal., 1884; Bull. U. S. Geol. Survey, No. 10, p. 27, pl. 3, figs. 1, 1 a-g.

—— HARLANI Green, 1834; Am. Jour. Sci., 1st series, vol. 35, p. 336. cf. *Paradoxides spinosus* Boeck.

—— HARLANI Green, 1835; Suppl. Mon. Tril. N. A., p. 14, cast 39.

—— HARLANI Rogers, 1856; Proc. Boston Soc. Nat. Hist., vol. 6, pp. 27, 40.

—— HARLANI Stodder, 1856; Proc. Boston Soc. Nat. Hist., vol. 6, p. 369.

—— HARLANI Rogers, 1856; Am. Jour. Sci., 2d series, vol. 22, p. 297.

PARADOXIDES.

—— *spinosus* Rogers, 1858; Geol. Pennsylvania, vol. 2, p. 816, fig. 590.
—— HARLANI Ordway, 1861; Proc. Boston Soc. Nat. Hist., vol. 8, p. 1, fig.
—— HARLANI Wal., 1884; Bull. U. S. Geol. Survey No. 10, p. 45, pl. 7, fig. 3; pl. 8, figs. 1, 1 a-c; pl. 9, fig. 1. Paradoxides zone. Braintree, Mass.
—— KJERULFI (Linrs.) Matt., 1886; Am. Jour. Sci., 3d series, vol. 31, p. 471. Referred by Holm (Geol. Füren. Stockholm Förhandl., vol. 9, 1887, häfte. 7, p. 1) to *Olenellus kjerulfi* Linrs., pls. 14, 15. Olenellus zone.
—— LAMELLATUS Hartt, 1868; Acadian Geology, p. 656, fig. 230. St. John, N. B.
—— LAMELLATUS Matt., 1882; Trans. Royal Soc. Canada, vol. 1, p. 105, pl. 9, figs. 16-18.
—— LAMELLATUS Matt., 1884; Trans. Royal Soc. Canada, vol. 2, p. 100, pl. 1, figs. 3, 4. St. John, N. B.
—— LAMELLATUS var. LORICATUS Matt., 1882; Trans. Royal Soc. Canada, vol. 1, p. 105, pl. 9, fig. 19. Portland, N. B.
—— LAMELLATUS Wal., 1884; Bull. U. S. Geol. Survey No. 30, p. 25, pl. 3, figs. 2, 2 a.
—— LAMELLATUS Matt., 1890; Trans. Roy. Soc. Canada, vol. 7, p. 135, pl. 11, fig. 9.
—— *macrocephalus*. (See *Olenellus thompsoni* Hall.)
—— MICMAC Hartt, 1868; Acadian Geology, p. 657, fig. 231. St. John, N. B.
—— MICMAC Matt., 1884; Trans. Royal Soc. Canada, vol. 2, p. 101, pl. 10, fig. 8.
—— MICMAC Matt., 1885; Trans. Roy. Soc. Canada, vol. 3, p. 80, pl. 7, fig. 18.
—— MICMAC var. PONTIFICALIS Matt., 1890; Trans. Roy. Soc. Canada, vol. 7, p. 136, pl. 11, fig. 8. Paradoxides zone. Portland, and Hanford Brook, St. Martins, N. B. Matthew refers the species described under *P. etcminicus* var. *pontificalis*, in 1882, to a variety of *P. micmac*.
—— ? *nevadensis*. (See *Olenoides nevadensis*.)
—— *quadrispinosus*. (See *Bathynotus quadrispinosus*).
—— TENELLUS Bill., 1874; Pal. Foss., vol. 2, p. 74, fig. 43. Chapel Arm, Trinity Bay, N. B.
—— TENELLUS Bill., 1882; Rep. Geol. Survey Newfoundland, p. 12, fig. 7. Paradoxides zone.
—— *thompsoni*. (See *Olenellus thompsoni* Hall.)
—— *triarthrus* Harlan; Trans. Geol. Soc. Pennsylvania, vol. 1, p. 264, pl. 15, fig. 5. (See *Triarthrus becki* Green.)
—— *vermontana*. (See *Olenellus* and *Mesonacis*.)
—— *walcotti* S. & F., 1888; Bull. Mus. Comp. Zool. Harvard Coll., vol. 16, p. 36, pl. 2, fig. 12. Attleborough, Mass. Referred by Walcott to *Olenellus walcotti*. Olenellus zone.

PELTURA.   PHACOPS.

PELTURA Milne-Edw., 1840; Crust., vol. 3, p. 344.
—— *holopyge*. (See *Bathynotus holopyge*.)
—— SCARABEOIDES (Wahl.) Matt., 1891; Trans. Royal Soc. Canada, vol. 9, p. 53, pl. 13, figs. 9 a–b. King Street and at Navy Island, St. John, also at Cape Breton.   Taconic.

PEMPHIGASPIS Hall, 1863; 16th Rep. New York State Mus. Nat. Hist., p. 221.
—— BULLATA Hall, 1863; 16th Rep. New York State Mus. Nat. Hist., p. 221, pl. 5 a, figs. 3–5. A species probably of the genus *Microdiscus*.

PHACOPS Emm., 1839; De Trilobites, etc., p. 18. The generic term *Trimerocephalus* McCoy has been used by authors for species like *Phacops volborthi*, compact in form; glabella inflated and expanded in front; lobes, except the basal ones, obscure; eyes small, and occupying the front portions of the head; genal angles rounded; pleuræ rounded; pygidium small, with few segments, and even border. *Acaste* Goldfuss for such species as *Phacops downingiæ* Murch. Diagnosis: Glabella not inflated nor greatly expanded in front; furrows distinct; eyes well developed, with numerous lenses; genal angles produced into spines; pleuræ rounded or truncate; pygidium small generally, with less than eleven segments; border even, but sometimes mucronate.

Syn., PHACOPS (Monorakos) Schm., 1866.
Calymene BUFO Green, 1832; Monthly Am. Jour. Geol., p. 559.
Calymene BUFO Green, 1832; Mon. Tril. N. A., p. 41, cast 10. The fossil is said to have been found in a dark grayish limestone in the State of New Jersey. The cast approaches so near *Phacops latifrons* Bronn that it may be the same species.
—— BUFO Hall, 1861; Des. New Species Foss., p. 65.
—— BUFO Hall, 1862; 15th Rep. New York State Cab. Nat. Hist., p. 93.
—— BUFO Hall, 1876; Illus. Dev. Foss. Crust., pl. 8, figs. 24, 25.
—— BUFO Hall, 1888; Pal. New York, vol. 7, p. 26, pl. 8, figs. 25, 26.
—— *bombifrons*. (See *Phacops cristata*.)
—— CACAPONA Hall, 1861; Des. New Species Foss., p. 68.   Cacapon River near its junction with the Potomac, Va.   Hamilton.
Syn., *Calymene bufo* Cast., 1843; Syst. Sil. de l'Amér., p. 21, pl. 2, figs. 1–4.
—— CACAPONA Hall, 1862; 15th Rep. New York State Cab. Nat. Hist., p. 96.
—— CACAPONA Hall, 1876; Illus. Dev. Foss. Crust., pl. 8, figs. 18–23.
—— CACAPONA Hall, 1888; Pal. New York, vol. 7, p. 27, pl. 8, figs. 19–24.
—— *callicephala*. (See *Dalmanites callicephala*.)
—— CRISTATA Hall, 1861; Des. New Species Foss., p. 67.
                                                                 Upper Helderberg.
Syn., *Phacops bombifrons* Hall, 1861; Des. New Species Foss., p. 67.
Phacops cristata Hall, 1862; 15th Rep. New York State Cab. Nat. Hist., p. 95.

PHACOPS.

>*Phacops cristata* Hall, 1876; Illus. Dev. Foss. Crust., pl. 6, figs. 1-17.
>
>*Phacops bombifrons* Hall, 1876; Illus. Dev. Foss. Crust., pl. 6, fis. 18-29.

—— CRISTATA Hall, 1888; Pal. New York, vol. 7, p. 14, pl. 6, figs. 1-31, 16-29, pl. 8 a, figs. 1-4. Oriskany Sandstone, Cayuga, Ontario, Upper Helderberg, Schoharie grit near Clarksville, Schoharie, N. Y., etc.

—— CRISTATA var. PIPA Hall, 1888; Pal. New York, vol. 7, p. 18, pl. 8 a, figs. 5-18. Waterville, N. Y., Le Roy, etc., N. Y.
Upper Helderberg.

—— *downingæ* Murch. (See *Phacops trisulcatus* Hall.)

—— HUDSONICA Hall, 1849; Pal. New York, vol. 3, p. 355, pl. 73, figs. 26-28. Becraft's Mt., near Hudson, N. Y.   Lower Helderberg.

—— ? *laticauda* Hall, 1847; Pal. New York, vol. 1, p. 248, pl. 64, fig. 6. (See *Dalmanites micrurus* Green.)

—— LOGANI Hall, 1859; Pal. New York, vol. 3, p. 353, pl. 73, figs. 15-25. Helderberg Mts., Becraft's Mt., Schoharie, Carlisle, Cherry Valley and other places.   Lower Helderberg.

>*Calymene* MACROPHTHALMA Green, 1832; Mon. Tril. N. A., p. 39, cast 9.
>
>*Asaphus* MEGALOPHTHALMUS Troost, 1834; Mem. Geol. Soc. France, vol. 3, p. 94, pl. 1, figs. 1-5. This species approaches very near to *Phacops hudsoni* Hall from the Lower Helderberg. My cabinet specimen from Tennessee has 5 lenses arranged into 5 vertical rows in each eye. *P. hudsoni* has only 4 lenses in each row.
>
>*Calymene nupera* Hall, 1843; Geol. New York, 4th Geol. Dist., p. 262, fig. 116.

—— *nupera* Emmons, 1860; Manual Geol., p. 149, fig. 138 (2).

—— *nupera* Hall, 1876; Illus. Dev. Foss. Crust., pl. 5, fig. 26.

—— *nupera* Hall, 1888; Pal. New York, vol. 7, p. 27, pl. 8, fig. 27. The type of this species was found in a loose block of sandstone in Chemung county. In all apparent features it agrees with *Phacops rana*.

—— ORESTES Bill., 1860; Canadian Naturalist, vol. 5, p. 65, fig. 10. Anticosti and Gaspé.   Cambrian.

—— PULCHELLUS Foerste, 1887; Bull. Denison Univ., vol. 2, p. 99, pl. 8, figs. 4, 20, 21.   Clinton.

—— PULCHELLUS Foerste, 1889; Proc. Boston Soc. Nat. Hist., vol. 24, p. 268.

—— RANA Green.

>*Somatrikelon megalomaton* McMurtrie, 1819; Sketches of Louisville and the Falls of the Ohio, p. 74, n. g. and n. sp. from the Falls of the Ohio.
>
>*Calymene bufo* var. RANA Green, 1832; Monthly Am. Jour. Geol., p. 559.   Hamilton.

CATALOGUE OF TRILOBITES. 335

PHACOPS.   PHILLIPSIA.
Calymene bufo var. RANA Green, 1832; Mon. Tril. N. A., p. 42, cast 11, 12.
Calymene bufo var. RANA Hall, 1843; Geol. New York, 4th Geol. Dist., p. 201, pl. 14.
Syn., *Calymene bufo* Owens, 1844; Geol. Expl. Iowa, Wisconsin and Illinois, p. 74, pl. 12, fig. 1.
Calymene bufo Emmons, 1860; Manual Geol., p. 138, woodcut 124, fig. 6.
—— RANA Hall, 1861; Description New Species Fossils, p. 65.
—— RANA Hall, 1862; 15th Rep. New York State Cab. Nat. Hist., p. 93, pl. 10, fig. 12.
—— RANA M. & W., 1868; Geol. Survey Illinois, vol. 3, p. 447, pl. 11, figs. 1 a-e.
—— RANA Nicholson, 1873; Pal. Province Ontario, p. 123, figs. 5, 6 a.
—— RANA Hall, 1876; Illus. Dev. Foss. Crust., pl. 6, figs. 14, 15; pl. 7, figs. 1-11; pl. 8, figs. 11-17.
—— RANA Whitf., 1882; Geol. Wisconsin, vol. 4, p. 339, pl. 26, figs. 17-19.
—— RANA Wal., 1884; Pal. Eureka Dist. Mon. U. S. Geol. Survey, vol. 8, p. 207.
—— RANA Hall, 1888; Pal. New York, vol. 7, p. 19, pl. 7, figs. 1-11; pl. 8, figs. 1-18; pl. 7 a, figs. 21-33. This species is common to the Hamilton group of New York, Indiana, Illinois, Iowa, Wisconsin, Maryland, Nevada and Province of Ontario, Canada.
—— RANA Clarke, 1888; Jour. Morphology, vol. 2, p. 253, pl. 21.
—— TRAGANUS Bill., 1863; Proc. Portland Soc. Nat. Hist., vol. 1, p. 124, pl. 1, figs. 26, 27. Square Lake, Maine. cf. *Phacops fecundus* Barr.   Silurian.
Calymene? TRISULCATUS Hall, 1843; Geol. New York, 4th Geol. Dist., p. 72, fig. 9. In the green shales at Rochester, N. Y.   Clinton.
—— TRISULCATUS Hall, 1852; Pal. New York, vol. 2, p. 300, pl. 66 a, figs. 3 a, b.

PHÆTHONIDES Aug., 1852; Pal. Scand., p. 21. (See *Proetus* and *Brachymetpus*.)

PHILLIPSIA Portl., 1843; Rep. Geol. Londonderry, etc., p. 315. Dr. Henry Woodward suggests the following brief diagnosis of the characters of the genera *Phillipsia* and *Griffithides* :

PHILLIPSIA.	GRIFFITHIDES.
1. Sides of the glabella nearly parallel.	Glabella pyriform.
2. Marked by either two or three short lateral furrows.	No short lateral furrows on the glabella.
3. Basal lobes continous with the glabella.	Basal lobes distinct from the glabella.
4. Eyes large, reniform.	Eyes small, suboval.

—— *bufo*. (See *Griffithides bufo*.)

PHILLIPSIA.

—— CLIFTONENSIS Shum., 1858; Trans. Acad. Nat. Sci. St. Louis, vol. 1, p. 227. Clifton Park, Kansas. Coal Measures.
—— CLIFTONENSIS Herr., 1887; Bull. Denison Univ., vol. 2, p. 11.
—— CLIFTONENSIS Vogd., 1887; Annals New York Acad. Sci., vol. 4, p. 84.
—— ? CONSORS Herr., 1889; Bull. Denison Univ., vol. 4, p. 53, pl. 1, figs. 16 a, b, c. Lodi, Ohio. Keokuk or Burlington.
—— CONSORS Herr., 1890; American Geologist, vol. 5, p. 254.
—— CONSORS Herr., 1891; Bull. Geol. Soc. America, vol. 2, p. 43, pl. 1, fig. 12.
—— ? coronata. (See *Cyphaspis coronata* Hall.)
—— doris (Hall) Winchell, 1865; Proc. Acad. Nat. Sci. Phila., 2d series, vol. 9, p. 133. (See *Proteus doris* Hall.)
—— HOWI Bill., 1863; Canadian Naturalist, vol. 8, p. 209, fig. Kennetook, Nova Scotia. Carboniferous.
—— HOWI Daw., 1868; Acadian Geology, p. 313, fig. 133.
—— HOWI Herr., 1887; Bull. Denison Univ., vol. 2, p. 63.
—— HOWI Vogd., 1887; Annals New York Acad. Sci., vol. 4, p. 91. cf. *Phillipsia meramecensis*.
—— INSIGNIS Winch., 1863; Proc. Acad. Nat. Sci. Phila., 2d series, vol. 7, p. 24. Burlington, Iowa. Burlington.
—— INSIGNIS Bill., 1863; Canadian Naturalist, vol. 8, p. 209.
—— INSIGNIS Herr., 1887; Bull. Denison Univ., vol. 2, p. 63.
—— INSIGNIS Vodg., 1887; Annals New York Acad. Sci., vol. 4, p. 87.
—— lævis. (See *Cyphaspis lævis*.)
—— MAJOR Shum., 1858; Trans. Acad. Sci. St. Louis, vol. 1, p. 226. Clinton Co., Missouri; Valley of the Verdigris River, 12 miles south of Lecompton, Kansas, Kansas City, and Bellevue, Neb.
Coal Measures.
—— MAJOR Meek, 1872; U. S. Geol. Survey Territories, Final Rep. Neb. p. 238, pl. 3, figs. 2 a, b, c.
—— MAJOR Herrick, 1887; Bull. Denison Univ., vol. 2, p. 60.
—— MAJOR Vogd., 1887; Annals New York Acad. Sci., vol. 4, p. 85, pl. 3, fig. 14.
—— MAJOR Hare, 1891; Kansas City Sci., vol. 8, p. 33, pl. 1, figs. 5 and 8 a-c.
—— MERAMECANSIS Shum., 1855; 1st and 2d Rep. Geol. Survey Missouri, p. 199, pl. B, fig. 9. Meramec River at Fulton, Missouri.
Archimides Limestone.
—— MERAMECANSIS Herr., 1887; Bull. Denison Univ., vol. 2, p. 59.
—— MERAMECANSIS Vogd., 1887; Annals New York Acad. Sci., vol. 4, p. 86, pl. 3, fig. 15.
—— MERAMECANSIS? Winch., 1863; Proc. Acad. Nat. Sci. Phila., 2d series, vol. 7, p. 24. Keokuk.
—— MERAMECANSIS Herr., 1888; Bull. Denison Univ., vol. 3, p. 28, pl. 11, fig. 3.

PHILLIPSIA.
Syn., *howi* (Bill.) Vogd., 1887; Annals New York Acad. Sci., vol. 4, p. 92.
    vindobonensis (Hartt) Vogd., Annals New York Acad. Sci., vol. 4, p. 92.
—— MERAMECANSIS Herr., 1889; Bull. Denison Univ., vol. 4, p. 54, pl. 1, fig. 6.
—— MERAMECANSIS Herr., 1891; Bull. Geol. Soc. America, vol. 2, p. 43, pl. 1, fig. 14. Curyahoga Shales.
—— *minuscula* Hall. (See *Cyphaspis minuscula*.)
—— MISSOURIENSIS Shum., 1858; Trans. Acad. Sci. St. Louis, vol. 1, p. 220. Lexington, Missouri.   Coal Measures.
—— MISSOURIENSIS Herr., 1887; Bull. Denison Univ., vol. 2, p. 59.
—— MISSOURIENSIS Vogd., 1887; Annals New York Acad. Sci., vol. 4, p. 86, pl. 3, figs. 1, 2, 14, 16.
—— NODACOSTATUS Hare, 1891; Kansas City Sci., vol. 8, p. 33, pl. 1, figs. 1 a–e and 7. Kansas City.   Upper Coal Measures.
——? *ornatus*. (See *Cyphaspis coranata*.)
—— (*Griffithides*) *portlocki* M. & W. (See *Griffithides portlocki*.)
—— PERANNULATA Shum., 1858; Trans. Acad. Sci. St. Louis, vol. 1, p. 296, pl. 11, fig. 10. Guadalupe Mts., New Mexico.   Permian.
—— PERANNULATA Vogd., 1887; Annals New York Acad. Sci., vol. 4, p. 84.
—— PRÆCURSOR Herr., 1888; Bull. Denison Univ., vol. 3, p. 29, pl. 12, fig. 1. Four miles west of Granville, Ohio.   Kinderhook.
—— PRÆCURSOR Herr., 1890; American Geologist, vol. 5, p. 254.
—— (*Proetus*) PRÆCURSOR Herr., 1889; Bull. Denison Univ., vol. 4, p. 54, pl. 1, fig. 1.   Waverly.
—— ROCKFORDENSIS Winch., 1865; Proc. Acad. Nat. Sci. Phila., 2d series, vol. 9, p. 133. Rockford, Indiana.   Kinderhook.
—— ROCKFORDENSIS Herr., 1887; Bull. Denison Univ., vol. 2, p. 62.
—— ROCKFORDENSIS Vogd., 1887; Annals New York Acad. Sci, vol. 4, p. 91.
—— SAMPSONI Vogd., 1888; Trans. New York Acad. Sci., vol. 7, p. 248, fig. Sedalia, Missouri.   Waverly.
—— (*Griffithides*) *sangamonensis* M. & W. (See *Griffithides sangamonensis*.)
—— (*Griffithides*) *scitula* M. & W. (See *Griffithides scitula*.)
—— SERRATICAUDATA Herr., 1889; Bull. Denison Univ., vol. 4, p. 52, pl. 1, figs. 8 a–d. Newark and Rushville, Ohio.   Keokuk.
—— SERRATICAUDATA Herr., 1890; American Geologist, vol. 5, p. 254.
—— *shumardi*. (See *Proetus auriculata*.)
—— STEVENSONI Meek, 1871; Rep. Regents Univ. W. Virginia, p. 73. Monongalia county, W. Virginia.   Chester.
—— STEVENSONI Vogd., 1887; Annals New York Acad. Sci., vol. 4, p. 88, pl. 3, fig. 6.
—— TUBERCULATA M. & W., 1870; Proc. Acad. Nat. Sci. Phila., p. 52. Kinderhook, Pike county, Illinois.   Burlington.

PHILLIPSIA.   PROETUS.

—— TUBERCULATA Vogd., 1887; Annals New York Acad. Sci., vol. 4, p. 92.
—— TRINUCLEATA Herr., 1887; Bull. Denison Univ., vol. 2, p. 64, pl. 1, figs. 23, 23 a, c, e, h; pl. 2, fig. 32; pl. 3, fig. 21. Coal Measures.
—— TRINUCLEATA Herr., 1889; Bull. Denison Univ., vol. 4, p. 49. Flint Ridge, Licking county, Ohio.   Coal Measures.
Proetus trinucleatus Vogd., 1888; Annals New York Acad. Sci., vol. 4, p. 81, pl. 2, figs. 7, 8, 9.
—— vindobonensis Hartt, 1868; Acadian Geol., p. 353. Windsor, Nova Scotia.   Carboniferous.
—— vindobonensis Herr., 1887; Bull. Denison Univ., vol. 2, p. 63.
—— vindobonensis Vogd., 1887; Annals New York Acad. Sci., vol. 4, p. 89. (cf. Phillipsia meramecansis Shum.)
Piliolites Cozzens, 1846; Annals Lyceum Nat. Hist. N. Y., vol. 4, p. 157.
—— ohioensis Cozzens, 1846; Annals Lyceum Nat. Hist. N. Y., vol. 4, p. 157, pl. 10, figs. 1 a, b. (cf. Proetus crassimarginatus?)
PROETUS Stein., 1831; Mém. Soc. Géol. France, vol. 1, p. 355, pl. 24, fig. 6. Syn., Æonia Burm., 1843; Organization Trilobites, p. 116 (Ray Soc. ed., p. 100). Burmeister separates Dalman's species Calymene concinna from the type of the genus Proetus, and gives it the new generic name of Æonia, with descriptions and figures of A. concinna, A. stokesii and A. verticalis.
Phæton Barr., 1846; Notice Prélim., p. 62. The author herein gives the new generic term Phæton for species of Proetus with fimbriated pygidia.
PRINOPELTIS Cord., 1847; Prodr., p. 121. New generic name proposed for the preoccupied term of Phæton. Pygidium with the pleuræ produced into spines forming a fimbriated circumference. Phaëtonellus Novak, 1890.
XIPHOGONIUM Cord., 1847; Prodr., p. 70. Proposed genus for certain species of Proetus, with nine segments in the thorax. Pleuræ of the pygidium unsegmented.
GONIOPLEURA Cord., 1847; Prodr., p. 80. This genus includes such species as Proetus clegantulus Loven, with twelve segments in the thorax.
FORBESIA McCoy, 1847; Sil. Foss. Ireland, p. 46. This author uses for the types of his genus Forbesia latifrons and F. stokesii, which is the same as that used by Barrande in 1846 for his genus Phæton and Angelin in 1878 for the genus Phætonides. Genal angles produced into long spines; pleuræ of the pygidium segmented; glabella furrows present and large tubercles terminating the neck furrow.
Gerastos Goldf., 1843; Neues Jahrbuch für Mineral., p. 557. This genus is a Proetus with G. lævigatus for its type, a species identical with Proetus curieri Stein.

PROETUS.

Trigonaspis and *Cylindraspis* Sand.; Verst. d. Rhein, Scht.-Syst., p. 30, pl. 3, 1850.

DECHENELLA Kays., 1880; Zeitschr. Deutsch.Geol.Gesell., Berlin, 1880, p. 703. For such species as *Proetus verticallis* and *P. haldemani*. The glabella has 4 pair of furrows, 10 thorax segments. I would prefer taking the older name of *Æonia* for this subgenus.

CELMUS Aug., 1852; Palæont. Scand., p. 23. Two pair of glabella furrows, 12 segments to the thorax.

—— ALARICUS Bill., 1860; Canadian Naturalist, vol. 5, p. 68, fig. 12. Anticosti. Hudson.

—— ALARICUS Bill., 1863; Geol. Canada, p. 219, fig. 230.

—— ANGUSTIFRONS Hall, 1861; Des. New Species Foss., p. 70. Schoharie, N. Y. Schoharie Grit.

—— ANGUSTIFRONS Hall, 1862; 15th Rep. New York State Cab. Nat. Hist., p. 98.

—— ANGUSTIFRONS Hall, 1876; Illus. Dev. Foss. Crust., pl. 20, figs. 1, 4, 7.

—— ANGUSTIFRONS Hall, 1888; Pal. New York, vol. 7, p. 91, pl. 20, figs. 1–5; pl. 22, figs. 1–3.

Phætonides ARENICOLUS Hall, 1888; Pal. New York, vol. 7, p. 134, pl. 25, figs. 12, 13. Schoharie, N. Y. Upper Helderberg.

—— *auriculatus*. (See *Proetus missouriensis* Shum.)

—— CANALICULATUS Hall, 1861; Des. New Species Foss., p. 73. Falls of the Ohio. Upper Helderberg.

—— CANALICULATUS Hall, 1862; 15th Rep. New York State Cab. Nat. Hist., p. 101.

—— CANALICULATUS Hall, 1876; Illus. Dev. Foss. Crust., pl. 20, figs. 10, 11.

—— CANALICULATUS Hall, 1888; Pal. New York, vol. 7, p. 107, pl. 20, figs. 10, 11; pl. 23, figs. 10, 11.

—— CLARUS Hall, 1861; Des. New Species Foss, p. 71. Stafford, Batavia and elsewhere in Genessee Co., N. Y. Upper Helderberg.

—— CLARUS Hall, 1862; 15th Rep. New York State Mus. Nat. Hist., p. 99.

—— CLARUS Hall, 1876; Illus. Dev. Foss. Crust., pl. 20, figs. 12–14.

—— CLARUS Hall, 1888; Pal. New York, vol. 7, p. 104, pl. 20, figs. 12–14; pl. 22, figs. 28–30.

—— CONRADI Hall, 1861; Des. New Species Foss., p. 69. Schoharie and Helderberg Mts., N. Y. Schoharie.

—— CONRADI Hall, 1862; 15th Rep. New York State Cab. Nat. Hist., p. 97.

—— CONRADI Hall, 1876; Illus. Dev. Foss. Crust., pl. 21, fig. 6 (not figs. 5, 8, 9, *P. crassimarginatus*); pl. 21, figs. 27, 28.

—— CONRADI Hall, 1876; Illus. Dev. Foss. Crust., pl. 20, figs. 5, 8, 9.

—— CONRADI Hall, 1888; Pal. New York, vol. 7, p. 89, pl. 20, fig. 9; pl. 21, figs. 27, 28; pl. 12, fig. 4.

Asaphus CORYCŒUS Con., 1842; Jour. Acad. Nat. Sci. Phila., vol. 8, p. 277, pl. 16, fig. 15. Lockport, N. Y. Niagara.

PROETUS.
 Asaphus CORYCŒUS Hall, 1843; Geol. New York, 4th Geol. Dist., pl. 19, fig. 3.
 —— CORYCŒUS Hall, 1852; Pal. New York, vol. 2, p. 315, pl. 67, fig. 15.
 Calymene CRASSIMARGINATUS Hall, 1843; Geol. New York, 4th Geol. Dist., p. 172, fig. 5.   Upper Helderberg.
 Piliolites ohioensis? Coz., 1846; Annals Lyceum Nat. Hist. New York vol. 4, p. 157, pl. 10, figs. 1 a, b.
 Calymene CRASSIMARGINATUS Owen, 1852; Rep. Geol. Wisconsin, Iowa and Minnesota, p. 623, pl. 3 a, fig. 6.
 —— CRASSIMARGINATUS Hall, 1859; 12th Rep. New York State Cab. Nat. Hist., p. 88.
 Phillipsia CRASSIMARGINATUS Bill., 1861; Canadian Journal, vol. vi, p. 362.
 —— CRASSIMARGINATUS Hall, 1861; Des. New Species Foss., p. 72. At Williamsville and other places in Western New York, in Canada West, Ohio and at the Falls of the Ohio.
 —— CRASSIMARGINATUS Hall, 1862; 15th Rep. New York State Cab. Nat. Hist., p. 100, pl. 10, fig. 10.
 —— CRASSIMARGINATUS Hall, 1876; Illus. Dev. Foss. Crust., pl. 20, figs. 20-31.
 Syn., *P. conradi*, pl. 20, figs. 5, 8, 9.
 —— CRASSIMARGINATUS Hall, 1888; Pal. New York, vol. 7, p. 99, pl. 20, figs. 6-8, 20-31; pl. 22, figs. 20-62, pl. 25, fig. 8.
 —— CURVIMARGINATUS Hall, 1888; Pal. New York, vol. 7, p. 94, pl. 22, figs. 13-19. Pendleton, Indiana.   Schoharie.
 —— *davenportensis* Barris. (See *Proetus prouti* Shum.)
 —— DEPHINULUS Hall, 1888; Pal. New York, vol. 7, p. 111, pl. 23, figs. 1, 2; pl. 25, fig. 6. Canandaigua, Ontario county, N. Y.
                 Upper Helderberg.
 —— DENTICULATUS Meek, 1877; U. S. Geol. Expl. 40th Par., vol. 4, p. 49, pl. 1, figs. 10 a, b.   Devonian.
 Phaëtonides (?) DENTICULATUS (Meek) Hall, 1888; Pal. New York, vol. 7, p. 139, pl. 25, figs. 14, 15. For other references, see *Proetus (Phæton) denticulastus* Meek. Steptoe Valley, Nev. Devonian (?).
 —— DETERMINATUS Foerste, 1887; Bull. Denison Univ., vol. 2, p. 91, pl. 8, figs. 2, 3, 3 a. Soldier's Home near Dayton, Ohio. Clinton.
 —— DORIS Hall, 1860; 13th Rep. New York State Cab. Nat. Hist., p. 112. Rockford, Indiana.   Goniatite limestone.
 Phillipsia DORIS Winch., 1865; Proc. Acad. Nat. Sci. Phila., 2d series, vol. 2, p. 62.
 Phillipsia DORIS Herr., 1887; Bull. Denison, Univ. vol. 2, p. 62.
 Phillipsia DORIS Vogd., 1887; Annals New York Acad. Sci., vol. 4, p. 90.
 —— ELLIPTICUS M. & W., 1865; Proc. Acad. Nat. Sci. Phila., p. 267. Jersey Co., Illinois.   Kinderhook.

CATALOGUE OF TRILOBITES. 341

PROETUS.
—— ELLIPTICUS M. & W., 1868; Geol. Survey Illinois, vol. 3, p. 460, pl. 14, fig. 3.
—— ELLIPTICUS Vogd., 1887; Annals New York Acad. Sci., vol. 4, p. 82, pl. 3, fig. 3.
—— FOLLICEPS Hall, 1888; Pal. New York, vol. 7, p. 101, pl. 23, figs. 3–8. Le Roy, Schoharie, Marbletown, N. Y., etc. Upper Helderberg.
—— granulatus Wetherby. (See *Griffithides granulatus*).
Phæthonides GEMMÆUS Hall, 1888; Pal. New York, vol. 7, p. 136, pl. 24, figs. 32–36. Canandaigua, N. Y., 18 Mile Creek, N. Y., etc.
Upper Helderberg.
—— HALDEMANI Hall, 1861; Des. New Species Foss., p. 74. Pennsylvania, Goniatite limestone, Manlius and Cherry Valley, N. Y. Hamilton.
—— HALDEMANI Hall, 1862; 15th Rep. New York State Cab. Nat. Hist., p. 102, pl. 10, fig. 6.
—— HALDEMANI Hall, 1876; Illus. Dev. Foss. Crust., pl. 21, figs. 7–9.
—— HALDEMANI Wal., 1884; Pal. Eureka Dist., Mon. U. S. Geol. Survey, vol. 8, p. 210.
—— HALDEMANI Herr., 1889; Bull. Denison Univ., vol. 4, p. 55, pl. 1, fig. 12.
—— HALDEMANI Hall, 1888; Pal. New York, vol. 7, p. 113, pl. 21, figs. 7–9; pl. 23, figs. 13–15.
Dechenella HADLEMANI Kayser, 1880; Zeitschr. Dutchsch. Geol. Gesell., Berlin, 1880, p. 770, pl. 27, fig. 9.
Dechenella HADLEMANI Tschernyschew, 1887; Mém. Comité Géol., vol. 3, No. 3, p. 14, pl. 1, fig. 9.
—— HALDEMANI Whiteaves, 1892; Contrib. Canadian Palæont, vol. 1, p. 246, pl. 31, figs. 6–8.
—— HESIONE Hall, 1861; Des. New Species Foss, p. 70. Schoharie, N. Y.
Schoharie.
—— HESIONE Hall, 1862; 15th Rep. New York State Cab. Nat. Hist., p. 98.
—— HESIONE Hall, 1876; Illus. Dev. Foss. Crust., pl. 20, figs. 15, 16.
—— HESIONE Hall, 1888; Pal. New York, vol. 7, p. 93, pl. 20, figs. 15, 16.
—— JEJUNUS Hall, 1888; Pal. New York, vol. 7, p. 124, pl. 25, fig. 7. Albany Co., N. Y. Hamilton.
—— JUNIUS Bill., 1863; Proc. Portland Soc. Nat. Hist., vol. 1, p. 122, pl. 1, fig. 23. Silurian.
—— LATIMARGINATUS Hall, 1888; Pal. New York, vol. 7, p. 97, pl. 22, figs. 7–12. Pendleton, Indiana. Schoharie.
—— LOGANENSIS H. & W., 1877; U. S. Geol. Expl., 40th Par., vol. 4, p. 264, pl. 4, fig. 33. Logan Cañon, Wahsatch Range, Utah.
Waverly.
—— LOGANENSIS Vogd., 1887; Annals New York Acad. Sci., vol. 4, p. 78.
—— LONGICAUDUS Hall, 1861; Des. New Species Foss., p. 80. Hamilton.
—— LONGICAUDUS Hall, 1862; 15th Rep. New York State Cab. Nat. Hist., p. 108, pl. 10, figs. 7–9.

PROETUS.
— LONGICAUDUS Hall, 1876; Illus. Dev. Foss. Crust., pl. 20, figs. 32-34.
— LONGICAUDUS Williams, 1881; Am. Jour. Sci., 3d series, vol. 21, p. 156.
— ? LONGICAUDUS Hall, 1888; Pal. New York, vol. 7, p. 131, pl. 20, figs. 32-34. Near Madison, Kansas, NE. of Des Moines, Iowa.
— MACROBIUS Bill., 1863; Proc. Portland Soc. Nat. Hist., vol. 1, p. 123, pl. 1, fig. 24. Silurian.
— MACROCEPHALUS Hall, 1861; Des. New Species Foss., pp. 77, 79. In Hamilton group, Genesco, Moscow, Pavilon, etc., N. Y.
— MACROCEPHALUS Hall, 1862; 15th Rep. New York State Cab. Nat. Hist., pp. 105, 107. Also var. a. *idem.*, p. 79, at Moscow and Bloomfield, and 18-Mile Creek, N. Y.
— MACROCEPHALUS Hall, 1876; Illus. Devonian Foss. Crust., pl. 21, figs. 10-21.
— MACROCEPHALUS Hall, 1888; Pal. New York, vol. 7, p. 116, pl. 21, figs. 10-21; pl. 23, fig. 30.
Calymene MARGINALIS Con., 1839; Ann. Rep. Pal. Dept. New York Geol. Survey, p. 66. In the Tully limestone near Ovid, Seneca county, N. Y.
— MARGINALIS Hall, 1861; Des. New Species Foss., p. 76. Hamilton.
— MARGINALIS Hall, 1862; 15th Rep. New York State Cab. Nat. Hist., p. 104.
— MARGINALIS Hall, 1876; Illus. Devonian Foss. Crust., pl. 21, figs. 24-28.
— MARGINALIS Wal., 1884; Pal. Eureka Dist. Mon. U. S. Geol. Survey, vol. 8, p. 210. This species is referred to *Proetus nevadæ* by Hall; Palæont. N. Y., vol. 7, p. 129.
— MARGINALIS Hall, 1888; Pal. New York, vol. 7, p. 122.
— MICROGEMMA Hall, 1888; Pal. New York, vol. 7, p. 109, pl. 22, figs. 33, 34. Le Roy, Canandaigua, Williamsville, N. Y.; Falls of the Ohio. Upper Helderberg.
Phillipsia MINUSCULA Hall, 1876; Illus. Devonian Foss. Crust., pl. 20, fig. 17. Schoharie, N. Y. Upper Helderberg.
— MINUTUS Herr., 1889; Bull. Denison Univ., vol. 4, p. 56, pl. 1, figs. 7 a, b. Kinderhook.
— MINUTUS Herr., 1890; American Geologist, vol. 5, p. 254.
— MISSOURIENSIS Shum., 1855; 1st and 2d Rep. Geol. Missouri, p. 196, pl. B, figs. 13 a, b. Hannibal, Louisiana and Chouteau Springs, Missouri; also Granville, Ohio. Waverly group.
Syn., *auriculatus* Hall, 1861; Des. New Species Foss., p. 79.
auriculatus Hall, 1862; 15th Rep. New York State Cab. Nat. Hist., p. 107.
Phillipsia (Proetus) auriculatus Herr., 1889; Bull. Denison Univ., vol. 4, p. 34, pl. 1, fig. 14.
Phillipsia shumardi Herr., 1887; Bull. Denison Univ., vol. 2, pp. 58, 69, pl. 7, fig. 14.

PROETUS.
— MISSOURIENSIS Vogd., 1887; Annals New York Acad. Sci., vol. 4, p. 75, pl. 3, fig. 1.
— MISSOURIENSIS Hall, 1888; Pal. New York, vol. 7, p. 133, pl. 23, fig. 32.
— MUNDULUS Whiteaves, 1892; Contrib. Canadian Palæont., vol. 1, p. 350, pl. 46, figs. 10, 11. Western shore of Dawson Bay, Lake Wiunipegosis, Manitoba. Devonian.
— NEVADÆ Hall, 1888; Palæont. N. Y., vol. 7, p. 129, pl. 23, fig. 19. cf. *P. rowi*, *P. angustifrons*, *P. clarus*, *P. marginalis* Conrad, also *P. prouti*. Comb's Peak, Eureka Dist., Nev.
— OCCIDENS Hall, 1861; Des. New Species Foss., p. 80. New Buffalo, Iowa. Hamilton.
— OCCIDENS Hall, 1862; 15th Rep. New York State Cab. Nat. Hist., p. 108.
— OCCIDENS Hall, 1876; Illus. Dev. Foss. Crust., pl. 21, figs. 22, 23.
— OCCIDENS Hall, 1888; Pal. New York, vol. 7, p. 130, pl. 21, figs. 22, 23.
— OVIFRONS Hall, 1888; Pal. New York, vol. 7, p. 110, pl. 22, figs. 31, 32. Canandaigua, N. Y. Upper Helderberg.
— PACHYDERMATUS Barrett, 1873; Am. Jour. Sci., 3d series, vol. 15, p. 370. Port Jervis, N. Y. Syn., *P. stokesi* (Hall) not Meek. Niagara.
— PARVIUSCULUS Hall, 1872; 24th Rep. New York State Cab. Nat. Hist., p. 223, pl. 8, fig. 14. Cincinnati, Ohio. Hudson.
— PARVIUSCULUS H. & W., 1875; Pal. Ohio, vol. 2, p. 109, pl. 4, fig. 18.
— PARVIUSCULUS Miller, 1874; Cincinnati Quart. Jour. Sci., vol. 1, p. 144.
— PEROCCIDENS H. & W., 1877; U. S. Geol. Expl. 40th Par., Pal., vol. 4, p. 262, pl. 4, figs. 28–32. Ogden and Logan Cañon, Utah. Waverly.
— PEROCCIDENS Vogd., 1887; Annals New York Acad. Sci., vol. 4, p. 79.
— PHOCION Bill., 1874; Pal. Foss., vol. 2, p. 63, fig. 31. Indian Gove, Gaspé. Devonian.
— PHOCION Hall, 1888; Pal. New York, vol. 7, p. 125, pl. 25, figs. 9, 10.
— PLANIMARGINATUS Meek, 1871; Proc. Acad. Nat. Sci. Phila., p. 89. Sylvania, Ohio. Corniferous.
— PLANIMARGINATUS Meek, 1873; Pal. Ohio, vol. 1, p. 233, pl. 23, figs. 3 a, b.
— PLANIMARGINATUS Hall, 1888; Pal. New York, vol. 7, p. 112, pl. 23, fig. 12.
— PROTUBERANS Hall, 1859; Pal. New York, vol. 3, p. 351, pl. 73, figs. 5–8. Schoharie Co., N. Y. Lower Helderberg.
— PROUTI Shum., 1863; Trans. Acad. Sci. St. Louis, vol. 2, p. 110. Davenport, Iowa. Hamilton.
Syn., *davenportensis* Barris, 1878; Proc. Davenport Acad. Sci., vol. 2, p. 287, pl. 11, fig. 8 (pl. 11, fig. 8, revised).

PROETUS.
— PROUTI Hall, 1888; Pal. New York, vol. 7, p. 126, pl. 23, figs. 16-18.
Daveuport, Iowa.
Calymene Le Row, 1837; Poughkeepsie Telegram, Nov. 22.
Calymene ROWI Green, 1838; Am. Jour. Sci., 1st series, vol. 33, p. 406.
Calymene ROWI Green, 1838; Annals Mag. Nat. Hist., 1st series, London, vol. 1, p. 79.
— ROWI Hall, 1861; Des. New Species Foss., p. 75. Fly Creek and other places in Otsego Co., N. Y., also in Scoharie Co., N. Y.
Hamilton.
— ROWI Hall, 1862; 15th Rep. New York State Cab. Nat. Hist., p. 103.
— ROWI Hall, 1876; Illus. Dev. Foss. Crust., pl. 21, figs. 2-6.
— ROWI Hall, 1888; Pal. New York, vol. 7, p. 119, pl. 21, figs. 2-6, 24-26; pl. 23, figs. 20-29.
Phillipsi shumardi Herr., 1887; Bull. Denison Univ., vol. 2, pp. 58, 69, pl. 7, figs. 14 a, b. (See *P. missouriensis* Shum.) Waverly.
Phillipsia shumardi Herr., 1888, Bull. Denison Univ., vol. 3, p. 29.
— SPURLOCKI Meek, 1872; Am. Jour. Sci., 3d series, vol. 3, p. 426. Cincinnati, Ohio. Hudson.
— SPURLOCKI Meek, 1873; Pal. Ohio, vol. 1, p. 161, pl. 14, fig. 12.
— SPURLOCKI Miller, 1874; Cincinnati Quart. Jour. Sci., vol. 1, p. 145. ·
— STENOPYGE Hall, 1888; Pal. New York, vol. 7, p. 110, pl. 22, fig. 27. Flint Creek, N. Y., North Caynga, Ontario. Upper Helderberg.
— STOKESI (Murch.) Hall, 1852; Pal. New York, vol. 2, p. 316, pl. 67, figs. 13, 14. Barrett refers this species to *P. pachydermatus.*
Niagara.
— SWALLOWI Shum., 1855; 1st and 2d Rep. Geol. Survey Missouri, p. 196, pl. B, figs. 12 a, b. Chouteau Springs, Missouri. Waverly.
Phillipsi SWALLOWI Herr., 1887; Bull. Denison Univ., vol. 2, p. 58.
Phillipsia TENNESSEENSIS Winch., 1869; Geol. Tennessee, p. 445. Hickman Co., Tenn. Waverly.
— TENNESSEENSIS Vogd., 1887; Annals New York Acad. Sci., vol. 4 p. 80.
— TUMIDUS Hall, 1888; Pal. New York, vol. 7, p. 113, pl. 23, fig. 9. Port Colborne and North Cayuga, Ontario. Upper Helderberg.
— *Phæthonides* VARICELLA Hall, 1888; Pal. New York, vol. 7, p. 135, pl. 24, figs. 29-31. Canandaigua, N. Y. Upper Helderberg.
— VERNEUILI Hall, 1861; Des. New Species Foss., p. 73. Williamsville, N. Y. Upper Helderberg.
— VERNEUILI Hall, 1862; 15th Rep. New York State Cab. Nat. Hist., p. 101.
— VERNEUILI Hall, 1876; Illus. Dev. Foss. Crust., p. 20, figs. 18, 19.
— VERNEUILI Hall, 1888; Pal. New York, vol. 7, p. 108, pl. 20, figs. 18, 19. Williamsville, Phelps and Canandaigua, also Schoharie Co., N. Y.

PROTYPUS.
PROTICHNITES Owen, 1852; Quart. Jour. Geol. Soc. London, vol. 8, p. 214.
—— ALTERNANS Owen, 1852; Quart. Jour. Geol. Soc. London, vol. 8, p. 221, pl. 8 a, fig. 7. Potsdam.
—— CARBONARIUS Daw., 1873; Am. Jour. Sci., 3d series, vol. 5. Coal Measures.
—— LATUS Owen, 1852; Quart. Jour. Geol. Soc. London, vol. 8, p. 218, pl. 11. Potsdam.
—— LINEATUS Owen, 1852; Quart. Jour. Geol. Soc. London, vol. 8, p. 220, pl. 8 a; pl. 13. Potsdam.
—— LOGANANUS Marsh, 1869; Am. Jour. Sci., 2d series, vol. 48, p. 46. Potsdam.
—— MULTINOTATUS Owen, 1852, Quart. Jour. Geol. Soc. London, vol. 8, p. 219, pl. 12. Potsdam.
—— OCTO-NOTATUS Owen, 1852; Quart. Jour. Geol. Soc. London, vol. 8, p. 217, pl. 10. Potsdam.
—— SEPTEM-NOTATUS Owen, 1852; Quart. Jour. Geol. Soc. London, vol. 8, p. 214, pl. 9. Potsdam.
PROTOLENUS Matt., 1892; Bull. Nat. Hist. Soc. New Brunswick, No. 10, p. 2.
—— ETEGANS Matt., 1892; Bull. Nat. Hist. Soc. New Brunswick, No. 10, p. 2, fig. Band b, Div. 1, St. John Group, Hanford Brook, N. B.
—— PARADOXIDES Matt., 1892; Bull. Nat. Hist. Soc. New Brunswick, No. 10, p. 3, fig. Band b, Div. 1, St. John Group, Hanford Brook, N. B.
PROTOPELTURA ACANTHURA (Ang.) Matt., 1891; Trans. Roy. Soc. Canada, vol. 9, p. 53; also var. *tetracanthura*, pl. 13, figs. 8 a-c. St. John.
PROTYPUS Wal., 1886; Bull. U. S. Geol. Survey No. 30, p. 211.
Ptychoparia? (subgen.?) CLAVATA Wal., 1887; Am. Jour. Sci., 3d series, vol. 34, p. 198, pl. 1, fig. 3.
—— CLAVATUS Wal., 1890; 10th Rep. U. S. Geol. Survey, p. 656, pl. 98, fig. 4. One mile east of North Greenwich, Washington county, N. Y. Taconic.
Angelina HITCHCOCKI Whitf., 1884; Bull. Am. Mus. Nat. Hist. N. Y., vol. 1, p. 148, pl. 14, fig. 13. Baintree, Mass. Taconic.
—— HITCHCOCKI Wal., 1886; Bull. U. S. Geol. Survey No. 30, p. 211, pl. 31, fig. 4.
—— HITCHCOCKI Wal., 1890; 10th Rep. U. S. Geol. Survey, p. 655, pl. 98, fig. 6.
Bathyurus SENECTUS Bill., 1861; Pamphlet published in adv. Pal. Foss. Canada, p. 15, figs. 20, 21.
Bathyurus SENECTUS Bill., 1861; Geol. Vermont, vol. 2, p. 953, figs. 359, 360. Moore's Corners, in St. Armand, to Saxe's Mills, in Highgate, Vermont. Taconic.
Bathyurus SENECTUS Bill., 1863; Geol. Canada, p. 286, figs. 298 a, b.
Bathyurus SENECTUS Bill., 1865; Pal. Foss., vol. 1, p. 15, figs. 20, 21.

PROTYPUS.   PTYCHASPIS.
—— SENECTUS Wal., 1886; Bull. U. S. Geol. Survey No. 30, p. 213, pl. 31,
    figs. 2, 2 a-c.
—·— SENECTUS Wal., 1890; 10th Rep. U. S. Geol. Survey, p. 655, pl. 98,
    fig. 7, 7 a-c. Olenellus zone.
—— SENECTUS var. PARVULUS. For other references see *Bathyurus parvulus*. Walcott 10th Rep. U. S. Geol. Survey, p. 568, refers it to
    a variety of *P. senectus*,
PTEROCEPHALIA Roemer, 1852; Kreid. von Texas, p. 93.
    Conocephalites (Pterocephalia) LATICEPS H. & W., 1877; U. S. Geol.
    Expl. 40th Par., vol. 4, p. 221, pl. 2, figs. 4-7. Pogonip Mt.,
    White Pine, Nev.
    Ptychoparia (Pterocephalia) LATICEPS Wal., 1884; Pal. Eureka Dist.
    Mon. U. S. Geol. Survey, vol. 8, p. 59. Richmond Mine Shaft,
    Ruby Hill, northeast side of cañon, and in the Hamburg belt,
    Nev.
    Ptychoparia (Pterocephalia) OCCIDENS Wal., 1884; Pal. Eureka Dist.
    Mon. U. S. Geol. Survey, vol. 8, p. 58, pl. 9, fig. 21. Hamburg
    Ridge, Nev.
—— SANCTI-SABÆ Roemer, 1849; Texas, p. 421. San Saba Shale, Texas.
—— SANCTI-SADÆ Roemer, 1852; Kreid. von Texas, p. 93, pl. 11.
PTYCHASPIS Hall, 1863; 16th Rep. New York State Cab. Nat. Hist., p. 170.
—— BARABUENSIS Winch., 1864; Am. Jour. Sci., 2d series, vol. 37, p. 230.
    Sauk Co., Wis.   Taconic.
    Arionellus CYLINDRICUS Bill., 1865; Pal. Foss., vol. 1, p. 406, fig. 385.
    Point Levis, Quebec.
    Conoceph (Ptychaspis) EXPLANATUS Whitf., 1880; Ann. Rep. Geol.
    Survey Wisconsin, 1879, p. 48. Hudson, Wis.   Taconic.
    Conoceph (Ptychaspis?) EXPLANATUS Whitf., 1882; Geol. Wisconsin,
    vol. 4, p. 181, pl. 1, figs. 27, 28.
    Dicellocephallus GRANULOSUS Owen, 1852; Rep. Geol. Survey Wisconsin, Iowa and Minnesota, p. 575, pl. 1, fig. 7.   Taconic.
—·— *granulosa* Hall, 1863; 16th Rep. New York State Cab. Nat. His., p.
    173, pl. 6. figs. 33-40. Trempaleau and Miniska. (See *P. striata*.)
—— *granulosa* Whitf., 1878; Ann. Rep. Geol. Survey Wisconsin, 1877,
    p. 34.
—— *granulosa* Whitf., 1882; Geol. Wisconsin, vol 4, p. 185, pl. 1, fig. 24.
—— GRANULOSA Whitf., 1882; Geol. Wisconsin, vol. 4, p. 185, pl. 1, fig. 24.
    Dicellocephalus MINISCAENSIS Owen, 1852; Rep. Geol. Survey Wisconsin, Iowa and Minnesota, p. 574, pl. 1, figs. 3, 12; pl. 1 a, figs.
    4, 5. Trempaleau and Miniska, Wis.
—— MINISCAENSIS Hall, 1863; 16th Rep. New York State Cab. Nat. Hist.,
    p. 171, pl. 6, figs. 41-46; pl. 10, figs. 21, 22.
—·— MINUTA Whitf., 1878; Ann. Rep. Geol. Survey Wisconsin, 1877, p. 55.
    At Robert's Shore, St. Croix Co., Wis.   Taconic.

PTYCHASPIS.   PTYCHOPARIA.
—— MINUTA Whitf., 1882; Geol. Wisconsin, vòl. 4, p. 186, pl. 1, figs. 25, 26.
—— MINUTA Wal., 1884; Pal. Eureka Dist., Mon. U. S. Geol. Survey, vol. 8, p. 60, pl. 10, fig. 23.
—— PUSTULOSA H. & W., 1877; U. S. Geol. Expl. 40th Par., Palæont, vol. 4, p. 223, pl. 2, fig. 27.   Taconic.
Dicellocephalus SESOSTRIS Bill., 1865; Pal. Foss., vol. 1, p. 198, fig. 184. Point Levis, Quebec.
—— SPECIOSUS Wal., 1879; 32d Rep. New York State Mus. Nat. Hist., p. 131.   Calciferous.
—— STRIATA Whitf., 1878; Ann. Rep. Geol. Survey Wisconsin, 1877, p. 55.   Taconic.
—— STRIATA Whitf., 1880; Ann. Rep. Geol. Survey Wisconsin, 1879, p. 51.
—— STRIATA Whitf., 1882; Geol. Wisconsin, vol. 4, p. 186.
Syn., *granulosa* Hall, 1863; 16th Rep. New York State Cab. Nat. Hist., p. 173, pl. 6, figs. 33–40. Not *P. granulosa* Owen.
Arionellus SUBCLAVATUS Bill., 1860; Canadian Naturalist, vol. 5, p. 315, figs. 15, 15 a. Point Levis, Quebec.
PTYCHOPARIA Cord., 1847; Prodr., p. 141, pl. 2, fig. 11.
Conocephalites ADAMSI Bill., 1861; Pamphlet published in adv. Pal. Foss., p. 12, fig. Highgate, Vermont.   Taconic.
Conocephalus Adams, 1848; Am. Jour. Sci., 2d series, vol. 5, p. 109.
Conocephalites ADAMSI Bill., 1861; Geol. Vermont, vol. 2, p. 950, fig. 355.
Conocephalus Bill., 1861; Am. Jour. Sci., 2d series, vol. 32, p. 231.
Conocephalus Bill., 1861; Canadian Naturalist, vol. 6, p. 324.
Conocephalites ADAMSI Bill., 1863; Geol. Canada, p. 286, fig. 294.
Conocephalites ADAMSI Bill., 1865; Pal. Foss., vol. 1, p. 12, fig. 15.
—— ADAMSI Wal., 1886; Bull. U. S. Geol. Survey No. 30, p. 195, pl. 26, figs. 1, 1 a-c.
Syn., *arenosus* (Bill.) Wal., 1886; Bull. U. S. Geol. Survey No. 30, p. 195.
Conoceph. arenosus Bill., 1861; Pamphlet published in adv. of Pal. Foss. Canada, vol. 1; also Pal. Foss., vol. 1, p. 15, fig. 18.
Conoceph. arenosus Bill., 1861; Geol. Vermont, vol. 2, p. 952, fig. 358.
Conoceph. arenosus Bill., 1863; Geol. Canada, p. 286, fig. 297.
Walcott (Bull. U. S. Geol. Survey No. 30, p. 195) refers this species to *Ptychoparia adamsi*.
—— ADAMSI Wal., 1890; 10th Rep. U. S. Geol. Survey, p. 649, pl. 96, figs. 1, 1 a. Olenellus zone.
—— *Eutoma?* AFFINIS, Wal., 1882; Pal. Eureka Dist., Mon. U. S. Geol. Survey, vol. 8, p. 54, pl. 10, fig. 12. Eureka Dist., Nev.
   Taconic. Cambrian.

PTYCHOPARIA.

 Crepicephalus (Bathyurus) ANGULATUS H. & W., 1877; U. S. Geol.
  Expl. 40th Par., Palæont., vol. 4, p. 220, pl. 2, fig. 28. Potsdam
  group of Pogonip Mt., White Pine, Nev.
——? ANGULATUS Wal., 1884; Pal. Eureka Dist., Mon. U. S. Geol. Survey,
  vol. 8, p. 269.
——? ANNECTANS Wal., 1884; Pal. Eureka Dist., Mon. U. S. Geol. Survey,
  vol. 8, p. 91, pl. 12, fig. 18. Hamburg Ridge, Nev.   Cambrian.
 Conoceph. ANTIQUATUS Salt., 1859; Quart. Jour. Geol. Soc. London,
  vol. 15, p. 554, fig. 2. Georgia, Vermont.   Taconic.
 Conoceph. ANATINUS Hall, 1863; 16th Rep. New York State Cab. Nat.
  Hist., p. 158, pl. 7, figs. 34, 35, and pl. 8, fig. 29. Lake Pepin, also
  Tremapleau, Wis.   Potsdam.
 Crepicephalus (Loganellus) ANYTUS H. & W., 1877; U. S. Geol. Expl.,
  40th Par., vol. 4, p. 219, pl. 2, figs. 19-21.
—— ANYTUS Wal., 1884; Pal. Eureka Dist., Mon. U. S. Geol. Survey, vol.
  8, p. 56, pl. 9, fig. 26. Schell Creek, Nev.   Taconic.
 Lisostracus ANYTUS Brögg., 1886; Geol. Fören. Stock. Förhandl., vol.
  8, p. 212.
—— ATTLEBORENSIS, Shaler & Foerste, 1888; Bull. Mus. Comp. Zool.
  Harvard College, vol. 19, p. 39, pl. 2, fig. 14. Attleborough, Mass.
——— ATTLEBORENSIS Wal., 1890; 10th Rep. U. S. Geol. Survey, p. 649, pl.
  95, fig. 2. Olenellus zone.
 Conoceph. aurora Hartt, 1868; Acadian Geology, p. 653. (See *Ptycho-*
  paria ouangondianus.)
 Conoceph. BILLINGSI Shum., 1861; Am. Jour. Sci., 2d series, vol. 32,
  p. 220. Morgan Creek, Burnet Co., Texas.   Taconic.
 Conoceph.? BINODOSUS Hall, 1863; 16th Rep. New York State Cab. Nat.
  Hist., p. 160, pl. 7, fig. 47. Osceola Mills, Wis.   Potsdam.
—— *Solenopleura?* BREVICEPS Wal., 1884; Pal. Eureka Dist., Mon. U. S.
  Geol. Survey, vol. 8, p. 49, pl. 10, fig. 9. Prospect Mt. group in
  the Hamburg shale north of Adams Hill, Nev.   Taconic.
—— BURNETENSIS Wal., 1880; Proc. U. S. Nat. Mus., vol. 13, p. 272, pl.
  21, fig. 1. Tatur Hill, Burnet Co., Texas.   Taconic.
· —— *Conocephalites* CALYMENOIDES Whitf., 1878; Ann. Rep. Geol. Survey
  Wisconsin, 1877, p. 52. Eau Claire, Wis.   Taconic.
 Conocephalites CALYMENOIDES Whitf., 1882; Geol. Wisconsin, vol. 4,
  p. 179, pl. 3, figs. 2-5.
—— CALYMENOIDES Wal., 1884; Pal. Eureka Dist., Mon. U. S. Geol. Survey,
  vol. 8, p. 48.
 Crepicephalus (Loganellus) CENTRALIS Whitf., 1880; Geol. Resources
  Black Hills Dakota, p. 341, pl. 2, figs. 21-24. Castle Creek, Black
  Hills, Dakota.   Taconic.
—— ? (subgenus?) *clavata* Wal., 1887; Am. Jour. Sci., 3d series, vol. 34,
  p. 198, pl. 1, fig. 3. 1¼ miles south of North Granville, etc.,
  Washington Co., N. Y. (See *Protypus clavata*.)   Taconic.

PTYCHOPARIA.
—— CONNATA Wal., 1890; Proc. U. S. Nat. Mus., vol. 13, p. 272, pl. 20, fig. 2. Eau Claire, Wis. Taconic.
Conoceph. CORDILLERÆ Rominger, 1877; Proc. Acad. Nat. Sci. Phila., p. 17, pl. 1, fig. 7. Mount Stephen, B. C. Taconic.
—— CORDILLERÆ Wal., 1888; Am. Jour. Sci., 3d series, vol. 36, p. 165.
Conoceph. DEPRESSUS Shum., 1861; Am. Jour. Sci., 2d series, vol. 32, p. 219. Clear Creek, Burnett Co., Texas. Potsdam.
Conoceph. DIADEMATUS Hall, 1863; 16th Rep. New York State Cab. Nat. Hist., p. 167, pl. 7, figs. 36–39; pl. 9, figs. 18–21 (?). Marine Mills, Wis. Potsdam.
? *Conoceph.* DEPRESSUS Shum., 1861; Am. Jour. Sci., 2d series, vol. 32, p. 219. Potsdam.
—— *Euloma?* DISSIMILIS Wal., 1884; Pal. Eureka Dist., Mon. U. S. Geol. Survey, vol. 8, p. 51, pl. 9, fig. 28. Secret Cañon, Prospect Mt. group, etc., Nev.
Conoceph. (*Arionellus?*) DORSALIS Hall, 1863; 16th Rep. New York State Cab. Nat. Hist., p. 222. Tremapleau, Wis. Potsdam.
Conoceph. ELEGANS Hartt, 1868; Acadian Geol., p. 650. St. John, N.B.
Conoceph. EOS Hall, 1863; 16th Rep. New York State Cab. Nat. Hist., p. 151, pl. 7, figs. 24, 25; pl. 8, figs. 2–9. Tremapleau, Wis. Potsdam.
—— EOS Wal., 1884; Pal. Eureka Dist., Mon. U. S. Geol. Survey, vol. 8, p. 51.
Conoceph. ERYON Hall, 1863; 16th Rep. New York State Cab. Nat. Hist., p. 157, pl. 7, figs. 10–16; pl. 8; figs. 16, 31. Tremapleau, Wis. Potsdam.
—— FITCHI Wal., 1887; Am. Jour. Sci., 3d series, vol. 34, p. 197, pl. 1, fig. 6. 2 miles south of North Granville, N. Y. Taconic.
——? FITCHI Wal., 1890; 10th Rep. U. S. Geol. Survey, p. 650, pl. 96, fig. 5. Olenellus zone.
Conoceph. formosus Hartt. (See *Ptychoparia robbi* Hartt.)
Conoceph. (*Ptychoparia*) *gallatinensis* Meek, 1873; 6th Ann. Rep. U. S. Geol. Survey Territories, 1872, p. 485. Referred by Walcott to *Ptychoparia oweni* Meek & Hayden. Potsdam.
Crepicephalus (*Loganellus*) GRANULOSUS H. & W., 1877; U. S. Geol. Expl. 40th Par., Palæont., vol. 4, p. 214, pl. 2, figs. 1, 2. 3. Eureka Dist., Nev. Taconic.
—— GRANULOSUS Wal., 1884; Pal. Eureka Dist., Mon. U. S. Geol. Survey, vol. 8, p. 57.
Crepicephalus (*Loganellus*) HAGUEI H. & W., 1877; U. S. Geol. Expl. 40th Par., Palæont., vol. 4, p. 210, pl. 2, figs. 14, 15.
—— HAGUEI Wal., 1884; Pal. Eureka Dist., Mon. U. S. Geol. Survey, vol. 8, p. 57. White Pine Dist., Nev. Taconic.
Lisostracus HAGUEI Brögg., 1886; Geol. Föreningens Stockholm Förhandl., vol. 8, p. 202.

PTYCHOPARIA.

Conoceph. halli Hartt. (See *Ptychoparia orestes* Hartt.)
Dicellocephalus HISINGERI Bill.; Matthew, Diffusion of the Cambrian fauna, 1893, p. 11, refers this species to the Conocephalites.
—— HOUSENSIS Wal., 1886; Bull. U. S. Geol. Survey, No. 30, p. 201, pl. 25, fig. 5. House Range, Antelope Springs, Utah. Taconic.
Conoryphe (Conocephalites) KINGII Meek, 1870; Proc. Acad. Nat. Sci. Phila., vol. 22, p. 63.
Conocoryphe (Conocephalites) KINGII Meek, 1877; U. S. Geol. Expl. 40th Par., Palæont., vol. 4, p. 20, pl. 1, fig. 4. House Range, Antelope Springs, Utah. Taconic.
Conocoryphe (Ptychoparia) KINGII Meek, 1873; 6th Ann. Rep. U. S. Geol. Survey Territories, 1872, p. 487.
Conocoryphe (Ptychyparia) KINGII White, 1876; U. S. Geog. and Geol. Surveys West 100th Mer., Palæont., vol. 4, p. 40, pl. 2, figs. 2 a, b, c.
Liostracus KINGII Brögg., 1886; Geol. Föreningens, Stockholm, Förhandl., vol. 8, p. 205.
—— KINGII Wal., 1886; Bull. U. S. Geol. Survey, No. 30, p. 193, pl. 27, figs. 4, 4 a.
—— LÆVICEPS Wal., 1884; Pal. Eureka Dist., Mon. U. S. Geol. Survey, vol. 8, p. 54, pl. 10, figs. 17, 18. White Pine Dist., Nev. Taconic.
Conocephalites (Pterocephalus) LATICEPS H. & W., 1877; U. S. Geol. Expl. 40th Par., Palæont., vol. 4, p. 221, pl. 2, figs. 4–7. West side Pogonip Mts., White Pine, Nev. Taconic.
Pterocephalus LATICEPS Wal., 1884; Pal. Eureka Dist., Mon. U. S. Geol. Survey, vol. 8, p. 59.
—— ? LINNARSSONI Wal., 1884; Pal. Eureka Dist., Mon. U. S. Geol. Survey, vol. 8, p. 47, pl. 9, figs. 18, 18 a. Prospect Mt. Group, Secret Cañon, etc., Nev. Taconic.
Liostracus PANOPE Wal., 1890; Proc. U. S. Nat. Mus., vol. 13, p. 275, pl. 21, fig. 13. Spring Creek Cañon, Black Hills, Dakota. Taconic.
—— LLANOENSIS Wal., 1890; Proc. U. S. Nat. Mus., vol. 13, p. 272, pl. 20, figs. 3, 4 and 5. Packsaddle Mt., Llano Co., Texas. Taconic.
Loganellus LOGANI Devine, 1863; Canadian Naturalist, vol. 8, p. 95, fig. Point Levis, Quebec.
Olenus? LOGANI Bill., 1865; Pal. Foss., vol. 1, p. 201, figs. 185, 186.
Syn., (*Loganellus*) *quebecensis* Bill.; *idem*, p. 203.
—— LOGANI Wal., 1884; Bull. U. S. Geol. Survey, No. 10, p. 36.
Crepicephalus (Loganellus) MACULOSUS H. & W., 1877; U. S. Geol. Expl. 40th Par., Palæont., vol. 4, p. 215, pl. 2, figs. 21, 25, 26 (?). Eureka, Nev. Taconic.
—— MACULOSUS Wal., 1884; Pal. Eureka Dist., Mon. U. S. Geol. Survey, vol. 8, pp. 269, 271.
—— METISENSIS Wal., 1890; 10th Rep. U. S. Geol. Survey, p. 651, fig. 68. Olenellus zone. In conglomerate at Metis, Quebec, Canada.

PTYCHOPARIA.

——? METRA Wal., 1890; Proc. U. S. Nat. Mus., vol. 13, p. 273, pl. 21, fig. 7. Tatur Hill, Burnet Co., Texas. Taconic.

Conoceph. MINOR Shum., 1863; Trans. Acad. Nat. Sci. St. Louis, vol. 2, p. 105. Bluffs of Mississippi near confluence of Black River, Wis.

(Undet.) trilobite Owen, 1848; Geol. Reconnoissance Chippewa Land Dist., p. 15, pl. 7, fig. 4.

Conoceph. MINOR Hall, 1863; 16th Rep. New York State Cab. Nat. Hist., p. 149, pl. 8, figs. 1–4. Trempaleau, Wis.

—— MINOR Wal., 1884; Pal. Eureka Dist., Mon. U. S. Geol. Survey, vol. 8, p. 91.

Conoceph. MINUTUS Bradley, 1860; Am. Jour. Sci., 2d series, vol. 30, p. 24, 3 figs. Keeseville, N. Y. Potsdam.

Conoceph. MINUTUS Bradley, 1860; Canadian Naturalist, vol. 5, p. 420, figs. 1–3.

Conoceph. MINUTUS Hall, 1863; 16th Rep. New York State Cab. Nat. Hist., p. 150, pl. 8, figs. 5–7.

—— MINUTUS Wal., 1884; Pal. Eureka Dist., Mon. U. S. Geol. Survey, vol. 8, p. 91.

—— MINUTUS Wall., 1886; Bull. U. S. Geol. Survey, No. 30, p. 21.

Conoceph. MISER Bill., 1861; New or little known Sil. Foss., p. 12, fig. 14. Straits of Belle Isle. Potsdam.

Conoceph. MISER Bill., 1861; Geol. Vermont, vol. 2, p. 950, fig. 354.

Conoceph. MISER Bill., 1863; Geol. Canada, p. 286, fig. 393.

Conoceph. MISER Bill., 1865; Pal. Foss., vol. 1, p. 12, fig. 14.

—— MISER Wal., 1886; Bull. U. S. Geol. Survey, No. 30, p. 199, pl. 27, fig. 2.

—— MISER Wal., 1890; 10th Rep. U. S. Geol. Survey, p. 651, pl. 96, fig. 8. Olenellus zone.

Crepicephalus (Loganellus) MONTANENSIS Whitf., 1876; Rep. Recon. Upper Missouri, p. 141, pl. 1, figs. 1, 2. Camp Baker, Montana. Potsdam.

—— MONTANENSIS Wal., 1884; Pal. Eureka Dist., Mon. U. S. Geol. Survey, vol. 8, pp. 53, 56.

—— MUCRONATUS Shaler & Foerste, 1888; Bull. Mus. Comp. Zool. Harvard Coll., vol. 16, p. 37, pl. 2, fig. 21. Attleborough, Mass. Taconic.

—— *Conoceph.* NASUTUS Hall, 1863; 16th Rep. New York State Cab. Nat. Hist., p. 155, pl. 7, figs. 3–9. Kickapoo, etc., Wis. Potsdam.

Conoceph. neglectus. (See *Ptychoparia tener.*)

Crepicephalus (Loganellus) NITIDUS H. & W., 1877; U. S. Geol. Expl. 40th Par., Palæont., vol. 4, p. 212, pl. 2, figs. 8–10. Eureka, Nev. Taconic.

Syn., *Crepicephalus (Loganellus) simulator* (H. & W.) Walcott.

—— NITIDUS Wal., 1884; Pal. Eureka Dist., Mon. U. S. Geol. Survey, vol. 8, p. 57.

PTYCHOPARIA.
—— OCCIDENTALIS Wal., 1884; Pal. Eureka Dist., Mon. U. S. Geol. Survey, vol. 8, p. 51, pl. 10, fig. 5. Eureka Dist., Nev. Taconic.
Conoceph. ORESTES Hartt, 1868; Acadian Geology, p. 649, fig. 225. Ratcliff's Millstream, St. John, N. B. Taconic.
Syn., Conoceph. halli (Hartt) Wal.; Bull. U. S. Geol. Survey, No. 10, p. 40.
Conoceph. thersites (Hartt) Wal.; Bull. U. S. Geol. Survey, No. 10, p. 40.
—— ORESTES var. THERSITES Wal., 1884; Bull. U. S. Geol. Survey, No. 10, p. 40.
Syn., orestes Wal., 1884; Bull. U. S. Geol. Survey, No. 10, p. 39, pl. 5, figs. 3, 3 a.
Solenopleura ORESTES Matt., 1887; Trans. Royal Soc. Canada, vol. 4, p. 154, pl. 2, figs. 4 a-e.
—— Pterocephalus OCCIDENTS Wal., 1884; Pal. Eureka Dist., Mon. U. S. Geol. Survey, vol. 8, p. 58, pl. 9, fig. 21. Cañon east of Hamburg Ridge, Nev. Taconic.
Conoceph. OUANGONDIANA Hartt., 1868; Acadian Geology, p. 651, fig. 226. Ratcliff's Millstream, St. John, N. B. Taconic.
—— OUANGONDIANA Wal., 1884; Bull. U. S. Geol. Survey, No. 10, p. 37, pl. 5, figs. 4, 4 a-f.
—— OUANGONDIANA var. AURORA (Hartt) Wal.; Bull. U. S. Geol. Survey, No. 10, p. 38, pl. 5, fig. 5.
Liostracus OUANGONDIANA Matt., 1887; Trans. Royal. Soc. Canada, p. 138.
—— OWENI (M. & H.) Wal., 1884; Pal. Eureka Dist., Mon. U. S. Geol. Survey, vol. 8, p. 55, pl. 10, figs. 3, 3 a. (See *Agraulus oweni* M. & H.) Big Horn Mts., Montana. Taconic.
The following species are referred by Walcott (Mon. U. S. Geol. Survey, vol. 8, p. 55) to *Ptychoparia oweni:*
Arionellus oweni M. & H., 1862; Am. Jour. Sci., 2d series, vol. 33, p. 74.
Agraulos oweni M. & H., 1864; Pal. Upper Missouri, p. 9, figs. a-c, pl. 1, fig. 4.
Conocoryphe (Ptychoparia) gallatinensis Meek, 1870; 6th Ann. Rep. U. S. Geol. Survey Territories, 1872, p. 485.
Crepicephalus (Loganellus) centralis Whit., 1880; Geol. Resources Black Hills, Dakota, p. 341, pl. 2, figs. 21-24.
Conoceph. OWENI Hall, 1863; 16th Rep. New York State Cab. Nat. Hist., p. 155, pl. 8, figs. 17-20. Marine Mills, Wis. Potsdam,
Conoceph. PATERSONI Hall; 16th Rep. New York State Cab. Nat. Hist., p. 159, pl. 7, figs. 45-46. Tremapleau, Wis. Potsdam.
—— *Liostracus* PANOPE Wal., 1890; Proc. U. S. Nat. Mus., vol. 13, p. 275, pl. 21, fig. 13. Spring Creek Cañon, Black Hills, Dakota.
Taconic.

PTYCHOPARIA.
Crepicephalus (Loganellus) PLANUS Whitf., 1877; Prelim. Rep. Pal. Black Hills, p. 11. Castle Creek, Black Hills, Dakota. Taconic.
Crepicephalus PLANUS Whit., 1880; Geol. Black Hills, Dakota, p. 343, pl. 2, fig. 20.
—— ? PERNASUTUS Wal., 1884; Pal. Eureka Dist., Mon. U. S. Geol. Survey, vol. 8, p. 49, pl. 10, figs. 8, 8 a, b. Cañon east side Hamburg Ridge, Nev. Taconic.
—— PERO Wal., 1890; Proc. U. S. Nat. Mus. vol. 13, p. 274, pl. 21, fig. 6. Morgan's Creek, Burnet Co., Texas.; Trempleau, Wis. Taconic.
Conoceph. PERSEUS Hall, 1863; 16th Rep. New York State Cab. Nat. Hist., p. 153, pl. 7, figs. 17-22; pl. 8, fig. 33. Opposite the mouth of the Chippewa on Mississippi River.
—— PIOCHENSIS Wal., 1866; Bull. U. S. Geol. Survey, No. 30, p. 201, pl. 26, fig. 2, 2 a, b; pl. 28, figs. 1, 1 a-e. Ely Mts. and western slope Highland Range, Nev. Taconic.
—— ? PROSPECTENSIS Wal., 1884; Pal. Eureka Dist., Mon. U. S. Geol. Survey, vol. 8, p. 46, pl. 9, fig. 20. Prospect Mt. on the east slope of Prospect Peak, Nev. Taconic.
—— ? PROSPECTENSIS Wal., 1886; Bull. U. S. Geol. Survey, No. 30, p. 202, pl. 27, fig. 5.
—— QUADRANGULARIS Whitf. The specific name of Conoceph. quadratus being preoccupied, Prof. R. P. Whitfield substitutes that of quadrangularis for this species. Eau Claire and at Ettrick, Wis.
Potsdam.
Conoceph. ? QUADRATUS Whitf., 1880; Ann. Rep. Geol, Survey, Wisconsin, 1879, p. 47.
Conoceph. ? QUADRATUS Whitf., 1882; Geol. Wisconsin, vol. 4, p. 180, pl. 1, figs. 15, 16. (See Ptychoparia quadrangularis Whitf.)
Conoceph. QUADRATUS Hartt, 1868; Acadian Geology, p. 654. At Coldbrook, St. John, N. B.
—— QUADRATUS Wal., 1884; Bull. U. S. Geol. Survey, No. 10, p. 39, pl. 5, fig. 1.
Conoceph. robbi Hartt, 1868; Acadian Geology, p. 648. Ratcliff's Millstream. St. John.
Syn., Conoceph. formosa (Hartt) Wal.; Bull. U. S. Geol. Survey, No. 10, p. 36.
—— robbi Wal., 1884; Bull. U. S. Geol. Survey, No. 10, p. 36, pl. 6, figs. 1, 1 a.
Solonopleura ROBBI Matt., 1887; Trans. Royal Soc. Canada, vol. 4, p. 153, pl. 2, figs. 3 a-l; fig. 4 a-e.
—— ROGERSI Wal., 1884; Bull. U. S. Geol. Survey, No. 10, p. 47, pl. 7, fig. 2. Braintree, Mass. Paradoxides zone.
Conoceph. SHUMARDI Hall, 1863; 16th Rep. New York State Cab. Nat. Hist., p. 154, pl. 7, figs. 1-2; pl. 3, figs. 19, 32. Kickapoo, Wis.
Potsdam.

PTYCHOPARIA.

— similis Wal., 1884; Pal. Eureka Dist., Mon. U. S. Geol. Survey, vol. 8, p. 52, pl. 10, figs. Secret Cañon Shale, Nev.
— similis var. robustus Wal.; *idem.*, p. 53, pl. 1, figs. 9, 9 a.
Conoceph. subcoronatus H. & W., 1877; U. S. Geol. Expl. 40th Par., Palæont., vol. 4, p. 237, pl. 2, fig. 1. Ute Peak, Wasatch Mts., Utah. Taconic.
— suada Wal., 1890; Proc. U. S. Nat. Mus., vol. 13, p. 274, pl. 21, fig. 9. Morgan's Creek, Burnet Co., Texas. Taconic.
— subcoronatus Wal., 1886; Bull. U. S. Geol. Survey, No. 30, p. 205, pl. 28, fig. 4. Ute Peak, Wasatch Range, Utah. Taconic.
— subcoronata Wal.; 10th Rep. U. S. Geol. Survey, p. 652, pl. 96, fig. 6. Olenellus zone.
Conoceph. teucer Bill., 1861; New or little known Sil. Foss., p. 12, fig. 16. Swanton, Vermont. Taconic.
Conoceph. teucer Bill., 1861; Geol. Vermont, vol. 2, p. 951, fig. 356.
Conoceph. teucer Bill., 1863; Geol. Canada, p. 286, fig. 295.
Conoceph. teucer Bill., 1865; Pal. Foss., vol. 1, p. 13, fig. 16.
— teucer Wal., 1886; Bull. U. S. Geol. Survey, No. 30, p. 197, pl. 26, fig. 3.
— teucer Wal.; 10th Rep. U. S. Geol. Survey, p. 652, pl. 96, fig. 3. Olenellus zone.
Conoceph. tener Hartt, 1868; Acadian Geol., p. 652. Coldbrook, N. B. Paradoxides zone.
Syn., Conoceph. neglectus (Hartt) Wal., 1884; Bull. U. S. Geol. Survey, No. 10, p. 41.
— tener Wal., 1884; Bull. U. S. Geol. Survey, No. 10, p. 41, pl. 5, figs. 6, 6 a, b.
Conoceph. thersites. (See *Ptychoparia orestus* Hartt.)
— trilineatus. (See *Atops trilineatus, Conocoryphe trilineatus, Conocephalites trilineatus, Calymene beckii, Triarthrus beckii.*)
Crepicephalus (Loganellus) unisulcatus H. & W., 1877; U. S. Geol. Expl. 40th Par., Palæont., vol. 4, p. 216, pl. 2, figs. 22, 23. Eureka, Nev. Taconic.
— unisulcatus Wal., 1884; Pal. Eureka Dist., Mon. U. S. Geol. Survey, vol. 8, p. 58. Eureka District, Nev.
—? urania Wal., 1890; Proc. U. S. Nat. Mus., vol. 13, p. 274, pl. 21, figs. 10 and 11. Packsaddle Mt., Texas. Taconic.
— vacuna Wal., 1890; Proc. U. S. Nat. Mus., vol. 13, p. 275, pl. 21, figs. 8–12. Spring Creek Cañon, Black Hills, Dakota. Taconic.
Conoceph. verucosus Whitf., 1884; Bull. Am. Mus. Nat. Hist., vol. 1, p. 146, pl. 14, fig. 12. Georgia, Vermont. Taconic.
Conoceph. vulcanus Bill., 1861; New or little known Sil. Foss., p. 14, fig. 17. Highgate, Vermont. Taconic.
Conoceph. vulcanus Bill., 1861; Geol. Vermont, vol. 2, p. 952, fig. 357.

PTYCHOPARIA. SHUMARDIA.

Conoceph. VULCANUS Bill., 1863; Geol. Canada, p. 286, fig. 290.
Conoceph. VULCANUS Bill., 1865; Pal. Foss., vol. 1, p. 14, fig. 17.
—— VULCANUS Wal., 1886; Bull. U. S. Geol. Survey, No. 30, p. 196, pl. 26, figs. 4, 4 a. cf. *Ptychoparia adamsi.*
—— VULCANUS Wal., 10th Rep. U. S. Geol. Survey, p. 653, pl. 96, figs. 4, 4 a. Olenellus zone.
Conoceph. WINONA Hall, 1863; 16th Rep. New York State Cab. Nat. Hist., p. 161, pl. 7, figs. 26–28. Mississippi River opposite mouth of Black River. Taconic.
Conoceph. ZENKERI Bill., 1860; Canadian Naturalist, vol. 5, p. 305, fig. 4. Point Levis, Quebec.
Conoceph. ZENKERI Bill., 1865; Pall. Foss., vol. 1, p. 398, fig. 375. Point Levis, Quebec.

REMOPLEURIDES Portl., 1843; Rep. Geol. Londonderry, etc., p. 254.
Syn., *Caphyra* Barr., 1846; Notice Prélim., p. 32.
Amphytryon Cord., 1847; Prodr., p. 228.
—— AFFINIS Bill., 1865; Pall. Foss., vol. 1, p. 325, fig. 313. Stanbridge Range 6, lot 20.
—— CANADENSIS Bill, 1865; Pal. Foss, vol. 1, p. 182, fig. 164. Front concession of the Township of Clarence. Chazy.
—— PANDERI Bill., 1865; Pal. Foss., vol. 1, p. 293, fig. 283. Table Head, N. F.
—— ? SCHLOTHEIMI Bill., 1865; Pal. Foss., vol. 1, p. 294, fig. 284. Table Head and Pistolet Bay, N. F.
Syn., *Paradoxides* or *Olenellus* Bill., 1863; Geol. Canada, pp. 871, 872.
—— STRIATULUS Wal., 1875; Cincinnati Quart. Jour. Sci., vol. 2, p. 347, fig. 27, 27 a b. Trenton Falls, N. Y. Trenton.

SAO Barr., 1846; Notice Prélim. Sys. Sil. Bohême, p. 13.
—— ? LAMOTTENSIS Whitf., 1886; Bull. Am. Mus. Nat. Hist. New York, vol. 1, p. 334, pl. 33, figs. 9–11. Isle La Motte.
Birdseye Limestone.

SHUMARDIA Bill., 1862; Pal. Foss., vol. 1, p. 92. Type, *Shumardia granulosa* Bill., from Point Levis, Quebec.
This generic name suggested for a minute Trilobite includes a specimen having an expanding glabella, with two anterior side nodules, similar to those of *Battus pusillus* Sars, the genus being allied to *Agnostus*, differing therefrom in having a regularly ribbed pygidium, similar to that of an *Asaphus*. Comparing *Shumardia granulosa* with *Conophrys salopiensis* Call. (Quart. Jour. Geol. Soc. London, vol. 33, p. 667, pl. 12, fig. 7), which Brögger (Sil. Etagen 2 und 3, p. 125) refers to as a synonym to *Battus pusillus*, we find a similar characteristic in the glabella. The pygidium of *Shumardia granulosa* has however a more semi-oval form, with a greater number of axis rings and an additional number of ribs on the side lobes.

SHUMARDIA. SOLENOPLEURA.

The head of *Shumardia granulosa* differs from *Battus pusillus* in having a groove, which runs from the anterior front of the glabella to the frontal margin of the head, a characteristic of some species of *Agnostidæ* (vid. Brögger, Sil. Etagen 2 und 3, pl. 1, fig. 10).

Mr. Billings remarks that "in some specimens this furrow has at the bottom a small triangular tubercle, giving the glabella the appearance of having a projecting augular process in the middle of its front." I am more inclined to view this triangular tubercle as a point of attachment for the muscles of the jaws, and not as a process for the attachment of antennæ.

Mr. E. Billings in the same book gives a description of a second species, which he calls *Shumardia glacialis*, figured p. 283, fig. 270. The species is represented with three glabella furrows, and should not be included in the genus *Shumardia*.

If these forms are not embryonic, I would advocate the acceptance of *Shumardia* amongst the *Agnostidæ* for such species as *Shumardia granulosa* Bill. and *Battus pusillus* Sars to replace the later generic term of *Conophrys* Call.

SOLENOPLEURA Ang., 1854; Palæont. Scand., p. 26. Type, *Solenopleura holometopa* Ang. Diagnosis: Glabella prominent, with three distinct furrows; fixed cheeks, elevated especially in the middle; front limb convex; occipital ring bearing a tubercle; genal angles pointed; thorax, pleuræ bluntly rounded; pygidium small with few segments; surface of shell granulated with or without scattered tubercles.

—— ACADIA Whiteaves, 1885; Trans. Royal Soc. Canada, vol. 3, p. 76, pl. 7, fig. 15. Porter's Stream, N. B. Taconic.

—— ACADIA Whiteaves, 1887; Trans. Royal Soc. Canada, vol. 4, p. 157, pl. 2, figs. 5 a, b, c.

—— ACADIA var. ELONGATA Matt., 1887; idem., p. 159, pl. 1, fig. 6.

—— BOMBIFRONS Matt., 1886; Trans. Royal Soc. Canada, vol. 2, p. 156, figs. 5, 5 a-b. Topsail Head, Conception Bay, N. F. Taconic.

—— BOMBIFRONS Wal., 1890; 10th Rep. U. S. Geol. Survey, p. 656, pl. 98, figs. 5, 5 a-b. Olenellus zone.

—— COMMUNIS Bill., 1874; Pal. Foss., vol. 2, p. 72. Chapel Arm, Trinity Bay.

—— COMMUNIS Matt., 1886; Trans. Royal Soc. Canada, vol. 2, p. 155, figs. 4, 4 a, b.

—— NANA Ford, 1878; Am. Jour. Sci., 3d series, vol, 15, p. 126. Troy, N. Y. Taconic.

—— NANA Wal., 1886; Bull. U. S. Geol. Survey, No. 30, p. 214, pl. 27, fig. 3.

—— ? NANA Wal., 1887; Am. Jour. Sci., 3d series, vol. 34, p. 196, pl. 1, figs. 1, 1 d.

SOLENOPLEURA. SPHÆROCORYPHE.
—— ? NANA Wal., 1890; 10th Rep. U. S. Geol. Survey, p. 658, pl. 98, figs. 1, 1 a-e, 2. Olenellus zone.
—— HARVEYI Wal., 1889; Proc. Nat. Mus., vol. 12, p. 45. Manuel's Brook, Conception Bay, N. F.
—— ? HARVEYI Wal., 1890; 10th Rep. U. S. Geol. Survey, p. 656, pl. 97, figs. 7, 7 a. Olenellus zone.
—— HOWLEYI Wal., 1889; Proc. Nat. Mus., vol. 12, p. 45. Manuel's Brook, Conception Bay, N. F.
—— ? HOWLEYI Wal., 10th Rep. U. S. Geol. Survey, p. 657, pl. 97, figs. 8, 8 a. Olenellus zone.
—— ORESTES (Hartt) Matt., 1887; Trans. Royal Soc. Canada, vol. 4, p. 154, pl. 2, figs. 4 a-e. Div. 1 c. at Portland, Simonds and St. Martin's, N. B. Taconic.
Syn., *robbi* (Hartt) Matt., 1887; Trans. Royal Soc. Canada, vol. 4, p. 153, pl. 2, figs. 3 a-l and figs. 4 a-e.
—— ? TUMIDA Wal., 1887; Am. Jour. Sci., 3d series, vol. 34, p. 196, p. 1, figs. 2, 2 a. Near North Greenwich, etc., Washington Co., N. Y. Taconic.
—— ? TUMIDA Wal., 1890; 10th Report U. S. Geol. Survey, p. 658, pl. 98, figs. 3, 3 a. Olenellus zone.
SPHÆRERROCHUS Beyr., 1845; Ueber einige böhmische Trilobiten, p. 19.
—— CANADENSIS Bill., 1866; Catalogue Sil. Foss. Anticosti, p. 64, fig. 21. South West Point, Div. 4.
—— MIRUS (Beyr.) Roemer, 1860; Die Sil. Fauna W. Tennessee, p. 81, pl. 5, fig. 20 Niagara.
—— PARVUS Bill., 1865; Pal. Foss., vol. 1, p. 180, fig. 161. Large Island, Mingan Island. Chazy. Trenton.
—— ROMINGERI Hall, 1862; Geol. Rep. Wisconsin, p. 434. Niagara.
Syn., *mirus* Hall (not Beyrich), 1867; 20th Rep. New York State Cab. Nat. Hist., p. 334, pl. 21, figs. 4–7. Milwaukee, Racine, Waukesha and Greenfield, Wis.; Bridgeport, Ill.
—— ROMINGERI Hall, 1867; 20th Rep. New York State Cab. Nat. Hist., p. 375, pl. 22, figs. 4–7.
—— ROMINGERI Hall, 1870; 20th Rep. New York State Cab. Nat. Hist., p. 425, pl. 21, figs. 4–7 (rev. ed.).
—— ROMINGERI Whitf., 1882; Geol. Wisconsin, vol. 4, p. 311, pl. 21, figs. 1–3.
—— ROMINGERI? Hall, 1876; 28th Rep. New York State Mus. Nat. Hist., p. 32, pl. 32, fig. 16.
—— ROMINGERI M. & W., 1875; Geol. Survey Illinois, vol. 6, p. 510, p. 24, fig. 4.
SPHÆROCORYPHE Ang., 1854; Palæont. Scand., p. 65.
—— ROBUSTUS Wal., 1875; Cincinnati Quart. Jour. Sci., vol. 2, p. 273, figs. 18 a, b. Trenton Falls, N. Y. Trenton.
—— SALTERI Bill., 1866; Catalogue Sil. Foss. Anticosti, p. 63. Junction Cliff, Div. 1.

STRENULLA Matt., 1886; Trans. Royal Soc. Canada, vol. 2, p. 154.
—— PARVUM Wal. For other references see *Anomocare parvum*.
—— STRENUUS Bill. For other references see *Agraulos strenuus*.
SYMPHYSURUS Goldf., 1843; Jahrb. für Mineral., 1843, p. 332.
—— ? GOLDFUSSI Wal., 1884; Pal. Eureka Dist., Mon. U. S. Geol. Survey, vol. 8, p. 95, pl. 12, fig. 16. Pogonip group, McCoy's Ridge, Eureka Dist., Nev.   Cambrian.
TELEPHUS Barr., 1852; Syst. Sil. Bohême, vol. 1, p. 890.
—— AMERICANUS Bill., 1865; Pal. Foss., vol. 1, p. 291, figs. 281. Table Head and Pistolet Bay, Portland Creek, N. F.
Terataspis Hall, 1863; 16th Rep. New York State Cab. Nat. Hist., p. 223. (See *Lichas* and *Acidaspis*.)
—— *grandis*. (See *Lichas* (*Terataspis*) *grandis* Hall.)
—— *eriops*. (See *Lichas* (*Conolichas*) *eriops* Hall.)
THALEOPS Con., 1843; Proc. Acad. Nat. Sci. Phila., vol. 1, p. 332.
—— *ovatus* Con. (See *Illænus ovatus*.)
TRIARTHRELLA Hall, 1863; 16th Rep. New York State Cab. Nat. Hist., p. 177.
—— AURORALIS Hall, 1863; 16th Rep. New York State Cab. Nat. Hist., p. 177, pl. 9, fig. 13. Lagrange Mt., Minnesota.   Taconic.
TRIARTHRUS Green, 1832; Monthly Am. Jour. Geol. vol, 1, p. 560 (June). Green, 1832; Mon. Tril. North America, p. 86.
Brongniartia CARCINODEA Eaton, 1832; Am. Jour. Sci., 1st series, vol. 22, p. 166 (March 6). Utica Slate, New York, etc.
Brongniartia CARCINODEA Eaton, 1832; Geol. Text Book, p. 33, pl. 1, fig. 3.
—— BECKII Green, 1832; Monthly Am. Jour. Geol., vol. 1, p. 560, pl. 1, fig. 3 (June). This species should take the older name of *carcinodea*. Eaton's paper was published March 6th, 1832, and Dr. Green's article June, 1832. A typical species of the Utica slate.
—— *beckii* Green, 1832; Mon Tril. N. A., p, 86, cast 34, pl. 1, fig. 3.
—— *beckii* Harlan, 1835; Trans. Geol. Soc. Penn., vol. 1, p. 205, pl. 15, fig. 6.
Syn., *Paradoxides triarthrus* Harlan, 1835; Trans. Geol. Soc. Penn., vol. 1, pl., p. 264, 15, fig. 5.
Paradoxides arcuatus Harlan, 1835; Trans. Geol. Soc. Penn., vol. 1, p. 265, pl. 15, figs. 1–3.
—— *beckii* Harlan, 1835; Med. Phys. Researches, p. 400, pl., fig. 6.
Syn., *Paradoxides triarthrus* Harlan, 1835; Med. Phys. Researches, p. 401, pl., fig. 5.
Paradoxides arcuatus Harlan, 1835; Med. Phys. Researches, p. 402, pl., figs. 1–3.
—— *beckii* Green, 1835; Suppl. Mon. Tril. N. A., p. vii.
Paradoxides beckii Hall, 1838; Am. Jour. Sci., 1st series, vol. 33, p. 142, fig. 2.
Syn., *Paradoxides eatoni* Hall, 1838; Am. Jour. Sci., 1st series, vol. 33, p. 142, fig. 2.
—— *beckii* Green, 1838; Am. Jour. Sci., 1st series, vol. 33, p. 344,

TRIARTHRUS. TRINUCLEUS.
—— *beckii* Mather, 1842; Geol. New York, 1st Geol. Dist., p. 390, fig. 24 (1).
—— *beckii* Emmons, 1842; Geol. New York, 2d Geol. Dist., p. 399, fig. 110 (1).
—— *beckii* Vanuxem, 1843; Geol. New York, 3d Geol. Dist., p. 57, fig. 8 (1).
—— *beckii* Hall, 1843; Geol. New York, 4th Geol. Dist., p. 504.
Calymene beckii Hall, 1847; Pal. New York, vol. 1, p. 237, pl. 64, figs. 2 a–e.
Calymene beckii Hall, 1847; Pal. New York, vol. 1, p. 250, pl. 66, figs. 2 a–k (not pl. 67, figs. 4 a–e, *Atops trilineatus*).
Syn.; *Atops trilineatus* (Emmons) Hall, 1847; Pal. New York, vol. 1, p. 237.
Calymene beckii Hall, 1848; Am. Jour. Sci., 2d series, vol. 5, p. 322, figs. 1, a, b, 2 a–g.
—— *beckii* Emmons, 1848; Am. Assoc. Adv. Sci., p. 17.
—— *beckii* Haldeman, 1848; Am. Jour. Sci., 2d series, vol. 8, p. 137.
—— *beckii* Emmons, 1855; Am. Geology, vol. 1, pt. 2, p. 214, pl. 15, fig. 12.
Calymene beckii Rogers, 1858; Geol. Survey Penn., vol. 2, p. 820, fig. 613.
—— *beckii* Barr., 1861; Soc. Géol. France, 2d series, p. 269, pl. 5, figs. 11, 12.
—— *beckii* Logan, 1863; Geol. Canada, p. 202, fig. 200.
—— *beckii* Miller, 1874; Cincinnati Quart. Jour. Sci., vol. 1, p. 146.
—— *beckii* Wal., 1876; Trans. Albany Inst., vol. 10, p. 23, pl. 2, figs. 1–16.
—— CANADENSIS Smith, 1861; Canadian Jour., vol. 6, p. 275. Hudson.
—— FISCHERI Bill., 1861; Pal. Foss., vol. 1, p. 291, fig. 280. Table Head, Pistolet Bay and Portland Creek, N. F.
—— GLABER Bill., 1859; Canadian Naturalist, vol. 4, p. 382. Lake St. Johns, Utica Slate.
—— GLABER Bill., 1863; Geol. Canada, p. 202, fig. 198.
—— SPINOSUS Bill., 1857; Geol. Survey Canada Rep. of Progress, 1853–56, p. 340. Township of Gloucester, County of Carleton, Canada, Utica Slate.
—— SPINOSUS Bill, 1859; Canadan Naturalist, vol. 4, p. 383.
—— SPINOSUS Bill., 1863; Geol. Canada, p. 202, fig. 199.
—— SPINOSUS Ami, 1882; Trans. Ottawa Nat. Club, vol. for 1882–83, p. 88, pl. 1, fig
Trimerus Green, 1832; Monthly Am. Jour. Geol., vol. 1, p. 560. (See *Homalonotus delphinocephalus* and *H. jacksoni*.)
TRINUCLEUS Lhwyd, 1698; Lithophyacii Brittannici Ichnographia, Epistola 1. Murch., Silurian System, 1839, p. 659. Sir R. I. Murchison has revived this old name of Lhwyd's, and all subsequent palæontologist have adopted it. Lhwyd's description meant no more than the general name of Trilobite of the more modern writers, and could not, except by courtesy, set aside Dr. Jacob Green's genus *Cryptolithus*.

TRINUCLEUS. ZACANTHOIDES.
—— BELLULUS Ulrich, 1878; Jour. Cincinnati Soc. Nat. Hist., vol. 1, p. 99, pl. 4, fig. 2.
Nuttainia CONCENTRICA Eaton, 1832; Geological Text Book, p. 34, pl. 1, fig. 2. Trenton and Hudson Groups of New York, etc.
Syn., *Trilobite* Bigsby, 1825; Annals Lyceum Nat. Hist. New York, vol. 1, pt. 2, p. 214, pl. 15, fig. 1.
Cryptolithus tessellatus Green, 1832; Monthly Am. Jour. Geol., vol. 1, pl. 1, fig. 4.
Cryptolithus tessellatus Green, 1832; Mon. Tril. N. A., p. 73, cast 38, pl. 1, fig. 4.
—— CONCENTRICA Hall, 1847; Pal. New York, vol. 1, pp. 249, 255, pl. 65, figs. 4 a, b; pl. 67, figs. 1 a–h.
—— CONCENTRICA Emmons, 1855; Am. Geol., vol. 1, pt. 2, p. 212, pl. 17, fig. 1.
—— CONCENTRICA Chapman, 1858; Canadian Jour., vol. 3, p. 514.
—— CONCENTRICA Bill., 1861; Geol. Vermont, vol. 1, p. 300, fig. 215.
—— CONCENTRICA Miller, 1874; Cincinnati Quart. Jour. Sci., vol. 1, p. 126.
ZACANTHOIDES Wal., 1887; Am. Jour. Sci., 3d series, vol. 36, p. 165.
—— EATONI Wal., 1889; Proc. U. S. Nat. Mus., vol. 12, p. 45. Washington county, N. Y.
—— EATONI Wal.; 10th Rep. U. S. Geol. Survey, p. 646, pl. 94, fig. 6. Olenellus zone.
Dicellocephalus? FLAGRICAUDA White, 1874; Prelim. Rep. Invt. Foss. Geog. and Geol. Survey West 100th Mer., p. 12.
Dicellocephalus? FLAGRICAUDA White, 1877; U. S. Geog. and Geol. Survey West 100th Mer., vol. 4, p. 60, pl. 3, figs. 8 a, b. Schellbourne, Schell Creek Range, Nevada. Taconic.
Olenoides FLAGRICAUDA Wal., 1886; Bull. U. S. Geol. Survey, No. 30, p. 185, pl. 25, fig. 4.
—— FLAGRICAUDA Wal., 1888; Am. Jour. Sci., 3d series, vol. 36, p. 165.
Olenoides LEVIS Wal., 1886; Bull. U. S. Geol. Survey, No. 30, p. 187, pl. 25, figs. 3, 3 a. Pioche, Nev. Taconic.
—— LEVIS Wal., 1888; Am. Jour. Sci., 3d series, vol. 36, p. 165.
—— LEVIS Wal.; 10th Rep. U. S. Geol. Survey, p. 646, pl. 94, figs. 5, 5 a. Olenellus zone.
Ogygia? SPINOSUS Wal., 1884; Pal. Eureka Dist., Mon. U. S. Geol. Survey, vol. 8, p. 63, pl. 9, fig. 22. Secret Cañon, Nev. Taconic.
Olenoides SPINOSUS Wal., 1886; Bull. U. S. Geol. Survey, No. 30, p. 184, pl. 25, figs. 6, 6 a.
Embolimus SPINOSUS (Rominger) Wal., 1888; Am. Jour. Sci., 3d series, vol. 36, p. 164.
—— SPINOSUS Wal., 1888; Am. Jour. Sci., 3d series, vol. 36, p. 165.
Olenoides TYPICALIS Wal., 1886; Bull. U. S. Geol. Survey, No. 30, p. 183, pl. 25, figs. 2, 2 a. Olenellus zone. Pioche, Nev. Taconic.
—— TYPICALIS Wal., 1888; Am. Jour. Sci., 3d series, vol. 36, p. 165.

PART III.

NON-TRILOBITIC GENERA AND SPECIES.

MEROSTOMATA.
EURYPTERIDÆ.

EURYPTERUS De Kay, 1825; Annals Lyceum Nat. Hist., vol. 1, p. 375.
  Silurian. Carboniferous.
DOLICHOPTERUS Hall, 1859; Palæont. New York, vol. 3, p. 414. Silurian.
ECHINOGNATHUS Wal., 1882; Am. Jour. Sci., 3d series, vol. 23, p. 213.
  Silurian.
ADELOPHTHALMUS Jordan and Meyer, 1854; Crust. Steinkohl. Saarb., p. 8.
  Carboniferous.
GLYPTOSCORPIUS Peach, 1880; Trans. Royal Soc. Edinb., vol. 30, p. 516.
  Carboniferous.
ARTHROPLEURA Jordan and Meyer, 1854; Crust. Steinkohl. Saarb., p. 13.
  Carboniferous.
EURYPTERELLA Matt., 1888; Trans. Roy. Soc. Canada, vol. 6, p. 60.
  Devonian.

PTERYGOTIDÆ.

PTERYGOTUS Agas., 1844; Mon. Poissons Foss., note, p. xix.
  Silurian. Devonian.
SLIMONIA Page, 1856; Advance Text-Book Geology, p. 135. Silurian.
STYLONURUS Page, 1856; Advance Text-Book Geology.
  Silurian. Devonian.

SYNZIPHOSURA.

ACANTHOTELSON M. and W., 1865; Proc. Acad. Nat. Sci. Phila., vol. 17. p. 46; Geol. Survey Illinois, vol. 2, 1866, p. 399. Carboniferous.

XIPHOSURA.
CYCLIDÆ.

CYCLUS De Kon., 1841; Mém. Acad. Sci. Belgique, vol. 14, p. 13.
  Carboniferous.
DIPELTIS Pack., 1885; American Naturalist, vol. 19, p. 293. Carboniferous.

BUNODIDÆ.

BUNODES Eichw., 1854; Beiträge Geol. u. Pal. Russ., p. 131. Silurian.
  cf. *Exapinurus* Nieszk., 1859; Archiv Nat. Liv-, Esth- u. Kurl., vol. 1, p. 380. Silurian.
HEMIASPIS H. W., 1865; Quart. Jour. Geol. Soc. London, vol. 21, p. 490.
  Silurian.

PSEUDONISCUS Nieszk., 1859; Archiv Nat. Liv-, Esth- u. Kurl., vol. 1, p. 381.   Silurian.
BUNODELLA Matt., 1888; Trans. Roy. Soc. Canada, vol. 6, p. 56.
Silurian.

LIMULIDÆ.

BELINURUS Koenig, 1825; Icones Foss., Sectiles, pl. 18, fig. 230.
Carboniferous.
PRESTWICHIA H. W., 1887; Quart. Jour. Geol. Soc. London, vol. 23, p. 32.
Carboniferous.
LIMULUS Müller, 1785; Entomostraca, etc., p. 124.   (Triassic.)
PROTOLIMULUS Pack., 1886; Mem. Natl. Acad. Sci., vol. 3, p. 150.
Carboniferous.
NEOLIMULUS H. W., 1868; Geol Mag., vol. 5, p. 1.   Silurian.

PHYLLOCARIDA.

I. CARAPACE UNIVALVE.

I. Flat or slightly convex shield.
 1. Neither sutured nor ridged along the back. Notched in front.
  A. Posterior border entire.
DISCINOCARIS H. W., 1866; Quart. Jour. Soc. Geol. London, vol. 22, p. 504.
Round shield, angular notch. Middle and Upper Silurian (Murch.)
Aspidocaris Reuss, 1867.   Triassic.
SPATHIOCARIS Clarke, 1882; Am. Jour. Sci., 3d series, vol. 23, p. 477. Angular notch.   Devonian.
PHOLADOCARIS H. W., 1882; Geol. Mag. decade 3, vol. 9, p. 388. Sinuous notch.   Devonian.
LISGOCARIS Clarke, 1882; Am. Jour. Sci., 3d series, vol. 23, p. 478. Oblong notch.   Devonian.
ELLIPSOCARIS H. W., 1882; Geol. Mag., decade 3, vol. 9, p. 444. Rounded notch.   Devonian.

  B. Posterior border slightly notched.
CARDIOCARIS H. W., 1882; Geol. Mag., decade 3, vol. 9, p. 386.

  C. Posterior border deeply notched.   Devonian.
PTEROCARIS Barr., 1872; Syst. Sil. Bohême, vol. 1, suppl., p. 464. Both notches angular. Lower Silurian (Murch.)
DIPTEROCARIS Clarke, 1883; Am. Jour. Sci., 3d series, vol. 23, p. 121. Both notches angular.   Cambrian. Devonian.
CRESCENTILLA Barr., 1872; Syst. Sil. Bohême, vol 1, suppl., p. 507. Both notches angular. Lower Silurian (Murch.)

 2. Ridged along the back.
DITHYOCARIS (Scouler) Portl.; Geol. Rep. Londonderry, etc., 1843, p. 313.
Ridges sometimes prickly.   Devonian and Carboniferous.
RACHURA Scudder, 1878. Only telson known.   Carboniferous.

3. Sutured along the back. Notched in front.

APTYCHOPSIS Barr., 1872; Syst. Sil. Bohême, vol. 1, suppl., p. 455. Angular notch. Middle and Upper Silurian (Murch.)

PELTOCARIS Salt., 1863; Quart. Jour. Geol. Soc. London, vol. 19, p. 87. Rounded notch. Lower Silurian (Murch.)

PINNOCARIS Eth., 1878; Proc. Royal Phys. Soc., Edinb., vol. 4, p. 167. Slightly notched. Lower Silurian (Murch.)

II. Folded shield, bent along the back so as to form an overarching carapace, or a pair of attached valves.

HYMENOCARIS Salt., 1852; Rep. 22d Meeting Brit. Adv. Sci., Trans. Sec., p. 56. Smooth. Lingula-flags. Taconic.

? CYTHEROPSIS TESTIS Barr, 1872. Not well known. (Lower Silurian Murch.)

PROTOCARIS Baily, 1872. Not well known. Uppermost Devonian or Lowest Carboniferous.

II. CARAPACE BIVALVED.

SACOCARIS Salt., 1868; Proc. Geol. Polytech. Soc., vol. 4, p. 588. Strongly emarginated behind. Lingula-flags.

1. Pod-like.

CARYOCARIS Salt., 1863; Quart. Jour. Geol. Soc. London, vol. 19, p. 139. Pod-like; elongate narrow, smooth. Arenig and Lingula-flags.

CERATIOCARIS McCoy, 1849; Annals Mag. Nat. Hist. London, 2d series, vol. 4, p. 412.
Pod-like, subovate, suboblong, striate. Tremadoc and Devonian in America.

PHYSOCARIS Salt., 1860; Annals Mag. Nat. Hist. London, 3d series, vol. 5, p. 159. Round. Upper Silurian (Murch.)

NOTHOZOE Barr., 1872; Syst. Sil. Bohême, vol. 1, suppl., p. 530. Oval. Lower Silurian (Murch.)

CRYPTOZOE Pack., 1888. Proc. Acad. Nat. Sci. Phil., 1888, p. 381. Suboblong. Carboniferous.

XIPHOCARIS Jones and Woodward, 1886. Only telson known. Silurian.

COLPOCARIS Meek, 1871; Proc. Acad. Nat. Sci. Phil., p. 334. Subovate; strongly emarginate at one end. Carboniferous.

2. Pod-like. Oculate.

EMMELEZOE J. & W., 1886; Rep. 56th Meeting Brit.Adv. Sci., p. 232.
Silurian.

III. With swellings in the antero-dorsal region, one of which on each valve may be ocular.

ECHINOCARIS Whitf., 1880; Am. Jour. Sci., 3d series, vol. 19, p. 34. Leperditioid. Segments spinose. Devonian.

ARISTOZOE Barr., 1872; Syst. Sil. Bohême, vol. 1, suppl., p. 474. Upper Silurian (Murch.)

OROZOE Barr., 1872; Syst. Sil. Bohême, vol. 1, suppl., p. 557. Upper Silurian (Murch.)

ELYMOCARIS Beecher, 1884; 2d Geol. Survey Penn., vol. PPP., p. 13.
Devonian.
TROPIDOCARIS Beecher, 1884; 2d Geol. Rep. Penn., vol. PPP, p. 15.
Wrinkled. Devonian.
PTYCHOCARIS Novák, 1885; Sitzungsb. böhm. Gesell., 1885, p. 343.
Wrinkled. Silurian.
PHASGANOCARIS Novák, 1886; Sitzungsb. böhm. Gesell., 1886, p. 498.
Only telson known. Silurian.
IV. With swellings in the antero-ventral region; ocular tubercle not apparent.
CALLIZOE Barr., 1872; Syst. Sil. Bohême, vol. 1, suppl., p. 503. Leperditoid. Silurian.
V. Conchiferoidal.
LINGULOCARIS Salt., 1866; Mem. Geol. Survey, vol. 3, pp. 252, 253, and 294.
Modioloid, and faintly ridged. Tremadoc.
STRIGOCARIS Vogd., 1889. Pod-shaped and concentrically marked.
Carboniferous.
SOLENOCARIS Young, 1868; Proc. Nat. Hist. Soc. Glasgow, vol. 1, p. 171.
? ORTHONOTELLA Ulrich, 1882. Oblong. Silurian.
MYOCARIS Salt., 1864; Quart. Jour. Geol. Soc. London, vol. 20, 1864, p. 292.
Quadrangular and ridged obliquely. Silurian or Devonian.

PINACARIDÆ.

MESOTHYRA Hall, 1888; Palæont. New York, vol. 7, pp. lvi, 187. Hamilton and Portage formations.

RHINOCARIDÆ.

RHINOCARIS Clarke, 1888; Palæont. New York, vol. 7, p. lviii. Anterior extremity of the head produced into a point. Devonian.

NOT CLASSIFIED.

PROTOCARIS Wal., 1885; Bull. U. S. Geol. Survey, No. 10, p. 50. Taconic.
ANOMALOCARIS Whiteaves, 1892; Canadian Rec. Sci., vol. 5, p. 205.
Taconic.
ROSTROCARIS Kinnear, 1887; Trans. Edinb. Geol. Soc., vol. 5, p. 417.
ARCHÆOCARIS Meek, 1871; Proc. Acad. Nat. Sci. Phila., p. 335.
Carboniferous.
DICTYOCARIS Salt., 1860; Annals Mag. Nat. Hist. London, 3d series, vol. 3, p. 101.
ACANTHOCARIS Peach, 1882; Trans. Royal Soc. Edinb., vol. 30, p. 511.

PHYLLOPODA.

LEAIA Jones, 1882; Mon. Fossil Estheria, p. 115. Quadrangular, ridged obliquely, and concentrically marked. Carboniferous.
ESTHERIA Rüppell, 1857; Mus. Senckenberg, vol. 2, p. 119. Like a bivalved mollusc and concentrically marked. Devonian. Carboniferous. Triassic. Rhætic. Jurassic. Cretaceous. Tertiary—Recent.
ESTHERIELLA Weiss, 1875; Zeitsch. Geol. Gesell. Berlin, vol. 27, p. 711.
SCHIZODISCUS Clarke, 1888; Palæont. New York, vol. 7, p. 207. Devonian.

BIVALVED ENTOMOSTRACA.

CYPRIDINIDÆ.

CYPRIDINA Milne-Edw., 1838; Lamarck's Anim. s. Vert., vol. 5, p. 178,
Carboniferous—Recent.
CYPRIDINELLA J. & K., 1874; Mon. Carboniferous Entom., p. 21.
Carboniferous.
SULCUNA Jones, 1873; Quart. Jour. Geol. Soc. London, vol. 29, p. 411.
Carboniferous.
CYPRIDELLA De Kon., 1841; Mém. Acad. Royal Bruxelles, vol. 14, p. 19.
Carboniferous.
BRADYCINETUS Sars, 1865; Oversigt af Norges Marine Ostracoder, p. 109.
Devonian—Recent.
PHILOMEDES Lilljeborg, 1853; Crust. in Scania Occurrentibus, p. 175.
Devonian—Recent.
RHOMBINA J. & K., 1874; Mon. Carboniferous Entom., p. 43.
Carboniferous.
CYPRELLA Dekon., 1841; Mem. Acad. Roy. Bruxelles, vol. 14.
Carboniferous.

ENTOMOCONCHIDÆ.

ENTOMOCONCHUS McCoy, 1839; Jour. Geol. Soc. Dublin, vol. 2, p. 91.
Carboniferous.
OFFA J. & K., 1874; Mon. Carboniferous Entom., p. 53. Carboniferous.

POLYCOPIDÆ.

POLYCOPE Sars, 1865; Oversigt af Norges Marine Ostracoder, p. 121.
Cambrian. Carboniferous—Recent.

ENTOMIDIDÆ.

ENTOMIS Jones, 1861; Mem. Geol. Survey Scotland, Expl. map 32, p. 137.
Silurian. Devonian. Carboniferous.
ENTOMIDELLA Jones, 1873. Annals Mag. Nat. Hist., London, 4th series,
vol. 11, p. 416. Silurian.
BOLBOZOE Barr., 1872; Syst. Sil. Bohême, vol. 1, suppl., p. 500. Étage
E. & G. Silurian.

CYTHERELLIDÆ.

CYTHERELLA Jones & Bosquet, 1848 and 1852. Silurian—Recent.
CYTHERELLINA Jones & Holl, 1869; Annals Mag. Nat. Hist., London, 4th
series, vol. 3, p. 215. Silurian.
CYTHERINA Lamarck, 1818; Animaux sans Vert., vol. 5, p. 125.
ÆCHMINA Jones & Holl, 1869; Annals Mag. Nat. His., London, 4th series,
vol. 3, p. 217. Silurian.

ISOXYSIDÆ.

ISOXYS Wal., 1891; 10th Rep. U. S. Geol. Survey, p. 625. Olenenus zone.
Taconic.

LEPERDITIDÆ.

LEPERDITIA Rouault, 1851; Bull. Soc. Géol. France, 2d series, vol. 8, p. 377.   Cambrian. Carboniferous.
ISOCHILINA Jones, 1858; Annals Mag. Nat. Hist., London, 3d series, vol. 1, p. 248.   Cambrian. Devonian.
JONESELLA Ulrich, 1890; Jour. Cin. Soc. Nat. Hist., vol. 13, p. 121.
LEPIDITTA Matt., 1885; Trans. Roy. Soc. Canada, vol. 3, p. 62. Paradoxides zone.   Taconic.

BEYRICHIDÆ.

BEYRICHIA McCoy, 1846; Synopsis Sil. Foss. Ireland, p. 57.
    Cambrian. Carboniferous.
BEYRICHIELLA J. & K., 1886; Geol. Mag., new series, decade 3, vol. 3, p. 438.   Carboniferous.
BEYRICHIOPSIS J. & K., 1886; Geol. Mag., decade 3, vol. 3, p. 434.
    Carboniferous.
KLŒDENIA Jones and Holl, 1886; Annals Mag. Nat. Hist., London, 5th series, vol. 17, p. 362.   Silurian.
HIPPONICHARION Matt., 1885; Trans. Royal Soc. Canada, vol. 3, p. 64. Paradoxides zone.   Taconic.
BEYRICHINA Matt., 1885; Trans. Roy. Soc. Canada, vol. 3, p. 65. Paradoxides zone.   Taconic.
ULRICHIA Jones, 1890; Quart. Jour. Geol. Soc. London, vol. 46, p. 543.
    Devonian. Carboniferous.
DEPRANELLA Ulrich, 1890; Jour. Cin. Soc, Nat. His., vol. 13, p. 117.
    Cambrian.
CTENOBOLBINA Ulrich, 1890; Jour. Cin. Soc. Nat. His., vol. 13, p. 108.
    Silurian. Devonian.
TETRADELLA Ulrich, 1890; Jour. Cin. Soc. Nat. His., vol. 13, p. 112.
    Cambrian.
HIPPA Barr., 1872, Syst. Sil. Bohême, vol. 1, suppl., p. 516.

PRIMITIDÆ.

PRIMITIA Jones and Holl, 1865; Annals Mag. Nat. Hist., London, 3d series, vol. 16, p. 416.   Cambrian. Carboniferous.
PRIMITOPSIS Jones, 1887; Notes on some Sil. Ostracoda from Gothland, p. 5.   Silurian. Devonian.
SCHMIDTELLA Ulrich, 1892; The Am. Geologist, vol. 10, p. 269.   Cambrian.

APARCHITES.

APARCHITES Jones, 1889; Annals Mag. Nat. Hist., 6th series, vol. 3, p. 385.   Cambrian. Silurian. Devonian.
EURYCHILLINA Ulrich, 1889; Contrib. Micro. Palæont., part 2, p. 52.
    Cambrian.
HALLIELLA Ulrich, 1890; Jour. Cin. Soc. Nat. Hist., vol. 13, p. 184,
    Devonian.

KIRKBYIDÆ.

KIRKBYA Jones, 1869; Annals Mag. Nat. Hist., London, 4th series, vol. 3, p. 223. Silurian. Permian.
STREPULA Jones and Holl, 1886; Annals Mag. Nat. Hist., London, 5th series, vol. 17, p. 403. Cambrian. Devonian.
TETRADELLA Ulrich, 1890; Jour. Cin. Soc. Nat. Hist., vol. 13, p. 11.
MOOREA J. and K., 1867; Annals Mag. Nat. Hist., London, 4th series, vol. 3, p. 225. Silurian. Carboniferous.
BOLLIA Jones and Holl, 1886; Annals Mag. Nat. Hist., London, 5th series, vol. 17, p. 360. Cambrian. Carboniferous.
PLACENTURA Jones and Holl, 1886; Annals Mag. Nat. Hist., London, 5th series, vol. 17, p. 407. Cambrian.

BURSULELLIDÆ.

BURSULELLA Jones, 1887; Notes on some Silurian Ostracoda from Gothland, p. 7. Silurian.
KYAMODES Jones, 1888; Annals Mag. Nat. Hist., London, 6th series, vol. 2, p. 295. Devonian.
BARYCHILINA Ulrich, 1889; Jour. Cin. Soc. Nat. Hist., vol. 13, p. 198. Devonian.
CYPROSIS Jones, 1881; Geol. Mag., decade 2, vol. 8, p. 338.
CYPROSINA Jones, 1881; Geol. Mag., decade 2, vol. 8, p. 338.
OCTONARIA Jones, 1887; Annals Mag. Nat. Hist., London, 5th series, vol. 19, p. 404. Silurian. Devonian.
ELPE Barrande, 1872; Syst. Sil. Bohême, vol. 1, suppl., p. 510.
THLIPSERA Jones and Holl, 1869; Annals Mag. Nat. Hist., 4th series, vol. 3, p. 213. Silurian.
PHREATURA J. and K., 1886; Quart. Jour. Geol. Soc. London, vol. 42, p. 507. Carboniferous.
PACHYDOMELLA Ulrich, 1890; Jour. Cin. Soc. Nat. Hist., vol. 13, p. 197. Devonian.

CYPRIDÆ.

AGLAIA Brady, 1867; Challenger Exped., Rep. on Ostracoda, p. 33.
CANDONA Baird, 1850; History Brit. Entomostraca, p. 159.
ARGILLŒCIA Sars, 1865; Oversigt af Norges Marine Ostracoder, p. 17.
MACROCYPSIS Brady, 1867; Intellectual Observ., vol. 12, p. 119.
BATHOCYPSIS Brady, 1880; Challenger Exped., Rep. on Ostracoda, p. 45.
BAIRDIA McCoy, 1844; Synopsis Carboniferous Foss. Ireland, p. 164.
PONTOCYPRIS Sars, 1865; Oversigt af Norges Marine Ostracoder, p. 13.

DARWINULIDÆ.

DARWINULA (Brady and Robertson) Jones, 1885; Quart. Jour. Geol. Soc., vol. 41, p. 346- Name Changed from preoccupied term Darwinella (Ann. Mag. Nat. Hist., 4th series, vol. 9, 1872, p. 50). Carboniferous.

CYTHERIDÆ.

CYTHERE Müller, 1785; Entomostraca, etc., p. 64.
XESTOLEBERIS Sars, 1865; Oversigt af Norges Marine Ostracoder, p. 66.
BYTHOCYTHERE Sars, 1865; Oversigt af Norges Marine Ostracoder, p. 82.
CARBONIA Jones, 1870; Geol. Mag., vol. 7, p. 218. Carboniferous.
YOUNGIA J. & K., 1886; Quart. Jour. Geol. Soc. London, vol 42, p. 507.
BERNIX Jones, 1884; Proc. Berwickshire Nat. Club, vol. 10, p. 316.
Carboniferous.

NOT CLASSIFIED.

CYTHEROPSIS (McCoy) Barr., 1872; Syst. Sil. Bohème, vol. 1, suppl., p. 508.
ZONOZOE Barr., 1872; Syst. Sil. Bohème, vol. 1, suppl., p. 554.
CARYON Barr., 1872; Syst. Sil. Bohème, vol. 1, suppl., p. 505.

ISOPODA.

ARTHROPLEURA Jordon, 1854; Ueber Crust. Steink. Saarrb., p. 13.
Carboniferous.
PRÆARCTURUS H. W., 1870; Trans. Woolhope Nat. Field Club, p. 266.
Old Red Sandstone.

CIRRIPEDÆ.

STROBILEPIS Clarke, 1888; Palæont. New York, vol. 7, p. 212. Devonian.
PALÆOCREUSIA Clarke, 1888; Palæont. New York, vol. 7, p. 210. Devonian.
PROTOBALANUS Whitf., 1888; Palæont. New York, vol. 7, p. lxii; Bull. Am. Mus. Nat. Hist., vol. 2, 1889, p. 66. Devonian.
TURRILEPAS H. W., 1865; Quart. Jour. Geol. Soc. London, vol. 21, p. 486.
Devonian.

DECAPODA.

ANTHRAPALÆMON Salt., 1861; Quart. Jour. Geol. Soc. London, vol. 17, p. 529. Carboniferous.
PALÆOPALÆMON Whitf., 1880; Am. Jour. Sci., 3d series, vol. 19, p. 40.
Upper Devonian.
CRANGOPSIS Salt., 1863; Quart. Jour. Geol. Soc. London, vol. 19, p. 80.
Carboniferous.
PSEUDO-GALATHEA Peach, 1882; Trans. Royal Soc. Edinb., vol. 30, p. 573.
Carboniferous.

SCHIZOPODA.

GAMPSONYX Meyer, 1847; Verhandl. Nat. Vereins Preuss., vol. 4, p. 86.
Carboniferous.
PALÆOCARIS M. & W., 1865; Proc. Acad. Nat. Sci. Phila., p. 48.
Carboniferous.
NECTOTELSON Brocchi, 1880; Bull. Soc. Géol. France, 3d series, vol. 8, p. 10.
PALÆORCHESTIA Zittel, 1885; Handb. Palæontologie, 1 Bd., 2 Abth., 4 Lieferung, p. 673, fig. 858. Carboniferous.

AMPHIPODA.

DIPLOSTYLUS Salt., 1863; Quart. Jour. Geol. Soc. London, vol. 19, p. 76.
Coal Measures.

STROMAPODA.

AMPHIPELTIS Salt., 1863; Quart. Jour. Geol. Soc. London, vol. 19, p. 75.
Devonian.

CATALOGUE OF NORTH AMERICAN NON-TRILOBITES.

ACANTHOTELSON M. & W., 1865; Proc. Acad. Nat. Sci. Phil., p. 46; Geol.
Survey Illinois, vol. 2, 1866, p. 299. Coal Measures.
—— EVENI M. & W., 1868; Am. Jour. Sci., 2d series, vol. 46, p. 28.
Mazon Creek, Grundy Co., Illinois. Coal Measures.
—— EVENI M. & W., 1868; Geol. Survey Illinois, vol. 3, p. 551, figs. a–d.
—— EVENI White, 1884; 13th Rep. Dept. Geol. Nat. Hist., Indiana, p. 177,
pl. 38, figs. 4–7.
—— EVENI Pack., 1886; Mem. Natl. Acad. Sci., vol. 3, p. 125.
—— *inequalis.* (See PALÆOCARIS TYPUS.)
—— ? MAGISTER Pack., 1886; Mem. Natl. Acad. Sci., vol. 3, p. 127, pl. 1,
fig. 2; pl. 2, figs. 4, 5. Mazon Creek, Grundy Co., Illinois.
Carboniferous.
—— STIMPSONI M. & W., 1865; Proc. Acad. Nat. Sci. Phil., vol. 17, p. 47.
Packard considers this species to be the young of *Acanthotelson
eveni.*
ÆCHMINA J. & H., 1887. Annals Mag. Nat. Hist., 5th series, vol. 3,
p. 217.
—— ABNORMIS Ulrich, 1890; Jour. Cincinnati Soc. Nat. Hist., vol. 13, p.
183, pl. 12, figs. 7, 7 a–b. Lockport, N. Y. Niagara.
—— BYRNESI (Miller) Jones, 1890; Quart. Jour. Geol. Soc., vol. 40, p. 12,
pl. 3, figs. 9–11. Cincinnati, Ohio. Cambrian.
For other references, see *Leperditia byrnesi* Miller.
—— MARGINATA Ulrich, 1890; Jour. Cincinnati Soc. Nat. Hist., vol. 13, p.
148, pl. 16, fig. 5. 18 Mile Creek, N. Y. Hamilton.
—— SPINOSA (Hall) Jones, 1890; Quart. Jour. Geol. Soc., vol. 46, p. 11,
pl. 3, figs. 4–8. Lockport, N. Y. Niagara.
For other references, see *Cytherina spinosa* Hall.
AMPHIPELTIS Salt., 1863; Quart. Jour. Geol. Soc. London, vol. 19, p. 75.
—— PARADOXUS Salt., 1863; Quart. Jour. Geol. Soc. London, vol. 19, p.
76, figs. 11, a b. St. John, N. B. Devonian.
—— PARADOXUS Dawson, 1868; Acadian Geol., p. 523, fig. 180.
—— PARADOXUS S. & W.; Chart of Fossil Crustacea, pl. 3, fig. 3.
ANOMALOCARIS Whiteaves, 1892; Canadian Rec. of Sci., vol. 5, p. 205.
—— CANADENSIS Whiteaves, 1892; Canadian Rec. of Sci., vol. 5, p. 206, fig.
Mount Stephen, British Columbia. Taconic.

ANTHRAPALÆMON. APTYCHOPSIS.

Anthraconectes M. & W., 1868; Am. Jour. Sci., 2d series, vol. 46, p. 21.
—— *mazonensis* M. & W. (See *Eurypterus mazonensis*.)
ANTHRAPALÆMON Salt., 1861; Quart. Jour. Geol. Soc. London, vol. 17, p. 529.
—— GRACILIS M. & W., 1865; Proc. Acad. Nat. Sci. Phila., vol. 17, p. 50. Mazon Creek, Grundy county, Illinois. Coal Measures.
—— GRACILIS M. & W., 1866; Geol. Survey Illinois, vol. 2, p. 407, pl. 32, figs. 4 a, b, c.
—— GRACILIS M. & W., 1868; Geol. Survey Illinois, vol. 3, p. 554, figs. a, b.
—— GRACILIS White, 1884; 13th Rep. Dept. Geol. Nat. Hist. Indiana, p. 180, pl. 38, figs. 8, 9.
—— GRACILIS Pack., 1885; American Naturalist, vol. 19, p. 880.
—— GRACILIS Pack., 1886; Mem. Natl. Acad. Sci., vol. 3, p. 135, pl. 4, figs. 1, 2, 3, 5, 6; pl. 7, figs. 3–6.
—— (*Palæocarabus*) HILLIANUS Daw., 1878; Geol. Mag., vol. 4, p. 56, fig. 1. Coal Measures.
—— (*Palæocarabus*) HILLIANUS Daw., 1878; Suppl. Acadian Geol., p. 55, fig. 10. South Joggins, Nova Scotia.
APARCHITES Jones, 1889; Annals Mag. Nat. Hist., 6th series, vol. 3, p. 385.
—— CONCINNUS Jones. This species was figured and described as *Cytheropsis concinnus*; Ann. Mag. Nat. Hist. Apl., 1858, p. 254, pl. 10, figs. 3, 4. Jones refers it to Aparchites; Contrib. Micro.Palæont., 1891, p. 99. Panquett's Rapids, Allumette Island, Ottawa River. Trenton.
—— INORNATUS Ulrich, 1890; Jour. Cincinnati Soc. Nat. His., vol. 13, p. 182, pl. 16. figs. 3, 3 a–b. Falls of the Ohio. Devonian.
—— METIS Jones, 1891; Contrib. Micro-Palæont., p. 3, p. 91, pl. 11, figs. 15 a–b. Athabasca River. Devonian.
—— METIS Whiteaves; Contrib. Canadian Palæont., vol. 1, p. 246.
—— MINUTISSIMUS (Hall) Ulrich, 1889; Contrib. Micro-Palæont., pt. 2, p. 49. For other references, see *Leperditia* (*Isochilina*) *minutissima* Hall. Stony Mountain, Manitoba. Hudson.
—— MUNDULUS Jones, 1891; Contrib. Micro-Palæont., pt. 3. p. 62, pl. 10, fig. 12 a–b. Province of Quebec. Chazy?
—— OBLONGUS Ulrich, 1890; Jour. Cincinnati Soc. Nat. Hist., vol. 13, p. 137, pl. 10, figs. 10, 10 a. Middletown, Ohio. Hudson.
—— UNICORNIS Ulrich, 1889; Contrib. Micro-Palæont., pt. 2, p. 50, pl. 9, fig. 11. For other references, see *Leperditia unicornis*. Stony Mountain, Manitoba. Hudson.
—— TYRRELLII Jones, 1891; Contrib. Micro-Palæont, pt. 3, p. 62, pl. 13, figs. 14 a, b, c. Black Island, Lake Winnipeg. Chazy.
APTYCHOPSIS Barr., 1872; Syst. Sil. Bohème, vol. 1, suppl., p. 436.

CATALOGUE OF NON-TRILOBITES. 371

ARCHÆOCARIS Meek, 1871; Proc. Acad. Nat. Sci. Phila., p. 335.
—— VERMIFORMIS Meek, 1871; Proc. Acad. Nat. Sci. Phila., p. 335. Danville, Ky. Waverly.
ARISTOZOE Barr., 1872; Syst. Sil. Bohême, suppl., vol. 1, p. 474.
—— ROTUNDATA Wal., 1887; Am. Jour. Sci., 3d series, vol. 34, p. 193, pl. 1, fig. 9. 2 miles southeast of North Greenwich, N. Y. Taconic.
—— ROTUNDATA Wal., 1890; 10th Rep. U. S. Geol. Survey, p. 627, pl. 70, fig. 3. Olenellus zone.
Leperditia TROYENSIS Ford, 1873; Am. Jour. Sci., 3d series, vol. 6, p. 138.
—— TROYENSIS (Ford) Wal., 1887; Am. Jour. Sci., 3d series, vol. 34, p. 193, pl. fig. 8. Troy and 1½ miles west North Greenwich.
Taconic.
—— TROYENSIS Wal., 1890; 10th Rep. U. S. Geol. Survey, p. 628, pl. 70, figs. 2, 2 a. Olenellus zone.
ARISTOZOE sp. Shaler and Foerste, 1888; Bull. Mus. Comp. Zoology, vol. 16, p. 35, pl. 2, fig. 18. Attleborough, Mass. Taconic.
BAIRDIA McCoy, 1844; Synopsis Carb. Fossils Ireland, p. 164.
—— ANTICOSTIENSIS Jones, 1890; Quart. Jour. Geol. Soc. London, vol. 46, p. 548, pl. 21, figs. 3, 3 a-b. English Head, Anticosti. Hudson.
—— CHESTERENSIS Ulrich, 1890; Jour. Cincinnati Soc. Nat. Hist., vol. 13, p. 210, pl. 17, figs. 6, 6 a-c, and 7, 7 a-b. Grayson Springs, Ky.
Chester.
—— LEGUMINOIDES Ulrich, 1890; Jour. Cincinnati Soc. Nat. Hist., vol. 13, p. 197, pl. 17, figs. 5, 5 a-c. 18 Mile Creek, N. Y. Hamilton.
BARYCHILINA Ulrich, 1890; Jour. Cincinnati Soc. Nat. Hist., vol. 13, p. 198.
—— PULCHELLA Ulrich, 1890; Jour. Cincinnati Soc. Nat. Hist., vol. 13, p. 199, pl. 13, figs. 4, 4 a-d. Falls of the Ohio. Devonian.
—— PUNCTO-STRIATA Ulrich, 1890; Jour. Cincinnati Soc. Nat. Hist., vol. 13, p. 199, pl. 13, figs. 1, 1 a-e, and 2; also var. *curta.*, p. 199, pl. 13, figs. 3, 3 a-c. Falls of the Ohio. Devonian.
BELINURUS Koenig, 1825; Icones. Foss. Sectiles, p. 18, fig. 230; Baily, Annals Mag. Nat. Hist., London, 3d series, vol. 11, 1863.
—— *danæ* M. & W. (See PRESTWICHIA DANÆ.)
—— LACŒI Pack., 1885; American Naturalist, vol. 19, p. 292. Mazon Creek, Illinois. Carboniferous.
—— LACŒI Pack., 1886; Mem. Natl. Acad. Sci., vol. 3, p. 149, pl. 5, fig. 5.
BEYRICHIA McCoy, 1846; Silurian Foss. Ireland, p. 58.
—— ÆQUILATERA Hall, 1860; Canadian Naturalist, vol. 5, p. 158, fig. 20.
Silurian.
—— ÆQUILATERA Dawson, 1868; Acadian Geology, p. 609, fig. 217.
—— ÆQUILATERA Jones, 1890; Quart. Jour. Geol. Soc., vol. 46, p. 18, pl. 2, fig. 6.
—— ÆQUILATERA Jones, 1891; Contrib. Micro-Palæont., pt. 8, p. 72, pl. 11, fig. 6. Arisaig, Nova Scotia.

BEYRICHIA.
—— ATLANTICA Bill., 1865; Pal. Foss., vol. 1, p. 300. Table Head, N. F.
—— BELLA Wal., 1883; Description New Species Trenton Group, p. 7, pl. 17, figs. 11, 11 a. Trenton Falls, N. Y. Trenton.
—— BELLA Wal., 1884; 35th Rep. New York State Mus. Nat. Hist., p. 213 pl. 17, figs. 11, 11 a.
—— BUCHIANA? Jones, 1890; Quart. Jour. Geol. Soc., vol. 46, p. 16, pl. 3, fig. 25. Cincinnati, Ohio. Cambrian.
—— CHAMBERSI Miller, 1874; Cincinnati Quart. Jour. Sci., vol. 1, p. 234, fig. 27. Cincinnati, Ohio. Hudson.
—— CHAMBERSI H. & W., 1875; Pal. Ohio, vol. 2, p. 104, pl. 4, figs. 11, 12.
—— CILIATA Emmons, 1855; American Geology, pt. 2, p. 219, fig. 74 c. Cincinnati, Ohio. Hudson.
—— CILIATA Miller, 1875; Cincinnati Quart. Jour. Sci., vol. 2, p. 351.
—— CILIATA Jones, 1890; Quart. Jour. Geol. Soc., vol. 46, p. 19, pl. 3, figs. 12-16; pl. 4, figs. 16-18.
Syn., *Beyrichia tumifrons* Hall. This species is used by Ulrich for the type of his new genus *Ctenobolbina*.
—— CINCINNATIENSIS Miller, 1875; Cincinnati Quart. Jour. Sci., vol. 2, p. 350, fig. 25. Weisburg, Ohio. Hudson.
—— CINCINNATIENSIS Wal., 1876; Trans. Albany Inst., vol. 10, p. 23.
—— CLARKEI Jones, 1890; Quart. Jour. Geol. Soc., vol. 46, p. 17, woodcut, fig. 2. Herkimer Co., N. Y. Lower Helderberg.
—— CLATHRATA Jones, 1858; Annals Mag. Hist., London, 3d series, vol. 1, p. 242, pl. 9, fig. 1. Niagara.
—— CLAVIGERA Jones, 1891; Contrib. Micro-Palæont, pt. 3, p 65, pl. 11, fig. 7, Quebec City; var. *clavifracta*, Quebec City, pl. 11, fig. 8.
—— DAGON Clarke,.1885; Bull. U. S. Geol. Survey, No. 16, p. 29, pl. 2, figs. 6, 7. Bristol Center, N. Y. Genesee.
—— DECORA Bill., 1866; Catalogue Sil. Foss. Anticosti, p. 67.
—— DIFFUSA Jones, 1890; Quart. Jour. Geol. Soc. London, vol. 46, p. 546, pl. 21, fig. 7. Jupiter River, Anticosti. Silurian.
—— DURYI Miller, 1874; Cincinnati Quart. Jour. Sci., vol. 1, p. 232, figs. 24, 25. Cincinnati, Ohio. Hudson.
—— FŒTOIDEA White and St. John, 1868; Trans. Chicago Acad. Sci., vol. 1, p. 126, figs. 11 a, b. Union and Page Co., Iowa. Niagara.
—— GRANULOSA Hall, 1876; 28th Rep. New York State Mus. Nat. Hist., Doc. Ed. Expl., pl. 32, fig. 4. Waldron, Decatur Co., Indiana.
Niagara.
—— GRANULOSA Hall, 1876; 28th Rep. New York State Mus. Nat. Hist., p. 186, pl. 32, fig. 4.
—— GRANULOSA Hall, 1883; 11th Annual Rept. Dept. Geol. Nat. Hist. Indiana, p. 331, pl. 34, fig. 4.
—— GRANULOSA Jones, 1890; Quart. Jour. Geol. Soc., vol. 46, p. 15, pl. 1, fig. 3. Schoharie county, N. Y.

BEYRICHIA.
- GRANULATA Hall, 1859; Pal. New York, vol. 3, p. 377, pl. 79 b, figs. 1 a-d. Lower Helderberg.
- HALLI Jones, 1890; Quart. Jour. Geol. Soc., vol. 46, p. 15, pl. 4, fig. 21. Utica, N. Y. Waterlime.
- HAMILTONENSIS Jones, 1890; Quart. Jour. Geol. Soc., vol. 46, p. 19, pl. 2, fig. 3. 18 Mile Creek, N. Y. Hamilton.
- JONESII Daw., 1868; Acadian Geol., p. 313, fig. 132. Carboniferous.
- KLOEDENI Jones, 1890; Quart. Jour. Geol. Soc. London, vol, 46, p. 538, pl. 21, figs. 1, 1 a-b. Ontario Co., N. Y. Devoni n.
- KOLMODINI Jones, 1890; Quart. Jour. Geol. Soc. London, vol. 46, p. 538, pl. 21, fig. 6. Clarke Co., Indiana. Devonian.
- (*Deparnella*) KALMODINI Ulrich, 1890; Jour. Cincinnati Soc. Nat. Hist., vol. 13, p. 190, pl. 14. figs. 1, 1 a-b. Falls of the Ohio. Hamilton.
- *lata*. (See *Bollia lata* Vanuxem.)
- *lithofactor* White and St.John; Prelim. Notice New Genera and Species Foss., May 8, 1867, p. 2. Changed to *Beyrichia petrifactor*, which see. Coal Measures.
- *lithofactor* var. *velata*; idem, p. 2. (See *Beyrichia petrifactor* var. *velata*.)
- *logani* Jones, 1858; Annals Mag. Nat. Hist., London, 3d series, vol. 1, p. 244, pl. 9, figs. 6-10. Chazy.
- *logani* Jones, 1858, Geol. Survey Canada, decade 3, p. 91, pl. 11, figs. 1-5. This with its varieties *reniformis*, fig. 1, and *leperditoides*, fig. 5, was referred to the genus *Primitia* in Ann. Mag. Nat. Hist., series 3, vol. 16, 1865, pp. 416-417.
- *logani* var. *reniformis* Jones, 1858; Geol. Survey Canada, decade 3, p. 91, pl. 11. fig. 1. (See *Primitia logani*.)
- *logani* var. *leperditoides* Jones, 1858; Geol. Survey Canada, decade 3, p. 91, pl. 11, fig. 5. (See *Primitia logani*.)
- LYONI Ulrich, 1890; Jour. Cincinnati Soc. Nat. His., vol. 13, p. 190, pl. 14, figs. 3, 3 a-c and 3. Falls of the Ohio. Devonian.
- - MACCOYANA Jones, 1855; Annals Mag. Nat. Hist., London, 2d series, vol. 16, p. 88, pl. 5, fig. 14. Silurian.
- MACCOYANA Jones, 1858; Annals Mag. Nat. Hist., London, 3d series, vol. 1, p. 252, pl. 10, fig. 15.
- MACCOYANA Jones, 1858; Geol. Survey Penn., vol. 2, p. 834, fig. 695.
- *mundula* Jones, 1855; Ann. Mag. Nat. History, 2d series, vol. 16, p. 90. (See *Primitia mundula*.)
- *notata*. (See *Klœdenia notata* (Hall) Jones.)
- NOVA SCOTICA J. & K., 1884; Geol. Mag., 5th series, vol. 1, p. 357, pl. 12, figs. 5, 6. Horton, Nova Scotia. Carboniferous.
- sp. Dawson, 1868; Acadian Geol., p. 256, fig. 78 c.
- (*Primitia*) OCCIDENTALIS Wal., 1884; Pal. Eureka Dist., Mon. U. S. Geol. Survey, vol. 8, p. 204, pl. 17, figs. 4, 4 a. Newark Mt., Eureka Dist., Nevada. Devonian.

BEYRICHIA.
— OCULIFERA Hall, 1871; Pamphlet Cincinnati Group, p. 8.   Cincinnati, Ohio.   Hudson.
— OCULIFERA Hall, 1872; 24th Rep. New York State Mus. Nat. Hist., p. 232, pl. 8, figs. 9, 10.
— OCULIFERA H. & W., 1875; Pal. Ohio, vol. 2, p. 103, pl. 4, figs. 9, 10.
— OCULIFERA Miller, 1874; Cincinnati Quart. Jour. Sci., vol. 1, p. 118.
— OCULIFERA Jones, 1890; Quart. Jour. Geol. Soc., vol. 46, p. 21, pl. 4, figs. 19 a, b, 20. Ulrich classes this species under the new genus *Tetradella*.
— OCULINA Hall, 1859; Pal. New York, vol. 3, p. 378, pl. 79 b, figs. 2 a-e. Schoharie Co., N. Y.   Lower Helderberg.
— OCULINA Jones, 1890; Quart. Jour. Geol. Soc., vol. 46, p. 16, pl. 1, fig. 4.   Schoharie Co., N. Y.
— ? (*Primitia*) PARALLELA Ulrich, 1890; Jour. Cincinnati Soc. Nat. Hist., vol. 13, p. 125, pl. 10, figs. 15, 15 a-c and d. Oxford, Ohio.
Cambrian.
— PARASITICA (Hall) Jones, 1890; Quart. Jour. Geol. Soc., vol. 46, p. 16, wood cut. For other references see *Leperditia parasitica*. Herkimer Co., N. Y.   Tentaculite limestone.
— PENNSYLVANICA Jones, 1858; Annals Mag. Nat. Hist., London, 3d series, vol. 1, p. 253, pl. 10, fig. 16-18.   Onondaga.
— PENNSYLVANICA Jones, 1858; Geol. Survey Penn., vol. 2, p. 834, fig. 696.   Near Barre Forge, Penn.
— *persulcata* Ulrich, 1879; Jour. Cincinnati Soc. Nat. Hist., vol. 2, p. 12, pl. 7, fig. 6. (See *Bollia persulcata*.) Covington, Ky. Hudson.
— PETRIFACTOR White and St. John; Trans. Chicago Acad. Sci., vol. 1, p. 125.
— PETRIFACTOR var. VELATA; *idem.*, p. 126. Near Pella, Marion Co., Iowa.   St. Louis.
— PLAGOSA Jones, 1858; Annals Mag. Nat. Hist., London, 3d series, vol. 1, p. 243, pl. 9, fig. 2.   Niagara.
— PUNCTULIFERA Hall, 1861; Des. New Species Fossils, p. 83. Ontario Co., N. Y.   Hamilton.
— PUNCTULIFERA Hall, 1862; 15th Rep. New York State Cab. Nat. Hist., p. 111.
— PUSTULOSA Hall, 1860; Canadian Naturalist, vol. 5, p. 157, fig. 19. Arisaig, Nova Scotia.   Silurian.
— PUSTULOSA Dawson, 1868; Acadian Geology, p. 608, fig. 216.
— PUSTULOSA Jones, 1890; Quart. Jour. Geol. Soc., vol. 46, p. 18, pl. 2, fig. 1 a-c. Referred to *Beyrichia tuberculata* Boll var. *pustulosa* Hall.
— *quadrilirata* H. & W., 1875; Pal. Ohio, vol. 2, p. 105, pl. 4, figs. 6, 7.
Hudson.
— *quadrilirata* Miller, 1875; Cincinnati Quart. Jour. Sci., vol. 2, p. 351. (See *Strepula quadrilirata*.)

BEYRICHIA.

—— QUADRIFIDA Jones, 1891; Contrib. Micro-Palæont., pt. 3, p. 66, pl. 11, fig. 19 a-b. Lorrett Falls, Quebec.
—— RADIATA Ulrich, 1890; Jour. Cincinnati Soc. Nat. Hist., vol. 13, p. 204, pl. 14, figs. 4, 4 a-b. Grayson Springs, Ky. Chester.
—— REGULARIS Emmons, 1855; American Geology, pt. 2, p. 219, fig. 74 b. Hudson.
—— REGULARIS Miller, 1875; Cincinnati Quart. Jour. Sci., vol. 2, p. 351. This species is placed as a synonym to *Bollia lata* by Jones (In Quart. Jour. Geol. Soc., vol. 46, p. 13).
—— RENIFORMIS Jones, 1858; Geol. Survey Canada, decade 3, p. 91, pl. 11, fig. 1. Referred to *Primitia logani.* Chazy.
—— RICHARDSONI Miller, 1874; Cincinnati Quart. Jour. Sci., vol. 1, p. 347, fig. 40. Wilmington, Clinton Co., Ohio. Hudson.
—— RUGULIFERA Jones, 1858; Annals Mag. Nat. Hist., London, 3d series, vol. 1, p. 242, pl. 9, fig. 4. Niagara.
—— SIGILLATA Jones, 1858; Annals Mag. Nat. Hist., London, vol. 1, p. 242, pl. 9, fig. 5. Niagara.
—— SIMPLEX Emmons, 1855; American Geology, vol. 1, pt. 2, p. 218, fig. 74 a.
—— SIMULATRIX Ulrich, 1890; Jour. Cincinnati Soc. Nat. Hist., vol 13, p. 205, pl., figs. 7, 7 a-b, cf. *B. foetoidea* W. and St. J. Grayson Springs, Ky. Chester.
—— *Cytherina* SPINOSA Hall, 1852; Pal. New York, vol. 2, p. 317, pl. 67, figs. 17-20. (See *Æchima spinosa.*) Lockport, N. Y. Niagara
—— *striato-marginatus* Miller, 1874; Cincinnati Quart. Jour. Sci., vol. 1, p. 233, fig. 26. 3 miles south of Osgood, Indiana. (See *Eurychilinda striato-marginatus.*) Hudson.
—— SUBQUADRATA Jones; Quart. Jour. Geol. Soc., vol. 46, p. 537, pl. 20, fig. 4. Ontario Co., N. Y. Devonian.
—— *symmetrica* Hall, 1852; Pal. New York, vol. 2, p. 317, pl. 67, fig. 16. (See *Bollia symmetrica*) Niagara.
—— TRICOLLINA Ulrich, 1890; Jour. Cincinnati Soc. Nat. Hist., vol. 13, p. 189, pl. 12, fig. 6. 18 Mile Creek, N. Y. Hamilton.
—— TRISULCATA Hall, 1859; Pal. New York, vol. 3, p. 381, pl. 79, figs. 5 a, b. Herkimer Co., N. Y. Lower Helderberg.
—— TRISULCATA Jones, 1890; Quart. Jour. Geol. Soc., vol. 46, p. 14, pl. 1, fig. 2.
—— TUBERCULATA (Kloeden) Jones, 1891; Contrib. Micro-Palæont., pt. 3, p. 74, pl. 11, fig. 3. Arisaig, Nova Scotia; var. *pustulosa*, pl. 11, fig. 2; var. *strictispiralis*, pl. 11, fig. 1; var. *nœtlingi*, pl. 11, figs. 4 a-b and 5.
—— TUMIFRONS Hall, 1871; Des. New Species Foss. Hudson River Group, p. 7. Hudson.
—— TUMIFRONS Hall, 1872; 24th Rep. New York State Mus. Nat. Hist., p. 231, pl. 8, fig. 11.

BEYRICHIA. BOLLIA.

—— TUMIFRONS H. & W., 1875; Pal. Ohio, vol. 2, p. 102, pl. 4, fig. 8.
—— TUMIFRONS Miller, 1874; Cincinnati Quart. Jour. Sci., vol. 1, p. 119. This species is classed by Jones, 1890; Quart. Jour. Geol. Soc., vol. 46, p. 19, as syn. *Beyrichia ciliata* Emmons.
—— VENUSTA Bill., 1866; Catalogue Silurian Fossils Anticosti, p. 68. East Point, Chaloupe River, Jumpers, etc., Div. 3, 4.
BEYRICHNONA Matt., 1885; Trans. Roy. Soc. Canada, vol. 3, p. 65.
—— PAPILIO Matt., 1885; Trans. Roy. Soc. Canada, vol. 3, p. 65, pl. 6, figs. 20 a-b. Paradoxides zone. Div. 1 b^2. Hanford brook, St. Martin's, N. B.
—— TINEA Matt., 1885; Trans. Roy. Soc. Canada, vol. 3, p. 66, pl. 6, figs. 21, 21 a-b. Paradoxides zone. Div. 1 b^2. Hanford brook, St. Martin's, N. B.
BOLLIA Jones and Holl, 1886; Annals Mag. Nat. Hist., 5th series, London, vol. 17, p. 360.
—— BILOBATA Jones, 1890; Quart. Jour. Geol. Soc., vol. 46, p. 540, pl. 20, fig. 12. Ontario Co., N. Y. Devonian.
—— HINDEI Jones, 1890; Quart. Jour. Geol. Soc. London, vol. 46, p. 540, pl. 20, fig. 5. 18 Mile Creek, N. Y. Devonian.
—— GRANIFERA Ulrich, 1890; Jour. Cincinnati Soc. Nat. His., vol. 13, p. 205, pl. 12, figs. 12, 12 a-b. Elizabethtown, Ky. St. Louis.
Agnostus LATUS Van., 1842; Geol. New York, 3d Geol. Dist., p. 80.
Clinton.
Beyrichia LATA Hall, 1852; Pal. New York, vol. 2, p. 301, pl. A 66, figs. 10 a-c. New Hartford, Oneida Co., N. Y.
Beyrichia LATA Jones, 1858; Notes on Palæozic Entomostraca, No. 2, p. 168, pl. 6, fig. 13.
—— LATA Jones, 1890; Quart Jour. Geol. Soc., vol. 46, p. 12, pl. 3, figs. 1-3.
—— OBESA Ulrich, 1890; Jour. Cincinnati Soc. Nat. Hist., vol. 13, p. 189, pl. 16, figs. 5, 5 a-b. Falls of the Ohio. Devonian.
—— PERSULCATA Ulrich, 1890; Jour. Cincinnati Soc. Nat. Hist., vol. 13, p. 116, fig. 3. Covington, Ky. Cambrian.
—— PUMILA Ulrich, 1890; *idem*, p. 117, pl. 12, figs. 1, 1 a-b. Weisburg, Indiana. Cambrian.
—— SEMILUNATA Jones, 1890; Quart. Jour. Geol. Soc. London, vol. 46, p. 548, pl. 21, figs. 9, 9 a-b. Junction Cliff. Anticosti. Hudson.
Beyrichia SYMMETRICA Hall, 1852; Pal. New York, vol. 2, p. 317, pl. 67, fig. 16. Lockport, N. Y. Niagara.
—— SYMMETRICA Jones, 1890; Quart. Jour. Geol. Soc., vol. 46, p. 12. This species is closely allied to *Bollia lata*.
—— UNGULA (Claypole MS.) Jones, 1889; American Geologist, vol. 4, p. 338, pl., figs. 10-13. Perry Co., near Bloomfield, Penn. Marcellus.
—— UNGULA Ulrich, 1890; Jour. Cincinnati Soc. Nat. His., vol. 13, p. 188, pl. 16. Falls of the Ohio.

BUMODELLA Matt., 1888, Trans. Roy. Soc. Canada, vol. 6, p. 56.
—— HORRIDA Matt., 1888; Trans. Roy. Soc. Canada, vol. 6, p. 56, pl. 4, fig. 8. Near Westfield Station, N. B. Silurian.
BYTHOCYPRIS Brady, 1880; Challenger Exped., Rep. Ostracoda, p. 45.
—— CYLINDRICA (Hall) Ulrich, 1889; Contrib. Micro-Palæont., pt. 2, p. 48, pl. 9, fig. 6. Stony Mountain, Manitoba. For other references, see *Leperditia (Isochilina) cylindrica* Hall.
Trenton and Hudson.
—— DEVONICA Ulrich, 1890; Jour. Cincinnati Soc. Nat. Hist., vol. 13, p. 196, pl. 17, figs. 1 a, b, c. Falls of the Ohio. Devonian.
—— FAVULOSA Jones, 1889; American Geol., vol. 4, p. 338, pl., figs. 2 a-c. Perry Co., near Bloomfield, Penn. Marcellus.
—— INDIANENSIS Ulrich, 1890; Jour. Cincinnati Soc. Nat. His., vol. 13, p. 197, pl. 16, figs. 11, 11 a, b, c. Falls of the Ohio. Devonian.
—— LINDSTRŒMII Jones, 1890; Quart. Geol. Jour. Soc., vol. 46, p. 548, pl. 21, figs. 11, a-c. Junction Cliff. Anticosti. Cambrian.
—— OBTUSA Jones, 1890; Quart. Jour. Geol. Soc. London, vol. 46, p. 549, pl. 21, figs. 4, 4 a-b. Hudson.
—— OVIFORMIS Jones, 1889; American Geol., vol. 4, p. 340, pl., figs. 3 a-c. Perry Co., Penn. Lower Helderberg.
—— PUNCTULATA Ulrich, 1890; Jour. Cincinnati Soc. Nat. His., vol. 13, p. 196, pl. 17, figs. 2, 2 a-b. Falls of the Ohio. Devonian.
CANDONA Baird, 1850; Hist. Brit. Entomostraca, p. 159.
—— ELONGATA J. & K., 1884; Geol. Mag., decade 3, vol. 1, p. 361, pl. 12, fig. 10. Joggins, Nova Scotia. Coal Measures.
CARBONIA J. & K., 1879; Annals Mag. Nat. Hist., London, 5th series, vol. 4, p. 28.
—— ? BAIRDIOIDES? J. & K., 1884; Geol. Mag., 5th series, vol. 1, p. 358, pl. 12, figs. 8 a-d.
Syn., *Bairdia* sp.? Dawson, 1868; Acadian Geol., p. 206, figs. 48 a (?) *Cythere? (Carbonia)* BAIRDIOIDES J. & K., 1879; Annals Mag. Nat. Hist., London, 5th series, vol. 4, p. 38, pl. 3, figs. 24, 25.
—— ? BAIRDIOIDES? J. & K., 1884; Geol. Mag., decade 3, p. 358, pl. 12, figs. 8 a d. Joggins, Nova Scotia.
—— ? ELONGATA J. & K., 1884; Annals Mag. Nat. Hist., London, 6th series, vol. 1, p. 361, pl. 12, fig. 10. Joggins, Nova Scotia.
Coal Measures.
—— ? ELONGATA J. & K., 1884; Geol. Mag., decade 3, p. 361, pl. 12, fig. 10.
—— FABULINA J. & K., 1879; Annals Mag. Nat. Hist,, London, 5th series, vol. 4, p. 31, pl. 2, figs. 1-9. Joggins, Nova Scotia.
Coal Measures.
—— FABULINA var. HUMILIS J. & K., 1879; Annals. Mag. Nat. Hist., London, 5th series, vol. 4, p. 31, pl. 2, figs. 11-14. Joggins, Nova Scotia. Coal Measures.

BUMODELLA.   CERATIOCARIS.
—— FABULINA Jones, 1889; Geol. Mag., decade 3, vol. 6, p. 270, figs. 1-4. Syn., *Cytherilla inflata* Dawson, 1868; Acadian Geol., p. 206, fig. 48 b.
—— FABULINA J. & K., 1889; Geol. Mag., decade 3, vol. 1, p. 358, pl. 12, figs. 9 a-e.
CERATIOCARIS McCoy, 1849; Annals Mag. Nat. Hist., London, vol. 4, p. 412.
—— ACULEATUS Hall, 1859; Pal. New York, vol. 3, p. 422*, pl. 80 a, fig. 10. Waterville, N. Y.   Waterlime.
—— ACUMINATUS Hall, 1859; Pal. New York, vol. 3, p. 422*, pl. 84, fig. 6. Near Buffalo, N. Y.   Waterlime.
—— ACUMINATUS Pohl., 1886; Bull. Buffalo Soc. Nat. Sci., vol. 5, p. 28, pl. 3, fig. 2.
—— *armatus* Hall. (See *Echinocaris punctata*.)
—— BEECHERI Clarke, 1881; Bull. U. S. Geol. Survey, No. 16, p. 44, pl. 2, fig. 1. Cashaque Creek, Livingston Co., N. Y.   Devonian.
—— BEECHERI Hall, 1888; Pal. New York, vol. 7, p. 104, pl. 31, fig. 3.
—— (*Colpocaris*) BRADLEYI Meek, 1871; Proc. Acad. Nat. Sci. Phila., vol. 23, p. 332. Danville, Ky.   Waverly.
Onchus DEWEYI Hall, 1852; Pal. New York, vol. 2, p. 320, pl. 71, figs. 1 a-d. Lockport and Rochester, N. Y.   Niagara.
—— (*Colpocaris*) ETYTROIDES Meek, 1871; Proc. Acad. Nat. Sci. Phila., vol. 23, p. 334. Danville, Ky.   Lower Carboniferous.
—— GRANDIS Pohl., 1881; Bull. Buffalo Soc. Nat. Sci., vol. 4, p. 19, fig. 5. Buffalo, N. Y.   Waterlime.
—— LONGICAUDUS Hall, 1863; 16th Rep. New York State Cab. Nat. Hist., p. 70, pl. 1, fig. 7 (not figs. 4, 5, 6). Bristol, N. Y.   Hamilton.
—— LONGICAUDUS Pack., 1883; 12th Annual Rep. U. S. Geol. Survey Territories, p. 450.
—— LONGICAUDUS J. & W., 1884; Geol. Mag., decade 3, vol. 1, p. 1.
—— LONGICAUDUS Eth., W. & J., 3d Rep. Comm. Foss. Phyllopoda, p. 35.
—— LONGICAUDUS Hall, 1888; Pal. New York, vol. 7, p. 163, pl. 31, fig. 1. Not *C. longicaudus* Clarke, 1885; Bull. U. S. Geol. Survey, No. 16, p. 20.
—— MACCOYANA Hall, 1859; Pal. New York, vol. 3, p. 421, pl. 84, figs. 1-5. Near Buffalo, N. Y.   Waterlime.
—— PUSILLUS Matt., 1888; Trans. Roy. Soc. Canada, vol. 6, p. 56, pl. 4, fig. 9. Cunningham Brook, Westfield Station, N. B.   Silurian.
——? *punctatus* Hall. (See *Echinocaris punctata*.)
—— SIMPLEX Clarke, 1885; Bull. U. S. Geol. Survey, No. 16, p. 43, pl. 2, fig. 2. Naples, N. Y.   Devonian.
——? SIMPLEX Hall, 1888; Pal. New York, vol. 7, p. 165, pl. 31, fig. 2.
——? SINUATA M. & W., 1868; Am. Jour. Sci., 2d series, vol. 46, p. 22. Waupecan Creek, Grundy Co., Illinois.   Coal Measures.
—— SINUATA M. & W., 1868; Geol. Survey Illinois, vol. 3, p. 540, fig. a.

COLPOCARIS. CYPRIDINA.
—— (*Solenocaris*) STRIGATA Meek, 1871; Proc. Acad. Nat. Sci. Phila., vol.
23, p. 335. Lower Carboniferous.
(*Solenocaris*) STRIGATA Meek, 1875; Pal. Ohio, vol. 2, p. 321, pl. 18,
figs. 4 a-c. Danville, Ky. (See *Strigocaris*.)
COLPOCARIS Meek, 1871. (See *Ceratiocaris (Colpocaris) elytroides* and
bradleyi.)
—— CHESTERENSIS Worthen, 1884; Bull. Illinois State Mus., p. 3, Chester,
Illinois. Chester limestone.
—— CHESTERENSIS Worthen, 1890; Geol. Illinois, vol. 8, p. 153, pl. 27, fig. 2.
CRYPTOZOE Pack., 1886; Proc. Am. Philos. Soc., vol, 23, p. 381.
—— PROBLEMATICUS Pack., 1886; Proc. Am. Philos. Soc., vol. 23, p, 382,
fig. 3, pl. Carboniferous.
CTENOBOLBINA Ulrich, 1890; Jour. Cincinnati Soc. Nat. His., vol. 13, p.
108.
—— ALATA Ulrich, 1890; Jour. Cincinnati Soc. Nat. His., vol. 13, p. 110,
pl. 7, figs. 4, 4 a-c. Cincinnati, Ohio. Hudson.
—— (*Bollia?*) ANTESPINOSA Ulrich, 1890; Jour. Cincinnati Soc. Nat. His.,
vol. 13, p. 187, pl. 15, figs. 9, 9 a-c. Falls of the Ohio. Devonian.
—— BISPINOSA Ulrich, 1890; Jour. Cincinnati Soc. Nat. Hist., vol. 13, p.
110, pl. 7, fig. 6. Cincinnati, Ohio. Hudson.
—— CILIATA (Emmons) Ulrich, 1890; Jour. Cincinnati Soc. Nat. Hist.,
vol. 13, p. 108, pl. 7, figs. 1, 1 a-b; also var. *curta*, idem, p. 109,
pl. 7, fig. 2, Cincinnati, Ohio; var. *emaciata*, idem, p. 109, pl. 7,
figs. 3, 3 a-c. Savannah, Illinois. For other references see
Beyrichia ciliata Emmons.
—— INFORMIS Ulrich, 1890; Jour. Cincinnati Soc. Nat. His., vol. 13, p.
107, pl. 15, figs. 6, 6 a-c. Falls of the Ohio. Devonian.
—— MINIMA Ulrich, 1890; Jour. Cincinnati Soc. Nat. Hist., vol. 13, p.
188, pl. 15, fig. 7. 18-Mile Creek, near Buffalo, N. Y. Hamilton.
—— PAPILLONA Ulrich, 1890; Jour. Cincinnati Soc. Nat. His., vol. 13, p.
186, pl. 15, figs. 8, 8 a-c. Falls of the Ohio. Devonian.
—— PUNCTATA Ulrich, 1890; Jour. Cincinnati Soc. Nat. His., vol. 13, p.
186, pl. 12, figs. 5, 5 a-c. Lockport, N. Y. Niagara.
—— TUMIDA Ulrich, 1890; Jour. Cincinnati Soc. Nat. Hist., vol. 13, p.
111, pl. 7, figs. 5, 5 a-b. McKinney's Station, Ky.
Middle bed Hudson group.
CYCLUS De Kon., 1841; Mém. Acad. Sci. Bruxelles, vol. 14, p. 18.
—— AMERICANA Pack., 1885; American Naturalist, vol. 19, p. 293. Mazon
Creek, Grundy Co., Illinois. Coal Measures.
—— AMERICANA Pack., 1886; Mem. Nat. Acad. Sci., vol. 3, p. 143, pl. 5,
figs. 1, 1 a; pl. 6, figs. 4. 4 a.
CYPRIDINA Milne-Edw., 1838; Lamarck's Anim. sans Vertb., vol. 5, p. 178.
—— HERZERI Ulrich, 1890; Jour. Cincinnati Soc. Nat. Hist., vol. 13, p.
209, pl. 14, figs. 9, 9 a-c. Richfield, Ohio. Waverly.

CYTHERE.   CYTHERINA.

CYTHERE Müller, 1785; Entomostraca, p. 64.
—— (*Beyrichia*) AMERICANA Shum., 1858; Trans. Acad. Sci. St. Louis, vol.
    1, p. 227.   Valley Verdigris River, Kansas.   Upper Coal Measures.
—— *carbonaria* Hall. (See *Leperditia carbonaria*.)
—— *cincinnatiensis* Meek. Danville, Ky. (See *Entomis cincinnatiensis*.)
                                                                                 Waverly.
—— CRASSIMARGINATA Winchell, 1862; Proc. Acad. Nat. Sci. Phil., p. 429.
                                                                                 Marshall.
—— IRREGULARIS Miller, 1878; Jour. Cincinnati Soc. Nat. Hist., vol. 1, p.
    106, pl. 3, figs. 7, 7 a. Cincinnati, Ohio.   Hudson.
—— NEBRASCENSIS Geinitz, 1866; Carb. und Dyas in Nebraska, p. 2, pl. 1,
    fig. 2. Nebraska City.   Coal Measures.
—— NEBRASCENSIS Meek, 1872; Rep. U. S. Geol. Survey Territories, Final
    Report Nebraska, p. 237, pl. 11, figs. 1, 2, 3 a-b (?).
—— NEBRASCENSIS Keyes, 1888; Proc. Acad. Nat. Sci. Phil., p. 222.
—— OHIONES Herrick, 1889; Bull. Den. Univ., vol. 4, p. 60, pl. 8, fig. 8;
    vol. 3, pl. 3, fig. 19. Newark, Rushville and Ashland Co., Ohio.
                                                                                 Waverly.
—— OHIONES Herrick, 1890; American Geologist, vol. 4, p. 254.
—— *okeni*. (See *Leperditia okeni*.)
—— SIMPLEX W. and St. J., 1867; Prelim. Notice New Species Foss., p. 3.
    At Sternberg's Mill, on Boone River, Iowa.   St. Louis.
—— SIMPLEX W. and St. J., 1868; Trans. Chicago Acad. Sci., vol. 1, p. 127.
CYTHERE sp. ? Dawson, 1868; Acadian Geology, p. 206, fig. 48 c.   Jones
    figures this species as *Leperditia okeni* (Münst.) var. *acuta*.   Geol.
    Mag., 1884, vol. 1, p. 357.   Carboniferous.
—— *sublævis*. (See *Leperditia sublævis*.)
—— *subrecta*. (See *Leperditia subrecta*.)
CYTHERELLA Jones and Bosquet, 1848 and 1852.
—— OVATIFORMIS Ulrich, 1890; Jour. Cincinnati Soc. Nat. Hist., vol. 13,
    p. 209, pl. 17, figs. 3 and 4 a-c. Caldwell, Ky.   Chester.
CYTHERINA Lamarck, 1818; Animaux sans Vertb., vol. 5, p. 125.
—— *alta*. (See *Leperditia alta*.)
—— CRENULATA Emmons, 1855; Am. Geology, vol. 1, p. 221, figs. 75 a-d.
    Middleville, Ohio.   Trenton.
—— *cylindrica*. (See *Isochilina cylindrica*.)
—— EMMONSI Vogd., 1889; Annals. N. Y. Acad. Sci., vol. 5, p. 13, to re-
    place *C. subcylindrica*. Middleville, Ohio.
—— *spinosa*. See *Beyrichia spinosa*.)
—— *subcylindrica* Emmons, 1855; Am. Geology, vol. 1, p 220, fig. 75 b.
    This name was used by Münster for a species of this genus.
    Change to *C. emmonsi*.   Trenton.
—— SUBELLIPTICA Emmons, 1855; Am. Geology, vol. 1, p. 220, figs. 75 a.
    One mile from Watertown, N. Y.   Trenton.

CYTHERELLINA. DIPTEROCARIS.

CYTHERELLINA J. & H., 1869; Annals Mag. Nat. His., London, 4th series, vol. 3, p. 215.
—— GRANDELLA Whitf., 1882; Bull. Am. Mus. Nat. Hist. New York, vol. 1, p. 94, pl. 9, figs. 28, 29. Spergin Hill, Indiana. *   Warsaw.
CYTHEROPSIS McCoy, 1855; Synopsis Classification Brit. Pal. Rocks, pl. 1 L, fig. 2. No definite description of the genus given. This is not now applied as a generic name to any fossil Ostacoda.
—— *concinna* Jones. (See *Primitia concinna*.)
—— *rugosa* Jones. (See *Cytherella rugosa*?)
—— SILIQUE Jones, 1858; Annals Mag. Nat. Hist., 3d series, London, vol. 1, p. 249, pl. 10, fig. 6. Probably of the genus *Macrocypsis*.
                                                                Trenton.
—— SILIQUE Jones, 1858; Geol. Survey Canada, decade 3, p. 99.
—— SILIQUE Jones, 1869; Annals Mag. Nat. Hist., London, 4th series, vol. 3, p. 216, pl. 14, figs. 1-6.
DEPRANELLA Ulrich, 1890; Jour. Cincinnati Nat. His., vol. 13, p. 117.
—— AMPLA Ulrich, 1890; Jour. Cincinnati Soc. Nat. Hist., vol. 13, p. 120, pl. 1, fig. 2. River gorge at High Bridge, Ky.   Chazy.
—— CRASSINODA Ulrich, 1890; Jour. Cincinnati Soc. Nat. His., vol. 13, p. 118, pl. 8, fig. 1, 1 a–c.   cf. *ampla*, *elongata*, *nitida* and *macra*. High Bridge, Ky.   (Birdseye.) Cambrian.
—— ELONGATA Ulrich, 1890; Jour. Cincinnati Soc. Nat. Hist., vol. 13, p. 121, pl. 8, figs. 5, 5 a–b. River gorge at High Bridge, Ky. Chazy.
—— MACRA Ulrich, 1890; Jour. Cinn. Soc. Nat. Hist., vol. 13, p. 119, pl. 8, figs. 4, 4 a–b. Near Lavergne, Tenn.
                                                (Glade Limestone.) Cambrian.
—— NITIDA Ulrich, 1890; Jour. Cinn. Soc. Nat. Hist., vol. 13, p. 119, pl. 8, figs. 3, 3 a–b. River gorge at High Bridge, Ky.   Chazy.
DIPELTIS Pack., 1885; Am. Naturalist, vol. 19, p. 293.
—— DIPLODISCUS Pack., 1885; Am. Naturalist, vol. 19, p. 293. Mazon Creek, Morris, Illinois.   Carboniferous.
—— DIPLODISCUS Pack., 1886; Mem. Natl. Acad. Sci., vol. 3, p. 145, pl. 5, figs. 2, 2 a.
DIPLOSTYLUS Salt., 1863; Quart. Jour. Soc. London, vol. 19, p. 76.
—— DAWSONI Salt., 1863; Quart. Jour. Geol. Soc. London, vol. 19, p. 77, fig. 6. Joggins, Nova Scotia.   Coal Measures.
DIPTEROCARIS Clarke, 1883; Am. Jour. Sci., 3d series, vol. 25, p. 121.
—— PENNÆ-DÆDALI Clarke, 1883; Am. Jour. Sci., 3d series, vol. 25, p. 122, fig. 1. Canadice, Ontario Co., N. Y.   Portage group.
—— PENNÆ-DÆDALI J. & W., 1884; Geol. Mag., decade 3, vol. 1, p. 349.
—— PENNÆ-DÆDALI Eth., W. & J., 1884; 2d Rep. Comm. Foss. Phyllopoda Pal. Rocks, p. 11.
—— PENNÆ-DÆDALI Eth., W. & J., 1885; 3d Rep. Comm. Foss. Phyllopoda Pal. Rocks, p. 3.

DIPTEROCARIS.   DOLICHOPTERUS.
—— PENNÆ DÆDALI Hall, 1888; Pal. New York, vol. 7, p. 200, pl. 35, fig. 24.
—— PENNÆ-DÆDALI Eth., W. & J., 1888; 6th Rep. Comm. Foss. Phyllopoda Pal. Rocks, p. 9.
—— PERCERVÆ Clarke, 1883; Am. Jour. Sci., 3d series, vol. 25, p. 123, figs. 4, 5. Dansville, N. Y.   Chemung.
—— PERCERVÆ J. & W., 1884; Geol. Mag., decade 3, vol. 1, p. 349.
—— PERCERVÆ Eth., W. & J., 1884; 2d Rep. Comm. Foss. Phyllopoda Pal. Rocks, p. 11.
—— PERCERVÆ Eth., W. & J., 1885; 3d Rep. Comm. Foss. Phyllopoda Pal. Rocks, p. 3.
—— PERCERVÆ Hall, 1888; Pal. New York, vol. 7, p. 202, pl. 35, figs. 20,21.
—— PERCERVÆ Eth., W. & J., 1888; 6th Rep. Comm. Foss. Phyllopoda Pal. Rocks, p. 9.
DIPTEROCARIS PRŒNE Clarke, 1883; Am. Jour. Sci., 3d series, vol. 25, p. 122, figs. 2, 3. Canadice, N. Y. Portage group and in Chemung at Haskinsville, N. Y.
—— PRŒNE J. & W., 1884; Geol. Mag., decade 3, vol. 1, p. 349.
—— PRŒNE Eth., W. & J., 1884; 2d Rep. Comm. Foss. Phyllopoda Pal. Rocks, p. 11.
—— PRŒNE Eth., W. & J., 1885; 3d Rep. Common Foss. Phyllopoda Pal. Rocks, p. 3.
—— PRŒNE Hall, 1888; Pal. New York, vol. 7, p. 201, pl. 35, figs. 25-27.
—— PRŒNE Eth., W. & J., 1888; 6th Rep. Comm. Foss. Phyllopoda Pal. Rocks, p. 9.
DITHYROCARIS (Scouler, 1843, MS.) Portl., 1843; Rep. Geol. Londonderry, etc., p. 313.
—— BELLI H. W., 1870; Geol. Mag., vol. 8, p. 106, pl. 3, figs. Gaspé, Canada.   Devonian.
—— BELLI Eth., W. & J., 1887; 5th Rep. Comm. Foss. Phyllopoda Pal. Rocks, p. 6.
—— BELLI Hall, 1888; Pal. New York, vol. 7, p. 194.
—— CARBONARIA M. & W., 1870; Proc. Acad. Nat. Sci. Phila., vol. 22, p. 55. Danville, Illinois.   Coal Measures.
—— CARBONARIA M. & W., 1873; Geol. Survey Illinois, vol. 5, p. 618, pl. 32, fig. 1.
—— CARBONARIA White, 1884; 13th Rep. Dept. Geol. Nat. Hist. Indiana, p. 178, pl. 39, fig. 2.
—— *neptuni* Hall. (See *Mesothyra neptuni*.)
Dolichocephala Claypole, 1883; Proc. Am. Philos. Soc. vol. 21, p. 236.
—— *locoana*, idem; p. 236, plate. (See *Stylonurus excelsior*.)
DOLICHOPTERUS (subgenus EURYPTERUS) Hall, 1859; Pal. New York, vol. 3, p. 414.
—— MACROCHEONS Hall, 1859; Pal. New York, vol. 3, p. 414*, pl. 83, fig. 1; pl. 83a, fig. 1. Near Buffalo, N. Y.   Waterlime.

ECHINOCARIS Whitf., 1880; Am. Jour. Sci., 3d series, vol. 19, p. 34.
—— CONDYLEPSIS Hall, 1888; Pal. New York, vol. 7, p. 173, pl. 19, figs. 14-17. Belmont, Alleghany Co., N. Y. Chemung.
—— MULTINODOSA Whitf., 1880; Am. Jour. Sci., 3d series, vol. 19, p. 38, fig. 8. (Plates with author's special edition only.) Paines' Creek, Le Roy, Ohio. Erie Shales.
—— MULTINODOSA Pack., 1882; Am. Naturalist, vol. 16, p. 952, fig. 10.
—— MULTINODOSA Pack., 1883; 12th Rep. U. S. Geol. Survey Territories, p. 451, fig. 71 B.
—— MULTINODOSA Beecher, 1884; 2d Geol. Survey Penn., vol. PPP, p. 5.
—— MULTINODOSA Eth., W. & J., 1885, 3d Rep. Comm. Foss. Phyllopoda Pal. Rocks, p. 35.
—— MULTINODOSA Hall, 1888; Pal. New York, vol. 7, p. 180, pl. 29, figs. 18-19. Paines' Creek, Le Roy, Lake Co., Ohio.
—— MULTINODOSA Eth., W. & J., 1888; 6th Rep. Comm. Foss. Phyllopoda Pal. Rocks, p. 8.
—— (*Ceratiocaris*) PUNCTATUS Hall, 1863; 16th Rep. New York State Cab. Nat. Hist., p. 74, pl. 1, fig. 8. Hamilton.
—— PUNCTATUS Whitf., 1880; Am. Jour. Sci., 3d series, vol. 19, p. 39.
Syn., *Ceratiocaris armatus* Hall, 1863; 16th Rep. New York State Cab. Nat. Hist., p. 72, pl. 1, figs. 1-3.
Ceratiocaris armatus Hall, 1876; Illus. Devonian Foss., pl 23, figs. 4, 5.
Certiocaris (Aristozoe) PUNCTATUS Hall, 1876; Illus. Devonian Foss., Expl. pl. 23, fig. 7.
—— PUNCTATUS Pack., 1882; Am. Naturalist, vol. 16, p. 952, fig. 12.
—— PUNCTATUS Pack., 1883; 12th Rep. U. S. Geol. Survey Territories, p. 450, fig. 70.
Syn., *armatus* Pack., 1883; 12th Rep. U. S. Geol. Survey Territories, p. 451.
—— PUNCTATUS Beecher, 1884; 2d Geol. Survey Penn., vol. PPP, p. 6, pl. 1, figs. 13-16; also fig. 1 in text, p. 4. At Delphi, Pratt's Falls, and the shore of Cayuga Lake, N. Y.
Syn., *armatus* J. & W., 1884; Geol. Mag., decade 3, vol. 1, p. 2, pl. 13, fig. 2.
armatus Eth., W. & J., 1885; 3d Rep. Comm. Foss. Phyllopoda Pal. Rocks, p. 35.
—— PUNCTATUS Hall, 1888; Pal. New York, vol. 7, p. 166, pl. 27, fig. 10; pl. 28, fig. 1-7; pl. 29, fig. 1-8. Delphi, near Fabius, Canandaigua Lake, etc., N. Y.
—— PUNCTATUS Eth., W & J., 1888; 6th Rep. Comm. Foss. Phyllopoda Pal. Rocks, p. 8.
—— PUSTULOSA Whitf., 1880; Am. Jour. Sci., 3d series, vol. 19, p. 38, pl. fig. 7, a plate with author's special edition. Le Roy, Ohio.
Erie Shales.

ECHINOCARIS. ECHINOGNATHUS.
—— PUSTULOSA Pack., 1883; 12th Rep. U. S. Geol. Survey Territories, p. 451.
—— PUSTULOSA Beecher, 1884; 2d Geol. Survey Penn., vol. PPP, p. 5.
—— PUSTULOSA J. & W., 1884; Geol. Mag., decade 3, vol. 1, p. 2, pl. 13, fig. 6.
—— PUSTULOSA Eth., W. & J., 1885; 3d Rep. Comm. Foss. Phyllopoda Pal. Rocks, p. 35.
—— PUSTULOSA Hall, 1888; Pal. New York, vol. 7, p. 178, pl. 29, figs. 9, 10. Le Roy, Lake Co., Ohio.
—— PUSTULOSA Eth., W. & J., 1888; 6th Rep. Comm. Foss. Phyllopoda Pal. Rocks, p. 8.
—— SOCIALIS Beecher, 1884; 2d Geol. Survey Penn., vol. PPP., p. 10, pl. 1, figs. 1-1 e. Warren, Penn. Chemung.
—— SOCIALIS Eth., W. & J., 1885; 3d Rep. Comm. Foss. Phyllopoda Pal. Rocks, p. 35.
—— SOCIALIS Hall, 1888; Pal. New York, vol. 7, p. 174, pl. 30, figs. 1-12.
—— SOCIALIS Eth., W. & J., 1888; 6th Rep. Comm. Foss. Phyllopoda Pal. Rocks, p. 8.
—— SUBLEVIS Whitf., 1880; Am. Jour. Sci., 3d series, vol. 19, p. 36, pl. figs. 4-6 (plate with the author's special edition only). Le Roy, Ohio. Erie Shales.
—— SUBLEVIS Pack., 1882; Am. Naturalist, vol. 16, p. 952, fig. 11.
—— SUBLÆVIS Pack., 1883; 12th Rep. Geol. Survey Territories, p. 450, fig. 71 b.
—— SUBLEVIS Beecher, 1884; 2d Geol. Survey Penn., vol. PPP, p. 5.
—— SUBLÆVIS J. & W., 1884; Geol. Mag., decade 3, vol. 1, p. 2, pl. 13, figs. 3, 4, 5.
—— SUBLÆVIS Eth., W. & J., 1885; 3d Rep. Comm. Foss. Phyllopoda Pal. Rocks, p. 35.
—— SUBLÆVIS Hall, 1888; Pal. New York, vol. 7, p. 176, pl. 29, figs. 11-13.
—— SUBLÆVIS Eth., W. & J., 1888; 6th Rep. Comm. Foss. Phyllopoda Pal. Rocks, p. 8.
—— WHITFIELDI Clarke, 1885; Bull. U. S. Geol. Survey, No. 16, p. 45, pl. 2, figs. 3, 4, Hatch Hill, Naples, N. Y. Portage.
—— WHITFIELDI Hall, 1888; Pal. New York, vol. 7, p. 172, pl. 29, figs. 20, 21.
—— WHITFIELDI Eth., W. & J., 1888; 6th Rep. Comm. Foss. Phyllopoda Pal. Rocks, p. 8.
Echinocaris wrightiana J. & W. 1884; Geol. Mag., decade 3, vol. 1, p. 395, pl. 13, figs. 1, 1 a-b.
ECHINOGNATHUS Wal., 1882; Am. Jour. Sci., 3d series, vol. 23, p. 213.
—— CLEVEVANDI Wal., 1882; Am. Jour. Sci., 3d series, vol. 23, p. 213, figs. 1, 2. North of Holland Patent, Oneida Co., N. Y. Utica.
—— (*Eurypterus?*) CLEVELANDI Wal., 1882; Am. Jour. Sci., 3d series, vol. 23, p. 151.

CATALOGUE OF NON-TRILOBITES. 385

ELPE. ESTHERIA.
ELPE TYRRELLII Jones, 1891; Contrib. Micro-Palæont., pt. 3, p. 93, fig. 8.
cf. *Cythere cincinnaticus* Meek. Dawson Bay. Lake Winnepegosis.
Devonian.
—— TYRELLII Whiteaves, 1892; Contrib. Can. Palæont., vol. 1, p. 346.
ELYMOCARIS Beecher, 1884; 2d Geol. Survey Penn., vol. PPP, p. 13.
—— CAPSELLA Hall, 1888; Pal. New York, vol. 7, p. 181, p. 31, fig. 4.
Tichinor's Glen, Canandaigua Lake, N. Y. Chemung.
—— SILIQUA Beecher, 1884; 2d Geol. Survey Penn., vol. PPP, p. 13, pl. 2, figs. 1, 2. Warren, Penn. Chemung.
—— SILIQUA Eth., W. & J., 1885; 3d Rep. Comm. Foss. Phyllopoda Pal., Rocks p. 35.
—— SILIQUA Hall, 1888; Pal. New York, vol. 7, p. 182, pl. 31, figs. 5, 6.
—— SILIQUA Eth., W. & J., 1888; 6th Rep. Comm. Foss. Phyllopoda Pal. Rocks, p. 8.
ENTOMIS Jones, 1861; Mem. Geol. Survey Scotland, Expl. Map. 32, p. 137; also, Annals Mag. Nat. Hist., London, 4th series, vol. 11, p. 413.
Cythere CINCINNATIENSIS Meek, 1871; Proc. Acad. Nat. Sci. Phila., vol. 23, p. 331. Cincinnati, Ohio. Hudson.
Cythere CINCINNATIENSIS Meek, 1873; Pal. Ohio; vol. 1, p. 158, pl. 14, figs. 1 a–d.
Cythere CINCINNATIENSIS Miller, 1874; Cincinnati Quart. Jour. Sci., vol. 1, p. 120.
—— CINCINNATIENSIS Jones, 1884; Annals Mag. Nat. Hist., London, 5th series, vol. 14, note p. 395.
—— MADISONENSIS Ulrich, 1890; Jour. Cincinnati Soc. Nat. Hist., vol. 13, p. 107, pl. 7, figs. 12, 12 a–b. Madison, Indiana. Cincinnati.
—— RHOMBOIDEA Jones, 1890; Quart. Jour. Geol. Soc., vol. 46, p. 10, pl. 2, figs. 9, 10 a, b. Seneca Lake, N. Y. Hamilton.
—— WALDRONENSIS Ulrich, 1890; Jour. Cincinnati Soc. Nat. Hist., vol. 13, p. 183, pl. 12, figs. 3, 3 a–b. Near Waldron, Ind. Niagara.
Equisetides wrightiana. This species is referred by W. & J. (3d Report Comm. Foss. Phyllopoda Brit. Assoc., 1885, p. 360) to *Echinocaris.* Prof. James Hall places it provisionally with *Stylonurus*, which see.
ESTHERIA Ruffell, 1857; Mus. Senckenberg, vol. 2, p. 119.
—— DAWSONI Jones, 1870; Geol. Mag., vol. 7, p. 220, pl. 9, fig. 15.
Carboniferous.
—— DAWSONI Jones, 1878; Geol. Mag., decade 2, vol. 2, p. 101, pl. 3, fig. 2.
—— sp. ? Dawson, 1868; Acadian Geology, p. 256, fig. 78 d.
—— DAWSONI J. & K., 1884; Geol. Mag., decade 3, vol. 1, p. 361, pl. 12, fig. 12. Horton, Nova Scotia.
—— PULEX Clarke, 1882; Am. Jour. Sci., 3d series, vol. 23, p. 476, pl., fig. 4. Hopewell, N. Y. Hamilton.
—— PULEX Pack., 1883; 12th Rep. U. S. Geol. Survey Territories, p. 355.

25

ESTHERIA. EURYPTERUS.
— PULEX Eth., W. & J., 1885; 5th Rep. Comm. Foss. Phyllopoda Pal. Rocks, p. 10.
— PULEX Hall, 1888; Pal. New York, vol. 7, p. 206, pl. 35, figs. 10, 11.
— PULEX Eth., W. & J., 1888; 6th Rep. Comm. Foss. Phyllopoda Pal. Rocks, p. 9.
Euproops Meek, 1867; Am. Jour. Sci., 2d series, vol. 34, p. 394. (See *Belinurus* and *Prestwichia*.) Mazon Creek, Grundy Co., Illinois.
EURYCHILINDA Ulrich, 1889; Contributions Micro-Palæont., pt. 2, p. 52.
— ÆQUALIS Ulrich, 1890; Jour. Cincinnati Soc. Nat. Hist., vol. 13, p. 129, pl. 9, figs. 5-8. Lebanon, Tenn. Birdseye and Chazy.
— GRANOSA Ulrich, 1890; Jour Cincinnati Soc. Nat. Hist., vol. 13, p. 128, pl. 9, figs. 9-12. Gorge of Kentucky River, at High Bridge, Ky. Chazy.
— LONGULA Ulrich 1890; Jour. Cincinnati Soc. Nat. Hist., vol. 13, p. 127, pl. 9, figs. 3, 3 a-b and 4. High Bridge, Ky. Birdseye.
— MANTIOBENSIS Ulrich, 1889; Contributions Micro-Palæont., pt. 2, p. 53, pl. 9, figs. 10, 10 a. Stony Mountain, Manitoba. Hudson.
— OBESA Ulrich, 1890; Jour. Cincinnati Soc. Nat. Hist., vol. 13, p. 129, pl. 9, fig. 13. High Bridge, Ky. Birdseye.
— RETICULATA Ulrich, 1890; Jour. Cincinnati Soc. Nat. Hist., vol. 13, p. 126, (type)—Minneapolis, St. Paul, Fountain, Minnesota.
Trenton.
— RETICULATA Jones, 1890; Quart. Jour. Geol. Soc., vol. 46, p. 539, pl. 20, figs. 13 a-b. Ontario Co., N. Y.
— RETICULATA Ulrich, 1889; Contributions Micro-Palæont., pt. 2, p. 52, pl. 9, figs. 9, 9 a. Trenton.
— STRIATO-MARGINATA (Miller) Ulrich, 1889; Contributions Micro-Palæont., pt. 2, p. 52. Osgood, Ind. For other references, see *Beyrichia striato-marginata*. Cambrian.
— STRIATO-MARGINATA Ulrich, 1890; Jour. Cincinnati Soc. Nat. Hist., vol. 13, p. 126, pl. 9, fig. 14. Madison, Ind.
— SUBRADIATA Ulrich, 1890; Jour. Cincinnati Soc. Nat. Hist., vol. 13, p. 126, pl. 9, figs. 1, 1 a-c and 2 a-c. Lebanon, Tenn.; Dixon, Illinois, and Minneapolis, Minn. Lower Trenton.
EURYPTERELLA Matt., 1888; Trans. Roy. Soc. Canada, vol. 6, p. 60.
— ORNATA Matt., 1888; Trans. Roy. Soc. Canada, vol. 6, p. 60, pl. 4, fig. 12. Lancaster, N. B. Devonian.
EURYPTERUS De Kay, 1825; Annals Lyceum Nat. Hist. N. Y., vol. 1, pt. 2, p. 375.
BEECHERI Hall, 1884; 2d Geol. Survey Pennsylvania, vol. PPP, p. 30, pl. 3, fig. 1. Warren, Penn. Chemung.
— BEECHERI Hall, 1888; Pal. New York, vol. 7, p. 156, pl. 27, fig. 5.
— BEECHERI Eth., W. & J., 1888; 6th Rep. Comm. Foss. Phyllopoda Pal. Rocks, p. 9.

EURYPTERUS.
—— BOYLEI Whiteaves, 1884; Pal. Foss., vol. 3, pt. 1, p. 42, pl. 7, fig. 3. Guelph formation of Ontario, Canada.
—— DEKAYI Hall, 1859; Pal. New York, vol. 3, p. 411*, pl. 82, fig. 1. Black Rock, N. Y.　　　　　　　　　　　　　　　　Waterlime.
—— ERIENSIS Whitf., 1881; Annals New York Acad. Sci., vol. 2, p. 196.
　　　　　　　　　　　　　　　　　　　　　　　Lower Helderberg.
—— GIGANTEUS Pohl., 1882; Bull. Buffalo Soc. Nat. Sci., vol. 4, p. 41, pl. 2, fig. 1. Buffalo, N. Y.　　　　　　　　　　　　　　Waterlime.
—— LACUSTRIS Harlan, 1834; Trans. Geol. Soc. Pennsylvania, vol. 1, p. 98, pl. 5, fig. 2. Williamsville, N. Y.　　　　　　Waterlime.
—— LACUSTRIS Harlan, 1835; Med. Phys. Researches, p. 297, pl., fig. 2.
—— LACUSTRIS Hall, 1859; Pal. New York, vol. 3, p. 407*, pl. 81, figs. 1-11; pl. 81 a, fig. 1; pl. 81 b, figs. 1-5; pl. 83 b, fig. 3; also fig. 5 in text p. 400*. Williamsville and Buffalo, N. Y.
—— LACUSTRIS var. ROBUSTUS Hall, 1859; Pal. New York, vol. 3, p. 410*, pl. 81 c, fig. 2. Near Bufialo, N. Y.
—— (*Dolichopterus*) MANSFIELDI Hall, C. E., 1877; Proc. Am. Philos. Soc., vol. 16, p. 621. Near Cannelton, Beaver Co., Penn. Carboniferous.
—— MANSFIELDI Hall, 1884; 2d Geol. Survey Pennsylvania, vol. PPP, p. 32, pl. 4, figs. 1-8; pl. 5, figs. 1-11; pl. 6, fig. 1; pl. 7, fig. 1; pl. 8, figs. 1-3.
—— (*Anthraconectes*) MAZONENSIS M. & W., 1868; Am. Jour. Sci., 2d series, vol. 46, p. 21. Mazon Creek, Grundy Co., Illinois.
　　　　　　　　　　　　　　　　　　　　　　　Coal Measures.
—— (*Anthraconectes*) MAZONENSIS M. & W., 1868; Geol. Survey Illinois, vol. 3, p. 544, figs.
—— (*Anthraconectes*) MAZONENSIS Hall, 1884; 2d Geol. Survey Pennsylvania, vol. PPP, p. 25, fig. 2; p. 27, fig. 3.
—— (*Anthraconectes*) MAZONENSIS White, 1884; 13th Ann. Rep. Dept. Geol. Nat. His. Indiana, p. 168, pl. 37, figs. 1, 2, 3.
—— MICROPHTHALMUS Hall, 1859; Pal. New York, vol. 3, p. 407*, pl. 80 a, fig. 7. Cazenovia, N. Y.　　　　　　　　　Lower Helderberg.
—— PACHYCHEIRUS Hall, 1859; Pal. New York, vol. 3, p. 412*, pl. 8, figs. 1-3. Black Rock four miles east of Bufialo, N. Y.　Waterlime.
—— PENNSYLVANICUS Hall, C. E., 1877; Trans. Am. Philos. Soc. Phila., vol. 7, p. 621. At Rooker Farm, Pithole City, Venango county, Penn.　　　　　　　　　　　　　　　　　　　Carboniferous.
—— PENNSYLVANICUS Hall, 1884; 2d Geol. Survey Pennsylvania, vol. PPP, p. 31, pl. 5, fig. 18.
—— POHLMANI Vogd., 1889; to take the place of the preoccupied name of *E. scorpionis* Grote and Pitt. Annals New York Acad. Sci., vol. 5, p. 22. Buffalo, N. Y.
—— POTENS Hall, 1884; 2d Geol. Survey Pennsylvania, vol. PPP, pl. 4, figs. 9, 10.　　　　　　　　　　　　　　　Carboniferous.

EURYPTERUS. HIPPONICHARION.

—— PROMINENS Hall, 1884; Proc. Am. Assoc. Adv. Sci., vol. 33, p. 420. Cayuga Co., N. Y. Clinton.
—— PROMINENS Hall, 1888; Pal. New York, vol. 7, p. 157, pl. 27, figs. 3, 4.
—— PULICARIS Salt., 1863; Quart. Jour. Geol. Soc. London, vol. 19, p. 78, figs. 9, 10. St. John, N. B. Coal Measures.
—— PULICARIS Daw., 1868; Acadian Geol., p. 523, figs. 179 a, b.
—— PUSTULOSUS Hall, 1859; Pal. New York, vol. 3, p. 413*, pl. 83 b, fig. 1. Near Buffalo. Waterlime.
—— REMIPES De Kay. Waterlime.
Fossil Fish Mitch., 1818; Am. Monthly Mag., vol. 3, p. 291.
—— REMIPES De Kay, 1825; Annals Lyceum Nat. Hist., vol. 1, p. 375, pl. 29.
—— REMIPES Harlan, 1834; Trans. Geol. Soc. Penn., vol. 1, p. 96, pl. 5.
—— REMIPES Harlan, 1835; Medical Phys. Researches, p. 297, pl., fig. 1.
—— REMIPES Milne-Edw., 1840; Hist. Nat. Crust., vol. 3, p. 422.
—— REMIPES Burm., 1843; Org. Trilobiten, p. 62 (Ray Soc. ed., p. 54).
—— REMIPES Bronn, 1837; Lethea Geognostica, vol. 1, p. 109, pl. 9^3, fig. 1; 2d ed., vol. 1, p. 61, pl. 9, fig. 1; also pl. 9^3, fig. 1.
—— REMIPES Salt., 1859; Quart. Jour. Geol. Soc. London, vol. 15, p. 255.
—— REMIPES Niesz., 1858; Der *Eurypterus remipes*, etc., pls. 1, 2, 3.
—— REMIPES Hall, 1859; Pal. New York, vol. 3, p. 404, pl. 80, figs, 1-12; pl. 80 a, figs. 1-6; pl. 83 b, fig. 2, also p. 403*, figs. 6, 7. Waterville, Westmoreland, Litchfield, N. Y.
—— REMIPES (numerous authors).
Eusarcus grandis Grote and Pitt, 1875; Bull. Buffalo Soc. Nat. Sci., vol. 3, p. 17. Buffalo, N. Y. Waterlime.
Referred by J. Pohlman (Bull. Buffalo Soc. Nat. Sci., vol. 5, p. 31) to *Eurypterus scorpionis*. An almost similar term, *E. scorpioides* was used for a species of this genus by H. Woodward in 1868. (See *E. pohlmani*.)
Eusarcus scorpionis Grote and Pitt, 1875; Bull. Buffalo Soc. Nat. Sci., vol. 3, pp. 1, 17, photographic plate. Waterlime.
Eusarcus SCORPIONIS (G. and P.) Pohl., 1886; Bull. Buffalo Soc. Nat. Sci., vol. 5, p. 30, pl. 2, fig. 3. Buffalo, N. Y.
Eusarcus grandis (G. and P.) Pohl., 1886; Bull. Buffalo Soc. Nat. Sci., vol. 5, p. 31. Referred to *Eurypterus pohlmani*.
—— STYLUS Hall, 1884; 2d Geol. Survey Penn., PPP, p. 34, pl. 5, figs. 12-15. Near Cannelton, Beaver Co., Penn. Carboniferous.

HALLIELLA Ulrich, 1890; Jour. Cincinnati Soc. Nat. Hist., vol. 13, p. 184.
—— RETIFERA Ulrich, 1890; Jour. Cincinnati Soc. Nat. Hist., vol. 13, p. 185, pl. 15, figs. 5, 5 a-e. Falls of the Ohio. Devonian.

HIPPONICHARION Matt., 1885; Trans. Roy. Soc. Canada, vol. 3, p. 60.
—— ERO Matt., 1885; Trans. Roy. Soc. Canada, vol 3, p. 64, pl. 6, figs. 19 a-b; Div. 1, b^1. Hanford Brook, St. Martin's, N. B. Paradoxides zone.

CATALOGUE OF NON-TRILOBITES. 389

HYMENOCARIS.

HYMENOCARIS Salt., 1853; 22d Rep. Brit. Assoc. Adv. Sci., Trans. of Sec., p. 56.

ISOCHILINA.

ISOCHILINA Jones, 1857; Mem. Geol. Survey Canada, decade 3, p. 97.
—— AMIANA Ulrich, 1890; Jour. Cincinnati Soc. Nat. Hist., vol. 13, p. 180, pl. 11, figs. 12 a-c. Ottawa, Canada. Chazy?
var. *insignis; idem.*, p. 181, pl. 181, pl. 11, fig. 13. Ottawa, Canada. Chazy?
—— AMPLA Ulrich, 1890; Jour. Cincinnati Soc. Nat. His., vol. 13, p. 179, pl. 11 f, figs. 8, 8 a-d. Nashville, Tenn. Trenton.
—— AMII Jones, 1891; Contrib. Micro-Palæont., pt. 3, p. 68, pl. 10, figs. 14. a, b. Lorette, Quebec. Trenton.
Leperditia (Isochilina) ARMATA Wal., 1883; Des. New Sp. Foss. Trenton Group, p. 7. Russia, Herkimer Co., N. Y.
Birdseye and Black River Limestone.
Leperditia (Isochilina) ARMATA Wal., 1884; 35th Rep. N. Y. State Mus. Nat. Hist., p. 213.
—— BELLULA Jones, 1891; Contrib. Micro-Palæont., p. 3, p. 92, pl. 11, figs. 16 a, b. Hay River, B. C. Devonian.
—— BELLULA Whiteaves, 1892; Contrib. Canadian Palæont., vol. 1, p. 246.
—— CRISTATA (Whitf.) Jones, 1890; Quart. Jour. Geol. Soc., vol. 46, p. 23, pl. 1, fig. 8. Cave Island, Vermont. For other references, see *Primitia cristata.* Calciferous.
Leperditia (Isochilina) CYLINDRICA Jones, 1858; Annals Mag. Nat. Hist., 3d series, vol. 1, p. 253; also, Geol. Survey Canada, decade 3, p. 101.
Leperditia (Isochilina) CYLINDRICA Hall, 1871; Description New Sp. Foss. Hudson River Group. Cincinnati, Ohio. Hudson.
Leperditia (Isochilina) CYLINDRICA Hall, 1872; 24th Rep. New York State Mus. Nat. Hist., p. 231, pl. 8, fig. 2; also, Pal. Ohio, vol. 2, 1875, p. 101, pl. 4, fig. 5. Jones (Quart. Jour. Geol. Soc., vol. 46, 1890, p. 8) refers this species to *Primitia minuta* Eichwald.
—— CYLINDRICA Miller, 1875; Cincinnati Quart. Jour. Sci., vol. 2, p. 351.
—— - DAWSONI Jones, 1891; Contrib. Micro-Palæont., p. 3, p. 92, fig. 7. Dawson Bay, Lake Winnipegosis. Devonian.
—— DAWSONI Whiteaves, 1892; Contrib. Canadian Palæont., vol. 1, p. 346.
—— FABACEA Jones, 1890; Quart. Jour. Geol. Soc., vol. 46, p. 22, pl. 2, fig. 11. 18-Mile Creek, N. Y. Hamilton.
Leperditia (Isochilina) GRACILIS Jones, 1858; Annals Mag. Nat. Hist., London, 3d series, vol. 1, p. 248, pl. 10, fig. 2; also, Geol. Survey Canada, decade 3, p. 98, pl. 2, fig. 15. White Horse Rapids, Isle Jesus, Canada. Trenton.
—— GRANDIS Jones, var. LATIMARGINATA Jones, 1891; Contrib. Micro-Palæont., pt. 3, p. 78, pl. 10, figs. 1 a-b-c, 2 a-b-c, and 3, 4. Saskatchewan River, east of Cedar Lake, etc. Silurian.

ISOCHILINA. JONESELLA.

—— GREGARIA (Whitf.) Jones, 1890; Quart. Jour. Geol. Soc., vol. 46, p. 22, pl. 1, figs. 9, 10. Cave Island, Ball's Bay, Lake Champlain, Vt. For other other references, see *Primitia gregaria*.
Calciferous.

—— JONESI Wetherby, 1881; Jour. Cincinnati Soc. Nat. Hist., vol. 4, p. 80, pl. 2, figs. 7, 7 a. Mercer Co., Ky. Subcarboniferous.

—— JONESI Ulrich, 1890; Jour. Cincinnati Soc. Nat. Hist., vol. 13, p. 179, pl. 11, figs. 9, 9 a–c. Near Harrodsburg, Ky. Trenton.

—— KENTUCKYENSIS Ulrich, 1890; Jour. Cincinnati Soc. Nat. Hist., vol. 13, p. 179, pl. 11, figs. 11, 11 a–d. Mercer Co., Ky., and Frankford, Ky. Birdseye.

—— LABELLOSA Jones, 1891; Contrib. Micro-Palæont., pt. 3, p. 69, pl. 10, figs. 16 a–b–c, 17 and 19. Quebec City. Chazy.

—— LINEATA Jones, 1890; Quart. Jour. Geol. Soc., vol. 46, p. 21, pl. 2, figs. 5 a, b, 8 a, b. Monteith's Point, Canandaigua, N. Y.
Hamilton.

Leperditia (Isochilina) MINUTISSIMA Hall, 1871; Des. New Species Foss. Hudson River Group, p. 7. Cincinnati, Ohio. Hudson.

Leperditia (Isochilina) MINUTISSIMA Hall, 1872; 24th Rep. New York State Mus. Nat Hist., p. 231, pl. 8, fig. 13.

Leperditia (Isochilina) MINUTISSIMA H. & W., 1875; Pal. Ohio, vol. 2, p. 102, pl. 4, fig. 5. (See *Aparchites minutissima*.)

Leperditia (Isochilina) OTTAWA Jones, 1858; Annals Mag. Nat. Hist., London, 3d series, vol. 1, p. 248, pl. 11, fig. 1. Northwest corner Township of L'Orignal, Canada. Chazy.

Leperditia (Isochilina) OTTAWA Jones, 1858; Geol. Survey Canada, decade 3, p. 97, pl. 9, fig. 14. A var. to this species is figured in Contrib. Micro-Palæont., pt. 3, p. 66, pl. 10, figs. 10 a b and 11 a b.

—— OTTAWA Jones, 1884; Annals Mag. Nat. Hist., London, 5th series, vol. 14, p. 345.

—— RECTANGULARIS Ulrich, 1890; Jour. Cincinnati Soc. Nat. Hist., vol. 13, p. 182, pl. 16, figs. 2 a–b. Falls of the Ohio. Devonian.

—— SAFFORDI Ulrich, 1890; Jour. Cincinnati Soc. Nat. Hist., vol. 13, p. 178, pl. 11, figs. 10 a–d. Nashville, Tenn. Trenton.

—— SEELYI (Whitf.) Jones, 1890; Quart. Jour. Geol. Soc., vol. 46, p. 22, pl. 1, fig. 7. For other references see *Primitia seelyi*. Shoreham, Vt.

—— SUB-NODOSA Ulrich, 1890; Jour. Cincinnati Soc. Nat. Hist., vol. 13, p. 177, pl. 11, figs. 7, 7 a–c. Perryville, Ky. Trenton.

—— WHITEAVESII Jones, 1891; Contrib. Micro-Palæont., pt. 3, p. 68, pl. 10, figs. 13 a–b. Lorette Falls, Quebec. Trenton.

JONESELLA Ulrich, 1890; Jour. Cincinnati Soc. Nat. Hist., vol. 13, p. 121.

—— CRASSA Ulrich, 1890; Jour. Cincinnati Soc. Nat. Hist., vol. 13, p. 123, p. 7, figs. 11, 11 a–c. Minneapolis, Minn. Trenton.

JONESELLA.

—— CREPIDIFORMIS Ulrich, 1890; Jour. Cincinnati Soc. Nat. Hist., vol. 13, p. 121, pl. 7, figs. 8, 9 a-b. Covington, Ky. For other references, see *Leperditia crepidiformis*. Cambrian.
—— DIGITATA Ulrich, 1890; Jour. Cincinnati Soc. Nat. Hist., vol. 13, p. 121, pl. 7, figs. 7, 7 a-b. Marion Co., Ky. Cambrian.
—— PEDIGERA Ulrich, 1890; Jour. Cincinnati Soc. Nat. Hist., vol. 13, p. 122, pl. 7, figs. 9, 9 a-b. Covington, Ky. Cambrian.

KLŒDENIA J. & H., 1886; Annals Mag. Nat. Hist., London, 5th series, vol. 17, p. 362.

Beyrichia NOTATA Hall, 1859; Pal. New York, vol. 3, p. 379, pl. 79 b, figs. 3 a, b, c. Herkimer Co., N. Y. Lower Helderberg.
Beyrichia NOTATA var. VENTRICOSA Hall, 1859; *idem*, p. 380, pl. 79 b, figs. 4 a, b, c. Herkimer Co., N. Y.
—— NOTATA Jones, 1890; Quart. Jour. Geol. Soc., vol. 46, p. 13, pl. 4, figs. 22, 23.
—— NOTATA var. VENTRICOSA Jones, 1890; Quart. Jour. Geol. Soc., vol. 46, p. 14, pl. 1, figs. 1 a, b; pl. 4, fig. 24.
—— PENNSYLVANICA Jones, 1889; American Geologist, vol. 4, p. 341, pl., figs. 5-9. Perry Co., Penn. Lower Helderberg.
—— SIMPLEX Jones, 1889; American Geologist, vol. 4, p. 338, pl., fig. 14. King's Mill, Perry Co., Penn. Upper Devonian.

KIRKBYA Jones, 1869; Annals Mag. Nat. Hist., 4th series, vol. 3, p. 223.
—— (*Barychilina?*) COSTADA Ulrich, 1890; Jour. Cinn. Soc. Nat. Hist., vol. 13, p. 208, pl. 18, fig. 2. Warsaw beds, Columbia, Ill. St. Louis.
—— LINDALLI Ulrich, 1890; Jour. Cinn. Soc. Nat. Hist., vol. 13, p. 207, pl. 18, figs. 6, 6 a-c. Warsaw beds, Columbia, Ill. St. Louis.
—— OBLONGA (J. & K.) Ulrich, 1890; Jour. Cinn. Soc. Nat. Hist., vol. 13, p. 206, pl. 18, figs. 4, 4 a-b and 5, 5 a b. Chester.
—— PARALLELA Ulrich, 1890; Jour. Cinn. Soc. Nat. Hist., vol. 13, p. 192, pl. 15, figs. 2, 2 a-b. Falls of the Ohio. Devonian.
—— SEMIMURALIS Ulrich, 1890; Jour. Cinn. Soc. Nat. Hist., vol. 13, p. 193, pl. 15, figs. 3, 3 a-b and 4, 4 a-c. Falls of the Ohio. Devonian.
—— SUBQUADRATA Ulrich, 1890; Jour. Cinn. Soc. Nat. Hist., vol. 13, p. 192, pl. 15, figs. 1, 1 a-c. Falls of the Ohio. Devonian.
—— TRICOLLINA (J. & K.) Ulrich, 1890; Jour. Cinn. Soc. Nat. Hist., vol. 13, p. 207, pl. 18, figs. 8, 8 a-b. Grayson Springs, Ky. Chester.
—— VENOSA Ulrich, 1890; Jour. Cinn. Nat. Hist., vol. 13, p. 208, pl. 18, figs. 3, 3 a-b. Grayson Springs, Ky. Chester.

LEAIA Jones, 1862; Mon. Foss. Estheriæ, p. 115.

Cypricardia LEIDYI Lea, 1855; Proc. Acad. Nat. Sci. Phila., vol. 7, p. 341, pl. 4. Near Pottsville, Penn. Coal Measures.
—— LEIDYI Jones, 1862; Mon. Foss. Estheriæ, p. 116, pl. 5, figs. 11, 12.
—— LEIDYI Daw., 1868; Acadian Geol., p. 256, fig. 78 c.

LEAIA.   LEPERDITIA.

—— LEIDYI Pack., 1882; 12th Rep. U. S. Geol. Survey Territories, p. 358, fig. 24.
    LEIDYI Jones, 1870; Geol. Mag., vol. 7, p. 219, pl. 9, fig. 11; also, Geol. Mag., decade 3, vol. 1, p. 361, pl. 12, fig. 13.
—— LEIDYI var. SALTERIANA Jones; Mon. Foss. Estheriæ; Pal. Soc., 1862, p. 119, pl. 1, fig. 2; also Geol. Mag., decade 3, vol. 1, 1884, p. 361, pl. 12, fig. 13. Horton, N. S.
—— TRICARINATA M. & W., 1868; Geol. Survey Illinois, vol. 3, p. 541, figs. B_1, B_2, B_3. La Salle Co., St. Clair Co., Vermilion Co., Illinois.
    Coal Measures.
—— TRICARINATA White, 1884; 13th Ann. Rep. Dept. Geol. Nat. Hist. Indiana, p. 167, pl. 39, figs. 10–13.
LEPERDITIA Rouault, 1851; Bull. Soc. Géol. France, 2d series, vol. 8, p. 377.
    Cytherina ALTA Conrad, 1842; Geol. Rep. New York, 3d Geol. Dist. (Vanuxem), p. 112. Abundant in the Tentaculite-limestone of Albany, Schoharie, Green, Herkimer, Oneida and Cayuga counties, N. Y.; also Greenfield, Ohio, Waubakee, Wisconsin, and Upper Silurian of Wellington Channel in the Artic Regions, etc. The syn. *L. jonesi* Hall is found in the Coralline-limestone of Schoharie and Herkimer counties of N. Y.   Lower Helderberg.
    Cytherina ALTA? Hall, 1852; Pal. New York, vol. 2, p. 338, pl. 78, figs. 2 a–d. Referred to *Leperditia jonesi* Hall, *idem.*, vol. 3, p. 372.
—— ÆQUILATERA Ulrich, 1892; The Amer. Geol., vol. 10, p. 365. High Bridge, Ky.   Birdseye.
—— ALTA Jones, 1856; Annals and Mag. Nat. Hist., London, 2d series, vol. 17, p. 88, pl. 7, figs. 6, 7.
—— ALTA Jones, 1858; Annals Mag. Nat. Hist., London, 3d series, vol. 1, p. 250, pl. 10, figs. 10, 11.
—— ALTA Hall, 1859; Pal. New York, vol. 3, p. 373, pl. 79 a, figs. 6 a–e.
—— ALTA Meek, 1873; Pal. Ohio, vol. 1, p. 187, pl. 17, figs. 2 a, b.
—— ALTA Jones, 1881; Annals Mag. Nat. Hist., London, 5th series, vol. 8, p. 346.
—— ALTA Whitf., 1882; Geol. Wisconsin, vol. 4, p. 323, pl. 25, figs. 8, 9.
—— ALTA Jones, 1884; Annals Mag. Nat. Hist., London, 5th series, vol. 14, p. 343.
—— ALTA Jones, 1890; Quart. Jour. Geol. Soc., vol. 46, p. 25, pl. 1, figs. 6 a, b.; also Contrib. Micro-Palæont., p. 84, pl. 13, figs. 10, 11 a and 11 b.
—— AMYGDALINA Jones, 1858; Annals Mag. Nat. Hist., London, 3d series, vol. 1, p. 341. One mile west of l'Original, Canada.   Chazy.
—— AMYGDALINA Jones, 1858; Geol. Survey Canada, decade 3, p. 97, pl. 11, figs. 18, 19.
—— AMYGDALINA Jones, 1881; Annals Mag. Nat. Hist., London, 5th series, vol. 8, p. 344, pl. 19, fig. 9.

LEPERDITIA.

—— AMYGDALINA Jones, 1884; Annals Mag. Nat. Hist., London, 5th series, vol. 14, p. 342.

—— ANGULIFERA Whitf., 1881; Annals New York Acad. Sci., vol. 2, p. 197. Greenville, Ohio. Lower Helderberg.

—— ANNA Jones, 1858; Annals Mag. Nat. Hist., London, 3d series, vol. 1, p. 347, pl. 9, fig. 8. Village St. Ann's, Canada. Calciferous.

—— ANNA Jones, 1858; Geol. Survey Canada, decade 3, p. 96, pl. 11, fig. 13.

—— ANTICOSTIANA Jones, 1884; Annals Mag. Nat. Hist., London, 3d series, vol. 14, p. 241. Trenton limestone, Indian Lorette, etc., Canada.

Syn., *canadensis* var. *anticostiana* Jones, 1858; Annals Mag. Nat. Hist., London, 3d series, vol. 1, p. 341. Hudson.

canadensis var. *anticostiana* Jones, 1858; Geol. Survey Canada, decade 3, p. 95, pl. 11, fig. 17.

—— ANTICOSTIANA Bill., 1866; Catalogue Sil. Foss., Anticosti, p. 68.

fabulites (Conrad) var. *anticostiana* Jones, 1881; Annals Mag. Nat. Hist., London, 5th series, vol. 8, p. 344, pl. 19, fig. 8.

—— ANTICOSTIANA Jones, 1891; Contrib. Micro-Palæont., pt. 3, pp. 98, 99.

—— APPRESSA Ulrich, 1890; Jour. Cinn. Soc. Nat. Hist., vol. 13, p. 176, pl. 11, figs. 5 a-d. Near the top of Trenton at Danville and Harrodsburg, Ky.

—— ARCTICA Jones, 1856; Annals Mag. Nat. Hist., London, 2d series, vol. 17, p. 87, figs. 1-5. Cape Hothan in Assistance Bay, etc. Silurian.

Syn., *balthica* var. *arctica* (Jones) Salt., 1852; Appendix Sutherland's Jour. Voyage Baffin's Bay, vol. 2, p. ccxxi, pl. 5, fig. 13.

balthica var. *artica* (Jones) Salt., 1853; Quart. Jour. Geol. Soc. London, vol. 9, p. 314.

—— ARGENTA Wal., 1886; Bull. U. S. Geol. Survey, No. 30, p. 146, pl. 8, fig. 5. Argenta in Big Cottonwood Cañon, Utah. Taconic.

—— BALTHICA (His.) Jones, 1891; Contrib. Micro-Palæont., pt. 3, p. 70; var. *primæva*, p. 70, pl. 10, fig. 18. Carleton Co., Ontario. var. *guelphica*, p. 80, pl. 13, figs. 12 a, b, and 13, a, b. c. Guelph limestone, Durham, Ontario.

—— BILLINGSI Jones, 1856; Annals Mag. Nat. Hist., London, 2d series, vol. 8, p. 345, pl. 15, fig. 9. Trenton.

—— BIVERTEX Ulrich, 1879; Jour. Cincinnati Soc. Nat. Hist., vol. 2, p. 11, pl. 7, figs. 5, 5 a. 1½ miles west of Covington, Ky. Hudson.

—— BIVIA White, 1874; Rep. Invert. Foss. U. S. Geog. and Geol. Survey W. 100th Mer., p. 11. Queen Spring Hill, Schell Creek Range, Nev. Cambrian?

—— BIVIA White, 1877; U. S. Geog. and Geol. Survey W. 100th Mer., Palæont., vol. 4, p. 58, pl. 3, figs. 7 a-d.

—— *byrnesi* Miller, 1874; Cincinnati Quart. Jour. Sci , vol. 1, p. 123, fig. 10. (See *Æchmina byrnesi*.) Hudson.

LEPERDITIA.

—— CÆCA Jones, 1891; Contrib. Micro-Palæont., pt. 3, p. 88, pl. 12, figs. 6, 7 and 9. Grand Rapids, Saskatchewan River, etc. Silurian.

—— CÆCIGENA Miller, 1881; Jour. Cincinnati Soc. Nat. Hist., vol. 2, p. 263, pl. 6, figs. 5, 5 a. Hudson.

—— CÆCIGENA Ulrich, 1890; Jour. Cincinnati Soc. Nat. Hist., vol. 14, p. 176, pl. 11, figs. 6, 6 a-d; also var. *frankfortensis*, p. 177. Frankfort, etc., Ky.

—— CANADENSIS Jones, 1858; Annals Mag. Nat. Hist., London, 3d series, vol. 1, p. 244, pl. 9, figs. 11-15. Chazy and Trenton.

—— CANADENSIS Jones, 1858; Geol. Survey Canada, decade 3, p. 92; var. *nana*, pl. 11, figs. 6, 7, 9, 10. This is the typical *Leperditia canadensis;* also var. *labrosa*, pl. 11, fig. 8; now *Leperditia labrosa*.

—— CANADENSIS Jones, 1884; Annals Mag. Nat. Hist., London, 5th series, vol. 14, p. 340.

—— *canadensis* var. *pauquettiana* Jones. (See *Leperditia louckana*.)

Cythere CARBONARIA Hall, 1858; Trans. Albany Inst., vol. 4, p. 33. Warsaw.

—— CARBONARIA Whitf., 1882; Bull. Am. Mus. Nat. Hist., New York, vol. 1, p. 94, pl. 9, figs. 24-27. Spergen Hill, near Bloomington, Indiana.

—— *capax* Saff.; Geol. Tennessee. (Not defined.)

—— CAYUGA Hall, 1861; Des. new sp. Foss., p. 83. Corniferous.

—— CAYUGA Hall, 1862; 15th Rep. New York State Cab. Nat. Hist., p. 111. Springport, near Lake Cayuga, N. Y. Corniferous.

—— CLAYPOLEI Jones, 1890; Quart. Jour. Geol. Soc., vol. 46, p. 25, pl. 3, figs. 17 a-c. Cincinnati, Ohio.

—— CONCINNULA Bill., 1865; Pal. Foss., vol. 1, p. 299. Point Rich, Table Head, N. F.

—— CREPIFORMIS Ulrich, 1879; Jour. Cincinnati Soc. Nat. Hist., vol. 2, p. 10, pl. 7, figs. 3, 3 a. Covington, Ky. This species is used by Ulrich for the type of the new genus Jonesella. Hudson.

Cytherina CYLINDRICA Hall, 1852; Pal. New York, vol. 2, p. 14, pl. 4, figs. 8 a, b. Medina, Oleans Co., N. Y. Medina.

—— DERMATOIDES Wal., 1887; Am. Jour. Sci., 3d series, vol. 34, p. 192, pl. 1, figs. 13, 13 a. One mile south of Greenwich, etc., Washington Co., N. Y. Taconic.

—— DERMATOIDES Wal.; 10th Rep. U. S. Geol. Survey, p. 626, pl. 80, figs. 1, 1 a. Olenellus zone.

—— (? *Primitia*) DORSICORNIS Ulrich, 1892; The American Geol., vol. 10, p. 267, pl. 9, figs. 24 to 26. Savannah, Illinois. Hudson.

—— EBENINA Dwight, 1889; Am. Jour. Sci., 3d series, vol. 38, p. 144, pl. 6, figs. 2-4. Near Stissing Station, N. Y. Taconic.

LEPERDITIA.

—— EXIGUA Jones, 1891; Contrib. Micro-Palæont., pt. 3, p. 90, pl. 12, fig. 10. Lake Winnipegosis, about 30 miles south of Long Point.
Devonian.
—— ? EXIGUA Whit., 1892; Contrib. Canadian Palæont., vol. 1, p. 347.
—— FABA Hall, 1876; 28th Rep. New York State Mus. Nat. Hist., Expl., pl. 32, figs. 1-3. Waldron, Decatur Co., Indiana. Niagara.
—— FABA Hall, 1879; 28th Rep. New York State Mus. Nat. Hist., p. 186, p. 32, figs. 1-3 (2d ed.).
—— FABA Hall, 1882; 11th Annual Rep. Dept. Geol. Nat. Hist. Indiana, p. 331, pl. 34, figs. 1-3.
Cytherina FABULITES Con., 1843; Proc. Acad. Nat. Sci. Phila., vol. 1, p. 332. Trenton.
—— FABULITES Jones, 1856; Annals Mag. Nat. Hist., London, 2d series, vol. 17, p. 89.
—— FABULITES Ulrich, 1890: Jour. Cincinnati Soc. Nat. Hist., vol. 13, p. 173, pl. 11, figs. 1, 1 a-d and 2. Lebanon, Lavergne, etc., Tenn.; High Bridge, Ky.; Dixon, Ill.; Beloit, Wis.; Minneapolis, Minn. A characteristic fossil of the lower portion of the series of limestones resting on the St. Peter's Sandstone.
—— FABULITES Jones, 1858; Annals Mag. Nat. Hist., London, 3d series, vol. 1, p. 341.
—— FABULITES Jones, 1881; Annals Mag. Nat. Hist., London, 4th series, vol. 8, p. 342.
—— FABULITES Whitf., 1883; Geol. Wisconsin, vol. 1, p. 60, fig. 5.
—— FABULITES Jones, 1884; Annals Mag. Nat. Hist., London, 5th series, vol. 14, p. 342.
—— FIMBRIATA Ulrich, 1892; The American Geologist, vol. 10, p. 268, pl. 9, figs. 34 to 36. Near Spring Valley, Minn. Hudson.
—— FONTICOLA, 1867; 20th Rep. New York State Cab. Nat. Hist., p. 335, p. 21, figs. 1-3. Fond du Lac, Wis. Niagara.
—— FONTICOLA, 1870; 20th Rep. New York State Cab. Nat. Hist. (rev. ed.), p. 428, pl. 21, figs. 1-3.
—— FRONTALIS Jones, 1890; Quart. Jour. Geol Soc., vol. 46, p. 547, pl. 21, figs. 8, a, b. Jumpers Anticosti. Silurian.
—— GERMANA Ulrich, 1892; The American Geologist, vol. 10, p. 266, pl. 9, figs. 16 to 18, Mineral Point, Wis. Birdseye.
—— GRANILABIATA Ulrich, 1892; The American Geologist, vol. 10, p. 267, pl. 9, figs. 31 to 33. St. Paul, Minn. Trenton.
—— GIBBERA Jones, 1856; Annals Mag. Nat. Hist., London, 2d series, vol. 17, p. 90, pl. 7, figs. 1-3. Niagara.
—— GIBBERA Jones, 1858; Annals Mag. Nat. Hist., London, 3d series, vol. 1, p. 250, pl. 10, figs. 7-9.
—— GIBBERA var. SCALARIS Jones, 1858; Geol. Survey Penn., vol. 2, p. 834, fig. 698. Williamsville, Penn.
—— *gracilis.* (See *Isochilini gracilis.*)

LEPERDITIA.

—— HISINGERI (Schm.) Jones, 1891; Contrib. Micro-Palæont., pt. 3, p. 82, pl. 13, figs. 1 and 9. Also, var. *fabulina*, pl. 10, figs. 5 and 7; pl. 12, fig. 15; pl. 13, figs. 2, 3 and 5; var. *gibbera*, pl. 13, fig. 4; var. *egena*, pl. 12, fig. 8.

—— HUDSONICA Hall, 1859; Pal. New York, vol. 3, p. 375, p. 79 a, figs. 7 a, b, c. Becraf's Mt., near Hudson, N. Y. Lower Helderberg.

—— HUDSONICA Jones, 1890; Quart. Jour. Geol. Soc., vol. 46, p. 24, pl. 1, figs. 5 a-c, 11 a-c; pl. 3, fig. 20 (?)

—— INFLATA Ulrich, 1892; The American Geologist, vol. 10, p. 265, pl. 9, figs. 12 to 15. High Bridge, Ky. Birdseye.

—— JONESI Hall, 1859; Pal. New York, vol. 3, p. 272, pl. 79 a, figs. 5 a e.
Lower Helderberg.

—— JONESI Jones, 1884; Annals Mag. Nat. Hist., London, 5th series, vol. 1, p. 343.

Syn., *alta* Hall, 1852; Pal. New York, vol. 2, p. 338, pl. 78, figs. 2 a-d.

—— JOSEPHIANA Jones, 1884; Annals Mag. Nat. Hist., London, 5th series, vol. 1, p. 341. Trenton.

Syn., *canadensis* var. *josephiana* Jones, 1858; Annals Mag. Nat. Hist., London, 3d series, vol. 1, p. 341.

canadensis var. *josephiana* Jones, 1858; Geol. Survey Canada, decade 3, p. 94, pl. 11, fig. 16.

fabulites var. *josephiana* Jones, 1881; Annals Mag. Nat. Hist., London, 5th series, vol. 8, p. 344, pl. 19, fig. 7; pl. 20, figs. 7, 8; also, p. 345, pl. 20, fig. 4 (?).

—— LABROSA Jones, 1858; described as *Leperditia canadensis* var. *labrosa*.

—— LINNEYI Ulrich, 1890; Jour. Cincinnati Soc. Nat. Hist., vol. 13, p. 174, pl. 11, figs. 3, 3 a-e. Near Harrodsburg, Perryville, Danville, etc., Ky. Trenton.

—— LOUCKIANA Jones, 1884; Annals Mag. Nat. Hist., London, 5th series, vol. 14, p. 340. Trenton.

Syn., *canadensis?* Jones, 1858; Annals Mag. Nat. Hist., London, 3d series, vol. 1, p. 245, pl. 9, fig. 16. Afterwards var. *louckiana* and *Leperditia canadensis* var. *pauquettiana*.

canadensis var. *louckiana* Jones, 1858; Geol. Survey Canada, decade 3, p. 93. pl. 11, fig. 11, now *Leperditia louckiana*.

fabulites var. *louckiana* Jones, 1881; Annals Mag. Nat. Hist., London, 5th series, vol. 8, p. 343.

Cypridina MARGINATA Keys., 1846; Wiss. Beob. auf einer Reise in das Petochora-Land, p. 288, pl. 11, fig. 16.

—— MARGINATA Jones, 1858; Annals Mag. Nat. Hist., London, 3d series, vol. 17, p. 91, pl. 7, figs. 11, 14, 15. See also Contrib. Micro-Palæont., pt. 3, p. 86, pl. 10, fig. 6 a, b, c. Lake Winnepegosis.

Syn., *Cypridina balthica* Eichw. (not *Cytherina balthica* Hisinger) Bull. Soc. Nat. Moscou, 1854, No. 1, p. 99, pl. 2, fig. 6.

LEPERDITIA.

— MILLEPUNCTATA Ulrich, 1892; The American Geologist, vol. 10, p. 268, pl. 9, figs. 37 to 39. One mile east of Fountain, Minn.
Trenton.
— (*Isochilina*) *minutissima* Hall. (See *Isochilina minutissima*.)
— *morgani* Saff.; Geol. Tennessee. (Not defined.)
— MUNDULA Ulrich, 1892; The American Geologist, vol, 10, p. 365. High Bridge, Ky. Birdseye. (Chazy?)
— *nana?* Jones, 1890; Quart. Jour. Geol. Soc., vol. 46, p. 27, pl. 4, fig. 4. This species is now included in *Leperditia canadensis*.
— NICLESI Ulrich, 1890; Jour. Cinn. Soc. Nat. His., vol. 13, p. 200, pl. 18, figs. 1, 1 a-e. Columbia, Moore Co., Illinois. St. Louis.
— ? OBSCURA Jones, 1891; Contrib. Micro-Palæont., pt. 3, p. 71, pl. 10, figs. 15 a, b, c. Lorette Falls, Quebec. Trenton.
— OKENI (Mün.) Daw., 1868; Acadian Geology, p. 256, fig. 78 b. This is figured by Jones as var. *Scotobur-digalensis* Hibbert.
— OKENI Jones, 1884; Geol. Mag., decade iii, vol. 1, p. 356, pl. 12, fig. 3. Horton, N. S. Lower Carboniferous.
— *ottawa*. (See *Isochilina ottawa*.)
— OVATA Jones, 1858; Annals Mag. Nat. Hist., London, 3d series, vol. 1, p. 252, pl. 10, fig. 14. Trenton.
— OVATA Jones, 1858; Geol. Survey Penn., vol. 2, p. 834, fig. 697. Potter's Fort, Penn's Valley, Penn.
— *pauquettiana*. (See *Leperditia canadensis* var. *pauquettiana*.)
— *parasitica* Hall, 1859; Pal. New York, vol. 3, p. 376, pl. 79 a, figs. 8 a, b. Referred by T. R. Jones to *Beyrichia parasitica*, which see.
Lower Helderberg.
— PARVULA Hall, 1859; Pal. New York, vol. 3, p. 376, pl. 79 a, figs. 9 a, b. Herkimer Co., N. Y. Lower Helderberg.
— PENNSYLVANICA Jones, 1858; Annals Mag. Nat. Hist., London, 3d series, vol. 1, p. 251, pl. 10, figs. 12, 13. Clinton.
— PENNSYLVANICA Jones, 1858; Geol. Survey Penn., vol. 2, p. 834, fig. 699. Near Barre Forge, Penn.
— PHASEOLUS (Hisinger) Jones, 1891; Contrib. Micro-Palæont., pl. 3, p. 85, pl. 13, figs. 7 and 8. Grand Rapids, Saskatchewan River.
— PHASEOLUS var. GUELPHICA, *idem.*, p. 86, fig. 5. Guelph limestone, Durham, Ontario.
— *punctulifera* Hall, 1860; 13th Rep. New York State Cab. Nat. Hist., p. 92. (See *Primitiopsis punctulifera*.) Hamilton.
— RADIATA Ulrich, 1879; Jour. Cincinnati Soc. Nat. Hist., vol. 2, p. 9, pl. 7, figs. 2 a, b. Cincinnati, Ohio. Utica.
— ROTUNDATA Wal., 1884; Pal. Eureka Dist,, Mon. U. S. Geol. Survey, vol. 8, p. 206, pl. 16, fig. 5. Richmond Mt., Eureka Dist., Nev.
Devonian.

LEPERDITIA.

—— SCALARIS (*Leperditia gibbera* var. *scalaris* Jones, 1858); Annals Mag. Nat. Hist. London, 3d series, vol. 1, p. 250, pl. 10, figs. 7-9.
Waterlime.
—— SELWYNII Jones, 1891; Contrib. Micro-Palæont., pt. 3, p. 89, pl. 12, figs. 1-5. Div. 2, Anticosti group, Jupiter River. Anticosti.
—— SENECA Hall, 1861; Des. New Species Foss., p. 84. Hamilton.
—— SENECA Hall, 1862; 15th Rep. New York State Cab. Nat. Hist., p. 112. Ontario Co., N. Y. Hamilton.
—— SENECA Jones, 1890; Quart. Jour. Geol. Soc., vol. 46, p. 23, pl. 1, figs. 13, 14.
—— SINUATA Hall, 1860; Canadian Naturalist, vol. 5, p. 158.
Upper Silurian.
—— SINUATA Daw., 1868; Acadian Geology, p. 609, fig. 217.
—— SINUATA Jones, 1990; Quart. Jour. Geol. Soc., vol. 46, p. 24, pl. 1, figs. 12 a, c. Arisaig, Nova Scotia.
—— SPINULIFERA Hall, 1861; Des. New Species Foss., p. 83. Corniferous.
—— SPINULIFERA Hall, 1862; 15th Rep. New York State Cab. Nat. Hist., p. 111. Ontario Co., N. Y. Hamilton.
Cythere SUBLÆVIS Shum., 1855; 1st and 2d Ann. Rep. Geol. Survey Missouri, p. 195, pl. B, fig. 15. Lower Magnesian.
—— SUBCYLINDRICA Ulrich, 1889; Contrib. Micro.Palæont., pt. 2, p. 49, pl. 9, figs. 4, 4 b. Stony Mountain, Manitoba. Cambrian.
—— SUBQUADRATA Jones, 1889; The American Geologist, vol. 4, p. 340, pl., figs. 4 a-d. Perry Co., Penn. Lower Helderberg.
—— SULCATA Ulrich, 1892; The American Geologist, vol. 10, p. 266, pl. 9, figs. 19 to 23; also var. VENTRICORNIS. High Bridge, Ky.
Birdseye.
——? SUBROTUNDA Ulrich, 1890; Jour. Cincinnati Soc. Nat. Hist., vol 13, p. 181, pl. 18, figs. 1, 1 a-c. Falls of the Ohio. Devonic.
—— *troyensis* Ford. (See *Aristozoe troyensis*.)
—— TUMIDA Ulrich, 1892; The American Geologist, vol. 10, p. 264. High Bridge, Ky. Birdseye.
—— TUMIDULA Ulrich, 1890; Jour. Cincinnati Soc. Nat. Hist., vol. 13, p. 175, pl. 11, figs. 4, 4 a-c. Danville, Ky. Trenton.
—— TURGIDA Bill., 1865; Palæozoic Foss. Canada, vol. 1, p. 299. Port aux Choix, Cape Norman, N. F.
—— UNICORNIS Ulrich, 1879; Jour. Cincinnati Soc. Nat. Hist., vol. 2, p. 10, pl. 7, figs. 4, 4 a, b. One and a half miles west of Covington, Ky. Referred by Prof. T. R. Jones to *Primitia unicornis*.
Hudson.
—— VENTRALIS Bill., 1865; Palæozoic Foss. Canada, vol. 1, p. 300. Bonne Bay, N. F.
—— WHITEAVESII Jones, 1891; Contrib. Micro-Palæont., pt. 3, p. 87, pl. 12, figs. 11-14, woodcut fig. 6. Cedar Lake. Silurian.

CATALOGUE OF NON—TRILOBITES. 399

LEPIDELLA. MESOTHYRA.
LEPIDELLA ANOMALA Matt., 1885; Trans. Royal Soc. Canada, vol. 3, p. 62, pl. 6, figs. 18, 18 a-b. Div. 1, c. Hanford Brook, St. Martin's, N. B. Paradoxides zone.
LEPIDITTA Matt., 1885; Trans. Royal Soc. Canada, vol. 3, p. 61.
—— ALATA Matt., 1885; Trans. Royal Soc. Canada, vol. 3, p. 61, pl. 6, figs. 16, 16 a. Paradoxides zone. Div. 1, c. Hanford Brook, St. Martin's, N. B.
—— CURTA Matt., 1885; Trans. Royal Soc. Canada, vol. 3, p. 62, pl. 6, fig. 17. Paradoxides zone. Div. 1, d. Porter's Brook, St. Martin's, N. B.
LEPIDOCOLEUS Faber, 1886; Jour. Cincinnati Soc. Nat. Hist., vol. 9, p. 15.
—— JAMESI (H. & W.) Faber, 1886; idem, vol. 9, p. 15, pl. 1, figs. a-f. For other references, see *Pumulites jamesi*. Hudson.
LISGOCARIS Clarke, 1882; Am. Jour. Sci., 3d series, vol. 23, p. 478.
—— LUTHERI Clarke, 1882; Am. Jour. Sci., 3d series, vol. 23, p. 478, pl. 1, fig. 5. Hopewell, Ontario Co., N. Y. Hamilton.
MACROCYPSIS Brady; Intellectual Observ., vol. 12, p. 119.
—— SILIQUA Jones, 1891; Contrib. Micro-Palæont., p. 98. This species was described and figured as *Cytheropsis siliqua* by Jones, 1858; Annals Mag. Nat. Hist., 3d series, vol. 1, p. 249, pl. 10, fig. 6; figured in a reversed position.
——(?) SUBCYLINDRICA Jones, 1890; Quart. Jour. Geol. Soc., vol. 46, p. 549, pl. 21, figs. 5, a, b. Anticosti.
MESOTHYRA Hall, 1888; Pal. New York, vol. 7, p. 187.
Dithyrocaris NEPTUNI Hall, 1876; Illus. Devonian Foss., pl. 23 (not pl. 22, figs. 1-5); pl. 23, figs. 1-3. Plainfield, N. Y. Hamilton.
Dithyrocaris NEPTUNI Pack., 1878; 12th Rep. U. S. Geol. Survey Territories, p. 452, fig. 73.
Dithyrocaris NEPTUNI Eth., W. & J., 1887; 5th Rep. Comm. Foss. Phyllopoda Pal. Rocks, p. 8.
—— NEPTUNI Hall, 1888; Pal. New York, vol. 7, p. 191, pl. 32, fig. 7; pl. 33, fig. 1.
—— NEPTUNI Eth., W. & J., 1888; 6th Rep. Comm. Foss. Phyllopoda Pal. Rocks, p. 8.
——— OCEANI Hall, 1888; Pal. New York, vol. 7, p. 187, pl. 32, figs. 1-6; pl. 33, figs. 4-7; pl. 34, figs. 1-5. Ithaca, N. Y. Portage.
Dithyrocaris neptuni Hall, 1876; Illus. Devonian Foss., pl. 22, figs. 1-5; pl. 33, figs. 1-3.
—— OCEANI Eth., W. & J., 1888; 6th Rep. Comm. Foss. Phyllopoda Pal. Rocks, p. 8.
—— STUMEA Hall, 1888; Pal. New York, vol. 7, p. 193, pl. 32, figs. 8, 9; pl. 34, fig. 2. Pratt Falls and Delphi, N. Y. Hamilton.
—— STUMEA Eth., W. & J., 1888; 6th Rep. Comm. Foss. Phyllopoda Pal. Rocks, p. 8.

MESOTHYRA. PALÆOCARIS.
—— (*Dithyrocaris?*) VENERIS Hall, 1888; Pal. New York, vol. 7, p. 193, pl
    33, fig. 3. East Bloomfield, Ontario Co., N. Y. Hamilton.
  Dithyrocaris? VENERIS Eth., W. & J., 1888; 6th Rep, Comm. Foss.
    Phyllopoda Pal. Rocks, p. 8.
MOOREA J. & K., 1867; Annals Mag. Nat. Hist., 4th series, vol. 3, p. 225.
—— BICONUTA Ulrich, 1890; Jour. Cincinnati Soc. Nat. Hist., vol. 13, p.
    191, pl. 16, figs. 4, 4 a–c. 18-Mile Creek, N. Y. Hamilton.
—— GRANNOSA Ulrich, 1890; Jour. Cincinnati Soc. Nat. Hist., vol. 13, p.
    206, pl. 12, figs. 9, 9 a, b. Grayson Springs, Ky. Chester.
—— KIRKBYI Jones, 1890; Quart Jour. Geol. Soc., vol. 46, p. 542, pl. 20,
    figs. 9, a, b; also figs. 10, a, b. Ontario Co., N. Y. Coniferous.
NEOLIMULUS H. W., 1868; Geol. Mag., vol. 5, p. 1.
NOTHOZOE Bar., 1872; Syst. Sil. Bohême, vol. 1, suppl., p. 536.
—— VERMONTANA Whitf., 1884; Bull. Am. Mus. Nat. Hist., vol. 1, p. 144,
    pl. 14, figs. 14, 15. Dunmore, near Middlebury, Vt. Taconic.
—— VERMONTANA Wal.; 10th Rep. U. S. Geol. Survey, p. 628, pl. 80, figs.
    4, 4 a, b. Olenellus zone.
OCTONARIA Jones, 1887; Annals Mag. Nat. Hist., 5th series, vol. 19, p. 404.
—— CLAVIGERA Ulrich; Jour. Cincinnati Soc. Nat. Hist., vol. 13, p. 195,
    pl. 16, figs. 7, 7 a, c. Falls of the Ohio. Devonian.
—— CURTA Ulrich, Jour. Cincinnati Soc. Nat. Hist., vol. 13, p. 195, pl. 12,
    figs. 4, 4 a, b. Lockport, N. N. Niagara.
—— LINNARSSONI Jones, 1890; Quart. Jour. Geol. Soc., vol. 46, p. 541, pl.
    20, figs. 7, a, b. Clarke Co., Indiana. Hamilton.
—— OVATA Ulrich, 1890; Jour. Cincinnati Soc. Nat. Hist., vol. 13, p. 194,
    pl. 16, figs. 6, 6 a, b. Falls of the Ohio. Devonian.
—— STIGMATA Ulrich, 1890; Jour. Cincinnati Soc. Nat. Hist., vol. 13, p.
    193, pl. 16, figs. 8, 8 a, b; also, var. *oblonga*, p. 194, p. 16, figs. 9,
    9 a, c, and var. *loculosa*, pl. 16, fig. 10. Falls of the Ohio.
       Denovian.
PACHYDONELLA Ulrich, 1890; Jour. Cincinnati Soc. Nat. Hist., vol. 13, p.
    197.
—— TUMIDA Ulrich, 1890; Jour. Cincinnati Soc. Nat. Hist., vol. 13, p.
    198, pl. 13, figs. 5, 5 a, c. Falls of the Ohio. Denovian.
PALÆOCARIS M. & W., 1865; Proc. Acad. Nat. Sci. Phila., vol. 17, p. 48.
—— TYPUS M. & W., 1865; Proc. Acad. Nat. Sci. Phila., vol. 17, p. 49.
    Mazon Creek, Grundy Co., Illinois. Coal Measures.
  Syn., *Acanthotelson inæqualis* M. & W., 1865; Proc. Acad. Nat. Sci.
    Phila., vol. 17, p. 48.
    Acanthotelson inæqualis M. & W., 1866; Geol. Illinois, vol. 2,
      p. 403, pl, 32, fig. 7, 7 a.
—— TYPUS M. & W., 1866; Geol. Illinois, vol. 2, p. 405, pl. 32, figs. 5 a–d.
—— TYPUS M. & W., 1868; Geol. Illinois, vol. 3, p. 552, figs. a, b.

PALÆOCARIS. PRESTWICHIA.
— TYPUS Peach, 1880; Trans. Royal Soc. Edinburgh, vol. 30, p. 85, pl. 10, figs. 10 g, 10 h.
— TYPUS Brocchi, 1880; Bull. Soc. Géol. France, vol. 8, p. 9, pl. 1, figs. 8-10.
— - TYPUS H. W., 1881; Geol. Mag., decade 2, vol. 8, p. 533, wood cut.
— TYPUS White, 1884; 13th Ann. Rep. Dept. Geol. Nat. Hist. Indiana, p. 179, pl. 38, figs. 1, 2, 3.
— TYPUS Pack., 1885; Am. Naturalist, vol. 19, p. 790.
— TYPUS Pack., 1886; Mem. Natl. Acad. Sci., vol. 3, p. 129, pl. 3, figs. 1-4; pl. 7, figs. 1, 2.
PALÆOCREUSIA Clarke, 1888; Pal. New York, vol. 7, p. 210.
— DEVONICA Clarke, 1888; Pal. New York, vol. 7, p. 210, pl. 36, figs. 24, 25. Le Roy, N. Y. Corniferous.
PALÆOPALÆMON Whitf., 1880; Am. Jour. Sci., 3d series, vol. 19, p. 40.
— NEWBERRYI Whitf., 1880; Am. Jour. Sci., 2d series, vol. 19, p. 41. Le Roy, Lake Co., Ohio. Erie shales.
— NEWBERRYI Whitf., 1880; Author's Edition pl., figs. 1, 2, 3.
— NEWBERRYI Hall, 1888; Pal. New York, vol. 7, p. 203, pl. 30, figs. 20-23.
PLACENTULA Jones & Hall, 1886; Annals Mag. Nat. Hist., 5th series, vol. 17, p. 407.
— INORNATA Ulrich, 1890; Jour. Cincinnati Soc. Nat. Hist., vol. 13, p. 124, pl. 10, figs. 14, 14 a-b. Covington, Ky. Cambrian.
— MARGINATA Ulrich, 1890; Jour. Cincinnati Soc. Nat. Hist., vol. 13, p. 124, pl. 10, figs. 13, 13 a-c. Cincinnati, Ohio. Cambrian.
Plumulites Barr. MS. in Sitz. Berichte K. Preuss. Acad. Wissensch., 1864, vol. 49, p. 215, note 2. (See the genus *Turrilepas*.)
POLYCOPE SUBLENTICULARIS Jones, 1890; Quart. Jour. Geol. Soc., vol. 46, p. 550, pl. 21, figs. 6 a, b. Anticosti.
PONTOCYPRIS? ACUMINATA Ulrich, 1890; Jour. Cincinnati Soc. Nat. Hist., vol. 13, p. 210, pl. 17, figs. 8, 8 a-b. Near Granville, Ohio.
Waverly.
— ? ILLINOISENSIS Ulrich, 1890; Jour. Cincinnati Soc. Nat. Hist., vol. 13, p. 107, pl. 11, figs. 16, 16 a-c. Savannah, Illinois. Hudson.
PRESTWICHIA H. W., 1867; Quart. Jour. Geol. Soc. London, vol. 23, p. 32
— DANÆ (M. & W.) Pack., 1886; Mem. Nat. Acad. Sci., vol. 3, p. 146, pl. 5, figs. 3, 39; pl. 6, figs. 1 a, 2, 2 a. Mazon Creek, Grundy Co., Ill. For [other references to this species, see *Euproops* and *Belinurus*.
— *eriensis* Williams. (See *Protolimulus eriensis*.)
— LONGISPINA Pack., 1886; Mem. Natl. Acad. Sci., vol. 3, p. 147, pl. 5, fig. 4. Pittston, Penn. Carboniferous.

26

PRIMITIA J. & H., 1865; Annals Mag. Nat. Hist., 3d series, London, vol. 16, p. 415.
—— ACADICA Matt., 1885; Trans. Roy. Soc. Canada, vol. 3, p. 66, pl. 6, figs. 2, 2 a-b. Div. 1, c, at Porter's Brook, St. Martin's, N. B. Paradoxides zone.
—— BILLINGSI Jones, 1890; Quart. Jour. Geol. Soc., vol. 46, p. 547, p. 21, fig. 10. West of Jupiter River, Anticosti. Clinton or Niagara.
—— CLARKEI Jones, 1890; Quart. Jour. Geol. Soc., vol. 46, p. 535, pl. 20, fig. 11. Ontario Co., N. Y. Devonian.
—— CENTRALIS Ulrich, 1890; Jour. Cincinnati Soc. Nat. Hist., vol. 13, p. 130, pl. 10, figs. 2 a-b. Cincinnati, Ohio. Utica.
—— CHESTERENSIS Ulrich, 1890; Jour. Cincinnati Soc. Nat. Hist., vol. 13, p. 201, pl. 14, figs. 7 a-c; also var. *caldwallensis*, pl. 14, figs. 7 a-c. Chester.
—— CINCINNATIENSIS (Miller) Ulrich, 1890; *idem*, p. 132, pl. 10, figs. 6 a-c. Weisburg and Clarkville, Ohio. Upper half of Cincinnati group.
—— *cristata* Whitf., 1889; Bull. Am. Mus. Nat. Hist., New York, vol. 2, p. 59, pl. 13, figs. 1, 2. (See *Isochilina cristata*.)
—— GLABRA Ulrich; Jour. Cincinnati Soc. Nat. Hist., vol. 13, p. 134, pl. 10, figs. 9 a-c. Oxford and Blanchester, Ohio. Cambrian.
—— GRANIMARGINATA Ulrich, 1890; *idem.*, p. 201, pl. 12, figs. 8 a-b. Grayson Springs Station, Ky. Chester.
—— *gregaria* Whitf., 1889; Bull. Am. Mus. Nat. Hist., New York, vol. 2, p. 58, pl. 13, figs. 3-5. (See *Isochilina gregaria*.)
—— IMPRESSA Ulrich, 1890; Jour. Cincinnati Soc. Nat. Hist., vol. 13, p. 131, pl. 10, figs. 3 a-c and 4 a-c. Savannah, Illinois. Cambrian.
—— LATIVIA Ulrich, 1889; Contrib. Micro-Palæont., pt. 2, p. 50, pl. 9, figs. 8, 8 a. Stony Mountain, Manitoba. Hudson.
Beyrichia logani var. *leperditioides* Jones, 1858; Annals and Mag. Nat. Hist., London, 3d series, vol. 1, p. 244, pl. 9, fig. 10.
Trenton.
Beyrichia logani var. *leperditioides* Jones, 1858; Geol. Survey Canada, decade 3, p. 91, pl. 11, figs. 1-5.
—— LEPERDITIOIDES Jones, 1884; Annals and Mag. Nat. Hist., 6th series, London, vol. 14, p. 345.
—— LOGANI Jones, 1891; Contrib. Micro-Palæont., pt. 3, p. 63, pl. Quebec City. Chazy.
—— MEDALIS Ulrich, 1890; Jour. Cincinnati Soc. Nat. Hist., vol. 13, p. 132, pl. 10, figs. 7 a, 7 b. Jefferson Co., Ky. Cambrian.
—— MILLERI Ulrich, 1890; Jour. Cincinnati Soc. Nat. Hist., vol. 13, p. 133, pl. 12, figs. 2, 2 a-c. Clarksville and Blanchester. Ohio.
Cambrian.
—— MUNDULA Jones, 1891; Contrib. Micro-Palæont., pt. 3, p. 64, var. *effossa*. Quebec City. Also var. *incisa*. For other references, see *Beyrichia mundula*. Lorette Falls, Quebec.

PRIMITIA.

—— MUNDULA Jones, 1889; American Geologist, vol. 4, p. 337, pl., figs. 1, 2, 10, 15. Perry Co., Penn. Upper Devonian.
—— NITIDA Ulrich, 1890; Jour. Cincinnati Soc. Nat. Hist., vol. 13, p. 135, pl. 8, fig. 7. Near Paris, Ky. Trenton.
—— NODOSA Ulrich, 1890; Jour. Cincinnati Soc. Nat. Hist., vol. 13, p. 134, pl. 10, figs. 11 a-b and 12 a-b. Cincinnati, Ohio.
Beyrichia (Primitia) OCCIDENTALIS Wal., 1884; Pal. Eureka Dist., Mon. U. S. Geol. Survey, vol. 8, p. 204, pl. 17, figs. 4, 4 a. Newark Mountain etc., Nevada. Devonian.
—— (Beyrichia?) PARALLELA Ulrich, 1889; Contrib. Micro-Palæont., pt. 2, p. 51, pl. 9, figs. 7-7 a. Oxford, Ohio. Hudson.
—— PENNSYLVANICA Jones, 1889; American Geologist, vol. 4, p. 339, pl., figs. 15 a, b. Perry Co., Penn. Marcellus.
—— PERMINIMA Ulrich, 1890; Jour. Cincinnati Soc. Nat. Hist., vol. 13, p. 131, pl. 8, fig. 7. Licking River, Covington, Ky. Utica.
—— RUDIS Ulrich, 1890; idem., p. 136, pl. 7, fig. 7. Covington, Ky.
Utica.
—— SCITULA Jones, 1891; Contrib. Micro-Palæont., pt. 3, p. 91, pl. 11, figs. 14 a-b. Hay River, which runs into Great Slave Lake, B. C. Devonian.
—— SCITULA Whit., 1892; Contrib. Canadian Palæont., vol. 1, p. 246.
—— SCULPTILIS Ulrich, 1890; Jour. Cincinnati Soc. Nat. Hist,, vol. 13, p. 136, pl. 8, fig. 6. Perryville, Ky. Trenton.
—— SEMINULUM Jones, 1890; Quart. Jour. Geol. Soc., vol. 46, p. 5, pl. 2, fig. 2. 18-Mile Creek, N. Y. Hamilton.
—— seelyi Whitf., 1889; Bull. Am. Mus. Nat. Hist., New York, vol. 2, p. 60, pl. 13, figs. 15-21. (See Isochilina seelyi.)
—— SIMULANS Ulrich, 1890; idem., p. 201, not figured. Chester, Ill.
Chester.
—— SUBSÆQUATA Ulrich, 1890; Jour. Cincinnati Soc. Nat. Hist., vol. 13, p. 202, pl. 14, figs. 8 a-b. Clayton P. O., Caldwell Co., Ky.
Chester.
—— UNICORNIS (Ulrich) Jones, 1890; Quart. Jour. Geol. Soc., vol. 46, p. 7, pl. 11, figs. 8-13. Cincinnati, Ohio. For other references, see Beyrichia unicornis Ulrich. The species is referred by Ulrich (Contrib. Micro-Palæont., pt. 2, p. 50) to Aparchites unicornis, pl. 9, fig. 11.
—— WALCOTTI Jones, 1890; Quart Jour. Geol. Soc., London, vol. 46, p. 543, fig 1. Thedford, Ontario, Canada. Referred to Kirkbya. walcotti Jones, 1891; Contrib. Micro-Palæont., pt. 3, p. 96, pl. 11, figs. 12 a, b.
—— WHITFIELDI Jones, 1890; Quart. Jour. Geol. Soc., vol. 46, p. 9, pl. 3; figs. 24 a, b. Cincinnati, Ohio. Cambrian.

PRIMITIOPSIS. PTERYGOTUS.

PRIMITIOPSIS Jones, 1887; Notes on some Sil. Ostracoda from Gothland, p. 5.
—— PUNCTULIFERA (Hall) Jones, 1890; Quart. Jour. Geol. Soc., vol. 46, p. 9, pl. 2, figs. 7 a, b, 12 a, b. Seneca Lake, etc., N. Y. For other references, see *Beyrichia punctulifera* Hall. Hamilton.
—— PUNCTULIFERA Jones, 1891; Contrib. Micro-Palæont., pt. 3, p. 95, pl. 11, figs. 11 a, b. Thedford, Ontario.
PROTOBALANUS Whitf., 1889; Bull. Am. Mus. Nat. Hist., New York, vol. 2, p. 66; Hall, 1888; Pal. New York, vol. 7, pp. lxii and 209.
—— HAMILTONENSIS (Whitf.) Hall, 1888; Pal. New York, vol. 7, p. 209, pl. 36, fig. 23. Avon, Genesee Co., N. Y.
—— HAMILTONENSIS Whitf., 1889; Bull. Am. Mus. Nat. Hist., New York, vol. 2, p. 67, pl. 13, fig. 22. Marcellus.
PROTOCARIS Wal., 1884; Bull. U. S. Geol. Survey, No. 10, p. 50.
—— MARSHI Wal., 1884; Bull. U. S. Geol. Survey, No. 10, p. 50, pl. 10. Georgia, Vermont. Olenellus zone. Taconic,
—— MARSHI Wal., 1886; Bull. U. S. Geol. Survey, No. 30, p. 148, pl. 15, fig. 1.
—— MARSHI Wal.; 10th Rep. U. S. Geol. Survey, p. 629, pl. 81. fig. 6. Olenellus zone.
PROTOLIMULUS Pack., 1866; Mem. Natl. Acad. Sci., vol. 3, p. 150.
Prestwichia ERIENSIS Williams, 1885; Am. Jour. Sci., 3d series, vol. 30, p. 45, 3 figs. Le Bœuf, Erie Co., Penn. Devonian.
—— ERIENSIS Pack., 1886; Mem. Natl. Acad. Sci., vol. 3, p. 150, figs. 11-13.
—— ERIENSIS Hall, 1888; Pal. New York, vol. 7, p. 153, pl. 26, figs. 1, 2.
PTERYGOTUS Agas., 1844; Mon. Poissons Fossiles, note, p. xix.
—— ACUTICAUDA Pohl., 1882; Bull. Buffalo Soc. Nat. Sci., vol. 4, p. 42, pl. 2, fig. 3. Buffalo, N. Y. Waterlime.
—— BILOBUS (Huxley and Salter) Pohl., 1886; Bull. Buffalo Soc. Nat. Sci., vol. 5, p. 27. Buffalo, N. Y. Waterlime.
—— BUFFALOENSIS Pohl., 1881; Bull. Buffalo Soc. Nat. Sci., vol. 4, p. 17' figs. 1, 2. Buffalo, N. Y. Waterlime.
—— BUFFALOENSIS Pohl., 1882; Bull. Buffalo Soc. Nat. Sci., vol. 4, p. 44, pl. 3, fig. 3.
—— BUFFALOENSIS Pohl., 1886; Bull. Buffalo Soc. Nat. Sci., vol. 5, p. 24, pl. 3, fig. 1.
—— CANADENSIS Daw., 1881; Canadian Naturalist, new series, vol. 9, p. 103, figs. 1, 2. Niagara.
—— COBBI Hall, 1859; Pal. New York, vol. 3, p. 417, pl. 83 b, fig. 4; pl. 84, fig. 8 (?). Near Buffalo, N. Y. Waterlime.
—— CUMMINGSI G. & P., 1875; Bull. Buffalo Soc. Nat. Sci., vol. 3, p. 18, fig. Buffalo, N. Y. Waterlime.

PTERYGOTUS.

— CUMMINGSI G. & P., 1878; Proc. Am. Assoc. Adv. Sci., 26th Meeting, p. 300, fig.
— GLOBICAUDATUS Pohl., 1882; Bull. Buffalo Soc. Nat. Sci., vol. 4, p. 42, pl. 2, fig. 2. Buffalo, N. Y. Waterlime.
— MACROPHTHALMUS Hall, 1859; Pal. New York, vol. 3, p. 418, pl. 80 a, figs. 8, 8 a. Wheelock's Hill, Lichfield, Herkimer Co., N. Y. Waterlime.
— MACROPHTHALMUS Pohl., 1882; Bull. Buffalo Soc. Nat. Sci., vol. 4, p. 44.
— OSBORNI Hall, 1859; Pal. New York, vol. 3, p. 419, pl. 80 a, fig. 9. Waterville, Oneida Co., N. Y.
— QUADRICAUDATUS Pohl., 1882, Bull. Buffalo Soc. Nat. Sci., vol. 4, p. 43, pl. 3, fig. 1. Buffalo, N. Y. Waterlime.

Rachura Scudder, 1878; Proc. Boston Soc. Nat. Hist., vol. 19, p. 296. Probably the same as the genus *Dithyocaris*.
— *venosa* Scudder, 1878; Proc. Boston Soc. Nat. Hist., vol. 19, p. 296. Coal Measures.
— (*Dithyocaris?*) *venosa* Eth., W. & J., 1888; 6th Rep. Comm. Foss. Phyllopoda, p. 4.

RHINOCARIS Clarke, 1888; Pal. New York, vol. 7, p. lviii.
— COLUMBINA Clarke, 1888; Pal. New York, vol. 7, p. 195, pl. 31, figs. 16-21. Vinegar Brook Glen, near Norton's Landing, on Cayuga Lake, N. Y. Hamilton.
— COLUMBINA Eth., W. & J., 1888; 6th Rep. Comm. Foss. Phyllopoda, etc., p. 8.
— SCAPHOPTERA Clarke, 1888; Pal. New York, vol. 7, p. 197, pl. 31, figs. 22, 23. Tichenor's Glen, Canandaigua Lake, N. Y. Hamilton.
— SCAPHOPTERA Eth., W. & J., 1888; 6th Rep. Comm. Foss. Phyllopoda Pal. Rocks, p. 8.

SCHIZODISCUS Clarke, 1888; Pal. New York, vol. 7, p. 207.
— CAPRA Clarke, 1888; Pal. New York, vol. 7, p. 207, pl. 25, figs. 1-9. Centerfield, N. Y. Hamilton.

SCHMIDTELLA Ulrich, 1892; The American Geologist, vol. 10, p. 269.
— CRASSIMARGINATA Ulrich, 1892; The American Geologist, vol. 10, p. 269, pl. 9, figs. 27 to 30. Mineral Point, Wis. Birdseye.

Solenocaris Meek, 1871; Proc. Acad. Nat. Sci. Philadelphia, vol. 23, p. 335. The name *Solenocaris* was used for a genus of fossil crustacea by Mr. J. Young, in 1868. Changed to *Strigocaris*.
— *ludovici* Worthen, 1884; Bull. Illinois State Cab. Nat. Hist., No. 2, p. 3. St. Louis.
Ceratiocaris (*Solenocaris*) *strigata* Meek, 1872; Proc. Acad. Nat. Sci. Philadelphia, vol. 23, p. 335. Waverly.
Ceratiocaris (*Solenocaris*) *strigata* Meek, 1875; Pal. Ohio, vol. 2, p. 321, pl. 18, figs. 4 a-c.

SPATHIOCARIS.   STYLONURUS.

SPATHIOCARIS Clarke, 1882; Am. Jour. Sci., 3d series, vol. 23, p. 477.
—— EMERSONI Clarke, 1882; Am. Jour. Sci., 3d series, vol. 23, p. 477, pl.,
    figs. 1-3. Bristol, Richmond and Naples, N. Y., etc. Hamilton.
—— EMERSONI Pack., 1883; 12th Rep. U. S. Geol. Survey Territories,
    p. 451.
—— EMERSONI Clarke, 1886; Neues Jarbuch für Mineral., p. 180.
—— EMERSONI W. & J., 1884; 2d Rep. Comm. Foss. Phyllopoda Pal.
    Rocks, p. 7.
—— EMERSONI Clarke, 1885; Bull. U. S. Geol. Survey, No. 16, p. 40.
    Naples, N. Y.
—— EMERSONI Eth., W. & J., 1885; 3d Rep. Comm. Foss. Phyllopoda Pal.
    Rocks, p. 3.
—— EMERSONI Hall, 1888; Pal. New York, vol. 7, p. 199, pl. 35, figs. 12-18.
—— EMERSONI Eth., W. & J., 1888; 6th Rep. Comm Foss. Phyllopoda Pal.
    Rocks, p. 8.
STREPULA J. & H., 1886; Annals Mag. Nat. Hist., 5th series, London, vol.
    17, p. 403.
—— LUNATIFERA Ulrich, 1889; Contrib. Micro-Palæont., pt. 2, p. 56, pl. 9,
    figs. 14, 14 b. Stony Mountain, Manitoba.   Hudson.
—— PLANTARIS Jones, 1890; Quart. Jour. Geol. Soc., vol. 46, p. 540, pl.
    20, figs. 8, a, b. 18-Mile Creek, N. Y.   Devonian.
—— QUADRILIRATA (H. & W.) Ulrich, 1889; Contrib. Micro-Palæont., pt.
    2, p. 54, pl. 9, fig. 12. Stony Mountain, Manitoba. For other
    references, see *Beyrichia quadrilirata*.
—— SIGMOIDALIS Jones, 1890; Quart. Jour. Geol. Soc., vol. 46, p. 11, pl.
    2, fig. 4. 18-Mile Creek, N. Y,   Hamilton.
STRIGOCARIS, name suggested to replace preoccupied generic name *Soleno-
    caris* (Meek) Vogd., 1889; Annals N. Y. Acad. Sci., vol. 5, p. 34.
    Solenocaris ST. LUDOVICI Worthen, 1884; Bull. Illinois State Cab. Nat.
    Hist., No. 2, p. 3.   St. Louis.
    Solenocaris ST. LUDOVICI Worthen, 1890; Geol. Illinois, vol. 8, p. 153,
    pl. 28, fig. 3. St. Louis limestone, South St. Louis, Mo.
    Ceratiocaris (Solenocaris) STRIGATA Meek, 1872; Proc. Nat. Acad. Sci.
    Phila., vol. 23, p. 335. Danville, Ky.   Waverly.
    Ceratiocaris (Solenocaris) STRIGATA Meek, 1875; Pal. Ohio, vol. 2, p.
    321, pl. 18, figs. 4 a-c.
STROBILEPSIS Clarke, 1888; Pal. New York, vol. 7, p. 212.
—— SPINIGERA Clarke, 1888; Pal. New York, vol. 7, p. 212, pl. 36, figs.
    20-22. Menteth's Point, Canandaigua Lake, N. Y.   Hamilton.
STYLONURUS Page, 1856; Advance Text-book Geology, p. 135.
—— EXCELSIOR Hall, MSS. Andes, Delaware Co., N. Y.   Catskill.
—— EXCELSIOR Martin, 1882; Trans. New York Acad. Sci., vol. 2, p. 8.
—— EXCELSIOR Hall, 1884; 36th Rep. New York State Mus., p. 77, pl. 5,
    fig. 1.

STYLONURUS. TROPIDOCARIS.
—— EXCELSIOR Hall, 1885; Proc. Am. Assoc. Adv. Sci., 33d Meeting, p. 421.
—— EXCELSIOR Hall, 1888; Pal. New York, vol. 7, p. 156, pl. 26; also p. 221, pl. 26 a. cf. *Dolichocephala lacoana* Claypole.
Equistides WRIGHTIANUS Daw., 1881; Quart. Jour. Geol. Soc., vol. 37, p. 303, pl. 12, fig. 10; pl. 13, fig. 20. Italy, Yates Co., N. Y.
Portage.
Equistides WRIGHTIANUS (Daw.) Wright; 35th Rep. New York State Mus. Nat. Hist., Expl., pl. 15, note, figs. 1, 2.
Echinocaris WRIGHTIANUS J. & W., 1884; Geol. Mag., decade 3, vol. 1, p. 3, pl. 13, figs. 1 a, b.
Echinocaris WRIGHTIANUS Eth., W. & J., 1885; 3d Rep. Comm. Foss. Phyl. Pal. Rocks, p. 35.
—— ? *Echinocaris* WRIGHTIANUS Hall, 1888; Pal. New York, vol. 7, p. 160, pl. 27, figs. 7-9.
TEDRADELLA Ulrich, 1890; Jour. Cincinnati Nat. Hist., vol. 13, p. 112.
—— CHAMBERSI (Miller) Ulrich, 1890; Jour. Cincinnati Society Nat. Hist., vol. 13, p. 112. For other references see *Beyrichia chambersi*.
—— LUNATIFERA Ulrich, 1890; Jour. Cincinnati Soc. Nat. Hist., vol. 13, p. 112. For other references see *Strepula lunatifera*.
—— OCULIFERA (H. & W.) Ulrich, 1890; Jour. Cincinnati Soc. Nat. Hist., vol. 13, p. 112, fig. For other references see *Beyrichia oculifera*. Cincinnati Hills, Ohio.
—— QUADRILIRATA (H. & W.) Ulrich, 1890; Jour. Cincinnati Soc. Nat. Hist., vol. 13, p. 112; (type) Birdseye limestone of Kentucky and Minnesota. For other references see *Beyrichia quadrilirata*.
—— SUBQUADRATA Ulrich, 1890; Jour. Cincinnati Soc. Nat. Hist., vol. 13, p. 113, fig. 2. Trenton Fall, N. Y. Trenton.
TROPIDOCARIS Beecher, 1884; 2d Geol. Survey Pennsylvania, vol. PPP, p. 15.
—— ALTERNATA Beecher, 1884; 2d Geol. Survey Penn., vol. PPP, p. 19, pl. 2, figs. 7, 8. Warren, Penn. Chemung.
—— ALTERNATA Eth., W. & J., 1885; 3d Rep. Comm. Foss. Phyllopoda Pal. Rocks, p. 35.
—— ALTERNATA Hall, 1888; Pal. New York, vol. 7, p. 186, pl. 31, figs. 14, 15.
—— BICARINATA Beecher, 1884; 2d Geol. Survey Penn., vol. PPP, p. 16, pl. 2, figs. 3-5. Warren, Penn. Chemung.
—— BICARINATA Eth., W. & J., 1885; 3d Rep. Comm. Foss. Phyllopoda Pal. Rocks, p. 35.
—— BICARINATA Hall, 1888; Pal. New York, vol. 7, p. 184, pl. 31, figs. 7-12.
—— HAMILTONENSIS Hall, 1888; Palæont. New York, vol. 7, p. 227, pl. 30, figs. 24-25. Foster's, Canandaigua Lake, N. Y. Hamilton.

TROPIDOCARIS. ULRICHIA.
—— INTERRUPTA Beecher, 1884; 2d Geol. Survey Penn., vol. PPP, p. 18,
pl. 2, fig. 5. Warren, Penn. Chemung.
—— INTERRUPTA Eth., W. & J., 1885; 3d Rep. Comm. Foss. Phyllopoda
Pal. Rocks, p. 35.
—— INTERRUPTA Hall, 1888; Pal. New York, vol. 7, p. 185, pl. 31, figs.
7-12.
TURRILEPAS H. W., 1865; Quart. Jour. Geol. Soc., vol. 21, p. 486.
Oploscolex Salt., 1873; Cat. Camb. and Sil. Foss. Woodwardian Mus.,
p. 129.
Plumulites Barr.; MS. in Sitz. Berichte K. Akad. Preuss. Wissensch.,
1864, vol. 49, p. 215, note 2.
—— CANCELLATUS Hall, 1888; Pal. New York, vol. 7, p. 216, pl. 36, fig. 2.
Canandaigua, N. Y. Corniferous.
Plumulites DEVONICUS Clarke, 1882; Am. Jour. Sci., 3d series, vol. 24,
p. 55, figs. 1, 2. Canandaigua, N. Y. Hamilton.
—— DEVONICUS Hall, 1888; Pal. New York, vol. 7, p. 215, pl. 36, fig. 1.
—— FLEXUOSUS Hall, 1888; Pal. New York, vol. 7, p. 215, pl. 36, fig. 1.
Canandaigua, N. Y. Corniferous.
—— FOLIATUS Hall, 1888; Pal. New York, vol. 7, p. 218, pl. 36, fig. 15.
Canandaigua, N. Y. Hamilton.
Plumulites JAMESI H. & W., 1875; Pal. Ohio, vol. 2, p. 106, pl. 4, figs.
1-3.
——? JAMESI Miller, 1875; Cincinnati Quart. Jour. Sci., vol. 2, p. 275, fig.
19. Referred by Faber (Jour. Cincinnati Soc. Nat. Hist., vol. 9,
1886, p. 15) to a new genus under the name of *Lepidocoleus*.
Plumulites NEWBERRYI Whitf., 1882; Annals New York Acad. Sci.,
vol. 2, p. 217. Sheffield and Birmingham, Erie Co. Ohio.
Hamilton.
—— NEWBERRYI Hall, 1888; Pal. New York, vol. 7, p. 219, pl. 36, figs.
16-19.
—— NITIDULUS Hall, 1888; Pal. New York, vol. 7, p. 218, pl. 36, fig. 4.
Canadaigua, N. Y. Hamilton.
—— SQUAMA Hall, 1888; Pal. New York, vol. 7, p. 217, pl. 36, figs. 5-8.
Canandaigua, N. Y. Hamilton.
—— TENER Hall, 1888; Pal. New York, vol. 7, p. 219, pl. 36, figs. 9-14.
Centerfield, N. Y. Hamilton.
ULRICHIA Jones, 1890; Quart. Jour. Geol. Soc., vol. 46, p. 543.
—— CONRADI *idem.*, p. 543, fig. 2. Thedford, Canada. Ontario, Canada.
—— CONRADI, Jones, 1891; Contrib. Micro-Palæont., pt. 3, p. 95, pl. 11,
fig. 13.
——? CONFLUENS Ulrich, 1890; Jour. Cincinnati Soc. Nat. Hist. vol. 13,
p. 203, pl. 12, figs. 11 a-b. Near Grayson Springs, Ky. Chester.
—— EMARGINATA Ulrich, 1890; Jour. Cincinnati Soc. Nat. Hist., vol. 13, p.
203, pl. 12, figs. 10, 10 a-c. Near Grayson Springs, Ky. Chester.

APPENDIX.

In the compilation of the Bibliography the following were inadvertently omitted:

Braum (F.) Petrefactenkunde der Natur. Sammulung zu Baireuth. Leipzig, 1840, 22 pls.

Calymene furcata n. sp.
This species is quoted on Munster's authority. I have not seen the work.

Krause (Aurel). Neue Ostrakoden aus märkischen Silurgeschieben.

In Zeitsch. Deutsch. Geol. Gesell., vol. 46, 1892, p. 383, pls. 21-22.

1, *Isochilina canaliculata* n. sp. 2, *Primitia plana* Krause var. *tuberculata* n. var. 3, *P. distans* Krause. 4, *P. elongata* Krause. 5, *P. corrugata* n. sp. 6, *P. plicata* n. sp. 7, *P. (Halliella) seminulum* Jones. 8, *P.* aff. *obliquipunctata* Jones. 9, *P. papillata* n. sp. 10, *Entomis obliqua* n. sp. 11, *Entomis (Primitia) flabellifera* n. sp. 12, *Primitia excavata* n. sp. 13, *P. (Ulrichia?) umbonata* n. sp. 14, *P. (Ctenobolbina?) globifera* n. sp. 15, *P. labrosa* n. sp. 16, *Entomis simplex* n. sp. 17, *E. auricularis* n. sp. 18, *E. plicata* n. sp. 19, *E. trilobata* n. sp. 20, *E. (Bursulella?) quadrispina* n. sp. 21, *Bollia minor* n. sp. 22, *B. major* n. sp. 23, *B. duplex* n. sp. 24, *Beyrichia dissecta* n. sp. 25, *B. mammillosa* n. sp. 26, *B. radians* n. sp. 27. *B. plicatula* n. sp. 28, *B. (Tedradella) harpa* n. sp. 29, *B. (Tetradella) carinata* n. sp. 30, *B. (Tedradella) signata* n. sp. 31, *B. (Ctenobolbina) rostrata* n. sp. 32, *B. (Ulrichia?) bidens* n. sp. 33, *Octonaria bifasciata* n. sp. 34, *Thlipsura v-scripta* var. *discreta* Jones. 35, *Æchmina bovina* Jones var. *punctata* n. var. 36, *Crustaceum* sp.?

Lea (Isaac). In the Proceedings Acad. Nat. Sci. of Philadelphia May, 1855, p. 341, Dr. Lea has described a small fossil found by Dr. Leidy in red sandstone at Tumbling Run Dam near Pottsville, Penn., as *Cypricardia leidyi*.

This species has been classed under the new genus of *Leaia* by Prof. T. Rupert Jones.

Lima (W. De). Note sur un nouvelle *Eurypterus* du Rothliegendes de Bussaco (Portugal).

In Communicacoes Comm. Trabalhos Geol. de Portugal, vol. 2, Facs. 2, 1892, p. 152, pl.

Eurypterus douvillei n. sp.

Matthew (G. F.) The diffusion and sequence of the Cambrian faunas.

In Trans. Royal Soc. Canada, Sec. 4, 1892.

Contains notes on some North American Palæozoic Crustacea.

Quenstedt (F. A.) Handbuche der Petrefactenkunde. Tübingen, 1852, with Atlas of 62 pls.; 2d ed., 1867, Atlas of 86 pls.; 3d ed., 1882–85, Atlas of 100 pls.

The author illustrates in the second edition the following Palæozoic Crustacea: *Limulus triloitoides* Buckl.

Plate 28—*Illænus crassicauda* Wahl. *Gerastos cornutus* Goldf., *G. lævigatus* Goldf. *Trill. expansus* Linné, *T. palifer* Beyr., *T. flabellifer* Goldf., *T. latifrons* Bronn, *T. sclerops* Dalm. *Entomolithus derbyensis* Martin. *Staurocephalus murchisoni* Barr. *Sphærexochus clavifrons* Dalm. *Trochurus speciosus* Beyr. *Tril. sternbergii* Boeck., *T. caudatus* Brünn, *T. hausmanni* Brong. *Lichas scabra* Beyr. *Odontopleura mirus* Barr. *Arges armatus* Goldf. *Tril. ceratophthalmus* Goldf. *Metopias verrucosus* Eichw.

Plate 29—*Tril. hoffi* Schloth, *T. senaria* Conrad. *Homalonotus lævicauda* n. sp. This species was described in the first edition, p. 294, pl. 23, fig. 9. *Homalonotus knighti* König. *Tril. sulzeri* Schloth, *T. striatus* Emm. *Eurypterus remipes* De Kay. *Harpes ungula* Sternb. *Tril. bohemicus* Boeck. *Battus integer* Beyr. *Tril. pisiformis* Linné. *Sao hirsuta* Barr. *Amphion polytoma* Dalm. *Agnostus tuberculatus* Kloden. *Cypris faba* Desm. *Cypridina edwardsiana* DeKon. *Cypris inflata* Murch. *Tril. ornatus* Sternb. *Cytherina balthica* His.

Plate 30—*Asaphas expansus* Linné. *Calymene blumenbachii* Brong. *Illænus crassicauda* Wahl. *Paradoxides spinosus* Wahl. *Amphion polytoma* Dalm. *Arethusina konincki* Barr. *Arionellus ceticephalus* Barr. *Cyphaspis burmeisteri* Barr. *Olenus truncatus* Brünn. *Zethus bellatula* Eichw., *Z. verrucosus* Pand. *Cryphaspis punctatus* Steinb. *Remopleurides radians* Barr. *Phacops sclerops* Dalm. *Dindymene haidingeri* Barr. *Chasmops odini* Eichw. *Agnostus nudus* Beyr., *A. granulatus* Barr. *Phillipsi kellii* Portl. *Phacops tettinensis* n. sp., *P. latifrons* Bronn. *Sphærexochus hemicranium* Kut. *Bumastus barriensis* Murch. *Acidaspis mira* Barr. *Asaphus cornutus* Pander. *Encrinurus punctatus* Brünn. *Ampyx portlocki* Barr. *A. nasutus* Dalm. *A. bucklandi* Barr. *Phillipsia granulifera* Phill. *Dionide formosa* Barr. *Himantopterus* sp.? Salt. *Bairdia subdeltoidea* Jones. *Estheria murchisoni* Jones. *Cypridina serrato-striata*. *Leperditia balthica* His.

Salter (J. W.) Description of Palæozic Crustacea and Radiata from South Africa.

In Trans. Geol. Soc. London, second series, vol. 7, 1845-1856, pp. 215-224, pl. 24.

Homalonotus herschelii Murch. *Phacops (Cryphæus) africanus* n. sp., *P. caffer* n. sp. *Phacops* sp. *Typhloniscus* n. g., *T. bainii* n. sp.

Shumard (B. F.) and Swallow (G. C.) Descriptions of new fossils from the Coal Measures of Missouri and Kansas.

In Trans. Acad. Sci. St. Louis, vol. 1, 1858, p. 198.

Phillipsia missouriensis Shum., *P. major* Shum., *P. cliftonensis* Shum. *Cythere (Beyrichia) americana* Shum.

Ulrich (E. O.) Contributions to Micro-Palæontology of the Cambro-Silurian Rocks of Canada. Part 2. Montreal, 1889.

Bythocypris cylindrica Hall. *Leperditia subcylindrica* n. sp. *Aparchites minutissimus* Hall. *A. unicornis* Ulrich, var. *Primitia lativia* n. sp., *P.? (Beyrichia) parallela* n. sp. *Eurychlina* n. g., *E. reticulata* n. sp., *E. manitobensis* n. sp. *Strepula quadrilirata* Hall, *S. lunatifera* n. sp.

Vanuxem (Lardner). Geology of New York. Part 3. Survey of the Third Geological District, Albany, 1842.

The author illustrates the following Trilobites: *Isotelus gigas* Dekay. *Triarthrus beckii* Green. *Hemicrupturus clintoni* n. sp. *Eurypterus remipes* Dekay. *Odontocephalus selenurus* Eaton. *Dipleura dekayi* Green.

Walchs (Johan Ernst Immanuel) Beyträg zur Naturgeschichte der Trilobiten.

In Der Naturforscher, vol. 9, Halle, 1776, p. 276, pl. 4, figs. 2, 3, 4.

Walcott (C. D.) Description of new genera and species of fossils from the Middle Cambrian.

In Proc. U. S. Natl. Mus., vol. 11, 1888, pp. 441–442.

Olenoides nevadensis Meek, *O. curticei* n. sp., *O.* sp. undet. *Karlia* n. g., *K. minor* n. sp., *K. stephenensis* n. sp. This name is used for *Menocephalus salteri* Rom. *Bathyuriscus (Kootenia) dawsoni* n. sp. for *Bathyurus? romingeri* Rom. *Ogygopsis* n. g. for *Ogygia klotzi* Rom.

―――― Description of new forms of Upper Cambrian fossils.

In Proc. U. S. Natl. Mus., vol. 13, 1890, pp. 266–279, pls. 20, 21.

Ptychoparia burnetensis n. sp., *P. connata* n. sp., *P. llanoensis* n. sp., *P.? metra* n. sp., *P. pero* n. sp., *P. suada* n. sp., *P. urania* n. sp., *P. vacuna* n. sp., *P. (Liostracus) panope* n. sp. *Agraulos? thea* n. sp., *A. saratogensis* n. sp. *Illænus dia* n. sp.

SUPPLEMENT TO CATALOGUE OF TRILOBITES.

AGRAULOS SARATOGENSIS Wal., 1890; Proc. U. S. Natl. Mus., vol. 13, p. 276, pl. 21, fig. 14. Saratoga Springs, N. Y. Taconic.
—— ? THEA Wal., 1890; Proc. U. S. Natl. Mus., vol. 13, p. 277, pl. 21, fig. 17. Oseola Mills, Wis. Taconic.
ILLÆNUS? DIA Wal., 1890; Proc. U. S. Natl. Mus., vol. 13, p. 277, pl. 20, fig. 9. Morgan's Creek, Burnet Co., Texas. Taconic.
KARLIA Wal., 1888; Proc. U. S. Natl. Mus., vol. 11, p. 445.
—— MINOR Wal., 1888; Proc. U. S. Natl. Mus., vol. 11, p. 445, Mount Stephens, Canada. Taconic.
—— STEPHENENSIS Wal., 1888; Proc. U. S. Natl. Mus., vol. 11, p. 445. Mount Stephens, Canada. For other references see *Mencephalus salteri* Rom. Taconic.
BATHYURISCUS (KOOTENIA) DAWSONI Wal., 1888; Proc. U. S. Natl. Mus., vol. 11, p. 446. For other references see *Bathyurus? romingeri*. Mount Stephens, Canada. Taconic.
OGYGOPSIS Wal., 1888; Proc. U. S. Natl. Mus., vol. 11, p. 446.
—— KLOTZI Wal., 1888; Proc. U. S. Natl. Mus., vol. 11, p. 446. For other references see *Ogygia klotzi* Rom. Mount Stephens, Canada. Taconic.

ERRATA.

Page 3, for 1891 read 1892.

Page 18, Bayle, line 13, for Barb. read Barr.

Page 19, Beyrich, line 4, for Boek read Boeck.

Page 32, Buch, line 7, for *varrucosus* read *verrucosus*.

Page 46, Dalman, line 2, for uwptäckter read upptäckter.

Page 46, Dalman, line 5, for Palæades read Palæader.

Page 59, Emmrich, line 26, for 7 read 9.

Page 63, Etheridge, 2d reference, for Pendland read Pentland.

Page 146, Maurer, 3d reference, add the following species: *Proetus catillus* n. sp., *P. nutilus* n. sp., *P. acutus* n. sp., *P. embryo* n. sp., *P. cephalotes* Cord. *Phacops* cf. *fecundus* var. *major* Barr.

Page 146, McCoy, 2d reference, line 6, for *coela* read *coelata*.

Page 147, line 16, for 1860 read 1862.

Page 198, Salter, 2d reference, lines 5 and 26, for corridensis read *corndensis*.

Page 220, Ulrich, line 13, for *Ctenolvina* read *Ctenobolvina*.

Page 286, line 18, for *Hemicrypturis* read *Hemicrupturus*.